Molecular and Cellular Mechanisms of H⁺ Transport

NATO ASI Series

Advanced Science Institutes Series

A series presenting the results of activities sponsored by the NATO Science Committee, which aims at the dissemination of advanced scientific and technological knowledge, with a view to strengthening links between scientific communities.

The Series is published by an international board of publishers in conjunction with the NATO Scientific Affairs Division

A Life Sciences B Physics	Plenum Publishing Corporation London and New York

C Mathematical and
   Physical Sciences
D Behavioural and
   Social Sciences
E Applied Sciences
	Kluwer Academic Publishers
	Dordrecht, Boston and London

F Computer and
   Systems Sciences
G Ecological Sciences
H Cell Biology
I Global Environmental
   Change
	Springer-Verlag
	Berlin Heidelberg New York
	London Paris Tokyo Hong Kong
	Barcelona Budapest

NATO-PCO DATABASE

The electronic index to the NATO ASI Series provides full bibliographical references (with keywords and/or abstracts) to more than 30000 contributions from international scientists published in all sections of the NATO ASI Series. Access to the NATO-PCO DATABASE compiled by the NATO Publication Coordination Office is possible in two ways:

- via online FILE 128 (NATO-PCO DATABASE) hosted by ESRIN,
  Via Galileo Galilei, I-00044 Frascati, Italy.

- via CD-ROM "NATO Science & Technology Disk" with user-friendly retrieval software in English, French and German (© WTV GmbH and DATAWARE Technologies Inc. 1992).

The CD-ROM can be ordered through any member of the Board of Publishers or through NATO-PCO, Overijse, Belgium.

Series H: Cell Biology, Vol. 89

Molecular and Cellular Mechanisms of H$^+$ Transport

Edited by

Barry H. Hirst

Department of Physiological Sciences
University of Newcastle upon Tyne
Medical School
Newcastle upon Tyne, NE2 4HH, U.K.

Springer-Verlag
Berlin Heidelberg New York London Paris Tokyo
Hong Kong Barcelona Budapest
Published in cooperation with NATO Scientific Affairs Division

Proceedings of the NATO Advanced Research Workshop on Molecular and Cellular Mechanisms of H+ Transport, held in York, U.K., July 27 to August 1, 1993.

QP
535
. H I
M65
1994

ISBN 3-540-58497-8 Springer-Verlag Berlin Heidelberg New York

Library of Congress Cataloging-in-Publication Data. Molecular and cellular mechanisms of H transport / edited by Barry H. Hirst. p. cm. – (NATO ASI series. Series H, Cell biology; vol. 89) "This volume comprises the proceedings of a NATO Advanced Research Workshop on Molecular and Cellular Mechanisms of H+ Transport held on 27th July-1st August, 1993, in York, England". Includes bibliographical references and index. ISBN 3-540-58497-8 1. Hydrogen ions–Metabolism–Congresses. 2. Biological transport, Active–Congresses. 3. Adenosine triphosphatase–Congresses. I. Hirst, Barry H., 1953- . II. NATO Advanced Research Workshop on Molecular and Cellular Mechanisms of H+ Transport (1993: York, England) III. Series. QP535.H1M65 1994 599'.0192–dc20 94-34255

© Springer-Verlag Berlin Heidelberg 1994
Printed in Germany

Typesetting: Camera ready by authors
SPIN 10101769   31/3130 - 5 4 3 2 1 0 - Printed on acid-free paper

PREFACE

This volume comprises the proceedings of a NATO Advanced Research Workshop on Molecular and Cellular Mechanisms of H^+ Transport held on 27th July - 1st August, 1993, in York, England. The Workshop and this volume reviews current knowledge of proton transport mechanisms in man, in organs, including the stomach, kidney and bone. Information on the gastric H^+, K^+-ATPase is compared with information on non-gastric transport H^+-ATPases, including similar P-type H^+, K^+-ATPases in addition to V-type and F-type H^+-ATPases in a variety of tissues. Contributions include information on similar H^+ pumps from plants and yeast where lessons of relevance to mammalian transport ATPases may be learnt. Contributions on other P-type ATPases, Na^+, K^+- and Ca^{2+}-ATPases, are included as is information on chimeric ATPases. The contributions then focus on other proton transport mechanisms, including proton/anion antiports, symports and channels. The final two areas of consideration, including regulation of proton and bicarbonate transport mechanisms, emphasise cellular mechanisms involved in translocating pumps to the secretory membranes, and a discussion of how specific organs have become adapted to resist environments of high acid concentration.

As with all such events, the success of the Workshop, as illustrated by the contributions in this volume, cannot be attributed to a single person but reflects the concerted efforts of organisers, contributors, participants and administrative help. I would like in particular to thank the members of the International Scientific Advisory Committee for this Workshop who provided incisive insight into both the topics and key contributors for the Workshop. The excellence and timeliness of the contributions in this volume are a reflection of their hard work. The Advisory Committee consisted of John G. Forte (Berkeley, U.S.A.), Eberhard Frömter (Frankfurt, Germany), Andrew Garner (United Arab Emirates), Miguel J.M. Lewin (Paris, France), Heini Murer (Zurich, Switzerland), George Sachs (Los Angeles, U.S.A.) and William Silen (Boston, U.S.A.). In particular, I would like to extend my thanks to John Forte and Andrew Garner who pushed forward the overall organisation of the meeting and oversaw its format.

The Workshop was made possible through generous sponsorship by the North Atlantic Treaty Organisation under its Advanced Research Workshop programme. In addition, important contributions were made by Astra Hässle (Sweden) and the Wellcome Trust (U.K.), as well as a contribution from SmithKline Beecham Pharmaceuticals (U.K.). The International Science Foundation (New York) provided sponsorship for two scientists from the former Soviet Union countries to enable them to attend the Workshop. The Federation of European Physiological Societies and the Biochemical Society Membrane Group (U.K.) generously sponsored named lectures.

The local arrangements during the Workshop were efficiently and willingly provided by Dr. Julian White and the staff of IFAB communications at the Institute for Applied Biology in the University of York.

Lastly, but by no means least I would like to acknowledge the particularly efficient help that I personally received from Mrs. Angela Bott in both organising the Workshop and corresponding with the participants, and latterly in the editing of this volume.

Barry H. Hirst

Participants at the NATO ARW on Molecular and Cellular Mechanisms of H⁺ Transport in York.

CONTENTS

Gastric H⁺,K⁺-ATPase

Gastric H^+, K^+-ATPase

Non-gastric transport-ATPases

Proton/anion antiports, symports and channels

Regulation of proton and HCO₃⁻ transport

Cellular resistance to acid

Author Index

Gene Structure and Regulation of Gastric Proton Pump

Masatomo Maeda
Department of Organic Chemistry and Biochemistry
The Institute of Scientific and Industrial Research
Osaka University
Mihogaoka 8-1
Ibaraki, Osaka 567
Japan

Summary

This article summarizes our recent molecular biological studies on the genes of the gastric proton pump (H^+/K^+-ATPase) and their cell-specific transcription. The exon/intron organizations of the genes for the H^+/K^+-ATPase α and β subunits are very similar to those of the corresponding subunits of Na^+/K^+-ATPase, suggesting that the α and β subunit genes, respectively, of the two ATPases were derived from common ancestors. In contrast to Na^+/K^+-ATPase subunits, the H^+/K^+-ATPase α and β subunits are expressed specifically in gastric parietal cells. Consistent with this, we found a gastric mucosal nuclear protein(s) recognizing a sequence motif in the 5'-upstream regions of the H^+/K^+-ATPase α and β subunit genes. This motif may be a binding site for a positive transcriptional regulator that functions specifically in parietal cells. We further demonstrated by cDNA cloning that novel zinc finger proteins that bind to this motif are present in the gastric mucosal layer. The characteristics of these proteins are described.

Introduction

The H^+/K^+-ATPase of gastric parietal cells is essential for acid secretion into the stomach and catalyzes electroneutral exchange of H^+ and K^+ coupled with ATP hydrolysis (Faller *et al*. 1982), maintaining a large pH difference of more than 5 units between the cytoplasm and gastric lumen. This membrane-intrinsic pump protein

NATO ASI Series, Vol. H 89
Molecular and Cellular Mechanisms
of H⁺ Transport
Edited by Barry H. Hirst
© Springer-Verlag Berlin Heidelberg 1994

consists of two subunits, α and β (Hall *et al.* 1990; Toh *et al.* 1990). The primary structures of the pig (Maeda *et al.* 1988a), rat (Shull and Lingrel, 1986), rabbit (Bamberg *et al.* 1992) and human (Maeda *et al.* 1990a; Newman *et al.* 1990) α subunits, and the pig (Toh *et al.* 1990), rat (Canfield *et al.* 1990; Maeda *et al.* 1991; Newman and Shull, 1991), mouse (Canfield and Levenson, 1991; Morley *et al.* 1992), rabbit (Reuben *et al.* 1990) and human (Ma *et al.* 1991) β subunits have been determined from cloned cDNAs and genomic DNAs. The catalytic α subunit (apparent molecular size, 100 kDa) is postulated to traverse the membrane several times, whereas the highly glycosylated β subunit may have a single transmembrane segment near its amino terminus. The molecular size of the β subunit was decreased from 60 - 85 kDa to 32 -35 kDa by glycopeptidase F treatment (Hall *et al.* 1990; Canfield *et al.* 1990). The pig enzyme is advantageous for biochemical studies on coupled ion transport and its regulation, because membrane vesicles enriched in this enzyme can be prepared from pig stomach easily and in large quantity (Maeda *et al.* 1988b). cDNA sequences of the pig α and β subunits have helped in construction of a membrane topological model of H^+/K^+-ATPase based on protein chemical studies (Tamura *et al.* 1989; Maeda *et al.* 1990b; Besancon *et al.* 1993). Acid secretion from parietal cells is regulated by extracellular signals such as acetylcholine, gastrin and histamine (Ueda *et al.* 1986). However, the molecular mechanisms of the intracellular events evoked by these signals are not well known.

The primary structures of the α and β subunits for H^+/K^+-ATPase are highly homologous to those of the corresponding subunits of Na^+/K^+-ATPase. In contrast to the wide distribution of Na^+/K^+-ATPase in various tissues (Sweadner, 1989), H^+/K^+-ATPase has been found only in the stomach (Canfield *et al.* 1990; Shull, 1990; Toh *et al.* 1990; Oshiman *et al.* 1991). Analysis of the gene structure of H^+/K^+-ATPase is the first step in understanding the cell-specific gene expression and its regulation by extracellular signals. Thus, we cloned H^+/K^+-ATPase subunit genes (Maeda *et al.* 1990a; Oshiman *et al.* 1991; Maeda *et al.* 1991). We then extended our studies to the identification of a DNA sequence motif in the control regions of the genes that is recognized by a gastric specific nuclear protein(s) (Maeda *et al.* 1991; Tamura *et al.* 1992). Furthermore, we cloned cDNAs for the candidate DNA binding proteins and examined their characteristics (Tamura *et al.* 1993).

Gene organization of gastric proton pump

H+/K+- and Na+/K+-ATPase subunit genes were derived from common ancestral genes ----- We screened a phage library constructed from human liver DNA with a cDNA of the pig H+/K+-ATPase α subunit (Maeda *et al.* 1988a) as a probe. One positive clone carried the entire genomic sequence, as the 5'- and 3'-flanking regions were found in a cloned 15-kb DNA (Maeda *et al.* 1990a). The coding region of the α subunit was separated by 21 introns (Fig. 1). The positions of intron insertions are essentially the same as those in the human Na+/K+-ATPase α2 (Shull *et al.* 1989) and α3 (Ovchinnikov *et al.* 1988) subunits . The first introns of the three enzymes are difficult to align, and, unlike in the Na+/K+-ATPase genes, the sixth exon in the H+/K+-ATPase gene is not separated by an intron. Furthermore, the ninth intron of the H+/K+-ATPase and the tenth

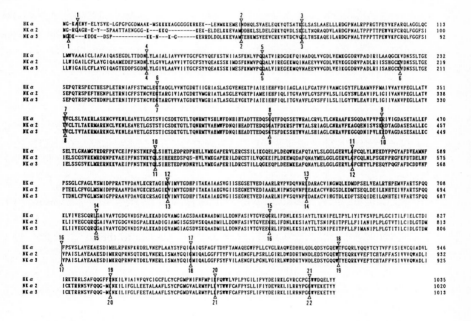

Fig. 1. Positions of introns in the α subunit genes of H+/K+- and Na+/K+-ATPases.
The primary structures and positions of introns (*closed triangles* with intron numbers) are indicated. HKα, human H+/K+-ATPase α subunit (Maeda *et al.* 1990); NKα2, human Na+/K+-ATPase α2 subunit (Shull *et al.* 1989); NKα3, human Na+/K+-ATPase α3 subunit (Ovchinnikov *et al.* 1988).

intron of the Na⁺/K⁺-ATPase α2 subunit are located two bases upstream of the position for the corresponding intron of the Na⁺/K⁺-ATPase α3 subunit. The other 19 exon/intron boundaries are located in the same positions in the three genes. The homology in the primary structures of the H⁺/K⁺- and Na⁺/K⁺-ATPase α subunits (about 60% identical residues) and also the similarity in the organizations of the above three genes suggest that they were derived from a common ancestral gene.

The close similarity of the two ATPases led us to suspect the presence of an additional subunit (β) of the H⁺/K⁺-ATPase, and later this β subunit was in fact identified as described above. We amplified a 88-bp segment of the gastric H⁺/K⁺-ATPase β subunit gene from rat liver chromosomal DNA by the polymerase chain reaction (PCR). One of the positive clones, isolated from a phage library using this segment as a probe, carried the entire reading frame (Maeda *et al.* 1991). The coding region of the β subunit gene is separated by 6 introns (Fig. 2). The positions of intron insertions of the related mouse Na⁺/K⁺-ATPase β2 subunit gene (Shyjan *et al.* 1991) are completely conserved, whereas the human β1 gene (Lane *et al.* 1989) was shown to have 5 introns, the positions 1, 2, and 5 being conserved. The similarities in intron/exon organizations and primary structures (30 - 40 % identical residues) suggest that the β subunit genes for H⁺/K⁺- and Na⁺/K⁺-ATPases were also derived from a common ancestor.

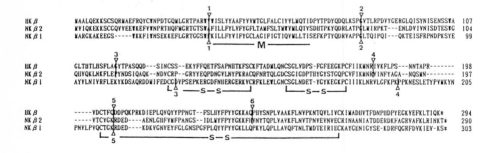

Fig. 2. Positions of introns in the β subunit genes of H⁺/K⁺- and Na⁺/K⁺-ATPases.
The primary structures and positions of introns (*closed triangles* with intron numbers) are indicated. HKβ, rat H⁺/K⁺-ATPase β subunit (Maeda *et al.* 1991); NKβ2, mouse Na⁺/K⁺-ATPase β2 subunit (Shyjan *et al.* 1991); NKβ1, human Na⁺/K⁺-ATPase β1 subunit (Lane *et al.* 1989). Positions of 3 S-S bridges identified in the Na⁺/K⁺-ATPase β1 subunit are also shown. The underline with **M** indicates the hypothetical transmembrane domain.

Role of the H$^+$/K$^+$-ATPase β subunit ----- Alignment of the H$^+$/K$^+$- and Na$^+$/K$^+$-ATPase β subunit sequences (Fig. 2) clearly showed the conservation of cysteine residues that form disulfide bridges in the Na$^+$/K$^+$-ATPase β1 subunit (Kirley, 1989). Furthermore, potential transmembrane segments were coded by the second exons of the three β subunit genes. These results suggest that the higher order structure of the H$^+$/K$^+$-ATPase β subunit is similar to that of Na$^+$/K$^+$-ATPase. The functional siginificance of the H$^+$/K$^+$-ATPase β subunit was demonstrated by an experiment on its expression in *Xenopus* oocytes (Noguchi *et al.* 1992). We injected messenger RNA for the α subunit of Na$^+$/K$^+$-ATPase into the oocytes together with that of the β subunit of H$^+$/K$^+$-ATPase. The Na$^+$/K$^+$-ATPase α subunit was assembled in the microsomal membranes with the H$^+$/K$^+$-ATPase β subunit, and became resistant to trypsin. These results suggest that the β subunit of H$^+$/K$^+$-ATPase has the role of forming a stable assembly with the catalytic subunit on the membranes similar to that of Na$^+$/K$^+$-ATPase.

Sequence motif for gastric DNA binding protein(s)

H$^+$/K$^+$-ATPase genes are transcribed only in the stomach ----- Immunoelectron microscopy has demonstrated that H$^+$/K$^+$-ATPase is located in gastric parietal cells (Toh *et al.* 1990; Morley *et al.* 1992). Consistent with this finding, we (Oshiman *et al.* 1991) and others (Canfield *et al.* 1990; Shull, 1990; Toh *et al.* 1990) found mRNAs of both subunits only in the stomach . Thus the expressions of both the α and β subunit are regulated at the transcriptional level, and a common gastric specific transcriptional regulatory factor(s) may recognize upstream sequence motifs located in the α and β subunit genes. The sequences of the 5'-upstream regions of the subunit genes are especially interesting, because, in general, the bindings of tissue- or cell-type specific proteins to the upstream sequences of genes are believed to be essential for the initiation of their transcription (Ptashne and Gann, 1990). As expected from their specific expression in the stomach, the 5' upstream sequences of the H$^+$/K$^+$-ATPase genes (Maeda *et al.* 1990a; Oshiman *et al.* 1991; Maeda *et al.* 1991) are different from those of the sodium pump (Ovchinnikov *et al.* 1988; Lane *et al.* 1989; Kano *et al.* 1989; Sverdlov *et al.* 1989; Shull *et al.* 1989; Kawakami *et al.* 1990a,b; Yagawa *et al.* 1992). Furthermore, PCR amplifications of the 5'-upstream sequences of the H$^+$/K$^+$-ATPase α subunit genes from gastric mucosa and liver of the same person (autopsy materials) demonstrated that they are identical and suggested that a specific trans-activating protein(s) participates in expression of the gastric gene (Tamura *et al.* 1992).

(G/C)PuPu(G/C)GAT(A/T)PuPy for gastric protein binding motif ----- As none of the sequence motifs except the TATA box (potential binding sites for TFIID) and CACCC sequences in the human and rat α subunit genes were commonly found in the rat β subunit sequence, it was important to demonstrate binding of a gastric specific nuclear protein(s) to the 5'-upstream sequences of the α and β subunit genes and to determine the recognition sequences for this protein(s). We first analyzed nuclear fractions from various rat tissues for protein binding to an upstream sequence (between -185 and +10 bp) of the rat β subunit gene (Maeda *et al.* 1991). Nuclear protein(s) from the stomach, but not the liver, brain, kidney, spleen or lung, bound to this sequence. Protein binding became undetectable in the presence of three unradiolabeled oligonucleotides carrying a GATAGC sequence. These oligonucleotides bound the stomach protein(s) in a mutually exclusive manner. The GATAGC sequences are located 30~60 bp upstream from potential TATA-boxes. Consistent with this, mRNA synthesis starts from the 3'-side of downstream TATA-box (Newman and Shull, 1991). Furthermore, the GATAGC sequences can be found in the mouse upstream sequence of the β subunit gene (Canfield and Levenson, 1991; Morley *et al.* 1992).

Comparison of the 5'-upstream sequences of the human and rat α subunit genes indicated the presence of highly conserved regions which may be important for specific

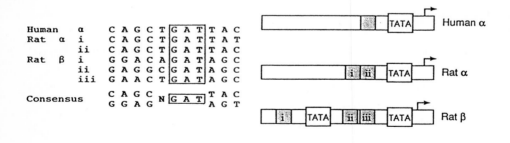

Fig. 3. Sequence comparison of the motifs in the α- and β-subunit genes.
The sequences of oligonucleotides that bind gastric nuclear protein [(G/C)PuPu(G/C)NGAT(A/T)PuPy] (Tamura *et al.* 1992) are compared (*left*). The positions of the motifs in the human α-, rat α- and rat β-subunit genes are shown schematically (*right*). Transcription start sites are shown by arrows.

transcription of the α subunit in gastric parietal cells (Oshiman *et al*. 1991). We found that gastric nuclear protein bound to the TAATCAGCTG sequences of α subunit genes immediately upstream (about 30 and 40 bp in the rat and human genes, respectively) of the potential TATA-boxes (Tamura *et al*. 1992). The binding of the protein to the sequence motifs in the α and β subunits was mutually competitive. From the sense-strand sequence of the binding motif in the α subunit gene, we concluded that (G/C)PuPu(G/C)GAT(A/T)PuPy is a core sequence motif for the gastric specific DNA binding protein. The positions of the sequence motifs in the upstream regions of the α and β subunit genes are shown schematically in Fig. 3.

It is also of interest to determine whether other genes specifically expressed in parietal cells have similar sequence motifs. We isolated a gene for the rat gastric intrinsic factor (Maeda, M., Mahmood, S. and Futai, M. in preparation). On sequence determination, we found similar sequence motifs upstream of the TATA-box of the gene. Analysis of the upstream region of the human H2 receptor gene is currently in progress (Maeda, M., Koike, T. and Futai, M. in preparation).

Novel gastric DNA binding proteins and their possible role(s)

The DNA sequence motif, (G/C)PuPu(G/C)NGAT(A/T)PuPy, common to human α, rat α and rat β subunit genes includes the (T/A)GATA(G/A) sequence recognized by the GATA-binding proteins (GATA-1, -2 and -3) (Zon *et al*. 1991), a group of zinc finger proteins implicated in the transcriptional regulation of erythroid-specific genes, endothelial determinants and others. The overlap of the sequences recognized by the gastric-specific and GATA-binding proteins led us to speculate that the gastric proteins might have similar zinc finger domains.

PCR amplification of cDNA from pig gastric mucosa demonstrated the presence of novel zinc finger proteins called GATA-GT1, GATA-GT2 and GATA-GT3; each had zinc finger sequences similar to previously characterized GATA-binding proteins. Subsequently, we obtained full length cDNA of GATA-GT1 and GATA-GT2 from rat stomach (Tamura *et al*. 1993). The proteins, GATA-GT1 and -GT2, have tandem zinc finger domains very similar to those of GATA-binding proteins. The zinc finger domains of GATA-GT1 and -GT2 were 70 - 85 % identical at the amino acid level with each other and with other GATA-binding proteins. We noted several potential protein kinase phosphorylation sites in the zinc finger region. In contrast, regions outside the zinc fingers shared essentially no similarities. GATA-GT1 and -GT2 were found to bind to the upstream sequence of the H^+/K^+-ATPase β gene. These proteins were expressed in the gastric mucosa, but not in the underlying muscular layer, and lower amounts were found in the intestine. This tissue distribution is quite different from that of GATA-1, -2 or -3. These results taken together clearly suggest that GATA-GT1 and -GT2 are involved in gene regulation specifically in the gastric epithelium and that they are two new members of the GATA-binding protein family. It will be interesting to determine how these proteins participate in transcriptional regulation of gastric proton pump (H^+/K^+-ATPase) genes.

Acknowledgements

This research was supported by grants from the Ministry of Education, Science and Culture of Japan, Osaka University, and a Human Frontier Science Program. I am grateful to the coworkers whose names appear in the references.

REFERENCES

Bamberg, K., Mercier, F., Reuben, M. A., Kobayashi, Y., Munson, K. B. and Sachs, G. (1992) cDNA cloning and membrane topology of the rabbit gastric H^+/K^+-ATPase α-subunit. Biochim. Biophys. Acta **1131**, 69-77.

Besancon, M., Shin, J. M., Mercier, F., Munson, K., Miller, M., Hersey, S. and Sachs, G. (1993) Membrane topology and omeprazole labeling of the gastric H^+, K^+-adenosinetriphosphatase. Biochemistry **32**, 2345-2355.

Canfield, V. A., Okamoto, C, T., Chow, D., Dorfman, J., Gros, P., Forte, J. G. and Levenson, R. (1990) Cloning of the H,K-ATPase β subunit: tissue-specific expression, chromosomal assignment, and relationship to Na,K-ATPase β subunits. J. Biol. Chem. **265**, 19878-19884.

Canfield, V. A. and Levenson, R. (1991) Structural organization and transcription of the mouse gastric H^+, K^+-ATPase β subunit gene. Proc. Natl. Acad. Sci. U.S.A. **88**, 8247-8251.

Faller, L., Jackson, R., Malinowska, D., Mukidjam, E., Rabon, E., Saccomani, G., Sachs, G. and Smolka, A. (1982) Mechanistic aspects of gastric ($H^+ + K^+$)-ATPase. Ann. N. Y. Acad. Sci. **402**, 146-163.

Hall, K., Perez, G., Anderson, D., Gutierrez, C., Munson, K., Hersey, S., J., Kaplan, J. H. and Sachs, G. (1990) Location of the carbohydrates present in the HK-ATPase vesicles isolated from hog gastric mucosa. Biochemistry **29**, 701-706.

Kano, I., Nagai, F., Satoh, K., Ushiyama, K., Nakao, T. and Kano K. (1989) Structure of the $\alpha 1$ subunit of horse Na,K-ATPase gene. FEBS Lett. **250**, 91-98.

Kawakami, K., Yagawa, Y. and Nagano, K. (1990a) Regulation of Na^+,K^+-ATPases: I. Cloning and analysis of the 5'-flanking region of the rat *NKAA2* gene encoding the $\alpha 2$ subunit. Gene (*Amst.*) **91** 267-270.

Kawakami, K., Yagawa, Y. and Nagano, K. (1990b) Regulation of Na⁺,K⁺-ATPases: I. Cloning and analysis of the 5'-flanking region of the rat *NKAB2* gene encoding the β2 subunit. Gene (*Amst.*) **91** 271-274.

Kirley, T. L. (1989) Determination of three disulfide bonds and one free sulfhydryl in the β subunit of (Na,K)-ATPase. J. Biol. Chem. **264**, 7185-7192.

Lane, L. K., Shull, M. M., Whitmer, K. R. and Lingrel, J. B. (1989) Characterization of two genes for the human Na,K-ATPase β subunit. Genomics **5**, 445-453.

Ma, J.-Y., Song, Y.-H., Sjostrand, S. E., Rask, L. and Mardh, S. (1991) cDNA cloning of the β subunit of the human gastric H,K-ATPase. Biochem. Biophys. Res. Commun. **180**, 39-45.

Maeda, M., Ishizaki, J. and Futai, M. (1988a) cDNA cloning and sequence determination of pig gastric (H⁺ + K⁺)-ATPase. Biochem. Biophys. Res. Commun. **157**, 203-209.

Maeda, M., Tagaya, M. and Futai, M. (1988b) Modification of gastric (H⁺ + K⁺)-ATPase with pyridoxal 5'-phosphate. J. Biol. Chem. **263**, 3652-3656.

Maeda, M., Oshiman, K., Tamura, S. and Futai, M. (1990a) Human gastric (H⁺ + K⁺)-ATPase gene: similarity to (Na⁺ + K⁺)-ATPase genes in exon/intron organization but difference in control region. J. Biol. Chem. **265**, 9027-9032.

Maeda, M., Tamura, S. and Futai, M. (1990b) Structure and chemical modification of pig gastric (H⁺ + K⁺)-ATPase. In Bioenergetics (Kim, C. H. and Ozawa, T. eds.), pp. 217-225. Plenum, New York.

Maeda, M., Oshiman, K., Tamura, S., Kaya, S., Mahmood, S., Reuben, M. A., Lasater, L. S., Sachs, G. and Futai, M. (1991) The rat H⁺/K⁺-ATPase β subunit gene and recognition of its control region by gastric DNA binding protein. J. Biol. Chem. **266**, 21584-21588.

Morley, G. P., Callaghan, J. M., Rose, B. R., Toh, B. H., Gleeson, P. A. and van Driel, I. R. (1992) The mouse gastric H,K-ATPase β subunit: gene structure and co-ordinate expression with the α subunit during ontogeny. J. Biol. Chem. **267**, 1165-1174.

Newman, P. R., Greeb, J., Keeton, T. P., Reyes, A. A. and Shull, G. E. (1990) Structure of the human gastric H,K-ATPase gene and comparison of the 5'-flanking sequences of the human and rat genes. DNA and Cell Biol. **9**, 749-762.

Newman, P. R. and Shull, G. E. (1991) Rat gastric H,K-ATPase β-subunit gene: intron/exon organization, identification of multiple transcription initiation sites, and analysis of the 5'-flanking region. Genomics **11**, 252-262.

Noguchi, S., Maeda, M., Futai, M. and Kawamura, M. (1992) Assembly of a hybrid from the α subunit of Na⁺/K⁺-ATPase and the β subunit of H⁺/K⁺-ATPase. Biochem. Biophys. Res. Commun. **182**, 659-666.

Oshiman, K., Motojima, K., Mahmood, S., Shimada, A., Tamura, S., Maeda, M. and Futai, M. (1991) Control region abd gastric specific transcription of the rat H⁺,K⁺-ATPase α subunit gene. FEBS Lett. **281**, 250-254.

Ovchinnikov, Y. A., Monastyrskaya, G. S., Broude, N. E., Ushkaryov, Y. A., Melkov, A.M., Smirnov, Y. V., Malyshev, I. V., Allikmets, R. L., Kostina, M. B., Dulubova, I. E., Kiyatkin, N. I., Grishin, A. V., Modyanov, N. N. and Sverdlov, E. D. (1988) Family of human Na⁺, K⁺-ATPase genes: structure of the gene for the catalytic subunit (αIII-form) and its relationship with structural features of the protein. FEBS Lett. **233**, 87-94.

Ptashne, M., and Gann, A. A. F. (1990) Activators and targets. Nature **346**, 329-331.

Reuben, M. A., Lasater, L. A. and Sachs, G. (1990) Characterization of a β subunit of the gastric H⁺/K⁺-transporting ATPase. Proc. Natl. Acad. Sci. U.S.A **87**, 6767-6771.

Shull, G. E. and Lingrel, J. B. (1986) Molecular cloning of the rat (H⁺ + K⁺)-ATPase. J. Biol. Chem. **261**, 16788-16791.

Shull, G. E. (1990) cDNA cloning of the β subunit of the rat gastric H,K-ATPase. J. Biol. Chem. **265,** 12123-12126.

Shull, M. M., Pugh, D. G. and Lingrel, J. B. (1989) Characterization of the human Na,K-ATPase $\alpha 2$ gene and identification of intergenic restriction fragment length polymorphisms. J. Biol. Chem. **264**, 17532-17543.

Shyjan A. W., Canfield, V. A. and Levenson, R. (1991) Evolution of the Na,K- and H,K-ATPase β subunit gene family: structure of the murine Na,K-ATPase $\beta 2$ subunit gene. Genomics **11**, 435-442.

Sverdlov, E. D., Bessarab, D. A., Malyshev, I. V., Petrukhin, K. E., Smirnov, Y. V., Ushkaryov, Y. A., Monastyrskaya, G. S., Broude, N. E. and Modyanov, N. N. (1989) Family of Na^+,K^+-ATPase genes: structure of the putative regulatory region of the α^+-gene. FEBS Lett. **244**, 481-483.

Sweadner, K. J. (1989) Isozymes of the Na^+/K^+-ATPase. Biochim. Biophys. Acta **988**, 185-220.

Tamura, S., Tagaya, M., Maeda, M. and Futai, M. (1989) Pig gastric $(H^+ + K^+)$-ATPase: Lys-497 conserved in cation transporting ATPase is modified with pyridoxal 5'-phosphate. J. Biol. Chem. **264**, 8580-8584.

Tamura, S., Oshiman, K., Nishi, T., Mori, M., Maeda, M. and Futai, M. (1992) Sequence motif in control regions of the H^+/K^+ ATPase α and β subunit genes recognized by gastric specific nuclear protein(s). FEBS Lett. **298**, 137-141.

Tamura, S., Wang, X.-H., Maeda, M. and Futai, M. (1993) Novel gastric DNA binding proteins recognize upstream sequence motifs of parietal cell-specific genes. Proc. Natl. Acad. Sci. U.S.A. **90**, in press.

Toh, B.-H., Gleeson, P. A., Simpson, R. J., Moritz, R. L., Callaghan J. M., Goldkorn, I., Jones, C., Martinelli, T. M., Mu, F.-T., Humphris, D. C., Pettitt, J. M., Mori, Y., Masuda, T., Sobieszczuk, P., Weinstock, J., Mantamadiotis, T. and Baldwin, G. (1990) The 60- to 90-kDa parietal cell autoantigen associated with autoimmune gastritis is a β subunit of the gastric H^+/K^+-ATPase (proton pump). Proc. Natl. Acad. Sci. U.S.A. **87**, 6418-6422.

Ueda, S., Oiki, S. and Okada, Y. (1986) Fluorescence-microscopical studies on acid secretion and Ca^{2+} mobilization in cultured parietal cells of rat stomach. Biomed. Res. **7**, 105-108.

Yagawa, S., Kawakami, K. and Nagano, K. (1992) Housekeeping Na,K-ATPase $\alpha 1$ subunit gene promoter is composed of multiple *cis* elements to which common and cell type-specific factors bind. Mol. Cell. Biol. **12**, 4046-4055.

Zon, L. I., Mather, C., Burgess, S., Bolce, M. E., Harland, R. M. and Orkin, S. H. (1991) Expression of GATA-binding proteins during embryonic development in *Xenopus laevis*. Proc. Natl. Acad. Sci. U.S.A. **88**, 10642-10646.

The Gastric H/K ATPase β Subunit Gene and Transcriptional Pathways in Acid-Secreting Epithelia of the Stomach and Kidney

Ian R. van Driel, Paul A. Gleeson, Seong-Seng Tan[1], and Ban-Hock Toh
Department of Pathology and Immunology
Monash University Medical School
Commercial Rd., Prahran, Vic, 3181
AUSTRALIA.

Gastric acid secretion and the gastric H/K ATPase

Stomach acid is produced by parietal cells and secreted into the gastric lumen for digestive and protective purposes. The gastric parietal cell has a highly developed acid secretory machinery that can maintain the pH of the gastric lumen at around pH 1 (Forte and Soll, 1993; Sachs et al., 1992; Faller et al., 1993; Rabon and Reuben, 1990). The central role of the gastric H/K ATPase in the secretion of gastric HCl is undisputed. The protein is located in the membranes of the parietal cell tubulovesicles and secretory canaliculi. The mechanism of action and structure of the gastric H/K ATPase have been extensively reviewed in this volume and elsewhere. The protein is composed of two subunits, a catalytic α and a β subunit. The existence of the α subunit has been known for some time. The protein is 95 kDa and has a large number of membrane spanning domains (probably 8) (Sachs et al., 1992; Green and Stokes, 1992). The β subunit was only recently discovered in a number of laboratories (Okamoto et al., 1990, Reuben et al. 1990, Shull, 1992, Toh et al, 1990). The 35 kDa protein core is highly glycosylated with N-linked glycans which may terminate in poly-N-acetyllactosamine structures (Callghan et al. 1990, 1992). The β subunit appears to be essential for membrane insertion, may contain intracellular trafficking signals and is required for functional activity (Horisberg et al., 1991; Gottardi and Caplan, 1993; Sachs et al., 1992)

Gene structure of the gastric H/K ATPase β subunit gene and evolutionary relationships

The gene for the α subunit of the gastric H/K ATPase belongs to a family of genes that encode "P-type" ATPases (1785, 1624). Different members of this family specify proteins with different ion motive activities. The H/K ATPase α subunit has the highest degree of similarity to the Na/K ATPase α subunits, and the genes encoding these proteins have very similar exon-intron structures (Lingrel et al.,

---

[1] *Department of Anatomy, The University of Melbourne, Parkville, Vic, 3052*

NATO ASI Series, Vol. H 89
Molecular and Cellular Mechanisms
of H⁺ Transport
Edited by Barry H. Hirst
© Springer-Verlag Berlin Heidelberg 1994

1990; Maeda et al., 1990; Newman et al., 1990). These two ATPases are also the only members of the gene family clearly demonstrated to have β subunits.

We (Morley et al., 1992) and others (Canfield and Levenson, 1991; Maeda and Oshiman, 1992) have isolated the genomic sequences encoding the gastric H/K ATPase β subunit. The mouse gene spans approximately 12 kbp and contains 7 exons. Two mammalian Na/K ATPases β subunits isoforms have been identified, β1 and β2, which are encoded by separate genes (Lane et al., 1989; Magyar and Schanchner, 1993). Lane *et. al.* (1989) have isolated a functional human β1 subunit gene (ATP1B) which is composed of 6 exons. The mouse Na/K ATPase β2 subunit (also known as the adhesion molecule on glia, AMOG) gene has also been analysed (Magyar and Schanchner, 1993). The gene structure is similar but not identical to the human Na/K ATPase β1 subunit gene, ATP1B (see below). Comparison of these β subunit genes has allowed a probable evolutionary relationship to be arrived at.

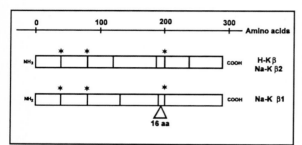

Fig. 1.   **Structure of P-type ATPase β subunit genes.**   The protein sequences of the mouse H/K ATPase β (H-K β), the mouse Na/K ATPase β2 (Na/K β2), and human Na/K ATPase β1 (Na/K β1) subunits are represented as rectangles. The positions at which introns interrupt the coding regions of the gene are indicated by a vertical line. Stars indicate that the positions of an intron are identical in all genes.

The placement of introns relative to the coding sequence of the mouse H/K ATPase β subunit gene is identical to the mouse Na/K ATPase β2 subunit gene (Fig. 1). In contrast, the structure of the human Na/K ATPase β1 subunit gene ATP1B is significantly different from these two genes. Although introns 1,2 and 5 interrupt the coding region in identical positions in all three genes, the introns 3 and 4 of the Na/K ATPase β1 subunit gene are found 17 and 7 codons distant from the corresponding introns in the other two genes. Furthermore, the H/K ATPase β and Na/K ATPase β2 genes have one additional intron not found in the Na/K ATPase β1 gene (Fig. 1).

This data strongly suggests the following genealogy for the ATPase β subunit genes (Fig. 2). As the positions of three introns are shared in all three genes it would appear that they have one common ancestor. The similarity between the Na/K ATPase β2 and H/K ATPase β subunit genes, and their dissimilarity with the Na/K ATPase β1 gene suggests that the divergence of the Na/K ATPase β1 lineage preceded the appearance of separate Na/K ATPase β2 and H/K ATPase β subunit genes. This scenario would be difficult to conclude from comparison of primary sequence data alone. The mouse Na/K ATPase β1 subunit has a higher level of sequence identity with the mouse Na/K ATPase β2 (42%) subunit than with the mouse H/K ATPase β subunit (33%). Perhaps of more significance, is that the Na/K ATPase β1 subunits from all species so far examined have a highly conserved insertion of 16 amino acids at position 190 of the mouse H/K ATPase β subunit genes that is not present in the other β subunit sequences, and the N-terminal 20 amino acids are quite dissimilar.

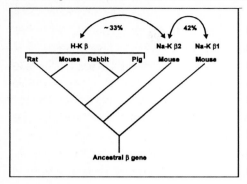

FIG. 2. **Genealogy of the ATPase β subunit genes.** A probable evolutionary pathway that has resulted in the present-day H/K ATPase β (H-K β), Na/K ATPase β1 (Na/K β1), and Na/K ATPase β2 (Na/K β2) subunit genes is shown. The percent amino acid identity between the proteins is shown. Length of lines does not represent evolutionary time.

It is of interest to consider the mechanisms that could have led to the observed differences in the structures of the β gene family. The presence of an extra intron in the H/K ATPase β and Na/K ATPase β2 subunit genes could be simply explained by intron loss in Na/K ATPase β1 subunit genes or intron gain in the ancestor of the H/K ATPase β and Na/K ATPase β2 subunit genes. Sequential intron deletion and insertion is unlikely to account for the positions of introns 3 and 4 of the Na/K ATPase β1 gene which are located at similar but not identical positions to the corresponding H/K ATPase β and Na/K ATPase β2 subunit introns. A mechanism of intron sliding, which is the result of the appearance of a new splice donor or acceptor site in the intron DNA adjacent to the existing splice site may explain this observation. Intron sliding would probably not lead to similarity in the sequence between the two exon/introns boundaries as the new DNA would be derived from non-coding regions. Comparing mouse H/K ATPase β subunit and the human Na/K ATPase β1 subunit proteins (Morley et al., 1992), 4 identities and 5 conservative substitutions are found in the 17 residues that separates the intron 3 positions (H-K ATPase residues 120-131). For the intron 4 positions, 2 identities, 3 conservative substitutions are present in 7 amino acids (H-K ATPase residues 176-182). An alternative explanation for the difference in the positions of intron 3 and 4 is that the region lying between the two closely-spaced introns may, at one time, have been encoded by a separate exon. Then loss of different introns in the two genes would lead to the present-day gene structures. This latter explanation *would* lead to the coding region between the two introns having sequence similarity.

An insertion of DNA encoding 4 amino acids that occurs at the 5' end of the exon 6 of the mouse and rat H/K ATPase β subunit genes, but not the pig and rabbit sequences, may have arisen as a result of intron sliding. The first nine amino acids residues encoded by mouse exon 6 are identical to the rat sequence in this region, arguing that the insertion is the result of an event that occurred prior to divergence of the mouse and rat species.

Coordinate expression of gastric H/K ATPase subunits during ontogeny

We have examined the expression of the gastric H/K ATPase subunits during ontogeny of the mouse (Morley et al., 1992; Pettitt et al., 1993). By examining the levels of α and β subunit protein with an ELISA we have confirmed that the levels of both subunits remain very low until approximately 10-15

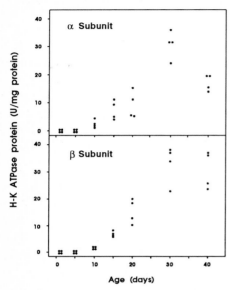

days after birth, and then increase in parallel to be reach adult levels by day 30. This increase in expression corresponds to the time at which the mucosa is morphologically and functionally reaching maturity. The stomach like other organs involved in digestion such as the pancreas, liver and salivary glands does not reach full functional development in mammals until weaning, a reflection of the need to digest more complex foods. We have further analysed expression of the H/K ATPase subunits using dual-colour immunofluorescence techniques in new born mice. Although the frequency of cells expressing protein was very low in these animals, we found that any cell that expressed one subunit also expressed the other, indicating that the mechanisms that control expression of the gastric H/K ATPase subunits is coordinated on a cellular level, and probably similar for both genes. The morphological development and expression of the gastric H/K ATPase subunits has also been examined. Pettitt *et al.* (1993) noted that identifiable secretory membranes were not visible in the parietal cells of foetal mice prior to the appearance of the H/K ATPase subunits.

Fig. 3. **Ontogeny of expression of the mouse gastric H/K ATPase subunits.** Levels of protein in mouse gastric membranes were determined by an ELISA. For details see Morley et al., 1992.

Functional analysis of the gastric H/K ATPase gene regulatory regions

One of the limiting factors in the investigation of parietal cell biology in general, and the investigation of parietal cell-specific transcription in particular, is the lack of cell lines with the differentiated phenotype of parietal cells. Thus, the standard functional approaches to identifying regions of genes that are involved in transcriptional regulation are inapplicable for genes expressed uniquely in parietal cells. Therefore, to functionally identify transcriptional control regions of the gastric H/K ATPase β subunit gene, we have turned to the use of transgenic mice (van Driel, *et al.*, unpublished). Up to 11 kbp of the 5' flanking sequences were placed 5' to an *E. coli lacZ* reporter gene. These hybrid genes where then introduced into the germline of mice, and the expression patterns of the *lacZ* reporter gene analysed by histochemical staining for β galactosidase.

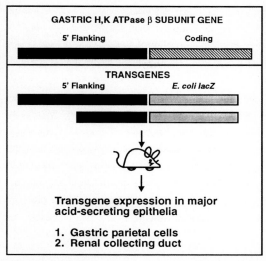

GASTRIC H,K ATPase β SUBUNIT GENE

5' Flanking   Coding

TRANSGENES

5' Flanking   E. coli lacZ

Transgene expression in major
acid-secreting epithelia

1. Gastric parietal cells
2. Renal collecting duct

Fig. 4. **Transgenes to assess the functional activity of the 5' flanking sequences of the mouse gastric H/K ATPase β subunit gene.** For further details see text.

We have found that transgenes containing either 11 kbp or 5 kbp of 5' flanking sequences direct expression of the reporter gene to the gastric mucosa and the kidney. In the gastric mucosa immunohistochemical staining with a parietal cell-specific antibody of gastric tissue sections revealed that the transgene was expressed only in the parietal cells and not other cell types. Hence, the 5' flanking sequences are able to specify parietal-cell specific expression within the mucosa.

A combination of morphological and histochemical observation indicated that the region in the kidney expressing the transgenes was the collecting ducts. Staining was observed in cortical and medullary collecting ducts and did not appear to be limited to a particular cell type.

Similarities between acid-secreting cells of the stomach and kidney

The gastric mucosa and the renal collecting ducts share the ability to acidify lumenal fluids. The excretion of urinary acid is a vital physiological function required for the maintenance of acid-base homoeostasis. There is strong evidence indicating that intercalated cells of the kidney collecting ducts are principally responsible for the acidification of the urine (Khadouri et al., 1991; Alexander and Schwartz, 1992). The acid secretory activity of the collecting duct intercalated cells is almost certainly mediated by more than one enzyme. It is well established that the vacuolar-type electrogenic H^{+}-ATPase is of some importance in this process (Gluck and Nelson, 1992). However, recent evidence suggests that a renal H/K ATPase is also involved. The renal H/K ATPase has been identified by enzymological (Khadouri et al., 1991; Doucet and Marsy, 1987; Wingo, 1989; Cheval et al., 1991) and immunochemical techniques (Wingo et al., 1990). Recent evidence confirmed that this enzyme is capable of vesicular acidification (Curran et al., 1992) and is probably also involved in K reabsorption (Wingo and Cain, 1993). The renal and gastric H/K ATPase enzymes appear to be distinct but several lines of evidence suggest that they are encoded by related genes Both enzymes are inhibited by similar compounds (Wingo, 1989; Cheval et al., 1991) and Wingo *et al.* (1990) demonstrated reactivity of some but not all anti-gastric H/K ATPase α subunit monoclonal antibodies with the intercalated cells of the cortex and outer medulla. We have partially purified a kidney membrane protein that binds to anti-gastric H/K ATPase β subunit monoclonal antibodies and hence is a candidate renal H/K ATPase β subunit (Callaghan, J.M. et al., unpublished). Recently, Crowson and Shull isolated a cDNA from the colon which encodes a P-type ATPase α subunit that hybridises to a mRNA in the kidney (Crowson and Shull, 1992). It is probable that the colon-derived cDNA is cross hybridising to a related renal cDNA.

In addition to the functional similarities, the renal collecting duct intercalated cells and the gastric parietal cells also have striking similarities in ultrastructure (Madsen and Tisher, 1985). They both possess cytoplasmic tubulovesicles that are adjacent to their apical surfaces. In both cell types, these vesicles have been implicated in the production of acid. Furthermore, as expected, there is a similarity in the enzymatic cohorts responsible for the production of acid including basal Cl^-/HCO_3^- exchangers, basal Na^+/H^+ exchangers, apical electrogenic H^+-ATPases, abundant carbonic anhydrase expression, and as mentioned H/K ATPase activity (see Hamm et al., 1991; Verlander et al., 1991; Forte and Soll, 1993; Doucet and Marsy, 1987; Wingo, 1989; Kriz and Kaissling, 1985).

The acid-secreting epithelia of the stomach and kidney contain similar transcriptional pathways

The expression of the gastric H/K ATPase 5' flanking sequence/*lacZ* transgenes in the kidney is probably the result of the gastric mucosa and the collecting ducts expressing common transcription factor activities. We hypothesise that the transcription factors activating the transgene in these cell types may be responsible for the expression of gene products that determine the common structural and functional properties of these cells, particularly acid secretion. Possible targets of these transcriptional activities are the related H/K ATPase genes found in these organs.

Observations similar to this have been made in other systems. Recently, work by Davis et al. (1992) demonstrated that the genes encoding trypsin I, elastase and amylase are normally expressed in pancreatic acinar cells as well as the chief cells of the stomach. The structure and function of the pancreatic acinar cells and gastric chief cells is very similar and distinctive as they are both zymogen granule-containing excretory cells. Previous to that investigation, Sweester et al. (1988) found that a transgene containing the regulatory regions of a gene expressed in intestinal enterocytes, "liver" fatty acid binding protein directed expression of a reporter gene to the proximal tubules of the kidney in addition to intestinal cells. These workers noted that the intestinal and renal cells expressing the "liver" fatty acid binding protein transgene shared a number of structural, functional and biochemical traits characteristic of highly polarised absorptive cell types. These findings and our work suggests that epithelia in different organs use similar transcriptional machinery to produce cells of given structural and functional properties. Similar embryological origins of the tissues involved in these observations can not fully account for these data. The gut and pancreas are derived from endoderm, whereas the renal collecting duct and renal nephron are mesodermal (Saxen, 1987).

Future investigations will attempt to identify the transcription factor proteins which activate transgene expression in the kidney and stomach, and further regulatory sequences of the gastric H/K ATPase β subunit gene that are required to limit expression to the gastric mucosa and silence expression in the kidney.

References

Alexander, E.A. and Schwartz, J.H. (1992). Regulation of acidification in the rat inner medullary collecting duct. Am. J. Kidney Diseases *18*: 612-618.

Callaghan, J.M., Toh, B.H., Pettitt, J.M., Humphris, D.C. and Gleeson, P.A. (1990) Poly-N-acetyllactosamine-specifc tomato lectin interacts with gastric parietal cells: Identification of a tomato-laectin binding $60\text{-}90\text{x}10^3$ Mr membrane glycoprotein of tubulovesicles. J.Cell Sci. *95*:563-576.

Callaghan, J.M., Toh, B.H., Simpson, R.J., Baldwin, G.S. and Gleeson, P.A. (1992) Rapid Purification of the Gastric H+/K+-ATPase Complex by Tomato-Lectin Affinity Chromatography. Biochem.J. *283*:63-68.

Canfield, V.A. and Levenson, R. (1991). Structural organization an transcription of the mouse gastric H^+,K^+-ATPase β subunit gene. Proc. Natl. Acad. Sci. U. S. A. *88*: 8247-8251.

Cheval, L., Barlet-Bas, C., Khadouri, C., Feraille, E., Marsy, S., and Doucet, A. (1991). K+-ATPase-mediated Rb transport in rat collecting tubule: modulation during K+ deprivation. Am. J. Pathol. *260*: F800-F805.

Crowson, M.S. and Shull, G.E. (1992). Isolation and Characterization of a cDNA Encoding the Putative Distal Colon H+,K+-ATPase - Similarity of Deduced Amino Acid Sequence to Gastric H+,K+-ATPase and Na+,K+- ATPase and messenger RNA Expression in Distal Colon, Kidney, and Uterus. J. Biol. Chem. *267*: 13740-13748.

Curran, K.A., Hebert, M.J., Cain, B.D., and Wingo, C.S. (1992). Evidence for the presence of a K-dependent acidifying adenosine triphosphatase in the rabbit medulla. Kidney Int. *42*: 1093-1098.

Davis, B.P., Hammer, R.E., Messing, A., and Macdonald, R.J. (1992). Selective Expression of Trypsin Fusion Genes in Acinar Cells of the Pancreas and Stomach of Transgenic Mice. J. Biol. Chem. *267*: 26070-26077.

Doucet, A. and Marsy, S. (1987). Characterisation of K-ATPase activity in distal nephron: stimulation by potassium depletion. Am. J. Physiol. *253:* F418-F423.

Faller, L.D., Smolka, A., and Sachs, G. (1993). The gastric H,K ATPase. In The enzymes of biological membranes, Vol. 3. A.N. Martonosi, ed. (New York: Plenum), pp. 431-438.

Forte, J.G. and Soll, A. (1993). Cell biology of hydrochloric acid secretion. In The Handbook of Physiology - The Gastrointestinal System III. S.G. Shultz, ed. (Bethesda.: American Physiological Society), pp. 207-228.

Gluck, S. and Nelson, R. (1992). The Role of the V-ATPase in Renal Epithelial H+ Transport. J. Exp. Biol. *172:*205 -218.

Gottardi, C.J. and Caplan, M.J. (1993). An Ion-Transporting ATPase Encodes Multiple Apical Localization Signals. J. Cell Biol. *121:* 283-293.

Green, N.M. and Stokes, D.L. (1992). Structural Modelling of P-Type Ion Pumps. Acta Physiol. Scand. *146*: 59-68.

Hamm, L.L., Weiner, I.D., and Vehaskari, V.M. (1991). Structural-functional characteristics of acid-base transport in rabbit collecting duct. Sem. Nephrol. *11*: 453-464.

Horisberg, J.-D., Jaunin, P., Reuben, M., Lasater, L.S., Chow, D.C., Forte, J.G., Sachs, G., and Geering, K. (1991). The H,K ATPase β-Subunit can act as a surrogate for the β-subunit of Na,K-pumps. J. Biol. Chem. *266*: 19131-19134.

Khadouri, C., Cheval, L., Marsy, S., Barlet-Bas, C., and Doucet, A. (1991). Characterization and control of proton-ATPase along the nephron. Kidney Int. *40:* S71-S78.

Kriz, W. and Kaissling, B. (1985). Structural organization of the mammalian kidney. In The kidney: Physiology and pathophysiology. D.W. Seldin and G. Giebisch, eds. (New York: Raven Press), pp. 265-306.

Lane, L.K., Shull, M.M., Whitmer, K.R., and Lingrel, J.B. (1989). Characterization of two genes for the Human Na,K-ATPase β Subunit. Genomics 5: 445-453.

Lingrel, J.B., Orlowski, J., Shull, M.M., and Price, E.M. (1990). Molecular Genetics of Na,K ATPase. Prog.Nuc.Acid.Res.Molec.Biol. 38: 37-89.

Madsen, K.M. and Tisher, C.C. (1985). Structural-functional relationships of H+-secreting epithelia. Fed. Proc. 44: 2704-2709.

Maeda, M., Oshiman, K.-I., Tamura, and Futai, M. (1990). Human gastric (H/K)-ATPase gene: Similarity to (Na/K)-ATPase genes in exon/intron organisation but difference in control region. J. Biol. Chem. 265: 9027-9032.

Maeda, M. and Oshiman, K.I. (1992). The rat H,K ATPase β subunit gene and recognition of its control region by gastric DNA binding protein. J. Biol. Chem. 266: 21584-21588.

Magyar, J.P. and Schanchner, M. (1993). Genomic structure of the adhesion molecule of glia (AMOG, Na/K-ATPase β2 subunit). Nucleic Acids Res. 18: 6695-6696.

Morley, G.P., Callaghan, J.M., Rose, J.B., Toh, B.H., Gleeson, P.A., and van Driel, I.R. (1992). The Mouse Gastric H/K ATPase β Subunit: Gene Structure and Co-Ordinate Expression with the α Subunit During Ontogeny. J. Biol. Chem. 267: 1165-1174.

Newman, P.R., Greeb, J., Keeton, T.J., Reyes, A.A., and Shull, G.E. (1990). Structure of the human H,K-ATPase gene and comparison of the 5'-flanking sequences of the human and rat genes. DNA Cell Biol. 9: 749-762.

Okamoto, C.T., Karpilow, J.M., Smolka, A. and Forte, J.G. (1990) Isolation and characterization of gastric microsomal glycoproteins. Evidence for a glycosylated β-subunit of the H^+/K^+-ATPase. Biochim.Biophys.Acta 1037:360-372.

Pettitt, J.M., Toh, B.H., Callaghan, J.M., Gleeson, P.A., and van Driel, I.R. (1993). Gastric parietal cell development: Expression of the H/K ATPase subunits coincides with the biogenesis of the secretory membranes. Immun. Cell Biol. 71: 191-200.

Reuben, M.A., Lasater, L.S. and Sachs, G. (1990) Characterization of a β-subunit of the gastric H^+/K^+-transporting ATPase. Proc.Natl.Acad.Sci.U.S.A. 87:6767-6771.

Rabon, E.C. and Reuben, M.A. (1990). The mechanism and structure of the gastric H,K ATPase. Annu. Rev. Physiol. 52: 321-344.

Sachs, G., Besancon, M., Shin, J.M., Mercier, F., Munson, K., and Hersey, S. (1992). Structural Aspects of the Gastric H,K-ATPase. J. Bioenerg. Biomembrane. 24: 301-308.

Saxen, L. (1987). Organogenesis of the kidney (Cambridge: Cambridge University).

Shull, G.E. (1990) cDNA Cloning of the β-Subunit of the Rat Gastric H,K-ATPase. J.Biol.Chem. 265:12123-12126.

Sweetser, D.A., Birkenmeier, E.H., Hoppe, P.C., McKeel, D.W., and Gordon, J.I. (1988). Mechanisms underlying generation of gradients in gene expression within the intestine: an analysis using transgenic mice containing fatty acid binding protein-human growth hormone genes. Genes. Dev. 2: 1318-1332.

Toh, B.H., Gleeson, P.A., Simpson, R.J., Moritz, R.L., Callaghan, J., Goldkorn, I., Jones, C.M., et al. (1990) The 60-90 kDa parietal cell autoantigen associated with autoimmune gastritis is a β subunit of the gastric H+/K+-ATPase (proton pump). Proc.Natl.Acad.Sci.U.S.A. 87:6418-6422.

Verlander, J.W., Madsen, K.M., and Tisher, C.C. (1991). Structural-functional features of proton and bicarbonate transport in the rat collecting duct. Sem. Nephrol. 11: 465-478.

Wingo, C.S. and Cain, B.D. (1993). The Renal H-K-ATPase: Physiological Significance and Role in Potassium Homeostasis. Annu. Rev. Physiol. 55: 323-347.

Wingo, C.S. (1989). Active proton secretion and potassium absorption in the rabbit outer medullary collecting duct: functional evidence for H+-K+-ATPase. J. Clin. Invest. 84: 361-365.

Wingo, C.S., Madsen, K.M., Smolka, A., and Tisher, C.C. (1990). H-K-ATPase immunoreactivity in cortical and outer medullary collecting duct. Kidney Int. 38: 985-990.

Functional Expression of Gastric H,K-ATPase in Sf9 Cells using the Baculovirus Expression System.

Corné H. W. Klaassen, Tom J. F. Van Uem, Mariëlle P. De Moel, Godelieve L. J.
De Caluwé, Herman G. P. Swarts and Jan Joep H. H. M. De Pont
Department of Biochemistry
University of Nijmegen
P.O.Box 9101
6500 HB Nijmegen
The Netherlands

The mechanism of action of H,K-ATPase is still only partly understood. Site-directed
mutagenesis and chimere production might give useful tools to enlarge our insight in
structure-function relationship of this enzyme. Functional expression *in vitro* is a
prerequisite for such studies. The baculovirus expression system makes advantage of
the high level expression of certain viral proteins not essential for viral replication in
a cell line from the fall armyworm *Spodoptera frugiperda* (Sf9 cells) [Vlak and
Keus,1990]. Two such proteins are polyhedrin and p10 protein. Both proteins can
reach expression levels of more than 30% in Sf9 cells, dependent on the stage of
infection. Replacing polyhedrin or p10 coding sequences by homologous
recombination between wild-type viral DNA and a transfer vector DNA has led to
the production of recombinant viruses expressing large amounts of recombinant
proteins. Many complex membrane proteins as well as cytosolic proteins have
successfully been expressed using this system [Parker et al.,1991; Janssen et al.,1991;
Fafournoux et al.,1991; Li et al.,1992; Smith et al.,1992; De Tomaso et al.,1993]. In
order to produce multiple subunit proteins, simultaneous expression of two or more
proteins has been made possible by using either one transfer vector containing two
or more promoters [Belyaev and Roy,1993] or by co-infection of Sf9 cells using a
mixture of two or more recombinant baculoviruses. We therefore decided to study
the *in vitro* synthesis of H,K-ATPase subunits using the baculovirus expression
system.

NATO ASI Series, Vol. H 89
Molecular and Cellular Mechanisms
of H⁺ Transport
Edited by Barry H. Hirst
© Springer-Verlag Berlin Heidelberg 1994

Both H,K-ATPase α-subunit and ß-subunit cDNA's were subcloned separately into baculovirus transfer vectors pAcDZ1 [Zuidema et al.,1990] and pAcAS3 [Vlak et al.,1990]. These vectors can be used for expression of recombinant proteins from the polyhedrin promoter and p10 promoter, respectively. Both transfer vectors also contain a ß-galactosidase (Lac Z) reporter gene expressed from the Drosophila HSP-70 promoter for convenient screening for recombinant viruses in plaque assays.

Figure 1 shows that infection of Sf9 cells with purified recombinant virus leads to expression of both H,K-ATPase α-subunit (left part) and ß-subunit (middle part).

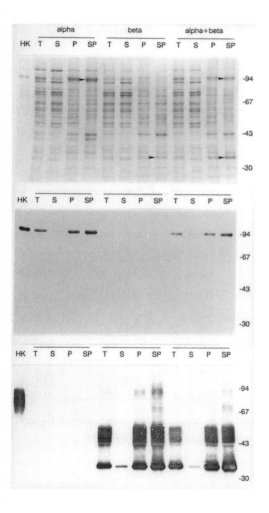

Figure 1. Analysis of recombinant proteins expressed in Sf9 cells. Top figure shows a Coomassie stained SDS-PAGE gel of protein samples (2.5 µg per lane) from cells infected for 48 h with virus expressing either the H,K-ATPase α-subunit, the ß-subunit or both alpha and beta subunits. T = total protein; S = supernatant and P = pelleted membranes after 1 h centrifugation of cell lysate at 100,000 x g;SP = pellet obtained by centrifugation of P over a discontinuous gradient of 20% and 40% sucrose. The arrows indicate the position of protein bands representing H,K-ATPase α-subunit (at 95 kDa) and unglycosylated ß-subunit (at 34 kDa) in the purified membrane fraction (SP). Middle figure shows Western blot of proteins in top figure identified by a MAb against the α-subunit (95-111) [Bayle et al.,1992]. Bottom figure shows a Western blot using a MAb against the H,K-ATPase ß-subunit (2G11) [Chow and Forte,1992]. The positions of molecular mass standards (in kDa) are indicated. [From Klaassen et al, 1993]

Expression of the α-subunit leads to formation of an extra protein band clearly visible in Coomassie stained SDS-PAGE gels, with an apparent molecular weight of 95 kDa, which is exactly the same size as that of native H,K-ATPase purified from pig gastric mucosa [Swarts et al.,1991]. This protein band is furthermore identified by Western blotting (Fig. 1, middle panel).

Expression of the ß-subunit also leads to production of an extra protein band visible in Coomassie blue stained gels and reactive with MAb's against the ß-subunit (Fig. 1, lower panel). The ß-subunit appears both as a single protein band of 34 kDa and partly as a smear of approximately 40-50 kDa. This smear can be explained by different glycosylation patterns of the protein produced in the insect cells since the rat ß-subunit contains 7 potential N-glycosylation sites and Sf9 cells are known to be capable of formation of N-linked sugars on peptide chains expressed by infection with recombinant baculoviruses [Vlak and Keus, 1990]. The protein band at 34 kDa could be identified as the naked core protein of the ß-subunit. Indeed when infected cells were incubated in the presence of 5 μg/ml tunicamycin, an inhibitor of the N-glycosylation process, the smear identified by the anti-ß-subunit antisera disappeared, leading to an increase of the immunoreactive protein band at 34 kDa (fig 2).

Although the *in vitro* produced ß-subunit in the absence of tunicamycin is glycosylated, glycosylation patterns were not as extensive as in the native enzyme which appeared on SDS-PAGE as a protein with an apparent molecular mass of 60-80 kDa, while the glycosylated ß-subunit as produced by Sf9 cells had a molecular mass between 40 and 50 kDa.

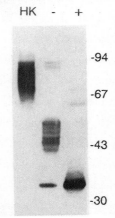

Figure 2. Western blot analysis of total protein samples from Sf9 cells infected with DZß virus expressing the H,K-ATPase ß-subunit. Protein samples (approximately 2.5 µg) were taken 48 h after infection. Cells were incubated in the presence (+) or absence (-) of 5 µg/ml tunicamycin. HK represents 500 ng of H,K-ATPase from pig gastric mucosa. The blot was incubated with monoclonal antibody 2G11 [Chow and Forte,1992] to detect expression of the ß-subunit. At the right, the positions of molecular mass standards (in kDa) are indicated. [From Klaassen et al, 1993]

The majority of H,K-ATPase subunits produced in insect cells were located in Sf9 membranes as demonstrated by immunofluorescence analysis (fig 3). This is furthermore supported by the finding that purified Sf9 membranes show an enrichment of each subunit by a factor 2.8 compared to total protein samples as calculated by ELISA (fig 4) and demonstrated by immunoblot (fig 1, lanes SP vs. T). Nearly all of the immunoreactive protein in the clarified cell lysate (lane T) can be recovered by centrifugation at 100,000 x g (lane P vs. S), indicating either membrane association or membrane localisation.

Since no apparent function for the ß-subunit of gastric H,K-ATPase is known, it was postulated that a presynthesized ß-subunit could be necessary for correct assembly with the α-subunit for transport and insertion into the cell membrane. Expression of either the α- or ß-subunit alone, however, leads to accumulation of recombinant protein subunits in Sf9 membranes, suggesting that one subunit is not essential for correct transport and insertion of the other subunit into Sf9 membranes. This seems to be a unique feature of Sf9 cells, since this has also been described for Na,K-ATPase [Blanco et al.,1993] while studies with other expression systems indicate that both with H,K-ATPase and with Na,K-ATPase the two subunits are necessary for transport and insertion into cell membranes [Gottardi and Caplan,1993; McDonough et al.,1990].

The above reported experiments clearly show that infection of Sf9 cells with viruses containing either the α- or the ß-subunit leads to formation of proteins, which react positively with antibodies against the respective subunits and are located in the plasma membrane fraction of these cells. We next studied the functional properties: $^{86}Rb^+$ uptake in Sf9 cells, ATPase activity and ATP-phosphorylation. Although these activities were present they could not be inhibited by SCH 28080, indicating that expression of either the α- or the ß-subunit alone did not lead to formation of an active H,K-ATPase.

We next infected Sf9 cells with a mixture of the two viruses in different ratios. Fig. 3 shows that under these conditions both α- and ß-subunits were expressed but in most cells one of either subunit was expressed primarily. A reason for this finding might be that a cell once infected by the one virus could have become less susceptible for infection by the second virus. Functional studies as described above

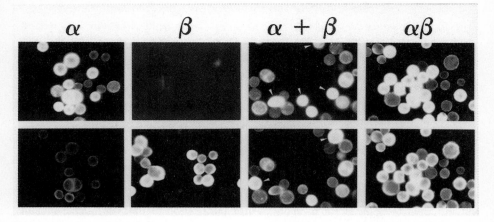

Figure 3. Immunohistochemistry of infected Sf9 cells expressing H,K-ATPase α- and /or β-subunits. Cells were infected with α-virus (α), β-virus (β), α-virus + β-virus (α + β) or αβ-virus (αβ). Upper half shows reaction with a polyclonal antibody against the α-subunit (565-585, gift of Michael Caplan) and lower half shows reaction of the same cells with MAb against β-subunit (2G11, Chow and Forte,1992). Arrows in α + β indicate cells expressing preferentially one subunit.

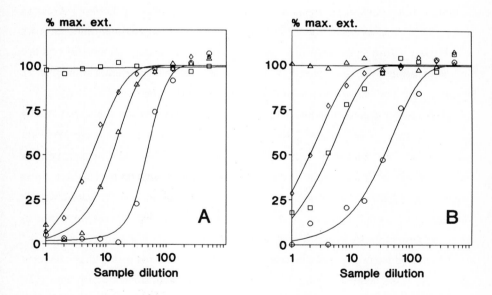

Figure 4. Inhibition ELISA of H,K-ATPase in Sf9 membranes. Samples represent pig H,K-ATPase (o), purified membranes from Sf9 cells infected with viruses expressing the α-subunit (△), the β-subunit (□) and both the α- and β-subunit (◊). A. ELISA with MAb 5B6 (van Uem et al, 1990). B. ELISA with MAb 2G11 (Chow and Forte,1992).

were again negative. We then decided to construct a single virus which expressed both subunits.

We used two successive recombination steps to produce a single recombinant baculovirus expressing both α-subunit and ß-subunit from the polyhedrin and p10 promoter, respectively. First the LacZ coding sequences were deleted from pAcDZ1 by digesting this transfer vector with EcoRI and XbaI. DNA ends were made blunt end and were religated. After cloning of the H,K-ATPase α-subunit cDNA into the BamHI cloning site, this transfer vector was used for recombination into the polyhedrin locus to produce a recombinant virus expressing the H,K-ATPase α-subunit from the polyhedrin promoter. After purification of this virus the ß-subunit cDNA was cloned into a transfer vector and used for recombination into the p10 locus of this virus generating a new virus expressing the α-subunit from the polyhedrin promoter as well as the ß-subunit from the p10 promoter and ß-galactosidase from the HSP70 promoter.

Infection of the purified virus in Sf9 cells led to the production of both α- and ß-subunits in the particulate fraction of these cells as is clear from both Western blotting (Fig. 1, lanes 9-12) and ELISA studies (Fig 4). In contrast to infection with two separate viruses immunohistochemical studies showed equal amounts of reactive antibodies against the α-subunit as well as the ß-subunit in individual cells (Fig 4), indicating that both subunits were indeed expressed in the same cell.

With these infected cells and membranes isolated thereof the functional studies were repeated. With either $^{86}Rb^+$ uptake studies or K^+-stimulated ATPase measurements no functional activity was found but with ATP-phosphorylation measurements positive results were obtained. Phosphorylation reactions were measured at 0.1 μM ATP in the presence of 1 mM $MgCl_2$, since at this very low ATP concentration phosphorylation is still possible and aspecific phosphorylation is relatively low. Only membranes containing both H,K-ATPase subunits exhibit a K^+ and SCH 28080 sensitive phosphorylation capacity (Fig.5), indicating that only cells expressing both subunits contain an active H,K-ATPase. When an ATP concentration of 2.0 μM instead of 0.1 μM was used the absolute SCH 28080 sensitive phosphorylation increased approximately twofold but phosphorylation levels not inhibitable by SCH 28080, increased even more (up to tenfold).

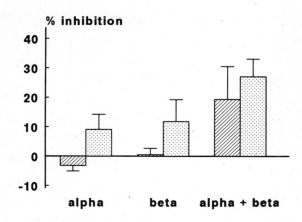

Figure 5. Phosphorylation capacities of purified Sf9 membranes. Cells were infected as described in Fig. 1. Phosphorylation capacities were measured in the presence of 0.1 μM ATP and 1.0 mM MgCl₂, with and without either 100 μM SCH 28080 (dashed) or 10 mM KCl (dotted). Values are expressed as percentage inhibition of control phosphorylation levels.

SCH 28080- and K⁺-sensitive phosphorylation capacities of Sf9 membranes containing both H,K-ATPase subunits are low compared to purified enzyme (1 pmol/mg versus 2 nmol/mg). Since estimates from ELISA experiments and Coomassie blue staining indicate that at least 5% of the membrane proteins in the infected cells are the two subunits of H,K-ATPase, this indicates that large amounts of produced subunits do not form a functional enzyme. This defect can be explained by several options. First it is yet unknown if all proteins produced are correctly folded in the Sf9 membranes, which is of course a prerequisite for the formation of an active enzyme. Secondly it is unknown if all α-subunits produced are also involved in formation of an α,ß heteroduplex molecule. Thirdly it could be possible that once some functional H,K-ATPase has been produced, the activity of the *in vitro* produced enzyme interferes with Sf9 cell metabolism, because of cellular alkalinization, inhibiting the formation of more functional enzyme. Fourthly it might be that the glycosylation pattern of the expressed ß-subunit, which is clearly different from that of the native subunit, is not optimal for full activity. Lastly it might be that native H,K-ATPase is not active as a single α,ß-oligomer but as a dimer or polymer of equimolar amounts of both α- and ß-subunits [Rabon et al.,1988]. If the latter is true, successful formation of large amounts of a functional H,K-ATPase *in vitro* would be even more dependent on the expression system used and difficult to establish. Nevertheless, because baculovirus infected Sf9 cells are capable of forming a functional H,K-ATPase we conclude that the baculovirus expression system can provide a useful method for studying the characteristics of gastric H,K-ATPase.

Acknowledgements: The authors wish to thank Dr. Gary Shull for his donation of H,K-ATPase cDNA clones, Dr. Just Vlak for the transfer vectors pAcDZ1 and pAcAS3, Sf9 cells and wild type AcNPV, Drs. Michael Caplan, John Forte and Annick Soumarmon for their antibodies, Dr. Björn Wallmark for his gift of SCH 28080 and Dr. Wim De Grip for his advice in the initial phase of this project. This work was sponsored by the Netherlands Foundation for Scientific Research (NWO) under grant no. 900-522-086.

Bayle D, Robert JC, Bamberg K, Benkouka F, Cheret AM, Lewin MJM, Sachs G, Soumarmon A (1992) Location of the cytoplasmic epitope for a K$^+$-competitive antibody of the (H+,K$^+$)-ATPase. J Biol Chem 267: 19060-19065

Belyaev AS, Roy P (1993) Development of baculovirus triple and quadruple expression vectors: co-expression of three or four bluetongue virus proteins and the synthesis of bluetongue virus-like particles in insect cells. Nucleic Acids Res 21: 1219-1223

Blanco G, Xie ZJ, Mercer RW (1993) Functional expression of the alpha2-isoforms and alpha3-isoforms of the Na,K-ATPase in baculovirus-infected insect cells. Proc Natl Acad Sci USA 90: 1824-1828

Chow D, Forte JG (1992) A monoclonal antibody, specific for the cytoplasmic domain of the β-subunit, inhibits gastric H,K-ATPase activity. FASEB J 6: A1187(Abstract)

De Tomaso AW, Xie ZJ, Liu GQ, Mercer RW (1993) Expression, targeting, and assembly of functional Na,K- ATPase polypeptides in baculovirus-infected insect cells. J Biol Chem 268: 1470-1478

Fafournoux P, Ghysdael J, Sardet C, Pouyssegur J (1991) Functional expression of the human growth factor activatable Na$^+$/H$^+$ Antiporter (NHE-1) in baculovirus-infected cells. Biochemistry 30: 9510-9515

Faller LD, Jackson R, Malinowska DH, Mukidjam E, Rabon E, Saccomani G, Sachs G, Smolka A (1982) Mechanistic aspects of gastric (H$^+$+K$^+$)-ATPase. Ann N Y Acad Sci 402: 146-163

Gottardi CJ, Caplan MJ (1993) An ion-transporting ATPase encodes multiple apical localization signals. J Cell Biol 121: 283-293

Janssen JJM, Mulder WR, De Caluwe GLJ, Vlak JM, De Grip WJ (1991) In vitro expression of bovine opsin using recombinant baculovirus: the role of glutamic acid (134) in opsin biosynthesis and glycosylation. Biochim Biophys Acta 1089: 68-76

Klaassen CHW, Van Uem TJF, De Moel MP, De Caluwé GLJ, Swarts HGP, de Pont JJHHM (1993) Functional expression of H,K-ATPase using the baculovirus expression system. FEBS Lett 329: 277-282.

Li ZP, Smith CD, Smolley JR, Bridge JHB, Frank JS, Philipson KD (1992) Expression of the cardiac Na$^+$-Ca^{2+} exchanger in insect cells using a baculovirus vector. J Biol Chem 267: 7828-7833

McDonough AA, Geering K, Farley RA (1990) The sodium pump needs its β subunit. FASEB J 4: 1598-1605

Parker EM, Kameyama K, Higashijima T, Ross EM (1991) Reconstitutively active G-protein-coupled receptors purified from baculovirus-infected insect cells. J Biol Chem 266: 519-527

Rabon EC, Gunther RD, Bassilian S, Kempner ES (1988) Radiation inactivation analysis of oligomeric structure of the H,K-ATPase. J Biol Chem 263: 16189-16194

Smith CD, Hirayama BA, Wright EM (1992) Baculovirus-mediated expression of the Na$^+$/glucose cotransporter in Sf9 cells. Biochim Biophys Acta 1104: 151-159

Swarts HGP, Van Uem TJF, Hoving S, Fransen JAM, De Pont JJHHM (1991) Effect of free fatty acids and detergents on H,K-ATPase - The steady-state ATP phosphorylation level and the orientation of the enzyme in membrane preparations. Biochim Biophys Acta 1070: 283-292

Van Uem TJF, Peters, W.H.M. De Pont, J.J.H.H.M. (1990) A monoclonal antibody against pig gastric H$^+$/K$^+$-ATPase, which binds to the cytosolic E$_1$.K$^+$ form. Biochim Biophys Acta 1023: 56-62

Vlak JM, Keus RJA (1990) Baculovirus expression vector system for production of viral vaccines. In: Viral Vaccines. Wiley-Liss, New York, pp 91-128

Vlak JM, Schouten A, Usmany M, Belsham GJ, Klinge-Roode EC, Maule AJ, Van Lent JWM, Zuidema D (1990) Expression of cauliflower mosaic virus gene I using a baculovirus vector based upon the P10 gene and a novel selection method. Virology 178: 312-320

Zuidema D, Schouten A, Usmany M, Maule AJ, Belsham GJ, Roosien J, Klinge-Roode EC, Van Lent JWM, Vlak JM (1990) Expression of cauliflower mosaic virus gene I in insect cells using a novel. Polyhedrin-based baculovirus expression vector. J Gen Virol 71: 2201-2209

Structural and Functional Integrity of H,K-ATPase Depends on its β-Subunit

John G. Forte and Dar C. Chow
Dept. of Molecular & Cell Biology
Univ. of California
Berkeley, Ca 94720, USA

H,K-ATPase compared to other P-Type ATPases

The primary gastric H^+ pump, the H,K-ATPase, belongs to the family of cation pump enzymes called P-type ATPases, which include the Na,K-ATPase and the Ca-ATPase (Jørgenson & Andersen 1988; Inesi & Kirtley 1992). This family of pump proteins all have the ability to form a phosphoenzyme intermediate, or E-P, in their catalytic cylcle, and predictably, they share a relatively high degree of sequence homology. Among the P-type ATPases, the H,K-ATPase and the Na,K-ATPase have the highest degree of homology (~60% identity), sharing at least two additional unique features: they are both cation exchange transporters, involving the cellular uptake of K^+ in exchange for the respective Na^+ or H^+ export from the cell; and both pumps consist of $\alpha\beta$-subunit heterodimers. The α-subunits of both Na,K- and H,K-ATPase have eight transmembrane segments, and ~70% of the 110 kDa mass is distributed within the cytoplasmic domain where ATP binding and phosphoenzyme formation occur. Thus the α-subunit is frequently called the catalytic subunit of the respective enzyme. The α-subunit of the Na,K-ATPase contains the ouabain binding site localized to the extracellular domain (Lingrel 1992). Although the H,K-ATPase does not bind ouabain, an omeprazole binding site has been localized to the extracellular domain of its α-subunit (Sachs et al. 1989). Further information on the structural and functional activities of the α-subunit of the H,K-ATPase can be found in the article of Sachs et al. in this volume.

The purpose of the present work is to review some of the structural and functional features of the much less studied β-subunit of these cation transporting ATPases, particularly the H,K-ATPase. The β-subunit of the H,K-ATPase is a glycoprotein with a short N-terminal cytoplasmic tail, a single transmembrane segment, and over 70% of the 34 kDa peptide mass is distributed on the extracellular side of the plasma membrane. In addition, the C-terminal extracellular domain is heavily N-glycosylated, contributing to an apparent molecular mass in the range of 60-80 kDa on SDS-PAGE.

When β-subunits from the H,K-ATPase are compared with various β-subunit isoforms of the Na,K-ATPase, a number of homologies are evident, including

NATO ASI Series, Vol. H 89
Molecular and Cellular Mechanisms
of H^+ Transport
Edited by Barry H. Hirst
© Springer-Verlag Berlin Heidelberg 1994

identical location of six cyteine residues within the extracellular domain (Canfield et al. 1990). This high degree of conservation for functionally divergent enzymes suggested some fundamental role for these cysteine residues, possibly stabilizing tertiary structure and contributing to the ion transport function. Indeed, studies on Na,K-ATPase demonstrated that the extracellular cysteine redidues of the β-subunit are oxidized as three disulfide bonds, and that reduction of the disulfides leads to inhibition of enzyme activity (Kawamura et al. 1984, 1985; Kirley 1989, 1990).

Disulfide bonds stabilize the β-subunit and the holoenzyme

Tests on the gastric H,K-ATPase showed that reduction of disulfides within the β-subunit also inhibited H,K-ATPase activity, as well as partial enzymatic reactions, such as K^+-stimulated p-nitrophenylphosphatase (pNPPase) activity (Chow et al. 1992). As indicated by Figure 1A and 1C, rather strong reducing conditions were required to inhibit enzyme function, suggesting that the essential disulfide bonds were highly stabilized by the structural order of the protein. Furthermore, K^+ and its congeners, Tl^+ and Rb^+, were shown to further stabilize the disulfide bonds in the β subunit, and thus protect H,K-ATPase activity from reductive inhibition (Fig. 1A and 1B). Binding of the K^+ congeners to the enzyme for these protective effects was in the same order and concentration range as for stimulating phosphatase activity of the enzyme (Forte et al. 1976). Since the hydrolysis of pNPP, or phosphatase activity, is catalyzed by the K^+ form of the enzyme, referred to as the E_2K conformer, this further supports a role for K^+ and the E_2K form in maintaing structural stability of the disulfide bonds. No change in the state of cysteine residues within the α subunit was observed over the time course of the treatment (Chow et al. 1992). These data strongly suggest that disulfide bonds in the extracellular domain of the β subunit are important for establishing a functional conformation of the H,K-ATPase, and conversely, that stability of the disulfide bonds is related to the conformational state of the H,K-ATPase, i.e. the disulfide bonds are more stable in the E_2K conformational state.

Stability of the H,K-ATPase to alcohols and denaturants

The functional and conformational stability of the H,K-ATPase can also be measured in response to treatment with denaturants, such as organic solvents and detergents. Dose-dependent inactivation by ethanol is shown in Figure 1D, along with the protective effects provided by KCl. Similar experiments were performed with a series of alcohols, including methanol, ethanol, 2-propanol, 1-propanol and butanol (Forte & Chow 1993). The measured IC_{50} values were inversely correlated with the respective oil/water partition coefficients. Several conclusions can be drawn from these data:

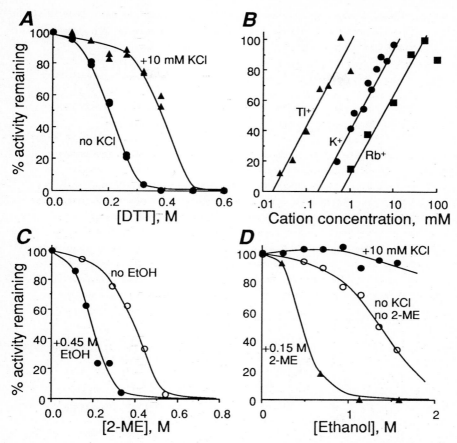

Figure 1. Inhibition of H,K-ATPasease activity by reducing agents and denaturants, and protective effects of K$^+$. Gastric microsomes were pre-incubated with the indicated concentration of reagent for 20 min at 44°C; samples were diluted 40-fold and K$^+$-stimulated pNPPase activity measured. A. Inhibition by dithiothreitol (DTT) and protection by 10 mM KCl. B. Protective effects of K$^+$-congeners. C. Inhibition by 2-mercaptoethanol (2-ME) in the absence and presence of 0.45 M ethanol. D. Inhibition by ethanol in absence and presence of protective reagents (KCl) and reducing reagents.

i) K$^+$ stabilizes the H,K-ATPase from inhibition by alcohol, as it did for reducing agents. ii) The potency of inhibition by alcohols is a function of their lipid solubility. iii) For all destabilizing agents tested, alcohols, detergents and reducing agents, the inhibition curves were sigmoidal, i.e. the H,K-ATPase resists inactivation at low concentrations, but is readily and irreversibly inactivated when a critical threshold is achieved. These latter relations are a feature of protein conformational stability. High sensitivity at critical threshold is thought to be due to a cooperative relationship among all bonding forces that stabilize the conformation of a protein.

Cooperativity of bonding forces and conformational stability

We further examined the issue of enzyme stability by measuring combined effects of denaturing and reducing conditions (Forte & Chow). In the presence of a totally non-toxic level of 3.5% (~0.45 M) ethanol, H,K-ATPase became much more sensitive to reducing reagent (Fig. 1C), and conversely, in the presence of reducing reagent, H,K-ATPase was more sensitive to alcohol (Fig. 1D), and these cooperative destabilizing effects were antagonized by K^+. In Figure 2 we have compared the measured reduction of disulfide bonds in the β subunit with the change in functional activity during various destabilizing conditions, with and without the protective effects of K^+. The high degree of correlation indicates that stability of the H,K-ATPase is contributed by non-covalent interactions within holoenzyme as well as disulfide bonds specifically within the β subunit, and that the holoenzyme is more stable in the E_2K conformational state.

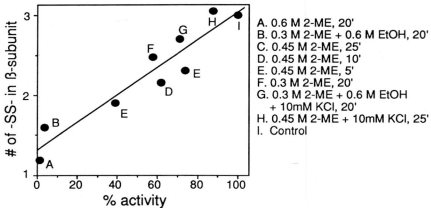

Figure 2. Relationship between number of remaining disulfide bonds (-SS-) in the β-subunit of H,K-ATPase and the residual enzyme activity after various destabilizing conditions. Microsomes were treated at 44°C with reagents and time indicated, then diluted for assay of % remaining enzyme activity (x-axis) and for titration of the % of remaining -SS- present in the β-subunit (y-axis) as described by Chow et al. (1992).

Inhibition of H,K-ATPase by an antibody against the β-subunit

In a recent survey of various monoclonal antibodies (Mab) directed against the β-subunit of the H,K-ATPase we found one that inhibited enzyme

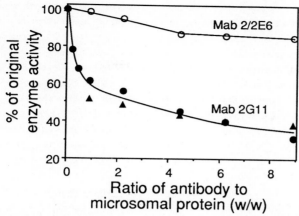

Figure 3. Titration of H,K-ATPase activity remaining after treatment with two different monoclonal antibodies against the β-subunit. Gastric microsomes were incubated with various amounts of purified Mab 2G11 or Mab 2/2E6 for 2 h (circles) or overnight (triangles), then assayed for K+ stimulated pNPPase. Data are normalized to % of specific activity of the untreated gastric microsomes.

activity (Chow & Forte 1993). As exemplified in Figure 3, inhibition of K^+-stimulated pNPPase by Mab 2G11 reached a maximum of about 50% when the amount of antibody protein was roughly equivalent to microsomal protein. Additional tests revealed that binding site for Mab 2G11 was definitively on the cytoplasmic domain of the β-subunit, and most likely within the first 36 amino acids from the N-terminus (Chow & Forte 1993).

There are two general explanations for the observed 50% reduction of enzyme activity by 2G11: i) 50% of the enzyme population might be bound and completely inhibited by 2G11, while the remainder was inaccessible to the antibody; or ii) 100% of the enzyme might be reduced to 50% of normal activity by 2G11. We attempted to distinguish between these two possibilities by studying the kinetics of K^+ activation. Gastric microsomes treated with affinity purified 2G11 were compared with control samples of microsomes, and after a 2 h incubation at 25°C, aliquots were assayed for ATPase and pNPPase activities at various concentrations of K^+.

Data for activation of ATPase activity by [K^+] are shown in Figure 4A in conventional form and as the double reciprocal tranformation. Treatment with Mab 2G11 reduced the V_{max} of ATPase activity without appreciable change in the apparent K_m for activation by K^+. Linear regression analysis revealed that for K^+-stimulated ATPase (N=3), V_{max} was decreased from 1.14 ±0.06 to 0.50 ±0.10 μmol/min/mg protein after treatment with 2G11 antibody, while the apparent K_m for K^+ activation of ATP hydrolysis was not changed (0.19 ±0.03 mM K^+, control; 0.20 ±0.03 mM K^+, 2G11-treated).

Parallel experiments to study the effect of 2G11 on the kinetics of K^+ activation of pNPPase activity are shown in Figure 4B. Best fits of the double reciprocal plots showed that V_{max} was decreased by Mab 2G11 (from 1.26 ±0.20 to 0.85 ±0.11 µmole/min/mg protein; N=4), but more interestingly, K_m was also changed. The apparent K_m for K^+ activation was 2.30 ±0.11 mM was increased about 2-fold to 4.67 ±0.45 mM for the 2G11-treated microsomes. Because of the significant change in the apparent K_m for K^+ activation of pNPPase activity, we can conclude that inhibition by Mab 2G11 is not simply the elimination of a population of enzymatic sites, but must also involve changes in the affinity of the enzyme for ligands.

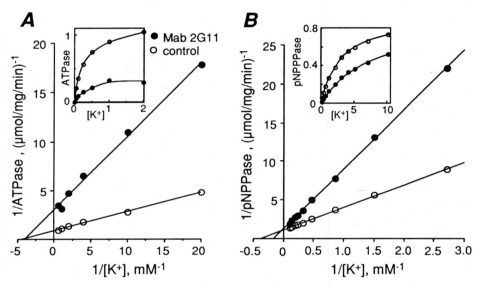

Figure 4. The effects of $[K^+]$ on the ATPase (A) and pNPPase (B) activities of control gastric microsomes (open circles) and 2G11-bound gastric microsomes (closed circles). Gastric microsomes were incubated with purified antibody for at least 2 h before enzymatic activity assays. Inset shows enzyme activity as a function of $[K^+]$; main graph is a Lineweaver-Burke transformation. Straight lines fit by method of least squares.

Conclusions

We have shown that the β-subunit of the H,K-ATPase is essential for the structural and functional stability of the holoenzyme. Reduction of disulfide bonds within the extracellular domain of the β-subunit leads to enzymic inactivation, including those steps known to be associated with the cytoplasmic domain of the α-subunit (Chow et al. 1992). On the other hand, binding of K^+ to its low affinity site within the cytoplasmic domain of the enzyme, not only alters conformational state of the α-subunit, but also changes the stability of the disulfide bonds within the β-subunit. Thus, the cytoplasmic domain of the α-subunit is conformationally linked to the extracellular domain of the β-subunit by some allosteric noncovalent interactions between the two subunits. The cooperative effects of reducing reagents and organic solvents in destabilizing enzyme function adds further support for association between the two subunits.

Studies with simple soluble enzyme systems have established that cooperative intramolecular interactions stabilize the structural and functional activity of the enzyme (Creighton 1990). For example, stability of disulfide bonds within the protein was quantitatively related to functional and conformational stability of the enzyme. The gastric H,K-ATPase is a much more complex system, being a membrane protein made up of distinct subunits. Organic solvents destabilize H,K-ATPase activity by interfering with noncovalent interactions within the protein or at the protein/lipid interface; at the same time they destabilize disulfide bonds in the β-subunit. Conversely, reduction of disulfide bonds in the β-subunit destabilizes the holoenzyme by making it more susceptible to denaturants. Thus, evidence for structural and functional cooperativity is presented for a complex heterodimeric membrane protein.

A monoclonal antibody specific for the short cytoplasmic domain of the β-subunit (Mab 2G11) inhibited the H,K-ATPase. Kinetic studies revealed that for p-nitrophenyl phosphate hydrolysis the affinity for K^+ was decreased by two-fold. The site for K^+ activation of pNPPase is known to be on the cytoplasmic side of the enzyme. Since the binding of K^+ at this cytoplasmic site also stabilizes disulfide bonds in the β-subunit and the structural integrity of the holoenzyme, we conclude that conformation of the β-subunit is integrally linked to structural and functional activities of the H,K-ATPase, especially the K^+-induced conformational states. These data further suggest that the β-subunit may be involved in some aspect of K^+-translocation (e.g. E_1K-E_2K transition) during the pumping cycle.

By analogy, the β-subunit of the Na,K-ATPase may also function in the same way, especially since the H,K-ATPase β-subunit has been shown to substitute for the Na,K-ATPase β-subunit in providing a functional holoenzyme (Horisberger et al. 1991). Coexpression of Na,K-ATPase α-subunit and H,K-ATPase β-subunit in yeast produced hybrids with altered ouabain binding affinity and different ouabain/K^+ antagonism compared to the Na,K-ATPase αβ complex (Eakle et al. 1992). Jaisser et al. (1992) reported that for Na,K-ATPase expressed in Xenopus oocytes, the K^+ activation constants depended on the isoform of the β-subunit. Finally, Capasso et al. (1992) have shown that a segment of the Na,K-ATPase β-subunit was protected against tryptic digestion by K^+ and its congeners; and this segment of β-subunit appeared to associate with the proteolyzed Na,K-ATPase remnant that retained the ability to occlude K$^+$. These observations are all consistent with the postulate that the β-subunits of these two closely related enzymes have some K$^+$-related role in the activities of their respective pumps.

References

Canfield VA, Okamoto CT, Chow D, Dorfman J, Gros P, Forte JG, Levenson R (1990) Cloning of the H,K-ATPase β-subunit: Tissue-specific expression, chromosomal assignment, and relationship to Na,K-ATPase β-subunits. J Biol Chem 265:19878-19884.

Chow DC, Browning CM, Forte JG (1992) Gastric H,K-ATPase activity is inhibited by reduction of disulfide bonds in the β-subunit. Am J Physiol (Cell Physiol 32) 263:C39-C46.

Chow DC and Forte JG (1993) Characterization of the β-sununit of the H,K-ATPase using an inhibitory monoclonal antibody. Am J Physiol (Cell Physiol 32) in press.

Creighton TE (1990) Protein folding. Biochem. J. 270:1-16.

Eakle KA, Kim KS, Kabalin MA, Farley RA (1992) High-affinity ouabain binding by yeast cells expressing Na$^+$,K$^+$-ATPase α subunits and the gastric H$^+$,K$^+$-ATPase β subunit. Proc Natl Acad Sci, USA. 89:2834-2838.

Forte JG and Chow DC (1993) Structural and functional significance of the gastric H,K-ATPase. In: Domschke W, Konturek S (eds) The Stomach. Springer-Verlag, Berlin, in press.

Forte JG, Ganser AL, Ray TK (1976) The K$^+$-stimulated ATPase from oxyntic glands of gastric mucosa. In: Kasbekar DK, Sachs G, Rehm W (eds) Gastric Hydrogen Ion Secretion. Dekker, New York, pp.302-330.

Horisberger JD, Jaunin P, Reuben MA, Lasater L, Chow DC, Forte JG, Sachs G, Rossier BC, Geering K (1991) The H,K-ATPase β-subunit can act as a surrogate for the β-subunit of the Na,K pump. J Biol Chem 26:19131-19134.

Inesi G, Kirtley MR (1992) Structural Features of Cation Transport ATPase. J Bioenerg Biomembr 24:271-284

Jaisser F, Canessa CM, Horisberger JD, Rossier BC (1992) Primary sequence and functional expression of a novel ouabain-resistant Na,K-ATPase. The beta subunit modulates potassium activation of the Na,K-pump. J Biol Chem 267:16895-16903.

Jørgensen PL, Andersen JP (1988) Structural basis for E1-E2 conformational transitions in Na,K-ATPase and Ca-ATPase. J Membrane Biol 103:95-120

Kawamura M, Nagano K (1984) Evidence for essential disulfide bonds in the β-subunit of (Na$^+$+K$^+$)-ATPase. Biochim Biophys Acta 774:188-192.

Kawamura M, Ohmizo K, Morohashi M, Nagano K (1985) Protective effect of Na$^+$ and K$^+$ against inactivation of (Na$^+$+K$^+$)-ATPase by high concentrations of 2-mercaptoethanol at high temperatures. Biochim Biophys Acta 821: 115-120.

Kirley TL (1989) Determination of three disulfide bonds and one free sulfhydryl in the β-subunit of (Na,K)-ATPase. J Biol Chem 264:7185-7192.

Kirley TL (1990) Inactivation of (Na$^+$,K$^+$)-ATPase by β-mercaptoethanol. J Biol Chem 265:4227-4232.

Lingrel JB (1992) Na,K-ATPase: isoform structure, function, and expression. J Bioenerg Biomembr 24:263-270.

Sachs, G., J. Kaunitz, J. Mendlein, and B. Wallmark. (1989) Biochemistry of gastric acid secretion. In: Handbook of Physiology - The Gastrointestinal System. Salivary, Gastric, Pancreatic, & Hepatobiliary Secretion. Bethesda, MD: Amer Physiol Soc Sect. 6, vol.III, chapt. 12, pp. 229-254.

The Topology of the α, β Subunits of the Gastric H/K ATPase

Jai Moo Shin, Krister Bamberg, Marie Besancon, Keith Munson, Frederic Mercier, Dennis Bayle, Steve Hersey* and George Sachs

UCLA and Wadsworth VAMC, Los Angeles
and
* Emory University, Atlanta

Supported by USVA SMI and NIH grant #'s 40615, 41301 and 14752.

INTRODUCTION

Knowledge of the secondary structure of the gastric H/K ATPase, a P type transport ATPase, allows deductions as to the ion transport pathway across the membrane spanning domain of the large, 1033 amino acid catalytic α subunit and as to the site of interaction between the smaller, glycosylated, ß subunit and the α subunit. The structure of this region of the enzyme also allows us to understand the molecular basis for inhibition of acid secretion by acid pump inhibitors, such as the substituted benzimidazoles and the K^+ competitive type of H/K ATPase inhibitor.

Secondary structure analysis of the H/K became possible after the primary sequence of a membrane spanning protein was available. Both subunits have been cloned from various species such as rat, rabbit or hog (1,2,3,4,5,6).

There are various methods available for determining secondary structure. The most commonly used method is to predict the location of membrane spanning α helices in the primary sequence using hydropathy plots (7). These plots are based on determining a moving average of hydrophobic, neutral and hydrophilic amino acids using a variety of scales. When the average is greater than a predetermined value for 11 or more amino acids, this sector is regarded as potentially membrane spanning. The most exact method is crystal structure derived from 2 or 3 dimensional crystals analysed by electron diffraction or X-ray.

Alternative methods use sites of labelling with cytoplasmic, extracytoplasmic or membrane directed chemical probes, determination of the

NATO ASI Series, Vol. H 89
Molecular and Cellular Mechanisms
of H⁺ Transport
Edited by Barry H. Hirst
© Springer-Verlag Berlin Heidelberg 1994

peptides remaining in the membrane following cleavage of the extramembranal domain (usually cytoplasmic), determination of the sidedness of epitopes for antibodies and development of putative membrane segement constructs for in vitro translation with and without microsomes. Most of these methods have been applied for analysis of the topology of the P type ATPases. In this short review we shall confine our attention to the gastric H/K ATPase enzyme. However, it is likely that what is true of the secondary structure of this enzyme will be applicable to the Na/K and Ca ATPases.

Four of these general methods have been applied in our laboratory in order to determine the number and nature of the membrane spanning segments of the α subunit. The first is to define the membrane segments as the residues left after cytoplasmic cleavage with trypsin ; the second is to define those sites labeled by extracytoplasmic reagents and the third is to define sidedness of antibody epitopes. In the fourth we have generated constructs of putative membrane spanning segments and determined, by in vitro translation in the presence and absence of microsomes, whether such sequences can act as membrane anchor or stop transfer sequences. We have performed a similar tryptic cleavage analysis of the ß subunit, and in addition have determined which region of the α subunit presumably interacts with the ß subunit, by biochemical methods and by deduction from the overlap between the 2 subunits of an epitope for a monoclonal antibody mAb 146-14 (8).

(a)Biochemical analysis of the H/K ATPase membrane topology

The hog enzyme is prepared as intact, cytoplasmic side out vesicles. Tryptic cleavage of the enzyme, followed by removal of the cytoplasmic fragments by washing, results in retention, in the membrane pellet, of the membrane spanning segments and their connecting extracytoplasmic loops. Since these are small peptides, they are often difficult to detect following separation on SDS-tricine gels. However, all of the segment/loop/segment sectors have cysteines in their sequence. This allows labeling of the cysteine residues with fluorescein maleimide, a fluorescent, covalent cysteine reagent. This manoeuvre allows ready identification of the fragments produced by trypsinolysis (9).

From these studies, 4 membrane spanning segment/extracytoplasmic loop/membrane spanning segment sectors were identified, corresponding to M1/M2 through M7/M8. Their locations are diagrammed in figure 1. Although the hydropathy plot predicts two additional membrane spanning segments at the C-terminal region of the enzyme, no evidence was obtained for this pair in spite of the fact that H9 is predicted to have 4 cysteine residues. It is possible that these cysteines are covalently modified so that F-MI is unable to react. Treatment with reducing agents or hydroxylamine to break disulfide bonds or thioester bonds respectively did not produce F-MI labeled fragments detectable in the membrane portion of the digest. Perhaps other covalent modification exists not cleaved by the treatments we have applied. In the absence of direct biochemical evidence for the membrane spanning nature of the 9th and 10th hydrophobic domains, they are placed tentatively as membrane or stalk associated but not membrane spanning in this figure and in figure 6. However, as discussed below, translation data do provide evidence that these sectors are able to act as membrane anchor or stop transfer sequences. These findings have not as yet been incorporated into our working model.

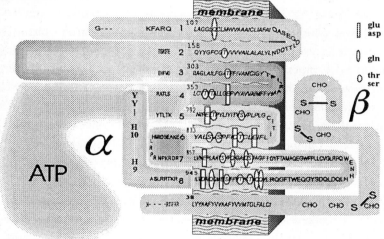

Figure 1: A two dimensional map of the α and ß subunits of the gastric H/K ATPase. This model shows the 8 definitive membrane spanning segments of the α subunit and the single membrane spanning segment of the ß subunit. Also shown is the close association between the ß subunit in the region of cys 161 to cys 178 and the M7 segment. Illustrated also are the H9 and H10 segments as stalk associated since these are absent by biochemical analysis. These may be membrane spans.

(b) Site of labelling with extracytoplasmic reagents

Figure 2 illustrates the K^+ competitive reagents used for locating their binding site on the H/K ATPase as well as defining some of the conformational changes that occur at that site. The K^+ competitive photoaffinity reagent, 3H-MeDAZIP, is derived as an analog of SCH 28080, an imidazopyridine (10). Figure 3 shows the structure of omeprazole, lansoprazole and pantoprazole which are all substituted benzimidazoles. These reagents are weak base, acid activated compounds, which form cationic sulfenamides in acidic environments. The sulfenamides then react with the SH group of cysteines in proteins to form relatively stable disulfides.

(b.i) K^+ competitive reagents

A series of K^+ competitive reagents have been developed, including imidazo-pyridines, that are known to react on the outside surface of the pump (10,11). 3H-MeDAZIP is an imidazopyridine that competitively inhibits before photolysis, and inhibits and binds in a saturable manner after photolysis. This shows that the compound binds covalently in the same region of binding of the non-photolysed compound. Cleavage of the enzyme with either trypsin or V8 protease showed that the sector binding this reagent was the M1/loop/M2 sector. Molecular modeling studies suggest that binding is to phe124 and asp136 (12). This region of the H/K ATPase is homologous to the region responsible for high affinity ouabain binding to the Na,K ATPase (13). These data substantiate the presence of the first two segments spanning the membrane in the α subunit of the H/K ATPase, as deduced from the trypsinolysis data.

Further, a fluorescent K^+ competitive arylquinoline, MDPQ, shows enhanced hydrophobicity of its environment with formation of the E2-P·I conformer of the enzyme, suggesting that the M1/loop/M2 segment moves cytoplasmically as the enzyme assumes the E2 conformation(14).

(b.ii) Site of labelling with substituted benzimidazoles

Another group of compounds, the substituted benzimidazoles, have been developed as covalent inhibitors of the functioning gastric H/K ATPase(15). The parent compound, timoprazole, was known to inhibit acid secretion independent of the pathway of stimulation. However, it showed toxicological side effects, namely on the thyroid and the thymus. As a coincidence, the first polyclonal antibody prepared against hog gastric H/K ATPase showed cross reactivity for both thyroid and thymus, for reasons that are still unclear(16). This similarity prompted an investigation as to the effect of timoprazole, and then later, picoprazole on the gastric H/K ATPase.

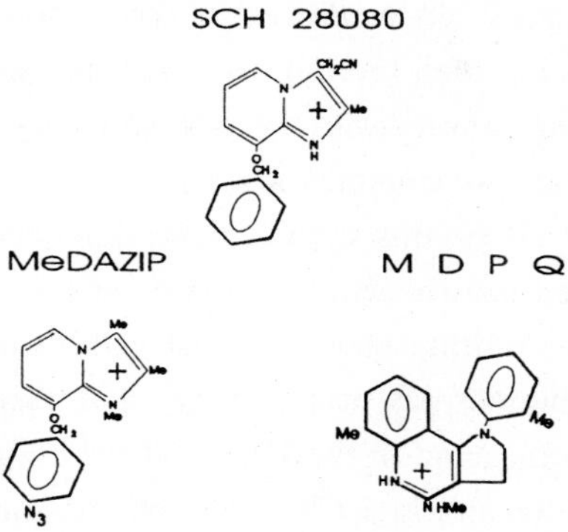

SCH 28080

MeDAZIP   M D P Q

Figure 2: The structure of some K$^+$ competitive reagents for the gastric H/K ATPase

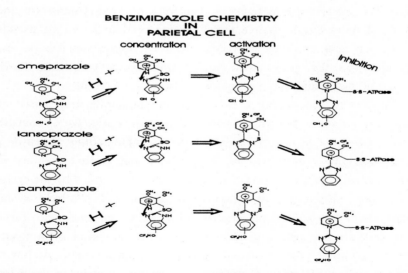

BENZIMIDAZOLE CHEMISTRY
IN
PARIETAL CELL

concentration   activation

Inhibition

omeprazole

lansoprazole

pantoprazole

Figure 3: The reaction mechanism for omeprazole, pantoprazole and lansoprazole, which involves pH gradient dependent accumulation, acid catalysed conversion to the sulfenamide and covalent reaction with certain cysteines of the α subunit.

Under our standard conditions, where the enzyme was assayed in the presence of nigericin, no inhibitory effect was observed at concentrations of 1-10 uM of the compounds. However, under acid transporting conditions, although the initial rate of acidification was almost the same, within a few minutes acid transport by the enzyme was inhibited. These results indicated that (a) these compounds were dependent on acidification for activity, and since acidification was occurring on the luminal face of the enzyme, this must be their site of inhibition (in the presence of nigericin there is no acidification, thus explaining the absence of activity under our standard assay conditions) and (b) the lag phase for inhibition showed that there was a time dependent conversion from the added compound to the active form, hence the substituted benzimidazoles were acting as prodrugs, and the rate limiting step was acid catalysed conversion to the active species (17). The compounds are weak bases and therefore concentrate in the acidic space, hence are targeted to the H/K ATPase both by accumulation and by acid conversion. These results in 1977-8 defined the target enzyme and the acid catalysed nature of the inhibition. The substitutions in both the pyridine ring and the benzimidazole ring in omeprazole eliminated the thyrotoxic and thymotoxic properties found for timoprazole.

It took some time to define the chemical mechanism, as shown, for example, for omeprazole in figure 3. The work of Arne Brandstrom and Per Lindberg at Astra Hassle showed that following protonation at the pyridine N, there was an acid catalysed conversion to a tetracyclic sulfenamide and the sulfenamide was the species that reacted with the ATPase (18). There are additional subtleties that were not appreciated until more recently.

The design criteria for a substituted benzimidazole drug based on the omeprazole mechanism, is a pKa of about 4.0, to ensure selective accumulation in the acid space of the parietal cell canaliculus; reasonable stability at neutral pH to reduce inappropriate activation of the prodrug and relative stability of the disulfide formed with the enzyme. The active compound formed by the cyclisation pathway, the sulfenamide, is a permanent cation, and relatively membrane impermeable even when the pH gradient disappears and hence once formed is targeted to the amino acids accessible only from the luminal or extracytoplasmic face of the pump. Moreover the sulfenamide is stable at acidic pH but highly unstable at neutral pH, again preventing reaction with proteins in the cytoplasm of the cell.

(b.iii) Omeprazole

Omeprazole requires acidification by the pump for its inhibitory action (19,20) and does not inhibit the ATPase under non-acid transporting conditions. The action of omeprazole following acidification is to react with available cysteine-SH groups. The sulfenamide does not break cys-S-S-cys bonds. The reaction with the cysteine-SH produces a disulfide, cys-S-S-X, which covalently inhibits the enzyme in terms of ATPase activity and acid transport. The reaction can be reversed by reducing agents such as DTT or mercaptoethanol, on the enzyme and in cells (21).

Using labelled omeprazole and by carrying out the inhibitory reaction in purified hog gastric vesicles under acid transporting conditions, the cysteines reacting in the ATPase were determined by trypsinolysis followed by SDS-PAGE separation of the tryptic fragments and sequencing the radioactive fragments. Of the 28 cysteines in the α subunit and the 9 in the ß subunit, the 2 cysteines reacting were found to be in position 813 or 822 and 892. These should be on or close to the extracytoplasmic face of the enzyme and are predicted to be in the M5/loop/M6 (cys 813 and 822) and M7/loop/M8 (cys 892) sector of the enzyme. This reagent therefore defines two additional pairs of membrane spanning segments. Since it does not react with the ß subunit, the 6 cysteines on this subunit predicted to be extracytoplasmic must be disulfide linked (9).

(b.iv)Other benzimidazoles

The development of omeprazole for clinical use was followed by the development of lansoprazole and pantoprazole as well. As illustrated in figure 3, the compounds share a common activation mechanism. They follow very similar chemistry, also forming cationic sulfenamides. They depend also on acid activation for ATPase inhibition and are equivalently displaced from the enzyme by disulfide reducing agents. However, they show significant selectivity in terms of the cysteines reacting on the H/K ATPase. Neither reacts stably with the ß subunit.

Lansoprazole, omeprazole and pantoprazole differ somewhat in their rate of inhibition of the H/K ATPase in isolated vesicles under acid transporting conditions. This is illustrated in figure 4. It can be seen that all 3 compounds show a lag in onset of inhibition for about 200 sec after acidification is initiated by the addition of ATP to vesicles in the presence of valinomycin, KCl and compound. Lansoprazole shows a more rapid inhibition than omeprazole and pantoprazole shows a slower inhibition than omeprazole.

Lansoprazole labels at three positions on the enzyme, at cysteine 321, at cysteine 813 or 822, and at cysteine 892 (22). Pantoprazole labels at both cys 813 and 822, with no labelling at either 892 or 321 (23). These cysteines are predicted to be in the M3/loop/M4 sector (cys 321) and in the M5/loop/M6 sector (cys 813,822). The labelling by lansoprazole provides additional evidence for the existence of the M3/M4 membrane spanning pair. Thus labelling by the K^+ competitive reagents and the benzimidazoles confirm the 8 membrane spanning segments deduced from tryptic hydrolysis but does not provide evidence for the additional pair predicted from hydropathy.

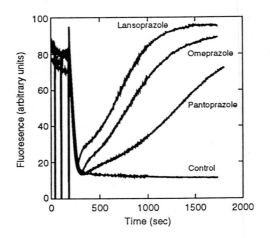

Figure 4: The inhibition of acidification by 20 uM lansoprazole, omeprazole and pantoprazole. The reaction is initiated by the addition of MgATP to a suspension of vesicles in 1uM acridine orange in the presence of valinomcyin 1 uM and KCl 150 mM in the absence and in the presence of the above compounds. A lag phase can be seen, followed by inhibition of the pH gradient, fastest for lansoprazole, slowest for pantoprazole.

Given that the cysteines at 813 or 822 are the only common site of reaction of these compounds, this site is probably the important one for enzyme inhibition, although the other sites may also contribute to the inhibition.

(b.v) Substituted benzimidazole selectivity

The selectivity of the different benzimidazoles is somewhat surprising, since the sulfenamide reacts with cysteine-SH groups in a diffusion limited reaction. Sterically the tetracyclic sulfenamide compounds are also very similar, being coplanar. There should be little difference in the rate of inhibition and no difference in the cysteines reacting with the sulfenamide derivatives of the different compounds. The explanation offered is that there is differential stabilisation of the protonated form of the benzimidazoles, prior to sulfenamide formation. Greater stabilisation allows access to deeper sites within the enzyme, lesser stabilisation reaction with more of the available cysteines accessible from the extracytoplasmic surface.

Following protonation of the pyridine N (with a pKa of 4), the formation of the tetracyclic compound depends on loss of this proton, and an increase in the nucleophilicity of the benzimidazole carbon, due to protonation of the benzimidazole N. The proton can come from free solution or from intramolecular transfer from the proton on the pyridine N. This latter pathway is the one illustrated in figure 3, but it is unlikely to be the exclusive route of protonation of the benzimidazole N since both reactions are extremely fast. The protonated species of the three drugs convert to the sulfenamide at somewhat different rates.

The protonated form is also probably stabilised by protein binding. The nature of the substituents could contribute considerably to the stabilisation and binding obtained. Contribution to binding from the fluorethoxy moiety on the pyridine of lansoprazole would be less than from the methoxy moiety on the pyridine and benzimidazole ring of omeprazole and these in turn less than the 2 methoxy moiety groups on the pyridine and the difluoromethoxy group on the benzimidazole of pantoprazole. Hence the latter drug would be the most likely to be stabilised by binding to the luminal surface of the ATPase α subunit.

The interpretation of the cysteine selectivity of pantoprazole is then that pantoprazole is significantly stabilised in its protonated form by binding within the extracytoplasmic domain of the enzyme. This allows access of the protonated form not only to cys 813 but also to cys 822. With formation of the sulfenamide, both cysteines in the M5/loop/M6 sector then react. The protonated form does not bind in the region of cys 892 or cys 321.

Omeprazole is less stabilised and converts rapidly to the sulfenamide after binding to the enzyme in its protonated form close to cys 813. Omeprazole reacts first with cys 813, thus preventing reaction with cys 822 by steric hindrance of access of the sulfenamide to this second cysteine in the M6 segment. Omeprazole also reacts with cys 892 which is probably in the vicinity of the omeprazole-H+ binding site.

Lansoprazole is the least stabilised by protein binding, and reacts with all the cysteines available from the extracytoplasmic surface, namely cys 321, 813 and 892. Perhaps this compound forms the sulfenamide before any of the protonated species is bound.

An additional point of interest is that following labelling with pantoprazole, but not omeprazole or lansoprazole, the digestion pattern in the region of the M5/loop/M6 sector changes, as if derivatising the second cysteine of this sector alters the site at which trypsin can attack. The position of cleavage moves from 776 to 783 and 792, implying that derivatising the second cysteine results in increased cytoplasmic exposure of the N-terminal end of M5. Since the reaction with benzimidazoles appears to stabilise an E2 conformation, this may imply a different conformation of this region of the enzyme in the E2 form. Thus from both the site of labelling with K^+ competitive reagents and from the results with pantoprazole, it may be that the membrane and extracytoplasmic domain of the H/K ATPase move cytoplasmically as the E2 conformation is achieved (23).

(c)Number of membrane spanning segments

There must be an even number of membrane spanning segments. The N terminal region of the enzyme in intact, cytoplasmic side out vesicles is rapidly cleaved at position 47 in the presence of trypsin at the cytoplasmic surface, showing that this site is cytoplasmic. The C-terminal 2 amino acids are tyr-tyr, and these can be iodinated in intact vesicles and the radioactivity reduced by carboxypeptidase digestion of the intact, iodinated vesicles. These data show that the C-terminal region is also cytoplasmic (24). Furthermore, an antibody generated against the C-terminal domain of the ATPase reacts cytoplasmically, not extracytoplasmically (25). The combination of biochemical methods provides direct evidence for at least 8 membrane spanning segments.

Further the most accessible site of attack by trypsin is the N terminal region of M7 and the C terminal region following M8, suggesting that these are placed somewhat peripheral to the other membrane segments (9), as predicted by the model discussed later.

(d)Antibody epitope studies on the α subunit

Various monoclonal antibodies have been raised against the H/K ATPase. With the sequence of both subunits now known, it is possible to define the epitopes for these antibodies, and by staining intact or permeable cells with either fluorescent or immunogold techniques, determine the sidedness of these epitopes.

Antibody 95, raised against the H/K ATPase, inhibits ATP hydrolysis in intact vesicles, and appears to be K competitive (26). Its epitope was defined by Western blotting of an E. coli expression library of fragments of the cDNA encoding the α subunit, as well as by Western analysis of tryptic fragments (27). The sequence recognised by this antibody was between amino acid positions 529 and 561. Since it inhibits intact vesicles this epitope must be cytoplasmic. Its epitope is close to the region known to bind the cytoplasmic reagent, FITC, in the loop between M4 and M5.

Antibody 1218 also raised against the H/K ATPase was determined to have its major epitope between amino acid positions 665 and 689 (25). This antibody is not inhibitory, but is known to have its epitope on the cytoplasmic surface of the enzyme and its predicted position is in the cytoplasmic loop between M4 and M5. A secondary epitope was also defined as being present between amino acid positions 853 and 946, using Western blots of tryptic digests. A synthetic peptide containing this second epitope was shown to displace 1218 from vesicles adsorbed to the surface of ELISA wells (27) confirming the presence of a second epitope for this monoclonal antibody. It was concluded from this work that this particular epitope was also cytoplasmic, but this is not likely based on the biochemical data where this epitope is found in the M7/loop/M8 sector and is predicted to be extracytoplasmic. Perhaps binding of the synthetic peptide to the antibody displaced it from the other, cytoplasmic, epitope in these experiments.

(e)Biochemical analysis of the ß subunit

The hydropathy prediction for the ß subunit is that there is a single rather long membrane spanning segment between amino acids 38 - 63, and that the rest of the protein is extracytoplasmic. In addition there are, dependent on species, 6 or 7 potential glycosylation sites (4,5). It is known that all the glycosylation sites are used (28) hence asn 220 in the hog sequence must be extracytoplasmic. Cleavage of lyophilised therefore permeabilised vesicles by trypsin followed by gel electrophoresis in the presence but not in the absence of DTT, produces a C-terminal fragment ala-gln-pro- beginning therefore at position 236 (25). A disulfide bridge must therefore connect this fragment to the N-terminal portion of the ß subunit. The C-terminal end of the disulfide is at position 262. This leaves little room for an addition membrane spanning α helix.

(e.i) Epitope for mAb 146-14

An antibody, mAb 146-14, was raised against intact rat parietal cells, and was shown to react extracytoplasmically using immunogold electron microscopy (8).

We initially determined the epitope for this antibody using Western analysis of fragments of the hog ATPase and of rat and rabbit enzyme. The antibody recognised a fragment of the hog α subunit, apparently beyond position 838 and before position 901, which was obtained by V8 protease digestion. It also recognised the α subunits of rat and rabbit. It did not recognise the ß subunit of hog enzyme, but did recognise the ß subunit of rat enzyme strongly and more weakly the ß subunit of rabbit enzyme. On Western blots the presence of reducing agents abolished its recognition of the ß subunit. Comparing sequences close to the disulfide bonds of the ß subunit shows that between cys 161 and cys 178 the hog enzyme has the non-conservative substitution of arg and pro for leu and val in the rat enzyme, probably explaining the lack of recognition of the hog ß subunit by this antibody and also the need for intact disulfide bonds in the rat enzyme (25).

Using the octamer walk technique, which involves synthesis of sequential peg immobilised octapeptides and ELISA assay of immunoreactivity, the α subunit sequence recognised by mAb 146-14 was between position 873 and 877, consistent with the results of Western analysis. The position of this sequence is predicted to be just at the extracytoplasmic face of M7. Expression of both subunits in baculovirus showed that mAb 146-14 reacted with the ß subunit expressed in SF9 cells, but not the α subunit. However on Western analysis, both subunits were recognised. Perhaps the epitope on the α subunit is too close to the membrane to bind the antibody in intact cells.

These data suggest that the region of the α subunit at the extracytoplasmic face of M7 is closely associated with the ß subunit in the region of the cys 161--cys

178 disulfide bridge, since the epitope for mAb 146-14 overlaps this region.

(e.ii)Effect of disulfide reduction on α subunit

It has been shown that reduction of the disulfides of the ß subunit inhibits the activity of the H/K ATPase (29). When these disulfide bonds are disrupted, the pattern obtained following tryptic digestion changes. The normal tryptic cleavage site at the N terminal end of M5 is at position 776. Following reduction, the cleavage site moves to position 792. This implies that the M5 segment moves cytoplasmically with reduction of the ß subunit. This change is similar in direction to that seen with labelling of both cys 813 and cys 822 with pantoprazole and the digestion pattern is also similar (23).

(e.iii)Site of association of the α and ß subunits

That the two subunits are closely associated is shown by the inability of detergents other than SDS to dissociate them and by the crosslinking between the two subunits by glutaraldehyde (30).

Tryptic digestion of intact inside out vesicles strips the cytoplasmic domain of the α subunit but leaves the ß subunit largely intact, since most of this subunit is extracytoplasmic and not accessible to trypsin. Furthermore this subunit is glycosylated in contrast to the α subunit. In fact the H/K ATPase has been purified using lectin columns based on binding of the ß subunit(31).

Following tryptic digestion, the membranes are precipitated by centrifugation and solubilised using non-ionic detergents such as NP-40 or $C_{12}E_8$. The soluble material is adsorbed to a WGA affinity column and eluted first with chromatography buffer and the eluate analysed for the peptides not retained along with the ß subunit. These are the first 3 membrane spanning pairs. After acetic acid elution, however, the ß subunit is eluted along with, quantitatively, the M7/loop/M8 sector. These data show that this region of the α subunit is tightly associated with the ß subunit such that non-ionic detergents are unable to dissociate it from the ß subunit. The shared epitope of mAb 146-14 also recognizes the region of the α subunit at the extracytoplasmic face of the M7 segment, a finding consistent with the association found by column chromatography (32).

When the tryptic digestion is carried out in the presence of KCl, a 20 kDa fragment is found, that represents the M7 segment to the C terminal end of the protein, as reflected by staining of this peptide with a C terminal antibody. Now when the association between the ß subunit and the membrane fragments is determined by WGA chromatography, not only is the 20 kDa fragment retained, but also the M5/loop/M6 segment. Thus absence of cleavage between M8 and H9 allows the additional retention of M5/loop/M6 (Shin JM unpublished observations).

(f) Molecular Biological analysis of the membrane structure of the H/K ATPase

The above data are derived entirely from biochemical methods. It seems possible, especially given the hydropathy analysis, that the H9 and H10 segments may be membrane spanning. The reasons for being unable to detect them in the membrane after trypsinolysis are unclear. Accordingly, we have developed a method allowing analysis of the membrane spanning properties of any putative membrane segment of the H/K ATPase.

The principle of this method is to construct a translation sequence containing a cytoplasmic N terminal domain, a variable region into which membrane spanning segment sequences can be inserted, and a C terminal region that contains N glycosylation sites. Translation in the presence of microsomes will result in glycosylation if the variable region contains an odd number of membrane spanning segments, and in no glycosylation if the variable region contains none or an even number of membrane spanning segments. The method is illustrated in figure 5.

For this method then, a plasmid encoding a fusion protein of the 100 N terminal amino acids from the H/K ATPase α subunit and the 180 C terminal amino acids from the ß subunit was assembled. The two subunits are joined by a variable region which must maintain the open reading frame. This method allows not only analysis of the membrane spanning properties of a given sequence, but also allows a determination of the minimum number of amino acids necessary for a sequence to act as either a membrane anchor or stop transfer sequence. One can also determine the role of positive charge distribution in determining the membrane associative properties of a given sequence.

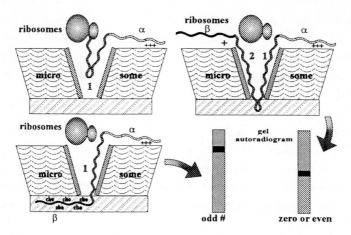

Figure 5. The conceptual basis for determining membrane segments by in vitro translation, in the presence and absence of microsomes. An odd number of membrane segments results in glycosylation, an even number or absence of membrane segments results in no glycosylation.

(f.i) M1/M2

In the presence of M1 ligated between the two subunit partial sequences, there is glycosylation of the translation product in the presence of microsomes. In the presence of M2 alone, there is also glycosylation of the product, but less is glycosylated proportionately as compared to the M1 domain. In the presence of both M1 and M2 there is no glycosylation detected. These data show that M1 and M2 can act as signal anchor sequences and that M2 can act as a stop transfer sequence after M1. These data are consistent with the conclusions derived from biochemical analyses as discussed above.

(f.ii)M3/M4

In the presence of M3 ligated between the partial sequences, there is glycosylation of the translation product. M4 alone can act as a signal anchor sequence only if the hydrophobic core terminates lys-arg. In the presence of M3 and M4 there is no glycosylation and in the presence of M1 and M4 there is also no glycosylation. Hence M3 acts as a signal anchor sequence and M4 as a stop transfer sequence, again consistent with the biochemical data.

(f.iii)M5/M6

Thus far we have not succeeded in demonstrating that this region of the enzyme contains membrane anchor or signal transfer sequences when translated on their own.

(f.iv)M7/M8

Various sequences containing M7 were ligated into the construct. So far, none of these have been able to act as a signal anchor sequence. M8 cannot act as a signal anchor sequence, but acts when in association with M1 as a stop transfer sequence.

(f.v)H9/H10

If H9 is inserted into the translation sequence there is glycosylation of the translation product. H10 does not act as a signal anchor sequence. If H9 and H10 are inserted into the variable region there is no glycosylation. If H10 is added also to M1 in the construct, there is also no glycosylation showing that the H10 sequence can act as a stop transfer sequence. The earlier data show that H 9 can perform as a membrane anchor sequence. These data are not consistent with the model derived from biochemical analysis, which did not reveal H9 or H10 as membrane spanning segments. This leads to the possible presence of an additional pair of membrane spanning segments. Another possibility is that in the structure of the intact enzyme, there is anchoring of the H9 and H10 segments to the stalk region of the protein, between M4 and M8, so that, although these are potentially membrane spanning, in the intact enzyme they are not membrane spanning. Further work is clearly necessary to become precise about the location of H9 and H 10 with respect to the membrane.

(g) A model of the secondary structure of the H/K ATPase based on biochemical analysis.

The above biochemical data can be summarised as showing that the α subunit has 8 membrane spanning segments and in this section we shall confine ourselves to an 8 membrane segment model, ignoring the data obtained from the translation of the H9 and H10 regions. The largest extracytoplasmic loop is between M7 and M8. The ß subunit has a single membrane spanning segment, a small N-terminal portion and most of this subunit is extracytoplasmic. There is close proximity between the extracytoplasmic face of M7 and the ß subunit, and a region of the ß subunit between cys 161 and cys 178. In addition, based on omeprazole labelling there is close proximity between cys 813 (822) and 892; cys 822 (813) is within the membrane domain, rather than on the surface, since only the longer lived pantoprazole intermediate is able to react with this cysteine.

MEMBRANE DOMAIN

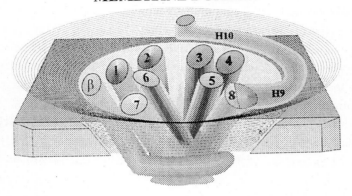

Figure 6: A possible 3 dimensional arrangement of the membrane and extracytoplasmic domain of the gastric H/K ATPase. Shown are the 8 segments defined by the biochemical methods outlined above, and the single membrane segment of the ß subunit. In addition, the M7/M8 segments are shown as lying peripheral to the central membrane domain which consists of 2 segments consisting of M1,2 and 6 and M 3,4 and 5. The H9 and H10 domains are shown as interacting with the stalk of the α subunit and not as membrane spanning segments. This speculative structure derives from a reinterpretation of the crystal structure at 14Å of the Ca^{2+} ATPase (33).

A conceptual 3 dimensional model which illustrates some these features is shown in Figure 6 and is derived from figure 1. The H9/10 segments have been placed above the plane of the membrane.

The model is also based on a re-interpretation of the 14 Å map of crystals of the sr Ca^{2+} ATPase that has been recently published (33). In this crystal, there of the sr Ca^{2+} ATPase that has been recently published (33). In this crystal, there are two longer membrane segments, connected by a long extracytoplasmic loop wrapped around the other extracytoplasmic loops. These segments sit external to the main membrane spanning domain, which exists as two clusters of membrane segments separating towards the extracytoplasmic face of the Ca^{2+} ATPase. The membrane domain is connected to the cytoplasmic domain by a stalk. A region of electron density is associated with the side of the stalk.

Our assumption is that given the characteristic hydropathy plot homology between the H/K ATPase and the Ca^{2+} ATPase, that the crystal structure of the α subunit of the two enzymes will be very similar. We interpret the peripheral membrane segments as being M7 and M8, rather than M7 and M1 (33), connected by a loop tightly associated with the other extracytoplasmic domains. The two clusters in the central membrane domain are M1/M2 and M3/M4 associated with M6 and M5 respectively. The 9th and 10th hydrophobic segments pass upward from the membrane back around the stalk into the cytoplasmic domain, and do not span the membrane, accounting for the electron density seen in this region of the Ca^{2+} pump.

This model is of course tentative, and will be revised as more biochemical, molecular biological and crystallographic data become available. Clearly several methods should be combined before the membrane helices can be defined.

REFERENCES

1.Shull, G.E. and Lingrel, J.B. (1986) J. Biol. Chem., 261, 16788-16791.
2.Maeda,M., Ishizaki,J., and Futai,M. (1988) Biochem. Biophys. Res. Commun., 157, 203-209.
3.Bamberg,K., Mercier,F., Reuben,M.A., Kobayashi,Y., Munson,K.B., and Sachs,G. (1992) Biochim. Biophys. Acta, 1131, 68-77
4.Reuben, M.A., Lasater, L.S. and Sachs, G. (1990) Proc. Natl. Acad. Sci. U.S.A., 87, 6767-6771
5.Toh,B.-H., Gleeson,P.A., Simpson,R.J., Moritz,R.L., Callaghan,J.M., Goldkorn,I., Jones,C.M., Martinelli,T.M., Mu,F.-T., Humphris,D.C., Pettitt,J.M., Mori,Y., Masuda,T., Sobieszczuk,P., Weinstock,J., Mantamadiotis,T., and Baldwin,G.S. (1990) Proc. Natl. Acad. Sci. U.S.A., 87, 6418-6422.
6.Newman PR; Shull GE. (1991) Genomics, 11(2), 252-62.
7.Kyte,J. and Doolittle,R.F. (1982) J. Mol. Biol., 157, 105-132.
8.Mercier,F., Reggio,H., Devilliers,G., Bataille,D., and Mangeat,P. (1989) J. Cell Biol., 108, 441-453.
9.Besancon, M., Shin, J. M., Mercier F., Munson, K., Miller, M., Hersey, S., and Sachs, G. (1993) Biochemistry 32, 2345-2355

10. Mendlein, J. and Sachs, G. (1990) J. Biol. Chem., 265, 5030-5036.
11. Munson, K.B. and Sachs, G. (1988) Biochemistry, 27, 3932-3938.
12. Munson, K.B., Gutierrez, C., Balaji, V.N., Ramnarayan, K., and Sachs, G. (1991) J. Biol. Chem., 266, 18976-18988.
13. Price, E.M., Rice, D.A., and Lingrel, J.B. (1990) J. Biol. Chem., 265, 6638-6641.
14. Rabon, E., Sachs, G., Bassilian, S., Leach, C., and Keeling, D. (1991) J. Biol. Chem., 266, 12395-12401.
15. Fellenius, E., Berglindh, T., Sachs, G., Olbe, L., Elander, B., Sjostrand, S-E., and Wallmark, B. (1981) Nature, 290, 159-161.
16. Fellenius E. Personal Communication
17. Lorentzon, P., Jackson, R., Wallmark, B., and Sachs, G. (1987) Biochim. Biophys. Acta, 897, 41-51.
18. Lindberg, P., Nordberg, P., Alminger, T., Brandstrom, A. and Wallmark, B. (1986) J. Med. Chem., 29, 1327-1329.
19. Wallmark, B., Brandstrom, A. and Lassen, H. (1984) Biochim. Biophys. Acta, 778, 549-558.
20. Keeling, D.J., Fallowfield, C., and Underwood, A.H. (1987) Biochem. Pharmacol. 36, 339-344.
21. Lorentzon, P., Eklundh, B., Brandstrom, A. and Wallmark, B. (1985) Biochim. Biophys. Acta, 817, 25-32.
22. Sachs, G., Shin, J.M., Besancon, M., and Prinz, C. (1993) Alimentary Pharmacology and Therapeutics, 7, 4-12,
23. Shin, J.M., Besancon, M., Simon, A., and Sachs, G. (1993) Biochim. Biophys. Acta, 1148(2), 223-33.
24. Scott, D.R., Munson, K., Modyanov, N., and Sachs, G. (1992) Biochim. Biophys. Acta, , 1112(2), 246-50.
25. Mercier, F., Bayle, D., Besancon, M., Joys, T., Shin, J.M., Lewin, M.J.M., Prinz, C., Reuben, A.M., Soumarmon, A., Wong, H., Walsh, J.H., and Sachs, G. (1993) Biochim. Biophys. Acta , 1149, 151-165
26. Bayle, D., Benkouka, F., Robert, J.C., Peranzi, G., and Soumarmon, A. (1991) Comp. Biochem. Biophys. 101B, 519-525.
27. Bayle, D., Robert, J.C., Bamberg, K., Benkouka, F., Cheret, A.M., Lewin, M.J.M., Sachs, G., and Soumarmon, A. (1992) J. Biol. Chem., 267, 19060-19065.
28. Treuheit, M.J., Costello, C.E., and Kirley, T.L. (1993) J. Biol. Chem., 268, 13914-13919.
29. Chow, D.C., Cathy, M. B., and Forte, J. G. (1992) Am. J. Physiol. 263, C39-C46.
30. Rabon, E.C., Bassilian, S., and Jakobsen, L.J. (1990) Biochim. Biophys.Acta, 1039, 277-289.
31. Okamoto, C.T., Karpilow, J.M., Smolka, A., and Forte, J.G. (1990) Biochim. Biophys. Acta. 1037, 360-372
32. Shin, J.M. and Sachs, G. (1993) J. Biol. Chem., submitted.
33. Toyoshima, C., Sasabe, H., and Stokes, D.L. Nature, 362, 469-471.

Synthetic Peptide Antibody Probes of Membrane Orientation of the Gastric H,K-ATPase

Adam Smolka,
Department of Medicine,
Division of Gastroenterology,
Medical University of South Carolina,
Charleston, SC 29425

Gastric acidification is mediated by H,K-ATPase, an integral membrane protein of the apical membrane of the gastric mucosal parietal cell. The enzyme is a heterodimer of α and β subunits, and belongs to the P-type class of membrane ATPases which form a phosphorylated intermediate, are inhibited by vanadate, and which oscillate between E_1 and E_2 conformations during the transport cycle. Other members of this class are Na,K-ATPase, Ca^{2+}-ATPase of plasma membrane and sarcoplasmic reticulum, H^+-ATPases of plants and fungi, and the K^+-ATPases of certain bacteria.

H,K-ATPase α subunit cDNAs of rat (Shull and Lingrel, 1986), pig (Maeda et al, 1988), human (Maeda et al, 1990), and rabbit (Bamberg et al. 1992) have been cloned and sequenced. The deduced primary structures are highly homologous, consisting of approximately 1033 amino acids, with subunit molecular weights of about 114,000 Daltons. Hydropathy analysis (Shull and Lingrel, 1986) predicts a cytosolic N-terminus (Smolka and Swiger, 1992) followed by four hydrophobic α-helices, a 52 kDa hydrophilic cytosolic domain which contains the phosphorylation site and FITC and pyridoxal phosphate binding sites, and finally four or six more hydrophobic α-helices preceding a cytosolically-oriented C-terminus (Scott et al, 1992).

H,K-ATPase β subunit cDNAs of rat (Shull, 1990; Canfield et al., 1990), pig (Toh et al. 1990), rabbit (Reuben et al, 1990), and human (Ma et al. 1991) have also been cloned and sequenced, and show substantial homology with Na,K-ATPase β_1 and β_2 subunits. The primary structures consist of 290-294 amino acids with a molecular weight of about 33,300 Daltons. Glycosylation at six or seven tripeptide consensus sequences results in an apparent molecular weight by SDS-PAGE of between 60 and 80 kDa. Hydropathy analysis of both H,K-ATPase and Na,K-ATPase β subunit sequences suggests a single transmembrane domain beginning approximately 40 residues from the N-terminus. While the transmembrane organization of H,K-ATPase β subunit has been amply confirmed (Okamoto et al, 1990; Hall et al. 1990), the

NATO ASI Series, Vol. H 89
Molecular and Cellular Mechanisms
of H⁺ Transport
Edited by Barry H. Hirst
© Springer-Verlag Berlin Heidelberg 1994

orientation of the β subunit with respect to the parietal cell apical membrane remains unclear.

In the case of the Na,K-ATPase, catalytic activity (ATP hydrolysis) has been shown to reside in the α subunit (Jorgensen, 1982), while functional activity such as vectorial ion transport and targeting to the appropriate cellular domain requires formation of an α/ß heterodimer (Ackermann and Geering, 1990). Despite detailed structural information for both subunits of Na,K-ATPase and H,K-ATPase, including identification of α subunit binding sites for physiological ligands and specific inhibitors such as omeprazole and SCH 28080 in the case of the H,K-ATPase, the mechanism of transmembrane ion transport remains obscure. The eventual acquisition of α and ß subunit crystals of sufficient quality to provide electron or X-ray diffraction patterns at 2 or 3 Å resolution will presumably facilitate mechanistic modelling of these enzymes; until that time, less direct approaches to clarification of secondary and tertiary structure of the P-type ATPases are necessary.

One such approach, based on site-specific polyclonal antibodies to the gastric H,K-ATPase, is discussed here. Using these reagents with membrane vesicles of known sidedness and a variety of immunochemical techniques has defined the orientation of the N-termini of both H,K-ATPase subunits, and has allowed documentation of some cytoplasmically-oriented tryptic cleavage sites which are sensitive to conformation of the H,K-ATPase.

Antibodies

Peptides corresponding to pig H,K-ATPase α subunit amino acids 1-17 (αN; MGKAENYELYQVELGPGC), and 1019-1034 (αC; GVRCCPGSWWDQELYY), and ß subunit amino acids 1-14 (ßN; MAALQEKKSCSQRMC) were synthesized with C-terminal cysteines being added to the N-peptides to allow directed coupling of carrier proteins. Desalting of peptides, blocking of free amino groups with citraconic anhydride, coupling to Keyhole limpet hemocyanin, and subsequent rabbit immunization were carried out as described (Smolka and Swiger, 1992). Whereas the antisera were shown by enzyme-linked immunosorbent assay (ELISA) to bind efficiently to their corresponding synthetic peptides, their use as topological probes required that they meet stringent criteria for recognition of H,K-ATPase.

Firstly, an appropriate tissue staining pattern was shown for all three antibodies (**Figure** 1). Pig acid secretory gastric mucosa was fixed for 1 hr (Bouin's), embedded in paraffin, sectioned (5 μm), and deparaffinized. Sections were treated with 3% H_2O_2 and primary antibody binding at 1:25 dilution was assessed using the peroxidase-anti-peroxidase technique and phase-contrast microscopy. ßN, αC, and αN antibodies (Ab) labelled pig gastric parietal cells exclusively (b,c and d); αC-Ab and ßN-Ab also labelled rat parietal cells (not shown).

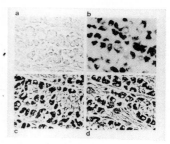

Labelling was distributed throughout the cytoplasm of parietal cells; with preimmune sera, no labelling was apparent in any cell type in either pig or rat gastric mucosa (a).

Secondly, inhibition of H,K-ATPase activity was shown by αN-Ab and ßN-Ab. Microsomes enriched in H,K-ATPase were prepared by differential and sucrose/Ficoll step gradient centrifugation of pig gastric mucosal homogenates as described (Rabon *et al*, 1988). ATPase activity in 1 µg aliquots of H,K-ATPase was measured in 50 µl 50 mM Tris-HCl (pH 7.2), 5 mM $MgCl_2$, 2 mM Na_2ATP, ± 20 mM KCl. Reactions proceeded for 15 min at 37°C, and P_i release was quantitated by colorimetric microassay (Henkel *et al*, 1988). When added to ATPase assay mixtures at room temperature 15 min before addition of ATP, α and ß subunit N-terminal antibodies inhibited K^+-stimulated ATPase activity by 55% and 40% respectively. αC-Ab had no effect on H,K-ATPase activity.

Thirdly, antibodies were shown to bind to proteins of appropriate molecular weight on polyvinylidene difluoride (PVDF) replicas of SDS-PAGE-resolved gastric microsomes (**Figure 2**). Separation of microsomal proteins by SDS-PAGE on 4%-20% acrylamide gels followed by Coomassie staining showed a prominent 94 kDa band

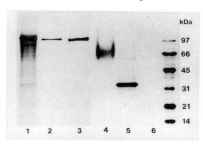

(H,K-ATPase α subunit) and numerous unknown bands of lower M_r (lane 1). Both α subunit antibodies reacted only with the 94 kDa band (lanes 2 and 3). ßN-Ab reacted with a 60-80 kDa band (lane 4). To confirm further the specificity of ßN-Ab for H,K-ATPase ß subunit polypeptide, rather than oligosaccharide substituents, microsomes were deglycosylated with N-glycosidase F, whereupon the ßN-Ab reacted with a 34 kDa band (lane 5), which corresponds to the ß subunit core peptide (Okamoto *et al*, 1990). Preimmune sera showed no reactivity with H,K-ATPase immunoblots (lane 6).

H,K-ATPase-Enriched Membranes

Having confirmed that our site-specific antibodies recognized gastric mucosal H,K-ATPase, we sought to validate our microsomal membrane preparation as an appropriate model for studies of H,K-ATPase topology. The sidedness of the vesicles is crucial, since ambiguity in this respect clouds interpretation of antibody binding data. Membrane vesicles enriched in H,K-ATPase are about 0.1 μm in diameter, and more than 90% of their protein migrates at 94 kDa by SDS-PAGE. Typical preparations of vesicles showed 8-fold stimulation of K^+-dependent ATPase activity in the presence of 10^{-5} M nigericin, a K^+/H^+ antiport ionophore. This degree of enzyme latency indicated that about 90% of the vesicles were oriented cytoplasmic side out, since the K^+ binding site on the α subunit is exposed on the luminal side of the parietal cell tubulovesicular membrane (Munson et al, 1991). Further confirmation of sidedness of this preparation comes from electron microscopy of the vesicles labelled with wheat germ agglutinin or Con A colloidal gold conjugates, which shows gold particles associated with the inner surface of vesicles, indicating a preferential lumen-inside (cytosol-out) vesicle orientation (Hall et al, 1990).

Immunoelectron Microscopy

In order to visualize directly the orientation of α subunit N- and C-termini, we labeled H,K-ATPase-enriched microsomes with antibodies, and then detected sites of membrane reactivity by electron microscopic localization of 15 nm gold-protein A conjugates. Microsomes were suspended in fixative (2% paraformaldehyde, 2.5% glutaraldehyde, 0.0625% picric acid, 0.1 M sodium cacodylate, pH 7.4) for 1.5 hr, washed in 0.1 M sodium cacodylate, dehydrated, and embedded in epoxy resin. 80 nm sections were dried onto 200 mesh nickel grids, rehydrated in 0.5% BSA/PBS for 30 min, and exposed to αN-Ab or αC-Ab or preimmune serum (50 μl, 1:25 dilutions) for 5 sec under microwave radiation. Grids were rinsed in PBS, immersed in gold-protein A conjugate (50 μl, 1:50 dilutions) for 2 hr, and rinsed in 0.5% BSA/PBS. After further rinsing in PBS and water, grids were stained with uranyl acetate and lead citrate, and then examined by electron microscopy. Micrographs (62,500 magnification) were overlaid with a translucent counting grid, and gold particles touching closed vesicular membrane profiles were counted. Particles were classified as a) touching the membrane inner surface; b) touching the membrane outer surface; c) lying directly on the membrane. Particles meeting criteria a, b, or c were expressed as % of total counts (a+b+c). **Figure 3** is a representative electron micrograph showing sites of αN-Ab reaction with gastric microsomes, and a bar graph depicting average particle counts for αN- and

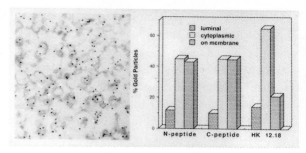

αC-terminal antibodies. From these data, we conclude that the majority of gold particles in both cases are in contact with the outer surface of microsomal vesicles, and that consequently, both termini of the H,K-ATPase α subunit are located on the cytoplasmic face of parietal cell tubulovesicular membranes.

Indirect Binding Assays

The solid phase enzyme-linked immunosorbent assay (ELISA) was used to measure interactions of site-specific antibodies with vesicular H,K-ATPase. Assays were carried out in 96 well polystyrene microplates at room temperature (Smolka and Swiger, 1992). In preliminary experiments, we showed that α and ß subunit site-specific antibodies reacted with H,K-ATPase-enriched microsomes adsorbed to ELISA plates. However, this interaction could not be interpreted topologically because of the unpredictable state of vesicles adsorbed to a polystyrene surface. We turned therefore to an indirect ELISA in which antibody binding to native, undenatured H,K-ATPase was measured by the decreased binding of the same antibody to the corresponding synthetic peptide or H,K-ATPase immobilized in polystyrene wells. Sub-saturating concentrations of antibodies were mixed with 0-5 µg freshly-prepared H,K-ATPase microsomes and transferred to microplate wells previously coated with competing antigen and blocked with BSA. After 1 hr incubation, antibody binding to the wells was measured by ELISA. As shown in **Figure 4**, αN-Ab and ßN-Ab were displaced from bound αN-peptide and bound H,K-ATPase respectively by native H,K-ATPase. Since vesicles are oriented

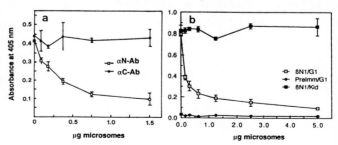

cytoplasmic side out, competition between vesicles and immobilized antigen again establishes that the α and ß subunit N-termini are displayed on the cytoplasmic side of gastric membranes. In contrast, antibody to the C-terminus of the α subunit was not displaced from bound C-peptide by native H,K-ATPase (**Figure 4a**), even when microsomes were solubilized in 0.12% $C_{12}E_8$ to expose putative intravesicular binding sites for this antibody; in native H,K-ATPase, the α subunit C-terminus is apparently not recognised by this antibody. The specificity of these competitions is shown by the failure of kidney microsomes to displace ßN-Ab from bound H,K-ATPase (**Figure 4b**).

Limited Tryptic Hydrolysis

Although αC-Ab does not react with native enzyme, its strong binding to denatured α subunit on PVDF membranes (**Figure 2**) allowed us to define a series of tryptic cleavage sites on the α subunit which are sensitive to E_1 or E_2 conformation of the enzyme. 100 μg H,K-ATPase was incubated at 37°C with 1.3 μg TPCK-treated trypsin (12,700 units/mg prot.) in 100 μl 250 mM sucrose, 20 mM Tris/Cl, pH 7.2, containing either 100 mM KCl (K^+ buffer, E_2 conformation) or 5 mM Na_2ATP, 1 mM EDTA, 90 mM choline Cl (ATP buffer, E_1 conformation). Reactions were stopped after 30 min by adding soybean trypsin inhibitor, followed by 10^5 g centrifugation for 1 hr. Pelleted membranes were resolved by SDS-PAGE and immunoblotted on PVDF with αC-Ab as described (Smolka *et al*, 1991). **Figure 5a** shows immunoreactive peptides resulting from tryptic hydrolysis of E_1-H,K-ATPase (ATP buffer) or E_2-H,K-ATPase (K^+ buffer). The peptide maps in both cases were quite dissimilar, showing that some α subunit sites are accessible to trypsin only in the E_1 form, and others only in the E_2 form. Three (*) immunoreactive tryptic peptides were microsequenced *in situ* by automated Edman degradation. These sequences allowed us to locate all conformation-sensitive αC-Ab immunoreactive membrane-bound tryptic peptides in the primary sequence; these are shown in **Figure 5b**. Microsequenced peptides (57 kDa, 37 kDa, and 18 kDa) were the most striking in terms of H,K-ATPase conformation. Arg-456 and Arg-854, the N-termini of the 57 kDa and 18 kDa peptides respectively, are accessible to trypsin only when α subunit assumes an E_2 conformation; Arg-672, the N-terminus of the 37 kDa peptide is equally accessible to trypsin

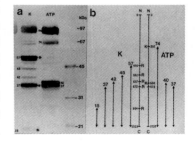

in both the E_1 and E_2 forms, but a nearby tryptic site at Lys-636 appears only in the E_1 form, together with another major tryptic site at Lys-291. Since the substrate for trypsin in these experiments was intact, cytoplasmic side out vesicles, we conclude that all the labelled cleavage sites are cytoplasmically oriented, including Arg-854, which hydropathy analysis has previously placed on the luminal side of the apical membrane (Shull and Lingrel, 1986).

Summary

We have shown that antibodies against synthetic peptides based on the N- and C-terminal amino acids of gastric H,K-ATPase α and ß subunits are useful probes of the topology of these and other subunit domains in the apical membrane of the parietal cell. Such antibodies are sometimes problematical because of limited reactivity with native protein; our antibodies react with H,K-ATPase in tissue sections, inhibit enzyme activity, and bind to α and ß subunits on immunoblots. We interpret immunoelectron micrographs of H,K-ATPase-enriched microsomes and indirect ELISAs of antibody binding to such vesicles in terms of cytoplasmically oriented N-termini in both the α and ß subunits. In addition, with limited tryptic hydrolysis of H,K-ATPase and immunorecognition of epitopic peptides, we have identified a series of cytoplasmically-oriented, conformation-sensitive cleavage sites. Current studies with these and other defined-epitope antibodies are focused on further clarification of secondary structure, especially variations in structure which may be induced by stimulation or inhibition of acid secretion.

Acknowledgements I thank Peter Zwarych and Daniel Massi for skilled technical assistance. This work was supported by NIH award DK 43138.

References

Ackermann U, Geering K (1990) Mutual dependence of Na,K-ATPase alpha and beta subunits for correct posttranslational processing and intracellular transport. FEBS Lett 269:105-108

Bamberg K, Mercier F, Reuben MA, Kobayashi Y, Munson KB, and Sachs G (1992) cDNA cloning and membrane topology of the rabbit gastric H,K-ATPase α-subunit. Biochim Biophys Acta, 1131:69-77

Canfield VA, Okamoto CT, Chow D, Dorfman J, Gros P, Forte JG, Levenson RL (1990) Cloning of the H,K-ATPase ß subunit. Tissue-specific expression, chromosomal assignment, and relationship to Na,K-ATPase ß subunits. J Biol Chem, 265:19878-19884

Hall K, Perez G, Anderson D, Guitierrez C, Munson K, Hersey SJ, Kaplan JH,

Sachs G. (1990) Location of the carbohydrates present in the H,K-ATPase vesicles isolated from hog gastric mucosa. Biochemistry, 29:701-706

Henkel RD, VandeBerg JL, Walsh RA (1988) A microassay for ATPase. Anal Biochem, 169:312-318

Jorgensen, PL (1982) Mechanism of the Na,K pump: protein structure and conformations of the pure Na,K-ATPase. Biochim Biophys Acta 694:27-68

Ma J-Y, Song Y-H, Sjöstrand SE, Rask L, Mårdh S (1991) cDNA cloning of the ß-subunit of the human gastric H,K-ATPase. Biochem Biophys Res Commun, 180:39-45

Maeda M, Ishizaki J, Futai M (1988) cDNA cloning and sequence determination of pig gastric (H^++K^+)-ATPase. Biochem Biophys Res Commun, 157:203-209

Maeda M, Oshiman K-I, Tamura S, and Futai M (1990) Human gastric (H^++K^+)-ATPase gene. J Biol Chem, 265:9027-9032

Munson KB, Gutierrez C, Balaji VN, Ramnarayan K, Sachs G (1991) Identification of an extracytoplasmic region of H,K-ATPase labeled by a K^+-competitive photoaffinity inhibitor. J Biol Chem, 266:18976-18988

Okamoto CT, Karpilow JM, Smolka A, Forte JG (1990) Isolation and charaterization of gastric microsomal glycoproteins. Evidence for a glycosylated ß-subunit of the H,K-ATPase. Biochim Biophys Acta, 1037:360-372

Rabon EC, Im WB, Sachs G (1988) Preparation of gastric H,K-ATPase. Methods Enzymol 157:649-654

Reuben MA, Lasater L, Sachs G (1990) Characterization of a ß subunit of the gastric H^+/K^+-transporting ATPase. Proc Natl Acad Sci USA, 87: 6767-6771

Scott DR, Munson K, Modyanov N, Sachs G (1992) Determination of the sidedness of the C-terminal region of the gastric H,K-ATPase α subunit. Biochim Biophys Acta, 1112:246-250

Shull GE (1990) cDNA cloning of the ß-subunit of the rat gastric HJ,K-ATPase. J Biol Chem, 265:12123-12126

Shull GE, Lingrel JB (1986) Molecular cloning of the rat stomach H,K-ATPase. J Biol Chem, 261:16788-16791

Smolka A, Alverson L, Fritz R, Swiger K, Swiger R (1991) Gastric H,K-ATPase topography: Amino acids 888-907 are cytoplasmic. Biochem Biophys Res Commun, 180:1356-1364

Smolka A, Swiger KM (1992) Site-directed antibodies as topographical probes of the gastric H,K-ATPase α-subunit. Biochim Biophys Acta, 1108:75-85

Toh B-H, Gleeson PA, Simpson RJ, Moritz RL, Callaghan JM, Goldkorn I, Jones CM, Martinelli TM, Mu F-T, Humphris DC, Pettitt JM, Mori Y, Masuda T, Sobieszczuk P, Weinstock J, Mantamadiotis Y, Baldwin GS (1990) The 60- to 90-kDa parietal cell autoantigen associated with autoimmune gastritis is a ß subunit of the gastric H^+/K^+-ATPase (proton pump). Proc Natl Acad Sci USA 87:6418-6422

Molecular Determinants of Subunit Assembly and Processing in Na,K-ATPase and H,K-ATPase

Käthi Geering, Philippe Jaunin, Ahmed Beggah and Frédéric Jaisser
Institute of Pharmacology and Toxicology
University of Lausanne
Rue du Bugnon 27
1005 Lausanne
Switzerland

Abstract

A chimeric approach has been used to determine the importance of the transmembrane and/or the ectodomain of β-subunits of Na,K-ATPase (βNK) or of H,K-ATPase (βHK) in the assembly with α-subunits (αNK or αHK), in the modulation of Na,K-pump activity and in the ER retention of unassembled βNK. Our results indicate that the transmembrane and the ectodomain in β-subunits cooperate for an efficient assembly with α–subunits as well as for the formation of functional Na,K-pumps with characteristic apparent K^+ affinities. Finally, a signal for ER retention of unassembled βNK is located in the ectodomain of the polypeptide.

Introduction

Na,K-ATPase and H,K-ATPase are the most closely related members of the P-type ATPase family (for review, see Lingrel, 1992; Sachs et al., 1992). Indeed, sequence homology is highest between the catalytic α-subunit of these two enzymes and, most interestingly, these two pumps are the only ATPases so far known that need a second subunit, a β-subunit, to be functionally active.

With respect to their membrane organization, the β-subunit of Na,K-ATPase (βNK) and H,K-ATPase (βHK) are very similar. They belong to the family of type II glycoproteins that are characterized by a short N-terminal cytoplasmic tail, one

NATO ASI Series, Vol. H 89
Molecular and Cellular Mechanisms
of H^+ Transport
Edited by Barry H. Hirst
© Springer-Verlag Berlin Heidelberg 1994

transmembrane region and a large extracytoplasmic domain containing several glycosylation sites and 3 disulphide bridges (see Table 1).

Significantly, the βHK can act at least partially as a surrogate for βNK to permit for the formation of functional Na,K-pumps at the plasma membrane (Horisberger et al., 1991a). The functional role of these β-subunits has long been an enigma, since indeed all main functional domains for binding of ATP, phosphate, cations or specific pharmacological inhibitors are located on the catalytical α-subunit. It is now well established that β-subunits play a crucial role in the posttranslational processing of the α-subunit, e.g. assembly of β-subunits is needed for the structural and functional maturation of the newly synthesized α–subunit and its transport to the plasma membrane (Jaunin et al., 1992). In addition, recent experimental evidence suggests that β-subunits might be modulators of the transport activity of Na,K-ATPase expressed at the plasma membrane (Eakle et al., 1992; Jaisser et al., 1992; Lutsenko and Kaplan, 1992; Schmalzing et al., 1992).

The structural determinants that mediate the various properties of the β-subunit are still poorly characterized. In this study, we have constructed chimera between βNK and βHK by exchanging their N-terminus plus transmembrane domains and their ectodomains and have analyzed these chimera after expression in the Xenopus oocyte in order to determine the respective importance of the transmembrane and/or the ectodomains of β-subunits in 1) the assembly with αNK or αHK, 2) the functional expression of Na,K-pumps at the cell surface and their activation by external K^+ and 3) the ER retention of unassembled βNK.

Material and Methods

The detailed description of the construction of the chimera between Xenopus βNK (Verrey et al., 1989) and rabbit gastric βHK (Reuben et al., 1990) is given in a forthcoming publication (P. Jaunin, F. Jaisser, A.T. Beggah, K. Takeyasu, P. Mangeat, B.C. Rossier, J.-D. Horisberger and K. Geering, submitted for publication). In brief, a Pvu II (384) restriction site was introduced by site directed mutagenesis in the cDNA of βNK at a position corresponding to an inherent Pvu II (282) site in the cDNA of βHK. This strategy permitted us to exchange the N- and the C-terminus of the two β-subunits and to produce, on the one hand, the chimera βNK/HK with the N-terminus plus the transmembrane domain of βNK and the ectodomain of βHK and, on the other hand, the chimera βHK/NK with the inversed arrangement (see Table I). cRNAs of β, Xenopus αNK (Verrey et al., 1989) and rabbit gastric αHK (Bamberg et al., 1992) templates were obtained by <u>in vitro</u> transcription and injected into Xenopus oocytes, as previously described (Geering et al, 1989). Oocytes were metabolically labelled with ^{35}S-methionine, then Triton extracts were prepared and βNK, βHK and β-chimera or αNK and αHK were immunoprecipitated with specific antibodies, as previously described (Jaunin et al., 1992). Na,K-pump activity was measured in Na^+ loaded oocytes as the outward current activated by addition of K^+ and maximal currents and the half activation constants ($K_{1/2}$) were determined as described (Horisberger et al., 1991b).

Results and Discussion

Involvement of the transmembrane and the ectodomain of the β-subunit in the assembly with αNK or αHK.

Fig 1 illustrates our previous finding (Geering et al., 1989) that αNK expressed alone in the Xenopus oocyte is rapidly degraded (Fig 1, lane 1). A similar observation is made with αHK (Fig 1, lane 6) indicating that, as αNK, αHK needs the β-subunit for its structural maturation. As expected, αNK is well stabilized by βNK (Fig 1, lane 2) and αHK by βHK (Fig 1, lane 8) permitting a significant accumulation of α-subunits. On the other hand, the assembly of αNK or αHK with the heterologous βHK (Fig 1, lane 3) or βNK (Fig 1, lane 7), respectively, is much less efficient. The differences in the association competence of the two β-subunits prompted us to produce chimeric proteins between βNK and βHK in order to assess the respective importance of the transmembrane and/or ectodomain in the efficient assembly of β-subunits with α-subunits. Surprisingly, both αNK and αHK are stabilized to a high extent by the chimera βNK/HK (Fig , lanes 4 and 9) and to a much lower extent by the chimera βHK/NK (Fig 1, lanes 5 and 10). The level of expression of αNK achieved with the different β-subunits at the protein level was closely paralleled by the number of functional Na,K-pumps at the cell surface as assessed by ouabain binding (Table I).

The data summarized in Table I imply that assembly of αNK with βNK is mainly determined by the transmembrane domain of βNK which is contained in the chimera βNK/HK while assembly of αHK with βHK is predominantly governed by the ectodomain of βHK. A more detailed structural analysis on the interaction sites in α - and β - subunits is needed to clarify these apparently contradictory results. The most likely explanation to reconcile the present observation is that both the transmembrane and the ectodomain of β-subunits are important for efficient assembly with α-subunits.

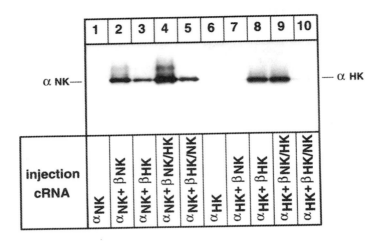

Fig 1: Cellular accumulation of αNK and αHK after assembly with βNK, βHK or with the chimera βNK/HK and βHK/NK. Oocytes were injected with 5 ng of αNK (lane 1) or 10 ng of αHK cRNA (lane 6) alone or together with 1 ng of βNK (lanes 2 and 7), 1 ng of βHK (lanes 3 and 8), 1 ng of βNK/HK (lanes 4 and 9) or 1 ng of βHK/NK cRNA (lanes 5 and 10). After a pulse of 16 h with ^{35}S-methionine and a chase of 72 h, Triton extracts were prepared and αNK (lanes 1-5) or αHK (lanes 6-10) were immunoprecipitated with specific antibodies.

An ER retention signal is located in the ectodomain of unassembled Xenopus βNK

ER retention of unassembled subunits of oligomeric proteins is a common mechanism and part of the cellular quality control (for review, see Hurtley and Helenius, 1989). α_1, β_1- and β_3-subunits of Xenopus Na,K-ATPase follow this rule when expressed alone in the Xenopus oocyte (Ackermann and Geering, 1990; Jaunin et al., 1992). On the other hand, β-subunits of H,K-ATPase can leave the ER without concomitant synthesis of α-subunits (Horisgerger et al., 1991). ER retention of βNK thus appears to be determined by specific structural characteristics of the protein. To determine the importance of the transmembrane and/or the ectodomain of βNK in ER retention, we analyzed the transport competence of the chimeric proteins between βNK and βHK in the absence of concomitant α-subunit synthesis. Similarly to βNK, the unassembled chimera βHK/NK with the ectodomain of βNK remains in its coreglycosylated ER form (Table 1). On the other hand, unassembled βHK and the chimera βNK/HK with the ectodomain of βHK become fully glycosylated, reflecting transport to a distal Golgi compartment (Table 1). These results indicate that ER retention of unassembled Xenopus βNK is mediated by a retention signal located in the ectodomain.

Table I: Summary of the properties of βNK, βHK and chimera βNK/HK and βHK/NK.

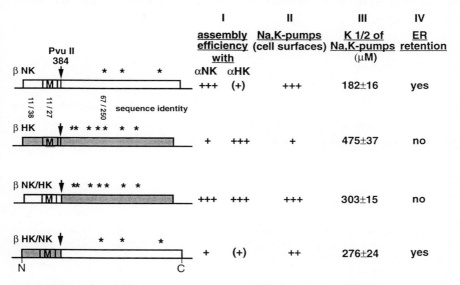

	I assembly efficiency with		II Na,K-pumps (cell surfaces)	III K 1/2 of Na,K-pumps (μM)	IV ER retention
	αNK	αHK			
β NK	+++	(+)	+++	182±16	yes
β HK	+	+++	+	475±37	no
β NK/HK	+++	+++	+++	303±15	no
β HK/NK	+	(+)	++	276±24	yes

I: Efficiency of assembly with α-subunits of Na,K-ATPase (αNK) or H,K-ATPase (αHK); II: Number of Na,K-pumps expressed at the cell surface as assessed by ouabain binding; III: Apparent K+ affinity (K1/2) of Na,K-pumps expressed at the cell surface (n=11-17); IV: Retention of unassembled β-subunits in the ER. For further details see text. With the exception of the results shown on the assembly efficiency of β-subunits with αHK, the table summarizes data from P. Jaunin, F. Jaisser, A. T. Beggah, K. Takeyasu, P. Mangeat, B. c. Rossier, J.-D- Horisberger and K. Geering, submitted for publication.

The transmembrane __and__ the ectodomain of β-subunits are important in the modulation of Na,K-pump activity.

Recent experimental evidence suggests that β-subunits can modulate the K+ activation of Na,K-pumps (Jaisser et al., 1992; Schmalzing et al., 1992). A similar conclusion can be drawn from the present study which shows that the apparent K^+ affinity ($K_{1/2}$) of Na,K-pumps composed of αNK-βNK complexes is significantly higher than of αNK-βHK complexes (Table 1). The analysis of the K^+ activation of Na,K-pumps composed of αNK and β- chimera reveals that the exchange of the N- and the C-terminal in βNK and βHK is not sufficient to re-establish the apparent K^+ affinity of either αNK-βNK or αNK-βHK wild type complexes. Indeed, the $K_{1/2}$ of the αNK- β-chimera complexes is similar and intermediate between the $K_{1/2}$ of αNK-βNK and αNK-βHK complexes (Table I). Hence, these results suggest that both the transmembrane and the ectodomain of β-subunits are involved in the modulation of the K^+ activation of Na, K- pumps.

Acknowledgements

The original cDNA of αHK and βHK and the αHK antibody were kindly provided by G. Sachs. The βHK antibody was a gift of J.G. Forte. This work was supported by the Swiss National Fund for Scientific Research, Grant No 3100-026241.89.

References

Ackermann U and Geering K (1990) Mutual dependence of Na,K-ATPase α- subunits and β-subunits for correct posttranslational processing and intracellular transport. FEBS Lett 269: 105-108.

Bamberg K, Mercier F, Reuben MA, Kobayashi Y, Munson KB and Sachs G (1992) cDNA cloning and membrane topology of the rabbit gastric H^+/K^+-ATPase α - subunit. Biochim Biophys Acta 1131: 69-77.

Eakle KA, Kim KS, Kabalin MA and Farley RA (1992) High-affinity ouabain binding by yeast cells expressing Na^+,K^+-ATPase α-subunits and the gastric H^+,K^+-ATPase β-subunit. Proc Natl Acad Sci USA 89: 2834-2838.

Geering K, Theulaz I, Verrey F, Häuptle MT and Rossier BC (1989) A role for the β- subunit in the expression of functional Na^+-K^+-ATPase in Xenopus oocytes. Am J Physiol 257: 851-858.

Horisberger JD, Jaunin P, Good PJ, Rossier BC and Geering K (1991 a) Coexpression of $α_1$ with putative $β_3$ subunits results in functional Na^+/K^+ pumps in Xenopus oocytes. Proc Natl Acad Sci USA 88: 8397-8400.

Horisberger JD, Jaunin P, Reuben MA, Lasater LS, Chow DC, Forte JG, Sachs G, Rossier BC and Geering K (1991 b) The H,K-ATPase β-subunit can act as a surrogate for the β-subunit of Na,K-pumps. J Biol Chem 266:19131-19134.

Hurtley SM and Helenius A (1989) Protein oligomerization in the endoplasmic reticulum. Annu Rev Cell Biol 5: 277-307.

Jaisser F, Canessa CM, Horisberger JD and Rossier BC (1992) Primary sequence and functional expression of a novel ouabain-resistant Na,K-ATPase. The β-subunit modulates potassium activation of the Na,K-pump. J Biol Chem 267: 16895-16903.

Jaunin P, Horisberger JD, Richter K, Good PJ, Rossier BC and Geering K (1992) Processing, intracellular transport and functional expression of endogenous and exogenous α-β_3 Na,K-ATPase complexes in Xenopus oocytes. J Biol Chem 267: 577-585.

Lingrel JB (1992) Na,K-ATPase - isoform structure, function, and expression. J Bioenerg Biomembr 24: 263-270.

Lutsenko S and Kaplan JH (1992) Evidence of a role for the Na,K-ATPase β-subunit in active cation transport. *In* Ion - Motive ATPases : Structure, Function and Regulation Scarpa A, Carafoli E and Papa S (eds) Ann NY Acad Sci 671 New York 147-155.

Reuben MA, Lasater LS and Sachs G (1990) Characterization of a β-subunit of the gastric H^+/K^+-transporting ATPase. Proc Natl Acad Sci USA 87: 6767-6771.

Sachs G, Besancon M, Shin JM, Mercier F, Munson K and Hersey S (1992) Structural aspects of the gastric H,K-ATPase. J Bioenerg Biomembr 24: 301-308.

Schmalzing G, Kroner S, Schachner M and Gloor S (1992) The adhesion molecule on glia (AMOG/β_2) and β_1 subunits assemble to functional sodium pumps in Xenopus oocytes. J Biol Chem 267: 20212-20216.

Verrey F, Kairouz P, Schaerer E, Fuentes P, Geering K, Rossier BC and Kraehenbuhl JP (1989). Primary sequence of Xenopus laevis Na^+,K^+-ATPase and its localization in A6 kidney cells. Am J. Physiol 256: F1034-F1043.

Assembly of Hybrid Pumps from Na,K-ATPase α Subunits and the H,K-ATPase β Subunit

Robert A. Farley[*][#] and Kurt A. Eakle[*]
Department of Physiology & Biophysics[*], and Department of Biochemistry[#]
University of Southern California School of Medicine
Los Angeles, CA 90033 USA

Introduction

The Na/K-ATPase (sodium pump) and the gastric H/K-ATPase (proton pump) are members of a structurally-related family of P-type ion transport proteins (Jorgensen and Andersen, 1988; Green and Stokes, 1992). The sodium pump and the proton pump are different from the other ion pumps, however, in that they require two subunits, α and β, for activity. The three isoforms of the α subunit of Na,K-ATPase (α1,α2,α3) all show ~85% amino acid sequence identity with one another, and the α subunit of H,K-ATPase is ~65% identical to Na,K-ATPase isoforms. In contrast, isoforms of the β subunit of these two pumps share only 30-35% amino acid sequence identity to one another. Although the importance of the β subunit for pump function is not known, the unique requirement of these ion pumps for a β subunit is intriguing since other pumps are capable of ATP hydrolysis and ion translocation with only an α subunit. In an effort to elucidate the role of the β subunit in Na,K-ATPase and H,K-ATPase function, the activities of hybrid pumps assembled from α subunits of Na,K-ATPase and the β subunit of gastric H,K-ATPase were investigated. Because the amino acid sequences of the β subunit of gastric H,K-ATPase (HKβ) and the β subunit of Na,K-ATPase (β1) are only about 30% identical, identification of differences in the biochemical properties of Na,K-ATPase and hybrid pumps may indicate functional features that depend upon the structure of the β subunit.

The β subunit of H,K-ATPase assembles into functional complexes with Na,K-ATPase α subunits: differences in K[+] antagonism of ouabain binding.

Yeast (*Saacharomyces cerevisiae*) do not contain endogenous sodium pumps, high-affinity ouabain binding sites or ouabain sensitive ATPase activities. We have previously shown that the heterologous expression of both α and β subunits of Na,K-ATPase in yeast results in the assembly of functional sodium pumps (Horowitz et al., 1990). Na,K-ATPase expressed in

NATO ASI Series, Vol. H 89
Molecular and Cellular Mechanisms
of H[+] Transport
Edited by Barry H. Hirst
© Springer-Verlag Berlin Heidelberg 1994

yeast cells is indistinguishable from Na,K-ATPase found in animal cells, and is characterized by high-affinity ouabain binding, and ouabain-sensitive enzymatic and transport activities (Horowitz et al., 1990; Wang and Farley, 1992; Scheiner-Bobis et al., 1991). An expression system in yeast has been designed in which separate plasmids with different selectable markers for yeast transformation are used to direct the heterologous expression of α and β subunits (Eakle et al., 1991). The synthesis of the β subunit is under the control of the inducible GAL1 promoter which is active only when yeast are grown in galactose and is repressed when cells are grown in glucose. This expression system permits the combinatorial expression of different α subunit - β subunit isoforms in yeast cells. Using this strategy, the Na,K-ATPase $\alpha 1$ or $\alpha 3$ subunit was expressed in yeast in combination with either the Na,K-ATPase $\beta 1$ or the gastric H,K-ATPase β subunit (Eakle et al., 1992). The binding of [^{3}H] ouabain was used to detect

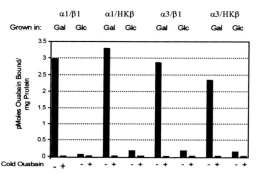

Figure 1. [^{3}H] Ouabain binding by α/β combinations. Yeast transformed with the expression plasmids indicated were grown either in galactose (Gal) or glucose (Glc) as indicated. A microsomal fraction of yeast membranes was prepared and [^{3}H] ouabain binding was tested as described (Eakle et al., 1992).

functional pumps in yeast membranes. Ouabain binds to the α subunit Na,K-ATPase after the enzyme is phosphorylated either by ATP in the forward catalytic direction in a reaction that depends on Na$^+$, or by inorganic phosphate in the reverse direction (Forbush and Hoffman, 1979; Yoda and Yoda, 1987). As shown in Figure 1, the gastric H,K-ATPase β subunit was assembled in the yeast cells with either the $\alpha 1$ or the $\alpha 3$ subunit of Na,K-ATPase into hybrid pumps that are capable of binding ouabain. The absence of ouabain binding to membranes prepared from cells grown in glucose confirms the requirement for a β subunit in these complexes. Scatchard analysis of ouabain binding to each hybrid pump complex indicated that the K_d values in this experiment were: 10.1 nM ($\alpha 1/\beta 1$), 10.2 nM ($\alpha 3/\beta 1$), 5.8 nM ($\alpha 1/HK\beta$), and 8.3 nM ($\alpha 3/HK\beta$). Thus, the structure of the β subunit appears to have little effect on the affinity of the pumps for ouabain.

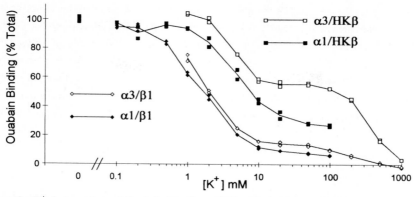

Figure 2. K$^+$ Antagonism of Ouabain Binding. [^{3}H] ouabain binding to yeast membranes from cells transformed with the indicated plasmids was titrated with increasing amounts of KCl. Duplicate points are shown with lines drawn through the mean. To correct for different levels of expressed proteins in different preparations, ouabain binding is normalized to the maximum binding (0 KCl) with non-specific binding (determined by addition of 1 mM unlabeled ouabain) subtracted.

Ouabain binding to Na,K-ATPase is antagonized by extracellular K$^+$ ions (Forbush, III, 1983). Although K$^+$ ions also prevent ouabain binding to pumps assembled with HKβ subunits (Eakle et al., 1992), the hybrid pumps are less sensitive to K$^+$ antagonism than pumps assembled from both Na,K-ATPase α subunits and β subunits (Figure 2). The K$_{0.5}$ for potassium antagonism of ouabain binding to the hybrid pumps is shifted from 1-2 mM for α1/β1 to 3-4 mM for α1/HKβ, and from 1-2 mM for α3/β1 to 8-10 mM for α3/HKβ. In addition, the titration of ouabain binding to the α3/HKβ complex by K$^+$ appears to distinguish between two K$^+$ binding sites. Non-specific effects due to high salt concentrations have not yet been excluded, nevertheless, a difference between the behavior of the α3/β1 complex and the α3/HKβ complex is observed even at high salt concentrations. The differences in the antagonism by K$^+$ of ouabain binding to Na,K-ATPase and to hybrid pumps indicate that the interaction between K$^+$ ions and the pumps is affected by changes in the structure of both the α subunit and the β subunit. Since it is likely that the ion binding sites reside largely, if not exclusively, within the structure of the α subunit, the influence of β subunit structure on ion binding may occur through a modification or distortion of the α subunit. To localize the domain of the β subunit that is responsible for the different sensitivities of Na,K-ATPase and hybrid pumps to antagonism of ouabain binding by K$^+$, chimeric β subunits were constructed in which the extracellular domains of the β1 and the HKβ subunits were exchanged.

A Bst EII restriction site was introduced into the cDNA encoding the Na,K-ATPase rat β1 subunit to generate β1B. This change introduces a conservative amino acid substitution

(Leu→Val). This site is located in the DNA at a point which is about 15 amino acids beyond the predicted transmembrane domain of the β subunit. The rat H,K-ATPase β subunit contains a Bst EII site in the equivalent location, and this site was used to exchange regions of the DNA between the two subunits. The polypeptide containing the cytoplasmic and transmembrane regions of Na,K-ATPase β subunit and the extracellular region of the H,K-ATPase β subunit is called NHβ1. The polypeptide containing the cytoplasmic and transmembrane regions of the H,K-ATPase β subunit and the extracellular region of the Na,K-ATPase β subunit is called

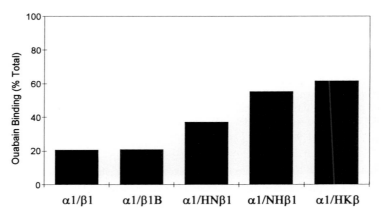

Figure 3. Antagonism of ouabain binding by 5 mM K⁺ for α1 + β-chimera hybrid pumps. [³H] ouabain binding was measured in the presence of 5 mM KCl, and normalized to % total binding as described in figure 2. Bars represent the average of two experiments done in duplicate. Variation between replicates is <1% of total binding in all cases.

HNβ1. The chimeric β subunits were co-expressed in yeast with Na,K-ATPase α1 and α3 subunits.

Figure 3 shows the antagonism by 5 mM KCl of ouabain binding to complexes formed from α1/β1, α1/β1B, α1/HKβ, α1/NHβ1, and α1/HNβ1. No difference is observed between the complexes containing β1 or β1B, indicating that the Leu→Val change in the β1 subunit has no effect on the interaction of the pump with K⁺ ions. Furthermore, the reduced apparent affinity of the α1/HKβ complex for K⁺ relative to α1/β1 appears to be due largely to differences in the structure of the extracellular region of the HKβ subunit and β1, since ouabain binding to the α1/NHβ1 complex was antagonized by K⁺ to a similar extent as α1/HKβ. The α1/HNβ1 complex showed an intermediate extent of antagonism of ouabain binding by K⁺, demonstrating that the transmembrane and/or cytoplasmic regions of the β subunit may also influence the structure of the K⁺ binding site.

The β subunit influences the Na⁺ dependence of ATP-dependent ouabain binding.

In sequential models of ion transport by Na,K-ATPase, the Na⁺ and K⁺ ions compete for the same sites, with the enzyme having a high affinity for Na⁺ and a low affinity for K⁺ in the E_1 conformation, and a high affinity for K⁺ and a low affinity for Na⁺ in the E_2 conformation. Since the structure of the β subunit can influence the interaction of the pumps with K⁺, we investigated the possibility that the interaction with Na⁺ may also be affected by the β subunit. To test this hypothesis, the binding of ouabain to the pumps was measured in an ATP-dependent reaction. In this reaction, we expect Na⁺ to be required to generate a phosphoenzyme which binds ouabain with high affinity. ATP-dependent ouabain binding was measured after a 3 minute incubation that was initiated by addition of membranes to the reaction components (35 mM imidazole pH 7.5, 5 mM $MgCl_2$, 100 μM Tris-ATP and different amounts of NaCl, all prewarmed to 37° C). The reaction was stopped by transfer to an ice/water bath. Yeast membranes, normally stored with 1 mM Na_2EDTA, were washed free of Na⁺ by diluting approximately 50 fold in Na⁺-free buffer, pelleting in the ultracentrifuge, resuspending the pellet in Na⁺-free buffer and re-pelleting. Measurements with a flame photometer confirmed that Na concentrations in the assay components were < 1 mM without added NaCl.

Figure 4 shows the ATP-dependent ouabain binding for hybrid pumps formed with α3 and either the β1, HKβ or the chimeric NHβ1.* Bars indicate 3 different concentrations of added Na⁺; 0 mM, 2 mM, and 20 mM. In the absence of added Na⁺, both α3/β1 and α3/HKβ show [3H] ouabain binding which is significantly above the background of non-specific binding (determined by the addition of excess unlabeled ouabain) in the assay. This phenomenon was also observed for α1/β1 and α1/HKβ, which show ~35% of total binding in the absence of added Na⁺. (data not shown) Control experiments indicate that removing ATP from the binding reaction reduces the 3H ouabain binding to background levels indicating that binding is dependent on Tris-ATP. Other controls suggest that only a small part of the phosphoenzyme formed in the absence of added Na⁺ can be attributed to phosphorylation of the pumps by inorganic phosphate.

At 2 mM Na⁺, the hybrid pump formed from the α3 and HKβ subunits shows enhanced [3H] ouabain binding compared to α3/β1 (Figure 4). This could reflect an increased affinity of the α3/HKβ hybrid pump for Na⁺, or alternatively, a change in the reaction kinetics which accelerates the formation of the ouabain binding phosphoenzyme conformation. This effect is

---

* HNβ1 when combined with α3 gives only very low levels of [3H] ouabain binding sites, suggesting a failure of these subunits to assemble properly. K.E. unpublished observations

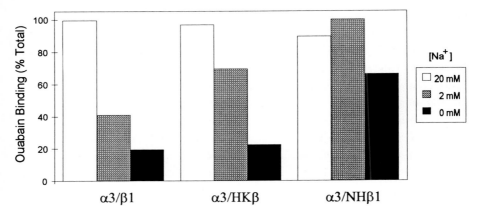

Figure 4. ATP-dependent Ouabain Binding. ATP-dependent [³H] ouabain binding was measured at different Na⁺ concentrations as described in the text. The data is normalized as % total binding and bars represent the mean of duplicate assays (variation < 1% of total binding).

unique to $\alpha3/HK\beta$; $\alpha1/HK\beta$ shows no increase in ouabain binding at 2 mM Na⁺ when compared to $\alpha1/\beta1$. (data not shown)

When $\alpha3$ is combined with the chimeric NHβ1 subunit, most of the [³H] ouabain binding (~65% of total binding) seen in this assay is no longer Na dependent (Figure 4). The binding is ATP dependent since removal of the Tris-ATP gives only background levels of [³H] ouabain binding. We see a similar effect when NHβ1 is combined with the $\alpha1$ isoform. (data not shown) It has been suggested that H⁺ or H$_3$0⁺ ions can be transported by the Na,K-ATPase (Hara and Nakao, 1986; Polvani and Blostein, 1988), and that Na⁺ ions can substitute for protons in the H,K-ATPase (Polvani et al., 1989). One possible explanation for the results obtained in the absence of Na⁺ ions, is that the chimeric NHβ1 subunit greatly enhances the ability of protons to substitute for sodium during phosphoenzyme formation. These data suggest that the structure of the β subunit may influence Na⁺ binding sites on the sodium pump and may affect phosphoenzyme formation.

Discussion

The experiments described in this report demonstrate that the β subunits of both Na,K-ATPase and gastric H,K-ATPase are able to assemble with the $\alpha1$ subunit and the $\alpha3$ subunit of Na,K-ATPase into hybrid complexes that are capable of binding ouabain. Since the binding of ouabain to Na,K-ATPase depends upon the catalytic activity of the pump, these results indicate that the hybrid pumps are functional. A similar observation has been reported by others in which the hybrid pumps were shown to catalyze K^+ transport in Xenopus oocytes (Horisberger et al., 1991; Noguchi et al., 1992). The amino acid sequences of the two β subunits are approximately 30% identical, and these observations indicate that sufficient structural conservation exists within the HKβ subunit to be recognized by and assembled with the Na,K-ATPase α subunits. The properties of the hybrid pumps are not identical to Na,K-ATPase, however, indicating that the structure of the β subunit can influence pump function. In particular, the structure of the β subunit appears to affect the interaction of the pumps with monovalent cations. The ability of K^+ ions to antagonize the binding of ouabain, and also the ability of Na^+ ions to participate in the transfer of phosphate from ATP to the protein, are influenced by the structure of the β subunit.

Acknowledgements

We thank J. Lingrel and G. Shull for providing the sheep Na,K-ATPase α1 subunit cDNA, E. Benz for the Na,K-ATPase α3 subunit cDNA, R. Mercer for the rat β1 cDNA, R. Levenson for the rat H,K-ATPase β subunit cDNA, J. Forte for antibodies against H,K-ATPase β subunit, and J. Kyte for antibodies against Na,K-ATPase. This work was supported by U.S. Public Health Service grants GM28673, GM9855, and HL39295, and grant DMB-8919336 from the National Science Foundation. KAE was supported in part by fellowship 983 F1-1 from the American Heart Association, Greater Los Angeles Affiliate.

References

Eakle, K.A., Horowitz, B., Kim, K.S., Levenson, R., and Farley, R.A. (1991). Expression and assembly of different α- and β-subunit isoforms of Na,K-ATPase in yeast. In The Sodium Pump: Recent Developments. J.H. Kaplan and P. DeWeer, eds. (New York: The Rockefeller University Press), pp. 125-130.

Eakle, K.A., Kim, K.S., Kabalin, M.A., and Farley, R.A. (1992). High-affinity ouabain binding by yeast cells expressing Na,K-ATPase α subunits and the gastric H,K-ATPase β subunit. Proc. Natl. Acad. Sci. USA *89*, 2834-2838.

Forbush, B. and Hoffman, J.F. (1979). Evidence that ouabain binds to the same large polypeptide chain of dimeric Na,K-ATPase that is phosphorylated from Pi.. Biochemistry *18*, 2308-2314.

Forbush, B., III (1983). Cardiotonic steroid binding to Na,K-ATPase. Current Topics in Membranes and Transport *19*, 167-201.

Green, N.M. and Stokes, D.L. (1992). Structural modelling of P-type ion pumps. Acta Physiol. Scand. *146*, 59-68.

Hara, Y. and Nakao, M. (1986). ATP-dependent proton uptake by proteoliposomes reconstituted with purified Na,K-ATPase. J. Biol. Chem. *261*, 12655-12658.

Horisberger, J., Jaunin, P., Reuben, M.A., Lasater, L.S., Chow, D.C., Forte, J.G., Sachs, G., Rossier, B.C., and Geering, K. (1991). The H,K-ATPase β subunit can act as a surrogate for the β subunit of Na,K-pumps. J. Biol. Chem. *266*, 19131-19134.

Horowitz, B., Eakle, K.A., Scheiner-Bobis, G., Randolph, G.R., Chen, C.Y., Hitzeman, R.A., and Farley, R.A. (1990). Synthesis and assembly of functional mammalian Na,K-ATPase in yeast. J. Biol. Chem. *265*, 4189-4194.

Jorgensen, P.L. and Andersen, J.P. (1988). Structural basis for E1-E2 conformational transitions in Na,K-pump and Ca-pump proteins. J. Membrane Biol. *103*, 95-120.

Noguchi, S., Maeda, M., Futai, M., and Kawamura, M. (1992). Assembly of a hybrid from the α subunit of Na,K-ATPase and the β subunit of H,K-ATPase. Biochem. Biophys. Res. Comm. *182*, 659-666.

Polvani, C., Sachs, G., and Blostein, R. (1989). Sodium ions as substitutes for protons in the gastric H,K-ATPase. J. Biol. Chem. *264*, 17854-17859.

Polvani, C. and Blostein, R. (1988). Protons as substitutes for sodium and potassium in the sodium pump reaction. J. Biol. Chem. *263*, 16757-16763.

Scheiner-Bobis, G., Eakle, K.A., Kim, K.S., and Farley, R.A. (1991). Expression of DNA for alpha, beta, and gamma polypeptides of Na, K-ATPase in yeast. Does gamma have a function in Na,K-ATPase activity?. In The Sodium Pump: Structure, Mechanism, and Regulation. J.H. Kaplan and P. DeWeer, eds. (New York: Rockefeller University Press),

Wang, K. and Farley, R.A. (1992). Lysine 480 is not an essential residue for ATP binding or hydrolysis by Na,K-ATPase. J. Biol. Chem. *267*, 3577-3580.

Yoda, S. and Yoda, A. (1987). Phosphorylated intermediates of Na,K-ATPase proteoliposomes controlled by bilayer cholesterol. Interaction with cardiac steroid. J. Biol. Chem. *262*, 103-109.

DCCD Inhibition of Cation Binding in the Gastric H,K-ATPase

Edd C. Rabon[+] and Kent Smillie
Department of Physiology
Tulane University Medical Center and
New Orleans Veterans Administration Center
1430 Tulane Ave
New Orleans, LA 70112
+Supported by NIH grant DK34286 and USVA

Introduction:

The H,K-ATPase (EC 3.6.1.3) is the plasma membrane ATPase dedicated to the active transport of H^+ across the gastric secretory epithelium (14,33,36). The H^+ pump functions as an H^+ for K^+ exchanger with lumenal K^+ actively exchanged for cytoplasmic H^+ (16,35,37). While the primary sequence of this enzyme and its structural organization within an a,ß-heterodimer have been elucidated, structural detail sufficient to identify residues involved in ligand binding sites within various domains is not presently available (17,22,28,40).

Structure/function studies utilizing chemical modification of amino acid residues have provided useful evidence of specific residues within functional domains of the H,K-ATPase. Notably, using ligand protection from functional inactivation as a criteria of domain involvement, several functionally essential residues within the nucleotide domain have been identified within the cytoplasmic loop bounded by the M4/M5 membrane spanning domains. While modification of the ATPase by FITC (12,13,18,19) pyridoxyl phosphate (23), and [4-(N-2-chlorethyl-N-methylamino)] benzoylamide-ATP (10) inhibits phosphoenzyme formation or ATP turnover in an ATP-protectable manner, this functional criteria does not distinguish between the loss of function due to the direct modification of a residue within a structural domain or a modification within a separate domain essential for overall activity. Thus site-specific mutagenes of these selected residues confirm the role of the essential ß-aspartyl phosphate within the catalytic center of the EP type ATPases, but also show that the FITC-labeled residue in the Na,K-ATPase and the Ca-ATPase do not participate directly in nucleotide binding (7,24,41).

NATO ASI Series, Vol. H 89
Molecular and Cellular Mechanisms
of H+ Transport
Edited by Barry H. Hirst
© Springer-Verlag Berlin Heidelberg 1994

Functional studies of partial reactions of the P type ATPase have been exploited as a more explicit means of identifying the functional consequences of structural modification within these enzymes. In the S.R. Ca-ATPase, Pick and Racker interpreted the Ca^{2+} protection of the Ca-ATPase as evidence that DCCD inhibition of the enzyme was due to the modification of residues within the cation binding site (31). Chemical modification studies using carboxyl reagents such as DEAC (1), DCCD (9,11,27,29,39) or EEDQ (34) have shown that modification of carboxyl residues within the P type ATPase family are protected by the transported ligand and thus provide similar evidence that carboxyl residues play an essential role in cation binding in diverse members of the P-type enzyme family.

In this report we present evidence that K^+ protects both K^+ stimulated turnover and cation binding in the H,K-ATPase from inactivation by the hydrophobic carboxyl reagent, DCCD. The inhibition is accompanied by the covalent incorporation of $[^{14}C]$-DCCD and suggests that the mechanism of inhibition involves the modification of carboxyl residues essential to cation binding. Sequence analysis of $[^{14}C]$-DCCD-labeled peptides shows that the label is incorporated into the C terminal transmembrane elements of the H,K-ATPase to suggest that carboxyl groups within these transmembrane elements are essential for cation binding activity.

Methods:

Preparation of membrane-bound H,K-ATPase; Gradient purified, microsomal H,K-ATPase was prepared, stored and resuspended by previously described methods (4).

Measurement of $^{86}Rb^+$ occlusion; $^{86}Rb^+$ occlusion was measured at 4°C in the digitonin-treated H,K-ATPase as previously described (32).

Tryptic hydrolysis and resolution of peptide fragments of the H,K-ATPase; The H,K-ATPase (1.0 mg/mL) was suspended in medium composed of 0.25 M sucrose and 50 mM TrisCl, pH 8.2. Trypsin was added at a protein/trypsin ratio of 20 (w/w) and the suspension digested for 30 min. at 37°C. The reaction was stopped by addition of trypsin inhibitor protein (TIP/trypsin = 15 (w/w)) and the suspension placed on ice. The membrane-bound protein was collected by centrifugation, resuspended and labeled with 0.1 mM fluorescein-5-maleimide (F-5-M) as reported by Besancon et al (2). The F-5-M-labeled peptides were resolved on a 16.5% discontinuous gel using the method of Schagger and von Jagow (38) and transferred to PVDF for measurement of $[^{14}C]DCCD$ incorporated and sequence analysis. The fluorescent bands of companion lanes were marked to provide boundaries for excision of intact bands or division into the 1 mm sections through the band region. [14C]-DCCD incorporation was measured in both intact bands and in 1 mm slices while sequence analysis was performed on selected, intact bands.

Stoichiometry of [14C]-DCCD incorporation; Aliquots containing 435 µg of the H,K-ATPase were suspended at 1.0 mg/mL protein in buffer composed of 20 mM Pipes/Tris, pH 7.4 ± 30 mM KCl. [14C]-DCCD was added (0.2 mM$_{final}$ and ≈ 30 cpm/ pmole) and incubated for time periods up to one hour. Periodically, aliquots containing 35 µg protein were quenched with addition to 1.0 mL of ice cold 10% Cl3CC00H and 0.15 mg/mL BSA. Membrane lipids were removed as described by Gorga (15) and the protein resuspended in 2% SDS. Covalently incorporated [14C]-DCCD was determined by liquid scintillation counting and the stoichiometry of incorporation calculated from the initial specific activity and the amount of protein reported by the BioRad system using γ-globulin standards.

Results:

Inhibition of ATPase activity; The H,K-ATPase was inhibited by DCCD concentrations ranging from .02 to 1.5 mM DCCD. In each case the time course of DCCD inhibition was described by a single inactivation constant, k', proportional to DCCD concentration and a residual factor inversely proportional to the DCCD concentration. In the experiment shown in figure 1, in K+ free media, DCCD rapidly inactivated approximately 90% of the measured activity of the H,K-ATPase over a time course; k' = 0.36±.04 min^{-1}. KCl protected the enzyme from activity loss. The upper trace shows that 80% of the control enzyme activity is retained over the same period sufficient to

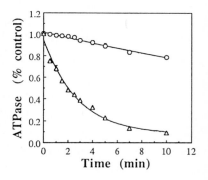

Figure 1.   *ATPase activity follow-ing treatment with 1.5 mM DCCD.* (O) K+-protected (Δ) Unprotected. Lines drawn to A exp$^{(-kx)}$ + res. For (Δ): k = .36±.04 min^{-1}, A = 0.87± .04, residual = .08±.03. For (0): k = 0.026 ±.001 min^{-1}, A = 1 and residual = 0.

inactivate 90% of enzyme activity in K+-free medium. 30 mM KCl decreased the rate of inhibition to; k_K' = .026±.001 min^{-1}. KCl also protected the enzyme from loss of ATPase activity at pH 6.1 where it decreased the time constant of the inactivation process from k′; 0.63±.04 to $k_K′$; 0.37±.06 min⁻¹. Since the ratio of the rate of DCCD inactivation of the unprotected to the KCl protected enzyme, k′/$k_K′$, was greater at pH 7.4, experiments were routinely performed at this pH to optimize the conditions for K⁺ protection.

Inactivation of ⁸⁶Rb⁺ occlusion and MDPQ fluorescence; The K⁺ protection of ATPase turnover suggests that inhibition is related to cation binding. ⁸⁶Rb⁺ binding to the H,K-ATPase was measured to provide a more specific functional test of DCCD interaction. Incubation with DCCD rapidly inhibits ⁸⁶Rb⁺ binding in the H,K-ATPase. The

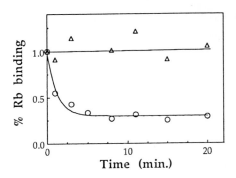

Figure 2   Rb occlusion following treatment with 1.5 mM DCCD (0) Rb+ occlusion in the unprotected enzyme (Δ) Rb+ occlusion of the K+ protected enzyme. Line drawn to; y = A exp⁽⁻$^{kx)}$ + residual. k = 0.84 ± .15 min⁻¹, A = 0.69±.05 and residual = 0.30 ± .02.

data in figure 2 show that ⁸⁶Rb⁺ binding was inhibited over a monoexponential time course where k; .84±.2 min⁻¹. A small fraction of the ⁸⁶Rb⁺ bound to the H,K-ATPase was resistant to DCCD and remained associated with the ATPase at the longest incubation periods measured. K⁺ fully protected the ⁸⁶Rb⁺ binding site. As shown in the upper trace, 10 mM KCl completely protected the cation binding site over the time period sufficient to inactivate the enzyme in the absence of the cation.

MDPQ, a fluorescent, K⁺ competitive inhibitor of the H,K-ATPase has been shown to be closely associated with cation binding on the extracytoplasmic side of the pump molecule (32). Its fluorescence intensity, enhanced with binding to the H,K-ATPase, is increased by enzyme phosphorylation so that both the K⁺-dependent response of the dephosphoenzyme and the ATP-dependent response of the phosphoenzyme can be evaluated following reaction with DCCD. The data in figure 3 show that DCCD inhibits both fluorescence responses with

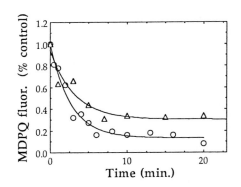

Figure 3 DCCD inactivation of MDPQ fluorescence. The H,K-ATPase was treated with 1.5 mM DCCD and added directly to 2.0 mL of the fluorescence assay buffer. Fluorescence was measured with addition of either (0) 0.1 mM ATP or (Δ) 10 mM KCl.

similar time courses where k; .38±.05 and .32±1 min⁻¹, respectively.

Stoichiometry of [¹⁴C] DCCD incorporation; Initial autoradiography of the [¹⁴C]DCCD-labeled membrane fraction revealed a membrane associated component migrating at the ion front. This component was eliminated by lipid extraction prior to the measurement of [¹⁴C]DCCD stoichiometry into the H,K-ATPase. The time dependence of [¹⁴C]DCCD incorporation is shown in figure 4. In K⁺ free medium, [¹⁴C]DCCD was incorporated into the H,K-ATPase at a rate k; 0.1±.03 min⁻¹. This rate is identical to the rate of inactivation of the H,K-ATPase at this DCCD concentration (k = 0.1±.01 min⁻¹ at 0.2 mM, data not shown). The [¹⁴C]DCCD incorporated into the H,K-ATPase was reduced 30% by KCl without change in the rate of incorporation. The maximal levels of [¹⁴C]DCCD incorporated in K⁺ and K⁺-free medium were 5.4±3 and 3.5±.4

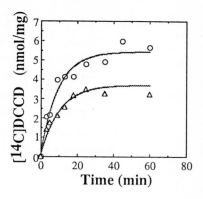

Figure 4   Stoichiometry of [¹⁴ C] DCCD incorporation. The H,K-ATPase was treated with [¹⁴C]DCCD (.17 mM @ 17 cpm/ pmole) in the presence, (Δ), and absence, (O), of 20 mM KCl. The lines are drawn to; y = C - Be⁻ᵏᵗ. For (Δ) C; 3.70±.26, B; 3.54±.39 nmoles/mg and k;.1±.03 min⁻¹. For (O) C; 5.44±.28, B; 5.18±.45 nmoles/mg and k;.1±.02 min⁻¹.

nmole [¹⁴C]DCCD/mg protein, respectively. This measurement provides a K⁺ specific component of [¹⁴C]DCCD incorporated into the H,K-ATPase of between 1 and 2 molecules of [¹⁴C]DCCD per mg protein. Additional experiment indicated that the loss of activity was reasonably proportional to the K⁺ sensitive DCCD incorporated where approximately 1.1 nmole DCCD/ mg protein were incorporated with complete inhibition of the ATPase.

Identification of [¹⁴C]DCCD labeled residues; Initially, the

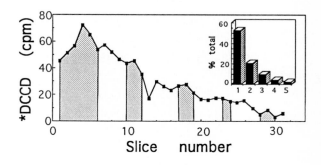

Figure 5   [¹⁴ C] DCCD incorporation into tryptic peptides of H,K-ATPase treated with [¹⁴ C] DCCD at 200 nmole/mg protein. Tryptic peptides were prepared from K⁺ free membranes resolved by SDS PAGE and transferred to PVDF. Points indicate [¹⁴ C] DCCD in peptide region revealed by F-5-M fluorescence (shaded area). Inset; % distribution of total cpms derived exclusively from the excised fluorescent bands.

[14C]DCCD-labeled enzyme was digested in KCl medium to obtain the membrane preparation enriched in a M_r; ≈23 kDa peptide (32). Autoradiography of the preparation resolved by SDS PAGE showed that the label was exclusively localized within the M_r; ≈23 kDa band (data not shown). Since this region encompasses the C-terminal end of the molecule from [857]Glu, conditions

Table I.   [14C]-labeled tryptic peptides

(A) M_r; 21 kDa	(B) M_r; 10.2 kDa
L (13.3/-)	L (2.4/ -)
V (16.9/10.9)	V (2.7/1.3)
N (15.9/8.6)	N (1.9/0.8)
E (21.7/9.5)	E (6.1/2.1)
P (19.4/2.6)	P (4.1/1.3)

Sequence obtained from PVDF transfer of tryptic peptides obtained in (A) KCl or (B) sucrose medium. ( ) ; pmol peptide yield per cycle/ pmole change from previous cycle.

were modified to obtain transmembrane peptides below M_r; 11 kDa. Figure 5 shows the distribution of the [14C]DCCD label into 1 mm slices throughout the M_r ≈ 10.2 - 5.7-kDa region and the overlap of this label with 5 fluorescently labeled peptide bands resolved within this region. While the [14C]DCCD is diminishingly distributed throughout this gel region it is evident that the peaks overlap the fluorescently labeled transmembrane regions and account for approximately 70% of the total label within this region. The inset provides a comparison of the distribution of [14C]DCCD within the labeled bands where 55% of the label is found within the largest peptide fragment resolved at M_r; ≈10.2 kDa with decreasing quantities present in the progressively smaller peptides at M_r; 8.8, 7.4, 6.8 and 5.7 kDa. The fluorescent bands at both M_r; ≈2 3 and 10.2 kDa were excised and submitted for microsequence analysis. A dominant sequence was resolved in each case along with minor residues of unidentifiable sequence. The sequence and yield obtained in each sequencing cycle is shown in Table I. The dominant peptide sequence obtained in each case was, leu-val-asp-glu-pro.

Discussion:

The purpose of this investigation was to provide biochemical evidence of essential carboxyl residues within the functionally defined cation binding domain of the H,K-ATPase. We have shown that chemical modification by DCCD, a hydrophobic carboxyl activating reagent, inhibits both K+ stimulated ATP turnover and [86]Rb+ binding in a K+ protectable manner. The time course of this functional inactivation is similar to the covalent incorporation of [14C]DCCD into the protein.

These criteria suggest that essential carboxyl residues are present in the cation

cation binding site of that enzyme (5,6).

Interestingly, DCCD also inhibits MDPQ fluorescence. This fluorescent, K^+ competitive inhibitor, is closely associated with the cation binding domain in that it is displaced by K^+, induces conformational changes in the nucleotide binding domain and like $^{86}Rb^+$ occlusion, is enhanced by tryptic digestion of the E_2-K enzyme. The M1/M2 sequence was not preferentially labeled by DCCD as might have been predicted from the absence of hydrophobic residues within the M1/M2 sequence though ^{131}Glu is labeled by the K^+-competitive photoaffinity label, MeDAZIP (26).

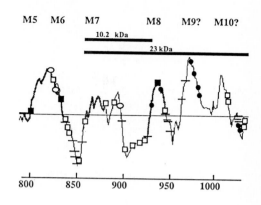

Figure 6 *Hydropathy model of the C-terminal H,K-ATPase.* Hydropathy profile from Kyte Doolittle analysis. Features marked include: ( ) omeprazole-labeled cysteine, ( ) cysteine, ( ) consensus Glu from SR Ca-ATPase implicated in cation binding, ( ) carboxyl groups, ( ) trypsin cleavage sites, ( ) peptides labeled with [^{14}C]-DCCD.

The lack of DCCD reactivity is also compatible with conclusions derived from the S.R. Ca-ATPase where mutation of 4 glutamate residues between M1 and M2 had no effect on Ca^{2+} transport activity (8). Our data do not eliminate the possibility that some label was incorporated into the M3/M4 transmembrane since approximately 30 % of the counts may overlap F-5-M fluorescent bands reported to contain M3/M4 sequence (2). Better peptide resolution and careful quantitation of the K^+ sensitive label will be necessary to resolve this question.

In summary, though it is difficult to eliminate the possibility that the modified residues are not within the K^+ binding site, the overall picture suggests that the covalent incorporation of 1 or 2 nmoles of DCCD specifically inhibits the Rb^+ binding, conformational changes involving cation binding and cation stimulated ATPase activity. Thus we suggest that the loss of K^+ dependent function results from the modification of an essential carboxyl residue(s) most likely within the M7/M8 transmembrane domain.

References:

1. Arguello, J.M. & Kaplan, J.H. (1991) *J. Biol. Chem. 266*, 14627-14635.
2. Besancon, M., Shin, J.M., Mercier, F., Munson, K., Miller, M., Hersey, S., & Sachs, G. (1993) *Biochemistry 32, No.9*, 2345-2355.
3. Brandl, C.J., Green, N.M., Korczak, B., & MacLennan, D.H. (1986) *C 44*, 597-607.
4. Chang, H., Saccomani, E., Rabon, E., Schackmann, R., & Sachs, G. (1977) *Biochim. Biophys. Acta 464*, 313-327.
5. Clarke, D.M., Loo, T.W., Inesi, G., & MacLennan, D.H. (1989) *Nature 339*, 476-478.
6. Clarke, D.M., Loo, T.W., & MacLennan, D.H. (1990) *J. Biol. Chem. 265*, 6262-6267.
7. Clarke, D.M., Loo, T.W., & MacLennan, D.H. (1990) *J. Biol. Chem. 265*, 22223-22227.
8. Clarke, D.M., Maruyama, K., Loo, T.W., Leberer, E., Inesi, G., & MacLennan, D.H. (1989) *J. Biol. Chem. 264*, 11246-11251.
9. De Ancos, J.G. & Inesi, G. (1988) *Biochemistry 27*, 1793-1803.
10. Dzhandzhugazyan, K. & Modyanov, N. (1985) in *The Sodium Pump* (Glynn, I.M. & Ellory, C., Eds.) pp 129-134, The Company of Biologists, Cambridge UK.
11. Famulski, K.S., Pikula, S., Wrzosek, A., & Wojtczak, A.B. (1990) *Cell Calcium. 11*, 275-280.
12. Farley, R.A. & Faller, L.D. (1985) *J. Biol. Chem. 260*, 3899-3901.
13. Farley, R.A., Tran, M.C., Carilli, C.T., Hawke, D., & Shively, J.E. (1984) *J. Biol. Chem. 259*, 9532-9535.
14. Forte, J.G. & Lee, H.C. (1977) *Gastroenterology. 73*, 921-926.
15. Gorka, F.R. (1985) *Biochemistry 24*, 6783-6788.
16. Gunther, R.D., Bassilian, S., & Rabon, E.C. (1987) *J. Biol. Chem. 262*, 13966-13972.
17. Hall, K., Perez, G., Sachs, G., & Rabon, E. (1991) *Biochim. Biophys. Acta 1077*, 173-179.
18. Jackson, R.J., Mendlein, J., & Sachs, G. (1983) *Biochim. Biophys. Acta 731*, 9-15.
19. Karlish, S.J.D. (1980) *J. Bioenerg. Biomembr. 12, Nos.3/4*, 111-136.
20. Karlish, S.J.D., Goldshleger, R., & Stein, W.D. (1990) *Proc. Natl. Acad. Sci. USA. 87*, 4566-4570.
21. MacLennan, D.H., Brandl, C.J., Korczak, B., & Green, N.M. (1985) *Nature 316*, 696-670.
22. Maeda, M., Ishizaki, J., & Futai, M. (1988) *Biochem. Biophys. Res. Commun. 157*, 203-209.
23. Maeda, M., Tagaya, M., & Futai, M. (1988) *J. Biol. Chem. 263*, 3652-3656.
24. Maruyama, K., Clarke, D.M., Fujii, J., Inesi, G., Loo, T.W., & MacLennan, D.H. (1989) *J. Biol. Chem. 264*, 13038-13042.
25. Mercier, F., Bayle, D., Besancon, M., Joys, T., Lewin, M., Prinz, C., Reuben, M., Soumarmon, A., Wong, H., Walsh, J., & Sachs, G. (1993) *Biochim. Biophys. Acta* .(in press)
26. Munson, K.B., Gutierrez, C., BALAJI, V.N., Ramnarayan, K., & Sachs, G. (1991) *J. Biol. Chem. 266*, 18976-18988.

27. Murphy, A.J. (1981) *J. Biol. Chem. 256*, 12046-12050.

28. Okamoto, C.T., Karpilo, J.M., Smolka, A., & Forte, A. (1990) *Biochim. Biophys. Acta 1037*, 360-372.

29. Pedemonte, C.H. & Kaplan, J.H. (1986) *J. Biol. Chem. 261*, 16660-16665.

30. Pedemonte, C.H. & Kaplan, J.H. (1990) *Am. J. Physiol. 258*, C1-C23.

31. Pick, U. & Racker, E. (1979) *Biochemistry 18*, 108-113.

32. Rabon, E., Smillie, K., Seru, V., & Rabon, R. (1993) *J. Biol. Chem. 268*, 8012-8018.

33. Rabon, E.C. & Reuben, M.A. (52) in *Annual Review of Physiology* (Hoffman, J.F., Ed.) pp 321-344,

34. Saccomani, G., Barcellona, M.L., & Sachs, G. (1981) *J. Biol. Chem. 256*, 12405-12410.

35. Sachs, G., Chang, H.H., Rabon, E., Schackmann, R., Lewin, M., & Saccomani, G. (1976) *J. Biol. Chem. 251*, 7690-7698.

36. Sachs, G., Spenney, J.G., & Rehm, W.S. (1977) *Int. Rev. Physiol. 12*, 127-171.

37. Schackmann, R., Schwartz, A., Saccomani, G., & Sachs, G. (1977) *J. Membr. Biol. 32*, 361-381.

38. Schagger, H. & von Jagow, G. (1987) *Anal. Biochem. 166*, 368-379.

39. Shani-Sekkler, M., Goldshleger, R., Tal, D., & Karlish, S.J.D. (1988) *J. Biol. Chem. 263*, 19331-19342.

40. Shull, G.E. & Lingrel, J.B. (1986) *J. Biol. Chem. 261*, 16788-16791.

41. Wang, K. & Farley, R.A. (1992) *J. Biol. Chem. 267*, 3577-3580.

Mapping of a Cytoplasmic Site of Gastric (H⁺,K⁺)-ATPase involved in the Transmission of K⁺-Activation across the Membrane

D. Bayle, J.C. Robert and A. Soumarmon
INSERM Unité 10,
Hôpital Bichat,
170 bd Ney,
75018 Paris, FRANCE

R. Brasseur
ULB Campus Plaine,
1050 Bruxelles,
BELGIUM.

H.G.P. Swarts, C.H.W. Klaassen, and J.J.H.M. De Pont,
Department of Biochemistry,
University of Nijmegen, 6500 HB
Nijmegen, The NETHERLANDS.

Monoclonal antibodies are useful tools to study the topology of membrane proteins. When in addition, monoclonal antibodies inhibit the protein function, the characterization of epitopes can provide insight in structure-function relationship.

Recently the groups from Nijmegen and Paris prepared inhibitory monoclonal antibodies against gastric (H⁺,K⁺)-ATPase which have close but not identical properties: Benkouka et al [1989] raised the monoclonal antibody 95-111 which inhibits the K⁺-stimulated ATPase activity at pH 7.0 for 60% and the K⁺-stimulated p-nitrophenylphosphatase activity for 25%. Bayle et al [1992], demonstrated that the inhibition of the ATPase activity is non-competitive towards ATP and competitive versus K⁺. Van Uem et al [1990] prepared the monoclonal antibodies 5-B6 and 4D10 [DePont et al, 1992]. The 5-B6 antibody inhibits 65% of the (H⁺,K⁺)-ATPase activity at pH 7.0 but does not inhibit the K⁺-stimulated p-nitrophenyl phosphatase activity. Inhibition is also non-competitive towards ATP but is uncompetitive towards K⁺.

The epitope for 95-111 was determined by a molecular biology approach [Bayle et al 1992]. The cDNA of rabbit α subunit was fragmented using restriction enzymes and the fragments were ligated to produce fusion proteins with ß galactosidase (PuEX vectors). The immunoreactivity of fusion proteins was determined by immunoblotting and the epitope was

NATO ASI Series, Vol. H 89
Molecular and Cellular Mechanisms
of H⁺ Transport
Edited by Barry H. Hirst
© Springer-Verlag Berlin Heidelberg 1994

mapped within the peptide sequence Cys_{529}-Glu_{561}. The epitope for 5-B6 was determined by proteolysis of the native membranous ATPase: well-defined trypsic fragments of the α-subunit of (H^+,K^+)-ATPase were prepared in the presence of either ATP or K^+ [Helmich de Jong et al, 1987]. In addition, papain cleaved a large part of the intracellular loop between the fourth and the fifth transmembrane segment of α into a water-soluble form [Saccomani et al 1987]. The epitope for 5-B6 was shown to lie between Ile_{456} and Gly_{620} [Van Uem et al 1991]. The location of epitope for 5-B6 was further shortened after formic acid treatment of the peptidic fragments. Since formic acid specifically cleaves the link between an aspartic acid and a proline and since formic acid treatment of the ATPase peptide destroyed the immunoreactivity, the epitope for 5-B6 was suggested to lie in the area of Asp_{507} and Asp_{510} where two Asp-Pro links exist.

When the molecular biology approach successfully used for mapping the 95-111 epitope was used with 5-B6, a fusion protein encoding for amino acids 500-620 of α was immunoreactive. To our surprise, a smaller fusion protein which encoded for the sequence between amino acids 529 and 561 of α was also immunoreactive. This was not in line with the mapping of 5-B6 epitope from formic acid experiments. Fusion proteins containing larger fragments of α (amino acids 112-529, 529-910, 270-666) were not immunoreactive with 5-B6 suggesting that they did not completely unfold in SDS and that the 5-B6 epitope was unaccessible for the mAb. We prepared new fusion proteins between glutathione transferase and the 421-700 and 343-554 large fragments of α. Both fusion proteins were immunoreactive with 5-B6 and 95-111 suggesting that the epitope for 5-B6 is mapped between amino acids 529 and 554. Therefore we suspected that the location (Asp_{507} Asp_{510}) of the epitope for 5-B6 assumed from formic acid treatment was wrong.

To investigate this problem further, we synthetized the peptide HTLEDPRDPRH corresponding to the sequence 503-513 of the α subunit of pig (H^+,K^+)-ATPase [Maeda et al 1988]. The synthesis was rather difficult because of the acid lability of the Asp-Pro bonds (Tesser et al, to be published). The synthetic peptide had no effect on the immunoreactivity of 5-B6 on native pig (H^+,K^+)-ATPase (ELISA assay) confirming that the epitope for 5-B6 is not located in the area of the two Asp-Pro bonds but rather lies between amino acids 524 and 554 of α. The most likely explanation is that formic acid treatment of papain-fragments of the ATPase did more than cleaving the Asp-Pro bonds; it could also convert asparagine and glutamine residues in aspartic and glutamic acids respectively. The possible candidates for that modification are Gln 537, 544 and 550.

The above results indicate that the epitopes for 95-111 and 5-B6 are both mapped between amino acids 529 and 554 of α. This sequence is on the cytoplasmic side of the ATPase, in the catalytic loop between M4 and M5. It is close to the FITC, i.e. the ATP binding site (Lys 517 (rat)). However, both antibodies are non-competitive inhibitors towards ATP indicating that they do not interfere with the binding of ATP. The close location of the epitope for 95-111 and 5B6 does not mean that the epitopes are equal. Competitive experiments show that 95-111 displaces peroxyidase labelled 5B6 from its binding site but only with a ten times higher concentration than unlabeled 5B6. However, the combination of 95-111 and 5B6 did not give more inhibition of the (H^+, K^+)-ATPase activity than 5B6 or 95-111 alone. Such an additivity was found by the combination of the two inhibitory mAbs 5B6 and 4D10 [De Pont et al, 1992], where the two epitopes are apparently completely separate. It was not found with the combination of 95-111 and 4D10.

The effect of mAb 95-111 is K^+-competitive with respect to the stimulation of ATPase and pNPPase and also K^+-competitive with respect to fluorescence quenching of bound FITC [Bayle et al 1992]. This means that the binding of 95-111 decreases the affinity for K^+. The inhibition by 5-B6 is uncompetitive with respect to K^+-stimulation of the ATPase. This means that the binding of 5-B6 increases the apparent affinity for K^+. Since K^+ binds to the extracytoplasmic side of the enzyme whereas the antibodies bind to the cytoplasmic side, we suggest an allosteric interaction across the membrane such that bindings of 95-111 and 5-B6 on the cytoplasmic catalytic loop change the conformation of the luminal site for K^+.

We must underline here that some experimental results remain unexplained: whereas, the group in Paris obtains competitive interactions between 95-111 and K^+, a mixed type of ATPase activity inhibition, mainly K^+-noncompetitive was found in Nijmegen. The reason for these discrepancies are not yet clarified. Medium composition could play an important role since medium pH modulates 5-B6 effects and since the binding of tertiary amines to the cytoplasmic side of (H^+,K^+)-ATPase competitively inhibits K^+-activation (Swarts et al submitted for publication).

The sequence between amino acids 529 and 554 carries both 5-B6 and 95-111 epitopes and is somehow functionally related to the luminal K^+-site of the ATPase with respect to the K^+-activation of dephosphorylation and the K^+-induced E1-E2 conformational changes. In recent data, we found that the accessibility of 95-111 epitope changes with E1 and E2: E1 has more sites for 95-111 than E2, whereas 5-B6 epitopes are almost constant (Bayle et al to be published) [Robert J.C. et al 1990]. This indicates that part of the aa 529-554 sequence moves in and out the ATPase as does Lys 517 (the FITC binding site) the environment of which is more hydrophobic in E1 than in E2. We calculated the three dimensional structure of this sequence using the protocol developed by R. Brasseur [1990] [Colson et al 1990], [Rosseneu et al 1990]. The 3-D structure is clearly amphipatic and as though could stand at the surface of the ATPase, the hydrophilic side facing the cytoplasm. Moreover, the structure is highly attractive for cations: due to its cytoplasmic location, it cannot be the activating K^+-site. If E2 binds luminal K^+, E1 binds cytoplasmic H^+. To our knowledge, the localization of H^+ binding site is unknown. We tested the effect of H^+ binding on the 3D modelized structure (Fig.1): protonation of carboxylic amino-acids (Glu and Asp) twists the structure and modifies the hydrophobicity pattern. This suggests that amino acids 529-554 could stand at the protein-cytoplasm interface and speculates that protonation has an active role in the E1-E2 movements. Of course, this is speculation since, if we have experimental evidence that the 529-554 sequence moves between E1 and E2, we have none that protonation is a driving force.

Figure 1 Modelization of the three dimensional structure of Ala519- Ala548 (rat (H^+,K^+)-ATPase) [Shull et al 1986]

unprotonated   **protonated**

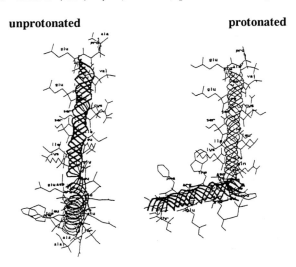

References

Bayle, D., Robert, J.C., Bamberg, K., Benkouka, F., Cheret, A.M., Lewin, M.J.M., Sachs, G. and Soumarmon, A. (1992) Location of the cytoplasmic epitope for a K^+-competitive antibody of the (H^+,K^+)-ATPase. J. Biol. Chem. 267, 19060-19065.

Benkouka, F., Péranzi, G., Robert, J.C., Lewin, M.J.M. and Soumarmon, A. (1989). A monoclonal antibody which inhibits (H^+,K^+)-ATPase activity but not chloride conductance. Biochimica Biophysica Acta, 987,205-211.

Brasseur, R. (1990) TAMMO: theoritical analysis of membrane molecular organization, in " Molecular description of biological membranes by computer-aided conformational analysis", R. Brasseur ed. CRC Press, vol I pp 203-220.

Colson, A.M., Turnpower, B.L. and Brasseur (1990) Insight into cytochrome b structure-function relationship by the combined approaches of genetics and computer modeling, in " Molecular description of biological membranes by computer-aided conformational analysis", Brasseur ed, CRC Press, vol II, pp 147-160.

DePont, J.J.H.H.M., Van Uem, T.J.F., Klaassen, C.H.W., Swarts, H.G.P. (1992) Inhibitory antibodies against gastric H,K-ATPase. Ann N Y Acad Sci 671, 446-448.

Helmich de Jong, M.L., Van Emst-De Vries, S.E., DePont, J.J.H.H.M. (1987) Conformational states of $(H^{++}K^+)$-ATPase studied using tryptic digestion as a tool. Biochim. Biophys. Acta, 905, 358-370.

Maeda, M., Ishizaki, J., Futai, M. (1988) cDNA cloning and sequence determination of pig gastric (H^+K^+)-ATPase. Biochem. Biophys. Res.Commun. 157, 203-209.

Robert, J.C., Benkouka, F., Bayle, D., Hervatin, F. and Soumarmon, A. (1990) (H^+,K^+)-ATPase contents of human, rabbit, hog and rat gastric mucosa. Biochimica Biophysica Acta, 1024, 167-172.

Rosseneu, M., De Loof, H., DeMeutter, J. Ruysschaert, J.M., Vanloo, B. and Brasseur, R. (1990) Identification and computer modeling of functional domains in plasma apolipoproteins, in " Molecular description of biological membranes by computer-aided conformational analysis", R. Brasseur ed. CRC Press, vol II pp 183.

Saccomani, G., Mukijdam, E. (1987) papain fragmentation of the gastric $(H^{++}K^+)$-ATPase Biochim. Biophys. Acta 912, 63-73.

Shull G.E., Lingrel J.B. (1986) Molecular cloning of the rat stomach (H^+,K^+)-ATPase. J. Biol. Chem. 261: 16788-16791.

Van Uem, T.J.F., Peters, W.H.M., DePont, J.J.H.H.M. (1990) A monoclonal antibody against gastric H^+/K^+-ATPase which binds the cytosolic $E1.K^+$ from. Biochim. Biophys. Acta, 1023, 56-62.

Van Uem, T.J.F., Swarts, H.G.P., DePont, J.J.H.H.M. (1991) Determination of the epitope for the inhibitory monoclonal antibody 5-B6 on the catalytic subunit of gastric Mg^{2+}-dependent H^+-transporting and K^+-stimulated ATPase Biochem. J. 280, 243-248.

New Approaches to Inhibition of Gastric Acid Secretion: Development and Mechanism of Omeprazole

Björn Wallmark
Preclinical R & D
Astra Hässle AB
431 83 Mölndal
Sweden

Introduction

With the exception of the secretory canaliculi of the parietal cell, the mammalian stomach is the most acidic compartment in the body, producing 140–160 mM hydrochloric acid. Fortunately, the oesophagus, the stomach and upper duodenum are well protected against acid attack and, in most healthy people, there is a balance between attacking and defending factors. Occasionally, this balance is disturbed, either by increased acid production, infection with *Helicobacter pylori,* or transient relaxation of the lower oesophageal sphincter. Whatever the precipitating factor, gastric acid is a key mediator of mucosal damage in diseases such as reflux oesophagitis and peptic ulcer. Furthermore, it is now well established that the most effective available methods of treating these diseases involve a reduction in gastric acid secretion.

Reduction of gastric acid secretion can be achieved surgically or pharmacologically. The former can involve removing part of the acid-secreting stomach or selective vagotomy of the nerves which stimulate acid secretion. Gastric/oesophageal surgery, however, is only performed in a small minority of patients, drug therapy being the more usual way to reduce acidity. Antacids, which neutralize acid that has already been secreted, provide only transient reduction of gastric acidity. Drugs that coat the mucosa in an attempt to protect it from exposure to acid and pepsin may also give some temporary relief of symptoms, but again they do not reduce acid secretion.

The secretion of gastric acid is driven by the enzyme, H^+,K^+-ATPase, the acid pump of the parietal cell, located in the oxyntic glands of the gastric mucosa. The parietal cell can be stimulated to secrete acid in a variety of ways – by the local release of histamine, by

NATO ASI Series, Vol. H 89
Molecular and Cellular Mechanisms
of H^+ Transport
Edited by Barry H. Hirst
© Springer-Verlag Berlin Heidelberg 1994

cholinergic nervous stimulation or by the antral release of gastrin. In addition, several other factors, such as epidermal growth factor, somatostatin and prostaglandins, may also have actions via endocrine or paracrine pathways (Berglindh and Sachs 1985). A number of drugs have been developed that act by inhibiting the secretion of acid from the parietal cell. The antisecretory agents available comprise anticholinergic agents, histamine H_2-receptor antagonists, antisecretory prostaglandins and the H^+,K^+-ATPase inhibitors.

The anticholinergic agents were developed to block the muscarinic effects of acetylcholine on the parietal cell (Feldman 1984). They were, however, found to be associated with side-effects, including blurred vision, dry mouth and more severe problems, such as urinary retention and the precipitation of glaucoma.

The discovery of a histamine H_2-receptor on the parietal cell led to the subsequent development of antagonists for this receptor (Black et al. 1972). The H_2-receptor antagonists proved to be more effective than other treatments available at the time and were associated with few side-effects. However, many patients with reflux oesophagitis, together with a smaller proportion of those with duodenal or gastric ulcer, remain symptomatic and unhealed, even after full courses of treatment with H_2-receptor antagonists.

Prostaglandins exert a weak inhibitory effect on gastric acid secretion (Robert 1984), probably by blocking the histamine-induced formation of cyclic AMP, but also possibly by an inhibitory action on gastrin secretion via the brain (Puurunen 1983). Stable prostaglandin analogues have been developed, and these are claimed to increase mucosal resistance by stimulating blood flow and the production of mucus and bicarbonate. These drugs have, however, proved only moderately effective in man.

The need for a more reliable treatment for acid-related diseases led Astra to embark on a drug project with the aim of finding new inhibitors of gastric acid secretion, and led to the discovery of the H^+,K^+-ATPase inhibitor, omeprazole.

Development of omeprazole

Research at Astra started in 1966 (for reviews on the development and mechanism of action of omeprazole see Lindberg et al. 1987, Lindberg et al. 1990), with an observation that some local anaesthetics reduced acid secretion when given orally to humans. The initial research led to a compound that was a very effective antisecretory compound in rats, but which had no effect in humans.

From 1972 onwards, testing was carried out in gastric fistula dogs. Pyridylthioacetamide (CMN131), originally investigated as an antiviral drug, was found to inhibit acid secretion, but was also found to be toxic to the liver. It was suspected that the functional group $CSNH_2$ was responsible for the liver toxicity of various compounds and was therefore replaced with other sulphur-containing groups. The development of this class of sulphur-containing compounds led to the synthesis of timoprazole, the parent compound of the pyridinylmethylsulphinyl benzimidazoles. Timoprazole proved to be effective in preventing acid secretion in dogs, but it also blocked the uptake of iodine into the thyroid gland, so was not taken forward for evaluation in humans. A number of analogues of timoprazole were synthesized, in order to try to separate the two effects. In 1977, one of the resultant compounds, picoprazole, was found not to influence iodine uptake by the thyroid. When tested in humans, it proved to be an effective antisecretory drug. Subsequently, the drug was found to act by inhibiting the newly-discovered enzyme H^+,K^+-ATPase (Fellinius et al. 1981, Wallmark et al. 1983, Ganser and Forte 1973).

Synthesis continued with the aim of finding an even more effective inhibitor of gastric acid secretion than picoprazole. Substituents with an electron-releasing effect, which elevated the pK_a value, were found to increase the effectiveness of the drug as an inhibitor of acid secretion. A methoxy group in the 4-position or a methyl group in the 3- or 5-positions strongly increases the basicity of a pyridine ring, so compounds including a 3,5-dimethyl-4-methoxy-substituted pyridine were synthesized. In 1979, omeprazole was finally selected as the optimal compound with respect to specific and highly effective inhibition of acid secretion.

Mechanism of action

Inhibition of H^+,K^+-ATPase

The site of action of omeprazole was confirmed by the intravenous administration of radio-labelled omeprazole into a mouse. The radiolabel was initially distributed to most organs and the blood. However, 16 hours after administration the drug was predominantly confined to the gastric mucosa. Light and electron microscopy showed the radiolabel to be present in the tubulovesicles and the secretory canalicular membranes of the parietal cell. These structures are also associated with the sites of acid secretion, and monoclonal antibodies against the gastric H^+,K^+-ATPase have been shown to bind exclusively to these membranes.

The proteins in a crude homogenate fraction of the radiolabelled gastric mucosa were separated by sodium dodecylsulphate polyacrylamide gel electrophoresis (SDS-PAGE). The radiolabel was confined to a protein of molecular weight of approximately 92,000 (Figure 1). As the gastric H^+,K^+-ATPase has been shown to contain a catalytically active peptide of 92kDa, a membrane fraction containing the H^+,K^+-ATPase was therefore purified from the crude homogenate. When this was subjected to SDS-PAGE, the 92 kDa peptide could now be clearly resolved from the rest of the proteins, and again the radiolabel was confined to this peptide region (Figure 1). These data clearly showed that omeprazole selectively binds to the H^+,K^+-ATPase in the oxyntic mucosa.

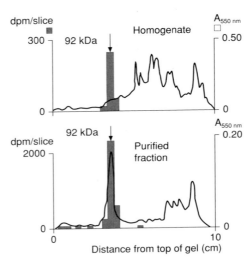

Figure 1. Binding of [3][H]omeprazole-derived radiolabel to gastric mucosal proteins. The proteins in the crude homogenate and in purified vesicles containing H^+,K^+-ATPase were separated by SDS-PAGE. After electrophoresis, the gel was stained and scanned for protein then sliced and the radiolabel quantified. dpm; disintegrations per minute. Reprinted from Biochem Pharmacol 37: Fryklund J, Gedda K, Wallmark B. Specific labelling of gastric H^+,K^+-ATPase by omeprazole; 2543–49 (©1988), with kind permission from Pergamon Press Ltd, Headington Hill Hall, Oxford OX3 OBW, UK.

A number of experimental findings show that inhibition of the enzyme H^+,K^+-ATPase by omeprazole parallels the inhibition of acid secretion. Omeprazole, given in submaximal doses to rats, inhibited both maximally-stimulated acid secretion and H^+,K^+-ATPase activity in a dose-dependent and parallel manner. The ED_{50} (the dose that gives 50% inhibition) was similar for inhibition of acid secretion and for H^+,K^+-ATPase activity. After blockade of acid secretion with omeprazole, both the H^+,K^+-ATPase activity and acid secretion returned to control levels at the same rate (Figure 2), suggesting that acid secretion is highly correlated to the activity of the gastric H^+,K^+-ATPase.

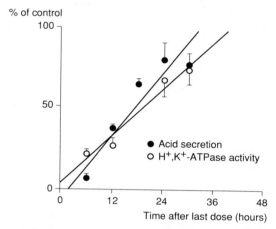

Figure 2. Recovery of acid secretion and H^+,K^+-ATPase activity following omeprazole administration. Omeprazole, 40μmol/Kg, was given once daily for three consecutive days. The effects on maximally stimulated acid secretion and K^+-stimulated ATPase activity were measured at the times indicated. The results are expressed as percentages of the secretory rates and K^+-stimulated ATPase activity obtained in control rats. Reproduced from Wallmark *et al.* (1985). J Biol Chem 260: 13681–4.

The gastric H^+,K^+-ATPase represents the final step in acid secretion, so inhibition of this enzyme is a highly effective and reliable way of controlling gastric acid secretion. By contrast, blockade of one receptor class involved in the stimulation of acid secretion can be bypassed by the action of agonists on other receptor classes. For example, food-stimulated acid secretion is mediated via cholinergic and gastrin pathways, which are unaffected or only partly affected by H_2-receptor antagonists.

Conversion of omeprazole to its active form

Omeprazole itself is not the active inhibitor of the H^+,K^+-ATPase, but is converted to an inhibitor within the acid compartments of the parietal cell, close to the enzyme. Omeprazole

(Figure 3, A) is a weak base and, at physiological pH, exists predominantly in its inactive base form, which can cross biological membranes. This form reaches the parietal cell from the blood and diffuses into the secretory canaliculus of the cell. The pH in the canaliculus is below 1, and omeprazole becomes protonated in this environment. In addition, the acid catalyses a molecular rearrangement of omeprazole, thereby transforming the drug from its inactive to its active form. The active form of the drug is a sulphenamide (Figure 3, B) which, being positively charged, does not pass readily through membranes and therefore remains within the acid compartment (secretory canaliculi) of the parietal cell.

In vitro, it has been shown that omeprazole is devoid of inhibitory activity and that transformation of omeprazole to the sulphenamide in acid is required in order to inhibit the H^+,K^+-ATPase. The sulphenamide reacts rapidly with mercaptans, such as beta-mercaptoethanol, to form a disulphide (Figure 3, C), which can be used as a model compound of the enzyme-inhibitor complex (Figure 3, D). During the inhibition of the H^+,K^+-ATPase, sulphenamide molecules bind to two cysteine-SH residues on the luminal face of the alpha subunit.

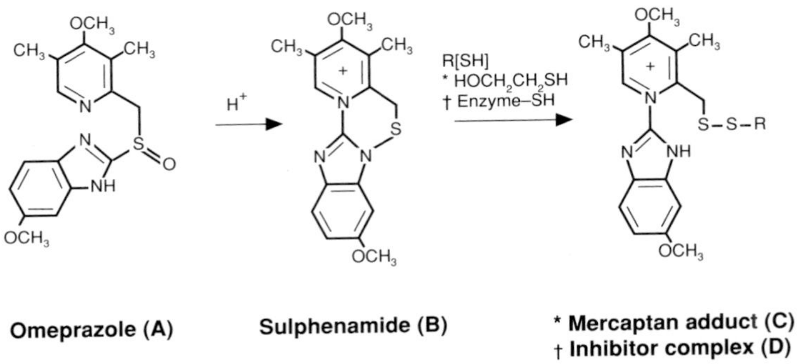

Omeprazole (A)   **Sulphenamide (B)**   *** Mercaptan adduct (C)**
                                                 † Inhibitor complex (D)

Figure 3. Chemical reactions of omeprazole leading to inhibition of H^+,K^+-ATPase. Reproduced from Lindberg et al. (1987). Trends Pharmacol Sci 8: 399–402.

In the isolated gastric mucosa or in gastric gland preparations, neutralization of the acid secretory canaliculi, by buffers that can permeate the membrane, blocks the acid-catalysed transformation of omeprazole to the sulphenamide and thus prevents the inhibition of acid secretion. Also, in vivo studies show omeprazole to be a more effective inhibitor of gastric acid secretion when the drug is administered during stimulation of acid secretion. Its efficacy is gradually decreased when the parietal cells are either in a basal state or are inhibited by either cimetidine or somatostatin at the time of omeprazole administration, demonstrating that acid-induced conversion of omeprazole is required for inhibition.

Advantages of the omeprazole mechanism

● Omeprazole, in its acid-activated form, binds specifically to the target enzyme, the gastric H^+,K^+-ATPase. This is the final step in acid secretion, and therefore inhibition of this stage provides reliable control of acid secretion, irrespective of the type of stimulus.

● The acid-activated form of omeprazole has a long duration of action because it forms a stable enzyme-inhibitor complex.

● Omeprazole has a highly specific mode of action. It is essentially stable and inactive at neutral pH, but is a weak base and therefore concentrates in acid regions. The H^+,K^+-ATPase is found only in the tubulovesicular and canalicular structures of the parietal cell, which are the only compartments of the body with an extremely low pH. Thus, omeprazole is concentrated in the acid canaliculus of the parietal cell, where it is rapidly transformed in the presence of acid into an active inhibitor of the enzyme. This active inhibitor is a cation and cannot easily penetrate the secretory membrane. It reacts very rapidly with mercapto groups (e.g. cysteines) on the luminal side of the target enzyme to form a covalently bound enzyme-inhibitor complex.

Clinical experience

Highly effective control of gastric acid leads to rapid symptom resolution and reliable healing in both peptic ulcer disease and reflux oesophagitis.

A meta-analysis of results from comparative studies (Blum 1990) has shown that omeprazole, 20 mg once daily, gives a healing rate for duodenal ulcer of 69% after 2 weeks and of 93% after 4 weeks. In gastric ulcer, the healing rates are 73% after 4 weeks and 91% after 8 weeks. Not only are these healing rates higher than those obtained with the H_2-receptor antagonists, but symptom resolution is also more rapid with omeprazole. Similarly, in patients with reflux oesophagitis, symptom resolution and complete healing are achieved in the majority of patients (Maton 1991). In addition, omeprazole is effective in healing and symptom resolution, even in patients who have remained unhealed after at least 3 months of treatment with H_2-receptor antagonists (Walan 1991). The massive hypersecretion of gastric acid in the few patients with Zollinger–Ellison syndrome has been difficult to treat in the past. Omeprazole has proved highly effective in this syndrome, controlling the hypersecretion in the majority of those treated (Lloyd-Davies *et al* 1986).

Omeprazole is the culmination of the search for improved control of gastric acidity. By specifically inhibiting the final step in gastric acid secretion, the enzyme H^+,K^+-ATPase in the gastric mucosa, omeprazole provides the physician with an effective treatment for acid-related diseases, with significant therapeutic gains over the H_2-receptor antagonists.

References

Berglindh T, Sachs G (1985) Emerging strategies in ulcer therapy: Pumps and receptors. Scand J Gastroenterol 20(suppl 108): 7–14

Black JW, Duncan WAM, Durant CJ, Ganellin CR, Parsons EM (1972) Definition and antagonism of histamine H_2-receptors. Nature 236: 385–90

Blum AL (1990) Treatment of acid-related disorders with gastric acid inhibitors: the state of the art. Digestion 47(Suppl 1): 3–10

Ganser AL, Forte JG (1973) K^+-stimulated ATPase in purified microsomes of bullfrog oxyntic cells. Biochim Biophys Acta 307: 169–80

Feldman M (1984) Inhibition of gastric acid secretion by selective and non-selective anticholinergics. Gastroenterology 86: 361–6

Fellenius E, Berglindh T, Sachs G *et al.* (1981) Substituted benzimidazoles inhibit gastric acid secretion by blocking H^+,K^+-ATPase. Nature 290:159–161

Fryklund J, Gedda K, Wallmark B (1988) Specific labelling of gastric H^+,K^+-ATPase by omeprazole. Biochem Pharmacol 37: 2543–49

Lindberg P, Brändström A, Wallmark B (1987) Structure-activity relationships of omeprazole analogues and their mechanism of action. Trends Pharmacol Sci 8: 399–402

Lindberg P, Brändstrom A, Wallmark B, Mattson H, Rikner L, Hoffman K-J (1990) Omeprazole: The first proton pump inhibitor. Med Res Rev 10: 1–54

Lloyd-Davies KA, Rutgersson K, Sölvell L (1986) Omeprazole in Zollinger-Ellison syndrome: four year international study. Gastroenterology 90: 1523 (Abstract)

Maton PN (1991) Drug therapy: Omeprazole. N Engl J Med 324: 965–75

Puurunen J (1983) Central inhibitory action of prostaglandin E_2 on gastric secretion in the rat. Eur J Pharmacol 91: 245–9

Robert A (1984) Prostaglandins: effects on the intestinal tract. Clin Physiol Biochem 2: 61–9

Walan A (1991) Efficacy and safety of omeprazole in the long-term treatment of patients with peptic ulcer or reflux oesophagitis resistant to H_2-receptor antagonists - Interim report. Hässle Clinical Report I-548/I-614. Mölndal, Sweden

Wallmark B, Larsson H, Humble L (1985) The relationship between gastric acid secretion and gastric H^+,K^+-ATPase activity. J Biol Chem 260: 13681–4

Wallmark B, Jareston B-M, Larsson H *et al.* (1983) Differentiation among the inhibitory actions of omeprazole, cimetidine and SCN on gastric acid secretion. Am J Physiol 245: G64–71

K$^+$-COMPETITIVE INHIBITORS OF THE GASTRIC (H$^+$+K$^+$)-ATPase.

David J. Keeling
Department of Cell Biology
Astra Hässle AB
431 83 Mölndal
Sweden.

Gastric acid secretion plays an important role in several diseases of the gastro-intestinal tract. The identification of the gastric (H$^+$+K$^+$)-ATPase as the final step in acid secretion (Ganser and Forte, 1973; Sachs *et al.*, 1976) has therefore led to great interest in the pharmacology of this acid pump. The (H$^+$+K$^+$)-ATPase is located in the acid-secreting parietal cells of the gastric mucosa, where it cycles between intracellular vesicles (in the resting cell) and the apical membrane (in stimulated cells). In the active state, the acid pump uses the free energy released from the hydrolysis of ATP to catalyze the electroneutral exchange of hydrogen ions for potassium ions across the cell membrane.

The first acid pump inhibitor to be approved for clinical use was omeprazole, which belongs to the class of compounds known as the pyridinmethylsulphinyl benzimidazoles. By virtue of a covalent interaction with cysteine residues on the extracellular face of the (H$^+$+K$^+$)-ATPase, omeprazole produces effective and long-acting inhibition of acid secretion, resulting in highly effective therapy in acid-related diseases of the gastrointestinal tract (Lindberg *et al.*, 1990). The intensive search for inhibitors of gastric acid secretion has identified additional classes of structures that are able to inhibit the gastric (H$^+$+K$^+$)-ATPase in a non-covalent manner (Figure 1). These include imidazopyridines (Kaminski *et al.*, 1985), guanidinothiazoles (LaMattina *et al.*, 1990), analogues of 4-(arylamino)-quinolines (Munson and Reevis 1982; Brown *et al.*, 1990) and a number of other organic amines including nolinium bromide (Nandi *et al.*, 1983; Im *et al.*, 1984; Nandi *et al.*, 1990). Many of these structures share the common property that the inhibition of (H$^+$+K$^+$)-ATPase activity is competitive with respect to the activating cation K$^+$ (Beil *et al.*, 1986, LaMattina *et al.*, 1990, Brown *et al.*, 1990, Nandi *et al.*, 1983). A potassium ion is transported into the parietal cell in exchange for the export of each H$^+$. Thus a high affinity potassium site exists on the extracellular face of the ion pump. In vesicular preparations of the (H$^+$+K$^+$)-

NATO ASI Series, Vol. H 89
Molecular and Cellular Mechanisms
of H$^+$ Transport
Edited by Barry H. Hirst
© Springer-Verlag Berlin Heidelberg 1994

ATPase (which are oriented inside-out such that the intravesicular space corresponds to the extracellular space of the parietal cell), K+ is able to stimulate ATPase activity by acting at this intravesicular (extracellular) site. The K+-competitive nature of the inhibition of ATPase activity has led to speculation that these inhibitors interact at the extracellular K+-site on the acid pump. The properties of these inhibitors will be discussed in relation to this proposition.

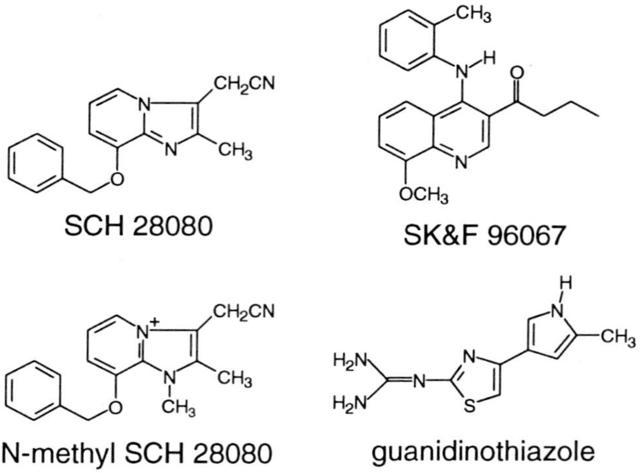

SCH 28080

SK&F 96067

N-methyl SCH 28080

guanidinothiazole

Figure 1. Structures of K+-competitive (H++K+)-ATPase inhibitors.

One of the most extensively studied K+-competitive inhibitors of the (H++K+)-ATPase is SCH 28080, an imidazopyridine identified originally by Schering Plough. This compound entered drug development and was shown to inhibit acid secretion in human volunteers (Ene *et al.*, 1982) before development was discontinued as a result of adverse effects in the liver (Kaminski *et al.*, 1987). SCH 28080 competitively inhibits K+-stimulated ATPase activity with a Ki value at neutral pH in the range 20 nM to 50 nM (Beil *et al.*, 1986; Keeling *et al.*, 1988). However, kinetic competition does not constitute proof of a common binding site. SCH 28080 is also a competitive inhibitor of the K+-stimulated paranitrophenylphosphatase (pNPPase) activity of the (H++K+)-ATPase. In this reaction, the activation by K+ occurs at the cytosolic face of the pump. Additional evidence is needed to identify on which side of the membrane the inhibitor binding site is situated. The N-methylated analogue of SCH 28080

(Figure 1) has proved a useful probe in studying inhibitor mechanisms. N-Methylation of the protonatable nitrogen, locks the inhibitor in a positively charged form, greatly reducing its membrane permeability. Effective inhibition of pNPPase activity by N-methyl SCH 28080 only occurs when the vesicle membrane barrier has been weakened by freeze-drying, providing evidence that the inhibitor binding site is on the extracellular side of the pump (Keeling *et al.*, 1988).

N-Methyl SCH 28080 has been further modified into a covalent probe of the (H^++K^+)-ATPase by the addition of an photoactivatable azido-group (Munson *et al.*, 1988). This compound specifically labels a region of the α-subunit of the (H^++K^+)-ATPase that is predicted, by hydrophobicity calculations, to form an extracellular domain between the first two membrane-spanning sections (M1/M2) of the primary sequence (Munson and Sachs, 1991). It remains to be seen if other extracellular domains of the protein also contribute to the binding site. It is interesting to note that the M1/M2 region of the closely-related (Na^++K^+)-ATPase is important in determining the affinity of inhibition by ouabain. However the binding sites are sufficiently different that there is virtually complete selectivity of these inhibitors for their respective ion pumps.

The binding of ^{14}C-SCH 28080 to the (H^++K^+)-ATPase is affected by the conformational state of the protein (Keeling *et al.*, 1989). The (H^++K^+)-ATPase alternates between at least two conformational states in its catalytic cycle, referred to as E1 and E2. In the E1 form of the protein, the ion binding site faces the cytosol. Hydrolysis of ATP occurs via the formation of a phosphoenzyme intermediate after which the ion binding site becomes accessible from the extracellular medium (E2 form). SCH 28080 binds with greatest affinity to the E2, phosphorylated form of the pump. The affinity of binding is the same as that for the inhibition of ATPase activity. However, in the absence of ATP, SCH 28080 can also bind to the dephosphorylated E2 form of the pump, albeit with approximately 10-fold lower potency. This corresponds to the affinity for inhibition of pNPPase activity by the pump, a reaction that does not require phosphorylation of the protein. SCH 28080 does not appear to bind to the E1 forms of the pump. Thus, conformational information from cytosolic, phosphorylation domains of the pump is transferred to the extracellular inhibitor binding site.

The detailed kinetics of SCH 28080 binding indicate a high (40 kJ/mol) activation energy for association and an equilibrium constant that is determined principally by entropy differences between the bound and unbound states (Keeling *et al.*, 1989). This suggests that hydrophobic interactions at the inhibitor binding site play an important role.

SCH 28080 is a protonatable weak base (pKa= 5.6) and exists either as a neutral or positively charged (protonated) form. By following affinity as a function of pH it can be shown that it is the positively charged, protonated form of the inhibitor that binds preferentially to the enzyme (Keeling *et al.*, 1988). This is consistent with the high affinity of the permanently charged N-methyl SCH 28080 and also with the role of the pump as a cation transporter. The preferential binding of the protonated form of SCH 28080 also helps to explain its effectiveness as an inhibitor of acid secretion in the intact tissue. In the active parietal cell, which secretes 160 mM HCl, the acidic pH of the medium bathing the extracellular face of the pump favours the accumulation of SCH 28080 by trapping it as the membrane-impermeable protonated form. Thus the active form of the inhibitor is accumulated at its site of action.

A second class of K^+-competitive (H^++K^+)-ATPase inhibitor is that of the 4-(arylamino)quinolines. This group of compounds was first described as inhibitors of acid secretion by Munson and Reevis (1982). Further medicinal chemistry in this area (Brown *et al.*, 1990; Leach *et al.*, 1992) resulted in the identification of SK&F 96067 (Figure 1) as a compound suitable to take forward into clinical trials (Ife *et al.*, 1992).

Many properties of the 4-(arylamino)quinolines are similar to those already described for the imidazopyridines. The low potency of membrane-impermeable analogues of 4-(arylamino)quinolines in intact gastric vesicles has indicated an extracellular binding site for this class of inhibitor (Brown *et al.*, 1990). At 7.0 pH, the Ki value for the K^+-competitive inhibition of ATPase activity by SK&F 96067 is 390 nM (Keeling *et al.*, 1991). This protonatable weak-base is also preferentially active in its protonated form, which can be accumulated at the site of action. It seems reasonable therefore to speculate that the two classes of compound bind to the same region of the protein. Consistent with this, competition studies between SK&F 96067 and SCH 28080 indicate that the two inhibitors may not bind to the enzyme at the same time (Keeling *et al.*, 1991).

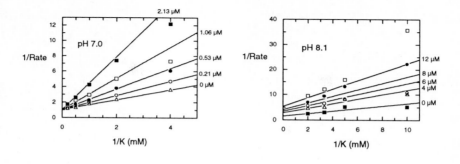

Figure 2. Inhibition pattern of SK&F 96067 as a function of pH.
Lineweaver Burk plots showing the change from competitive inhibition at pH 7.0
to non-competitive (mixed) at pH 8.1. Experiments were performed using
lyophilized gastric vesicles containing (H^++K^+)-ATPase prepared from pig
gastric mucosa. The rate of K^+-stimulated ATPase activity (IU/mg protein) is
shown as a function of K^+ concentration and defined concentrations of
SK&F 96067 as indicated. Data analysis yielded the inhibitory constants defining
the effects on the slope (Kis) and intercept (Kii) of the Lineweaver Burk plot. At
pH 7.0 Kis= 0.49 μM. At pH 8.01 Kis= 4.9 μM, Kii= 4.8 μM. Adapted from
Keeling *et al.* (1991) Biochem. Pharmacol. 42:123-130 with kind permission from
Pergamon Press Ltd, Headington Hill Hall, Oxford OX3 0BW, UK.

However at pH values above 7.0, the inhibition pattern of SK&F 96067 changes
from purely competitive to non-competitive (Figure 2). This indicates that the
binding of inhibitor and K^+ to the (H^++K^+)-ATPase are no longer mutually
exclusive and that a ternary complex is possible between pump, K^+ and inhibitor.
Thus under these conditions SK&F 96067 cannot be considered to occupy the
entire K^+-site. At these higher pH values, SK&F 96067 (pKa= 6.5) is present
predominantly in the non-protonated, neutral form. It is perhaps this form that is
able to bind simultaneously with K^+ as suggested Figure 3.

A number of analogues of SK&F 96067 have been found to be fluorescent
(Brown *et al.*, 1990; Leach *et al.*, 1992). Using one such analogue, it has been
shown that phosphorylation of the cytosolic domains of the (H^++K^+)-ATPase,
results in increased hydrophobicity at the luminal inhibitor binding site (Rabon *et
al.*, 1991). This is further evidence for concerted conformational changes on both
sides of the membrane as the pump goes through the ion-translocation cycle.

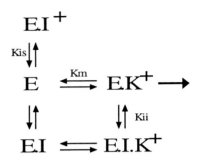

Figure 3. A kinetic scheme for the inhibition of ATPase activity by SK&F 96067. At pH 7.0 or lower, the only detectable interaction is between the protonated inhibitory (I^+) and the (H^++K^+)-ATPase (E) as defined by the competitive inhibitory constant Kis. Above pH 7.0 both inhibitor and K^+ can bind to the enzyme simultaneously as defined by the additional inhibitory constant Kii. It has been assumed that it is the non-protonated form of the inhibitor (I) that binds in this case, such that the net charge bound to the site does not exceed one unit. The binding of I alone to the (H^++K^+)-ATPase is also shown. This form could contribute to the experimentally determined Kis particularly at higher pH. Reprinted from Keeling *et al.* (1991) Biochem. Pharmacol. 42:123-130 with kind permission from Pergamon Press Ltd, Headington Hill Hall, Oxford OX3 0BW, UK.

In summary SCH 28080 binds in its cationic form to the extracellular face of the (H^++K^+)-ATPase. Binding is optimal when the enzyme adopts the phosphor-ylated E2 conformation which binds K^+ in the ion translocation cycle. This is consistent with (but not proof that) it binds to the K^+-transporting site of the (H^++K^+)-ATPase. However, the cation binding site of the (H^++K^+)-ATPase, or indeed that of the other closely related P-type ion pumps, has yet to be defined in detail. Rubidium occlusion studies on the (Na^++K^+)-ATPase suggest that the membrane spanning sections at the C-terminus of the α-subunit form an important part of the cation binding site (Karlish *et al.*, 1990). It is not yet known whether the M1/M2 region also contributes to the cation binding site. SK&F 96067 shares many of the properties of interaction with SCH 28080. However at high pH a second mode of binding is possible that is not competitive with K^+ indicating that the "K^+-competitive" inhibitors do not always occupy the entire cation binding site of the (H^++K^+)-ATPase.

Despite considerable effort in the field of non-covalent (H^++K^+)-ATPase inhibitors, no compound has yet been developed successfully for clinical use. It remains to be seen whether non-covalent (H^++K^+)-ATPase have a role to play in the treatment of acid-related diseases' of the gastrointestinal tract.

References.

Beil W, Hackbarth I, Sewing K-Fr (1986) Mechanism of gastric antisecretory effect of SCH 28080. Br J Pharmacol 88:19-23.

Brown TH, Ife RJ, Keeling DJ, Laing SM, Leach CA, Parsons ME, Price CA, Reaville DR, Wiggall KJ (1990) Reversible inhibitors of the gastric (H^+/K^+)-ATPase. 1. 1-aryl-4-methylpyrrolo[3,2-c]quinolines as conformationally restrained analogues of 4-(arylamino)quinolines. J Med Chem 33:527-533.

Ene MD, Khan-Daneshmend T, Roberts CJC (1982) A study of the inhibitory effects of SCH 28080 on gastric secretion in man. Br J Pharmac 76:389-391

Ganser AL, Forte JG (1973) Ionophoretic stimulation of K^+-ATPase of oxyntic cell microsomes. Biochem Biophys Res Commun 54:690-696

Ife RJ, Brown TH, Keeling DJ, Leach CA, Meeson ML Parsons ME, Reaville DR, Theobald CJ, Wiggall KJ (1992) Reversible inhibitors of the gastric (H^+/K^+)-ATPase. 3. 3-Substituted-4-(phenylamino)quinolines. J Med Chem 35:1413-3422

Kaminski JJ, Bristol JA, Puchalski C, Lovey RG, Elliott AJ, Guzik H, Solomon DM, Conn DJ, Domalski MS, Wong S-C, Gold EH, Long JF, Chiu PJS, Steinberg M, McPhail AT (1985) Antiulcer agents. 1. Gastric antisecretory and cytoprotective properties of substituted imidazo[1,2-a]pyridines. J Med Chem 28:876-892

Kaminski JJ, Perkins DG, Frantz JD, Solomon DM, Elliott AJ, Chiu PJS, Long JF (1987) Antiulcer agents. 3. Structure-activity-toxicity relationships of substituted imidazo[1,2-a]pyridines and a related imidazo[1,2-a]pyrazine. J Med Chem 30:2047-2051

Karlish SJD, Goldshleger R and Stein WD (1990) A 19-kDa C-terminal tryptic fragment of the α-chain of Na/K-ATPase is essential for occlusion and transport of cations. Proc Natl Acad Sci 87:4566-4570

Keeling DJ, Laing SM, and Senn-Bilfinger J (1988) SCH 28080 is a lumenally acting, K^+-site inhibitor of the gastric $(H^+ + K^+)$-ATPase. Biochem Pharmacol 37:2231-2236

Keeling DJ, Taylor AG, and Schudt C (1989) The binding of a K^+ competitive ligand, 2-methyl,8-(phenylmethoxy)imidazol(1,2-a)pyridine 3-acetonitrile, to the gastric (H^+,K^+)-ATPase. J Biol Chem 264:5545-551

Keeling DJ, Malcolm RC, Laing SM, Ife RJ and Leach CA (1991) SK&F 96067 is a reversible, lumenally acting inhibitor of the gastric (H^++K^+)-ATPase. Biochem Pharmacol 42:123-130

LaMattina JL, McCarthy PA, Reiter LA, Holt WF, Yeh L-A (1990) Antiulcer agents. 4-substituted 2-guanidinothiazoles: reversible, competitive, and selective inhibitors of gastric H^+,K^+-ATPase. J Med Chem 33:543-552

Leach CA, Brown TH, Ife RJ, Keeling DJ, Laing SM, Parsons ME, Price CA, Reaville DR, Wiggall KJ (1992) Reversible inhibitors of the gastric (H^+/K^+)-ATPase. 2. 1-Arylpyrrolo[3,2-c]quinolines: Effect of the 4-substituent. J Med Chem 35:1845-1852

Lindberg P, Brändström A, Wallmark B, Mattsson H, Rikner L, Hoffman K-J (1990) Omeprazole: The First Proton Pump Inhibitor. Med Res Reviews 10:1-54

Munson HR Jr, Reevis SA, (1982) US Patent 4,343,804

Munson KB and Sachs G (1988) Inactivation of H^+,K^+-ATPase by a K^+-competitive photoaffinity inhibitor. Biochem 27:3932-3938

Munson KB, Gutierrez C, Balaji VN Ramnarayan K, Sachs G (1991) Identification of an extracellular region of H^+,K^+-ATPase labeled by the K^+-competitive photoaffinity label. J Biol Chem 266:18976-88

Nandi J, Wright MV, Ray TK (1983) Mechanism of gastric antisecretory effects of nolinium bromide. Gastroenterol 85:938-945

Rabon E, Sachs G, Bassilian S, Leach C, Keeling D (1991) A K^+-competitive fluorescent inhibitor of the H^+,K^+-ATPase. J Biol Chem 266:12395-12401

Sachs G, Chang HH, Rabon E, Schackman R, Lewin M Saccomani G (1976) A non-electrogenic H^+ pump in plasma membranes of hog stomach. J Biol Chem 251:7690-7698

Bioactive Peptide Inhibitors and An Endogenous Parietal Cell Protein: Interactions With the Gastric H^+/K^+ ATPase

John Cuppoletti, Pingbo Huang and Danuta H. Malinowska
Department of Physiology and Biophysics
University of Cincinnati College of Medicine
231 Bethesda Avenue
Cincinnati, Ohio 45267-0576

The gastric H^+/K^+ ATPase is responsible for HCl secretion by the parietal cell. A variety of amphipathic polypeptide inhibitors of the gastric H^+/K^+ ATPase, including melittin from *Apis mellifera*, mastoparan from *Vespula lewisii*, and synthetic model polypeptides have been identified (Cuppoletti *et al*, 1989; Cuppoletti, 1990). In functional studies, the effects of melittin and mastoparan on ATP hydrolysis, para-nitrophenylphosphatase (pNPPase) activity, steady-state levels of phosphoenzyme, and enzyme conformation measured with fluorescein isothiocyanate (FITC)-labelled enzyme have been compared and contrasted (Cuppoletti *et al*, 1992). In structural studies, photoaffinity analogs of melittin and mastoparan were employed to demonstrate direct interactions between melittin and mastoparan, and the alpha subunit of the H^+/K^+ ATPase (Cuppoletti *et al*, 1989; Cuppoletti *et al*, 1992). In the search for intracellular proteins which might bind to the H^+/K^+ ATPase by virtue of having epitopes in common with these insectotoxins, antibodies raised against melittin were reacted with parietal cell proteins (Cuppoletti *et al*, 1993). These studies resulted in the identification of a parietal cell "melittin-like" protein which reversibly associated with gastric H^+/K^+ ATPase-containing (apical) membranes when parietal cells were in the stimulated state of HCl secretion, but it was in the cytosol when the cells were in the resting state. The question addressed in the present manuscript is whether a definable binding site for these bioactive polypeptides exists on the gastric H^+/K^+ ATPase, and whether the binding site (or sites) exhibits overlap for the various polypeptides which interact with it. Evidence is presented for overlapping binding sites for various amphipathic helical polypeptides and the "melittin-like" protein on the H^+/K^+ ATPase.

Materials and Methods

Membrane vesicles enriched in the gastric H^+/K^+ ATPase from secreting (histamine plus diphenhydramine-injected) or non-secreting (cimetidine-injected) rabbit gastric mucosa were prepared as previously described (Cuppoletti & Sachs,

NATO ASI Series, Vol. H 89
Molecular and Cellular Mechanisms
of H⁺ Transport
Edited by Barry H. Hirst
© Springer-Verlag Berlin Heidelberg 1994

1984). Preparation of $[^{125}I]$azidosalicylyl melittin ($[^{125}I]$Mel) and photoaffinity labelling studies were carried out as previously described (Cuppoletti et al, 1989). $[^{125}I]$azidosalicylyl mastoparan ($[^{125}I]$Mast) was prepared as described for $[^{125}I]$Mel. Preparation of purified parietal cells and immunoanalysis were carried out as previously described (Cuppoletti et al, 1993). Measurement of ATP hydrolysis, pNPPase activities, and determination of protein concentration were carried out as described in detail elsewhere (Cuppoletti and Sachs, 1984). Labelling of H^+/K^+ ATPase with FITC and measurement of FITC fluorescence changes were carried out as described (Cuppoletti & Abbott, 1990). Antimelittin antibody was the kind gift of Dr. John Dedman of the Department of Physiology and Biophysics, University of Cincinnati College of Medicine.

Results

Effects of amphipathic helical polypeptides on the gastric H+/K+ ATPase. Melittin inhibited ATP hydrolysis non-hyperbolically, with a $K_{0.5}$ of less than 0.5 μM, and a Hill coefficient of approximately 2. Melittin also inhibited pNPPase activity, reversed K^+-dependent changes in fluorescence of FITC-labelled H^+/K^+ ATPase and reduced steady state levels of phosphorylation of the H^+/K^+ ATPase (Huang et al, 1993b). Mastoparan was inhibitory to K^+-stimulated ATPase activity with a $K_{0.5}$ of 6 μM with a Hill coefficient of approximately 1. Mastoparan also reversed K^+ dependent changes in FITC fluorescence, and reduced steady state levels of phosphorylation of the H^+/K^+ ATPase, but did not affect pNPPase activities (Huang et al, 1993b).

Evidence for direct interaction of melittin with the gastric H^+/K^+ ATPase. In order to demonstrate direct interactions of melittin with the H^+/K^+ ATPase, a radioactive azidosalicylyl melittin $[^{125}I]$ASMel was prepared and used to label the proteins in gastric vesicle preparation which contain the H^+/K^+ ATPase. As shown in Figure 1A, the alpha subunit was shown to be the receptor for $[^{125}I]$ASMel in these studies. Similar labelling of the alpha subunit of the Na^+/K^+ ATPase was also observed, and this labelling was prevented by K^+ (Cuppoletti and Abbott, 1990). The beta subunits of these pumps were not labelled by $[^{125}I]$ASMel.

Site of interaction of melittin within the primary structure of the H^+/K^+ ATPase. $[^{125}I]$ASMel-labelled H^+/K^+ ATPase, when proteolysed by trypsin in the presence of K^+, yielded labelled fragments with mobilities on sodium dodecyl sulfate gel polyacrylamide gel electrophoresis (SDS-PAGE), similar to

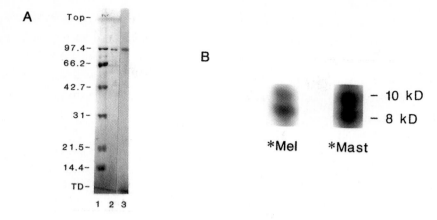

Figure 1. (A) Interaction of [^{125}I]Azidosalicylyl Melittin with the Alpha Subunit of the H$^+$/K$^+$ ATPase. Coomassie blue stained protein pattern (lane 2) shows the H$^+$/K$^+$ ATPase alpha subunit from hog and the corresponding autoradiogram (lane 3) shows that the alpha subunit is photoaffinity labelled with [^{125}I]ASMelittin. Molecular weight markers are shown in lane 1. (From Cuppoletti *et al*, 1989). **(B) Melittin and Mastoparan Binding Sites on the Alpha Subunit of the H$^+$/K$^+$ ATPase: V$_8$ Protease Fragmentation Study.** Limiting V$_8$ protease fragments of purified hog gastric H$^+$/K$^+$ ATPase alpha subunit labelled with [^{125}I]ASMelittin and [^{125}I]ASMastoparan, respectively. Both melittin and mastoparan peptides react with fragments of 8 and 10 kDa.

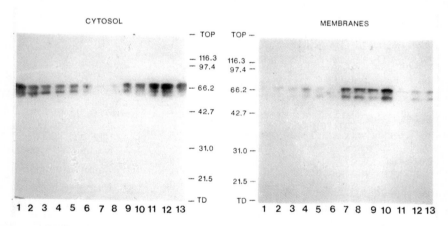

Figure 2. Identification of a Stimulus-Associated "Melittin-Like" Protein in the Gastric Parietal Cell. Immunoblots using antimelittin antibody of cytosol and membrane fractions from purified rabbit parietal cells are shown. Lane 1, basal state; lanes 2-6, 5, 15, 30, 45 & 60 min with cimetidine; lanes 7-11, 5, 15, 30, 45 & 60 min with histamine; lanes 12 and 13 are 45 & 60 min after addition of a 10-fold molar excess of cimetidine to cells previously treated with histamine for 30 min. The immunoreactive 67 kDa protein reversibly redistributes from cytosol in non-secreting cells to membranes in secreting cells. (From Cuppoletti *et al*, 1993).

fragments previously identified by others as deriving from the cytosolic domain (Helmich-deJong *et al*, 1987). These studies demonstrated that melittin did not bind non-specifically to the transmembrane domains of the H^+/K^+ ATPase, but rather to a definable region. In addition, these studies suggested that melittin binding to the H^+/K^+ ATPase was limited to a small region (or regions) of the H^+/K^+ ATPase.

It was therefore of interest to determine whether mastoparan, with a different primary sequence but similar structure, bound to overlapping sites on the H^+/K^+ ATPase. V_8 protease treatment of the labelled H^+/K^+ ATPase was carried out. The results (**Figure 1B**) showed that labelling was restricted to two fragments of the H^+/K^+ ATPase, whether the protein was labelled with the photoaffinity analog of melittin or mastoparan (Huang et al 1993b). In more recent studies (manuscript in preparation), we have demonstrated that the mastoparan and melittin binding sites overlap, and the mastoparan binding sites have been sequenced and localized to a region smaller than the peptide probes (approximately 7 amino acids) within the cytosolic domain (Huang et al 1993a, 1993b). These studies have established for the first time that there is a definable polypeptide binding region on the H^+/K^+ ATPase. Since this binding region is in the cytoplasmic domain of the H^+/K^+ ATPase, the question is raised whether this site has physiological significance. Perhaps it represents an intra/inter-molecular binding site within the gastric H^+/K^+ ATPase, as occurs with calmodulin regulation of the sarcoplasmic reticulum Ca^{2+} ATPase (Enyedi *et al*, 1989). In the search for endogenous proteins of the gastric parietal cell which may bind to the H^+/K^+ ATPase, by virtue of having determinants in common with melittin, a polyclonal antimelittin antibody which was reported to react with an amphipathic helical region of myosin light chain kinase was employed (Cuppoletti *et al*, 1993). This antibody was of interest since both melittin and myosin light chain kinase bind to calmodulin at overlapping sites (Kaetzel & Dedman, 1988).

Endogenous "melittin-like" protein of the gastric parietal cell. Purified parietal cells in the basal state, treated with cimetidine (non-secreting cells), histamine (secreting cells), or histamine followed by an excess of cimetidine were homogenized, and cytosol and membrane fractions of the cells were prepared. The proteins of the fractions were separated by SDS-PAGE, transblotted, and subjected to immunoanalysis using antimelittin antibody. The results are shown in **Figure 2**. Of all the proteins which were present, only a 67 kDa protein (doublet) exhibited immunoreactivity. In the basal state and with cimetidine treatment, this protein was present in the cytosol, but absent in the membrane fraction. However, upon treatment with histamine, the protein disappeared from the cytosol and appeared in the membranes. Subsequent treatment of histamine-treated cells with excess cimetidine resulted in loss of the protein from the membranes and re-appearance of the protein in the cytosol. Thus, the parietal cell contains a 67 kDa protein which reacts with antimelittin antibody, and exhibits stimulus-dependent association with components the membranes. As shown in **Figure 3**, the

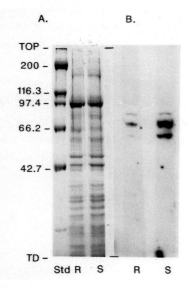

Figure 3. Presence of the "Melittin-Like" Protein in Ficoll-Purified, H^+/K^+ ATPase-Containing Membrane Vesicles from Stimulated Rabbit Gastric Mucosae. Immunoblots using antimelilittin antibody of membranes from non-secreting or resting (R) and secreting (S) rabbit gastric mucosae. A is protein pattern and B is corresponding autoradiogram showing a differential enrichment of the 67 kDa immunoreactive protein in H^+/K^+ ATPase-containing membranes from secreting gastric mucosae. (From Cuppoletti *et al*, 1993)

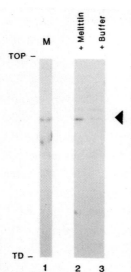

Figure 4. Displacement by Melittin of the "Melittin-Like Protein" from Stimulated Membranes: Evidence for an Overlapping Binding Site. Immunoblots using antimelittin antibody of stimulated rabbit gastric membranes (lane 1) were treated with 100 µg/ml melittin (lane 2), or buffer alone (lane 3). Resultant supernatants obtained by centrifugation show enrichment of the 67 kDa "melittin-like" protein (arrowhead) when the membranes were treated with melittin.

immunoreactive 67 kDa protein was present in the most highly purified preparations of H^+/K^+ ATPase-containing membrane vesicles from secreting rabbit gastric mucosae which are available, and was virtually absent from similar preparations from non-secreting rabbit gastric mucosae. These results suggest that the 67 kDa "melittin-like" protein binds to a component of the H^+/K^+ ATPase-containing membrane, which we speculate, from our studies with melittin, to be the gastric H^+/K^+ ATPase alpha subunit (Cuppoletti, et al, 1993). To determine whether there is overlap of the binding site of this protein with the binding site for melittin, we attempted to displace the "melittin-like" protein from H^+/K^+ ATPase-containing membranes using melittin.

Displacement of the "melittin-like" protein from rabbit gastric parietal cell membranes. As shown in **Figure 4**, treatment of H^+/K^+ ATPase-containing membranes with buffer which contained inhibitory concentrations of melittin, but not buffer alone, resulted in displacement of the 67 kDa "melittin-like" protein into a soluble fraction. These results suggested that melittin and the "melittin-like" protein bind to overlapping sites.

Discussion

Melittin and mastoparan are polypeptides of dissimilar sequence but similar amphipathic alpha helical structure which inhibit ATP hydrolysis and partial reactions of the gastric H^+/K^+ ATPase. Uniquely, mastoparan does not affect the pNPPase activity of the gastric H^+/K^+ ATPase, reminiscent of the stimulatory effect of covalent modification with thiomerosal on the H^+/K^+ ATPase (Forte et al, 1981), and the Na^+/K^+ ATPase (Askari et al, 1979). Using photoaffinity-labelled analogs of these bioactive polypeptides, it was demonstrated that they both interacted with the gastric H^+/K^+ ATPase in the cytoplasmic domain of the alpha subunit (Cuppoletti, 1990; Huang et al, 1993a, 1993b). Previous reports of proteins reacting with various ion pumps have appeared. Thus, ankyrin reacts with a small fraction of H^+/K^+ ATPase (Smith et al, 1993), phospholamban interacts with the sarcoplasmic reticulum Ca^{2+} pump (James et al, 1989), and calmodulin interacts with the plasmalemmal Ca^{2+} pump, which in turn exhibits intramolecular interactions with similar sites (Enyedi et al, 1989). Melittin interacts with all ion pumps tested, and has been demonstrated to be approximately stoichiometric with the Na^+/K^+ ATPase (Cuppoletti et al, 1989; Cuppoletti & Abbott, 1990, Cuppoletti & Malinowska, 1992). As demonstrated in proteolytic fragmentation studies, photoaffinity analogs of melittin and mastoparan interact with the gastric

H^+/K^+ ATPase alpha subunit in the cytoplasmic domain, and the sites for binding melittin and mastoparan overlap. This suggests that particular regions of the gastric H^+/K^+ ATPase cytoplasmic domain are involved in either intra- or inter-molecular interactions with polypeptides or polypeptide domains.

Our studies using antimelittin antibody resulted in the identification of a 67 kDa immunoreactive protein which contains determinants in common with melittin, and exhibits stimulus-dependent association with the parietal cell apical membrane, suggesting that intermolecular interactions between the protein and the H^+/K^+ ATPase may occur. Direct evidence for interaction of the "melittin like" protein with the alpha subunit has not yet been obtained and the precise site of interaction of this protein with the gastric H^+/K^+ ATPase is not known at present. Nevertheless, displacement of the "melittin-like" protein and other circumstantial evidence suggests that this protein's site of interaction overlaps with the binding site for other bioactive polypeptides on the alpha subunit of the H^+/K^+ ATPase. The role of the "melittin-like" protein is not known. However since it interacts with H^+/K^+ ATPase-containing vesicles in the stimulated state, it could alter the function of the H^+/K^+ ATPase *per se*, the function of associated K^+ or Cl^- channels which are essential for HCl secretion, or it could serve a role in the localization of the H^+/K^+ ATPase to the plasma membrane as has been suggested for gastric ezrin (Hanzel *et al*, 1991). Exploration of the possibility of direct interaction of this novel "melittin-like" protein with the gastric H^+/K^+ ATPase alpha subunit and whether this protein affects the function of the H^+ pump will be obtained from reconstitution studies, which are in progress. We speculate that similar proteins may be identified which bind to other ion pumps and alter their function or localization.

Acknowledgements

This work was supported by NIH grants DK43816 and DK43377 (JC & DHM) and NSF Research Opportunities for Women Career Advancement Award DCB9109605 (DHM). PH was supported by NIH Training Grant HL07571.

References

Askari A, Huang W, Henderson GR (1979) Na,K-ATPase: Functional and structural modifications induced by mercurials. In Na,K-ATPase: Structure and Kinetics Skou JC & Norby JG (eds), Academic Press, New York

Clarke MA, Conway TM, Shorr RGL, Crooke ST (1987) Identification and isolation of a mammalian protein which is antigenically and functionally related to the phospholipase A_2 stimulatory peptide melittin. J Biol Chem 262 : 4402-4406.

Cox JA, Comte M, Fitton JE, DeGrado WF (1985) The interaction of calmodulin with amphipathic peptides. J Biol Chem 260 : 2527-2534.

Cuppoletti J (1990) [^{125}I]Azidosalicylyl melittin binding domains: evidence for a polypeptide receptor on the gastric H$^+$/K$^+$ ATPase. Archs Biochem Biophys 278 : 409-415

Cuppoletti J, Abbott AJ (1990) Interaction of melittin with the Na$^+$/K$^+$ATPase: evidence for a melittin-induced conformational change. Archs Biochem Biophys 283 : 249-257

Cuppoletti J, Malinowska DH (1992) Interaction of polypeptides with the gastric (H$^+$ + K$^+$)ATPase: melittin, synthetic analogs, and a potential intracellular regulatory protein. Mol Cell Biochem 114 : 57-63

Cuppoletti J, Sachs G (1984) Regulation of gastric acid secretion via modulation of a chloride conductance. J Biol Chem 259 : 14952-14959

Cuppoletti J, Blumenthal KM, Malinowska DH (1989) Melittin inhibition of the gastric H$^+$/K$^+$ ATPase and photoaffinity labeling with [^{125}I]azidosalicylyl melittin. Archs Biochem Biophys 275 : 263-270

Cuppoletti J, Chernyak BV, Huang P, Malinowska DH (1992) Structure-function relationships in the interaction of amphipathic helical polypeptides with the gastric H$^+$/K$^+$ ATPase. Ann NY Acad Sci 671 : 443-445

Cuppoletti J, Huang P, Kaetzel MA, Malinowska DH (1993) Stimulus-associated protein in gastric parietal cell detected using antimelittin antibody. Am J Physiol 264 (Gastrointest Liver Physiol 27 : G637-G644

Enyedi A, Vorherr T, James P, McCormick DJ, Filoteo AG, Carafoli E (1989) The calmodulin binding domain of the plasma membrane Ca^{2+} pump interacts both with calmodulin and with another part of the pump. J Biol Chem 264 : 12312-12321

Forte JG, Poulter JL, Dykstra R, Rivas J, Lee HC (1981) Specific modification of gastric K$^+$ stimulated ATPase activity by thimerosal. Biochim Biophys Acta 644 : 257-265

Hanzel D, Reggio, H, Bretscher A, Forte JG, Mangeat P (1991) The secretion-stimulated 80K phosphoprotein of parietal cells is ezrin, and has properties of a membrane cytoskeletal linker in the induced apical microvilli. EMBO J 10 : 2363-2373.

Helmich-de Jong ML, van Emst-de Vries SE, de Pont JJHHM (1987) Conformational states of the (K$^+$ + H$^+$)-ATPase studied using tryptic digestion as a tool. Biochim Biophys Acta 905 : 358-370

Huang P, Chernyak B, Malinowska DH, Cuppoletti J (1993a) Identification of two melittin binding sites on the alpha subunit of hog H$^+$/K$^+$ ATPase. FASEB J 7(4ptII) : A574

Huang P, Chernyak BV, Malinowska DH, Cuppoletti J (1993b) Identification and sequence of a polypeptide binding site on the gastric H$^+$/K$^+$ ATPase. In preparation for submission to Biochemistry

James P, Inui M, Tada M, Chiesi M, Carafoli E (1989) Nature and site of phospholamban regulation of the Ca^{2+} pump of sarcoplasmic reticulum. Nature (Lond) 342 : 90-92

Kaetzel MA, Dedman JR (1988) Affinity purified melittin antibody recognizes the calmodulin binding domain on calmodulin target proteins. J Biol Chem 263 : 3726-3729

Smith PR, Bradford AL, Joe EH, Angelides KJ, Benos DJ, Saccomani G (1993) Gastric parietal cell H$^+$/K$^+$ ATPase microsomes are associated with isoforms of ankyrin and spectrin. Am J Physiol (Cell Physiol) 264 : C63-C70

Zhang K, Wang ZQ, Gluck S (1992) Identification and partial purification of a cytosolic activator of vacuolar H$^+$-ATPases from mammalian kidney. J Biol Chem 267 : 9701-9705

The Gastric H/K -ATPase: The Principle Target in Autoimmune Gastritis.

Paul A. Gleeson, Ban-Hock Toh, Frank Alderuccio and Ian R. van Driel
Department of Pathology and Immunology,
Monash University Medical School, Alfred Hospital,
Melbourne, Vic. 3181,
AUSTRALIA

Introduction

The immune system is programmed to defend the body against invading microbes by recognising antigens displayed on the surface of the organisms. At the same time, the immune system should not respond to "self" antigens. This ability to discriminate between "foreign" and "self" is known as immunological tolerance. Autoimmune disease develops when "tolerance" to self antigens breaks down. Currently, little is known about how tolerance can be broken, although it is generally thought that this may occur in genetically susceptible individuals exposed to certain environmental "triggers". Furthermore, little is known about the early events which initiate the immune response to self antigens and in particular the way in which self antigens are recognised by T cells to activate the self-destructive autoimmune process. While in full blown autoimmune disease, the autoimmune response is clearly recognising a number of different antigenic determinants, it is possible that the early response may be directed towards one or only a few antigenic determinants.

Autoimmune diseases fall into two broad groups: those affecting multiple organs in the body (systemic autoimmune diseases) and those in which the diseases are restricted to a single organ (organ-specific autoimmune diseases). The latter group of diseases traditionally comprises autoimmune diseases of the endocrine glands and the stomach. The organ-specific autoimmune disease of the stomach (also known as autoimmune gastritis or chronic atrophic gastritis (Type A)) gives rise to a condition in man known as pernicious anaemia. This gastric autoimmune disease is an excellent model

NATO ASI Series, Vol. H 89
Molecular and Cellular Mechanisms
of H⁺ Transport
Edited by Barry H. Hirst
© Springer-Verlag Berlin Heidelberg 1994

for the study of the organ-specific autoimmune diseases. The condition is characterised by the loss of the acid-secreting parietal cells and the enzyme-secreting chief cells from the lining of the stomach, an accumulation of lymphocytes in this location and the presence of autoantibodies to parietal cells in the circulation (Strickland and Mackay, 1973). The anaemia is the result of the lack of intrinsic factor, a component normally secreted by parietal cells and which is necessary for vitamin B12 absorption in the small bowel. A genetic basis for this disease is suggested by the familial occurence of pernicious anaemia and by the presence of parietal cell autoantibodies, and associated autoimmune gastritis, in 20-30% of relatives of patients with the disease.

In order to understand the genesis of these autoimmune diseases it is critical to have a precise knowledge of the self antigens which are targeted by the body's own immune system, especially those target antigens responsible for the initial induction of tissue injury. Further, the basis for the loss of tolerance to these molecules must be determined. The occurence of circulating autoantibodies in patients with these diseases provides a convenient reagent for the identification of the target antigens, irrespective of whether or not the autoantibodies themselves are the cause of the tissue injury. We adopted this philosophy and have used the parietal cell autoantibodies present in the circulation of patients with autoimmune gastritis to identify the target antigen(s) in this disease.

The target antigen in pernicious anaemia is the gastric H/K-ATPase

The major function of the gastric parietal cells is to produce hydrochloric acid, an activity mediated by the gastric H/K-ATPase (proton pump) localised to the intracellular membranes of parietal cells. These intracellular membranes have the configuration of "tubulovesicles" in resting cells which are replaced by secretory "canaliculi" when the cells are stimulated to produce acid. The circulating autoantibodies found in patients with pernicious aneamia are parietal cell-specific, and show strong staining of the membranes of the intracellular canaliculi and tubulovesicles, as detected by immunofluorescense and immunoelectron microscopy (Hoedemaker and Ito, 1970; Callaghan et al., 1990). Further, these parietal cell autoantibodies specifically react with two major gastric membrane antigens, a 95 kDa component and a 60-90 kDa glycoprotein. From a variety of biochemical and molecular studies these autoantigens have been shown to be physically associated and identified as the 95 kDa α-subunit and the 60-90 kDa glycoprotein β subunit of the gastric H/K-

ATPase (Karlsson et al., 1988; Gleeson and Toh, 1991; Jones et al., 1991; Toh et al., 1990; Toh et al., 1992).

We have examined the reactivity of 26 parietal cell autoantibody-positive sera and all sera immunoblot both subunits under appropiate but mutually exclusive conditions (Callaghan et al., 1993). For example, the reactivity of anti-parietal cell autoantibodies with the 95 kDa α-subunit is optimal when the SDS-PAGE is carried out with samples which are reduced but not boiled. Whereas, reactivity with the 60-90 kDa β subunit is optimal with samples which are boiled but not reduced (Fig. 1). Autoantibody binding to the β subunit is critically dependent on the presence of a full complement of N-linked glycans since the partially deglycosylated protein failed to interact with the autoantibody (Fig. 1).

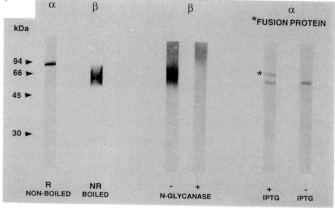

Fig. 1. Reactivity of pernicious anaemia serum with the gastric H/K ATPase by immunoblotting. Reactivity of pernicious anaemia serum with α and β subunits from reduced (R) non-boiled, and non-reduced (NR) boiled samples of gastric membranes, respectively. Autoantibody reactivity with the β subunit is sensitive is to removal of carbohydrate with N-glycanase. Autoantibodies react with a bacterial fusion protein containing residues 506-961 of the α-subunit. Asterisk indicates the 70 kDa glutathionine-S-transferase /α subunit fusion protein induced by treatment of bacteria with isopropyl–β-D-thiogalactoside (IPTG).

The lack of autoantibody binding to either reduced β-subunit or deglycosylated β-subunit suggests that the β-subunit binding sites of these parietal cell autoantibodies consist of either peptide sequences whose three dimensional structure is critically dependent on the presence of correctly-processed glycans, or composed of both glycan and peptide. In addition, these studies also suggest that the antibody binding sites are on the lumenal domain of the β subunit where there are 6 potential N-linked glycosylation sites and 6 conserved cysteines capable of forming disulphide bonds. In the case of the α subunit, the reactivity of parietal cell autoantibodies with a bacterial fusion protein incorporating the predominantly cytoplasmic domain of this molecule suggests that the antibody binding sites are in this region of the molecule. As this portion of the molecule also includes the catalytic domain, our findings are consistent with those of Burman et al. (1989) who showed that parietal cell autoantibodies were able to deplete H/K ATPase acitivity *in vitro*. However, whether the autoantibodies to the proton pump plays a role in the destruction of the parietal cells in the lining of the stomach in the human disease is not known.

A mouse model of gastric autoimmunity

To study the role of the proton pump in the immunopathology of gastritis we have adopted an animal model of this autoimmune disease (Fig. 2). Surgical removal of the thymus (thymectomy) in the narrow window of 2-4 days after birth in mice induces organ-specific autoimmune diseases (Kojima and Prehn, 1981). These autoimmune diseases are ideal experimental models since they show the cardinal features of human organ-specific autoimmunity with selective destruction of target cells, lymphocytic infiltration of the affected organ and circulating organ-specific autoantibodies. The organ targeted for autoimmune disease depends on the genetic background; BALB/c and AJ strains give the highest frequency of autoimmune gastritis. These mice have gastric lesions similar to those of the human disease and have circulating parietal cell autoantibodies which we have shown react with the two subunits of the gastric proton pump (Jones et al., 1991). Thus the target antigens recognised by the mouse autoantibodies are identical to those recognised in the human disease.

Are the parietal cell autoantibodies pathogenic in this animal model? Apparently not as the mouse disease cannot be transferred to a healthy individual by these autoantibodies. However, the disease can be transferred by T cells (Sakaguchi et al., 1990), thus the disease is clearly mediated through cells. The pathogenic T cells that cause autoimmune gastritis by adoptive transfer are a subset of "helper" T cells bearing a surface molecule designated CD4. On the other hand, "cytotoxic" T cells do not seem to have a role in this disease since depletion of this T cell population does not reduce the capacity of the remaining T cells to transfer disease (Smith et al., 1992). As T cells recognise different antigenic determinants than those recognised by antibodies, the specificity of the auto-reactive T cells that cause the tissue damage must be defined. However, the antigen specificity of the pathogenic T cells in murine autoimmune gastritis has not yet been established, although a T cell delayed hypersensitivity response to an enriched parietal cell population has been demonstrated. In addition, we have shown by *in vitro* proliferation assays that T cells specific for the gastric proton pump are present in mice with gastric autoimmunity (Martinelli et al., unpublished observations).

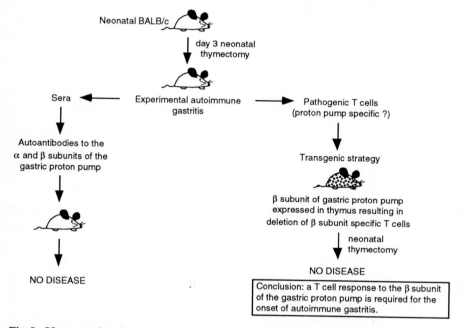

Fig. 2. Mouse model of autoimmune gastritis and role of the H/K ATPase in the genesis of disease.

Transgenic expression of organ-specific autoantigens

The studies described above have revealed that both human and mouse autoimmune gastritis, like many other autoimmune diseases, has multiple autoantibody targets, i.e. α and β subunits of the gastric H/K ATPase and intrinsic factor. However, for most autoimmune diseases it is not known if these molecules recognised by autoantibodies are also T cell-targets or whether the immune response to these antigens is involved in disease induction. Furthermore, immune responses in autoimmunity are usually most clearly defined at later stages of disease when the pathology is well established. The predominant specificities of lymphocytes at this phase of disease may be quite different from the specificities that initiated autoimmunity during events which may have occurred months or even years before.

The thymus is a primary lymphoid organ specialised for the development of T cells. During maturation of T cells within the thymus, autoreactive T cells are very efficiently destroyed by a process known as clonal deletion. Significantly, the gastric H/K ATPase is a peripheral autoantigen and is not expressed in the thymus of normal individuals. Hence, T cells which recognise the gastric H/K-ATPase are not deleted within the thymus but seeded into the periphery. We have exploited the central role of the thymus in tolerance induction to examine whether autoreactive T cells directed against the gastric H/K ATPase are responsible for the induction of autoimmune gastritis. This has been achieved by the generation of transgenic mice where the anti-proton pump T cells are specifically silenced or tolerised within the thymus (Fig. 2) (Alderuccio et al., 1993).

Previous studies showed that clonal deletion of developing T cells within the thymus can be directed by MHC class II-positive cells (Blackman et al., 1990). Therefore, we have produced transgenic mice that express the β-subunit of the H/K ATPase under the control of the MHC class II promoter (βH/K-transgenic), resulting in widespread tissue expression of the β-subunit, including the thymus. Four lines of transgenic mice were produced in which expression of the gastric H/K ATPase β-subunit was driven by the 5' flanking region of the MHC class II gene (van Ewijk et al., 1988). H/K ATPase β-subunit mRNA expression was found in all tissues examined from βH/K-transgenic mice, including the thymus. We reasoned that during T cell development, autoreactive T cells specific for the H/K ATPase β-subunit would be tolerised within the thymuses of βH/K-transgenic mice. Thus, autoimmune gastritis would be prevented if T cells of this specificity are required for the onset of disease (Fig. 2).

We neonatally thymectomised mice from four independent βH/K transgenic mouse lines. Transgenic littermates failed to develop anti-parietal cell autoantibodies and mononuclear cell infiltrates in the gastric mucosa (Fig. 3). In contrast, the incidence of autoimmune gastritis in thymectomised non-transgenic littermates was at the expected frequency, as defined by the presence of cellular infiltrates in the gastric mucosa. The same pattern of disease prevention was found in all the 4 independent βH/K-transgenic lines, indicating that the phenomenon observed was reproducible and not associated with the insertion of the βH/K-transgene within a particular genetic locus. Neonatally thymectomised BALB/c mice also develop oophoritis (Kojima et al., 1981). The incidences of oophoritis in both βH/K-transgenic mice and non-transgenic littermates were identical (Fig. 3). This finding indicated that the immune system of transgenic mice was still capable of initiating an autoimmune response to ovarian antigens. Thus the prevention of thymectomy-induced autoimmune disease in β-H/K-transgenic mice was gastric specific. From this data we conclude that a T cell response to the H/K ATPase β-subunit is an absolute requirement for the development of autoimmune gastritis (Alderuccio et al., 1993). Furthermore, it appears that the immune response to the other molecular target in this disease, the H/K ATPase α-subunit is either insufficient to cause disease, or a secondary event to an anti-β–subunit response. However, a causitive role for this molecule in disease induction has not yet been formally addressed.

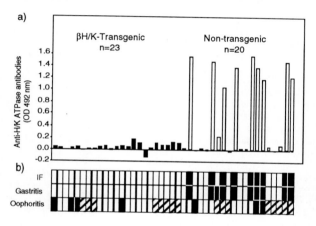

Fig. 3. **H/K ATPase autoantibodies, gastritis and oophoritis in neonatally thymectomised transgenic mice. a)** Mice were thymectomised at day 3 and transgenic status determined at weaning. Sera were collected at 3-5 months and approximately half the βH/K-transgenic and non-transgenic mice were killed for histological examination. Anti-H/K ATPase antibodies were detected by ELISA. **b)** Immunofluorescence (IF) to detect anti-parietal cell autoantibodies and histological examination of stomachs and ovaries of mice from **a**. Mice are represented in the same order as **a** and are shown directly below the corresponding ELISA readings. The presence of parietal cell autoantibodies, gastritis or oophoritis is indicated by a black box. Striped boxes indicate male mice excluded from analysis of oophoritis. Reproduced from the Journal of Experimental Medicine, 1993, 78, (in press) by copyright permission of the Rockefeller University Press.

Our findings here argue that an immune response to the β-subunit of the gastric H/K ATPase is essential for the development of autoimmune gastritis. The ability to identify the causative antigens in autoimmune diseases has ramifications in developing successful immunotherapeutic strategies. This approach, directed towards the identification of T cell targets in autoimmune gastritis, is applicable to other autoimmune diseases which will help to expand our understanding of the mechanisms and molecular basis of autoimmune diseases.

REFERENCES

Alderuccio F, Toh BH, Tan SS, Gleeson PA, and van Driel IR (1993) An autoimmune disease with multiple molecular targets abrogated by the transgenic expression of a single autoantigen in the thymus. J. Exp. Med. *178* (in press).

Blackman M, Kappler J, and Marrack P (1990) The role of the T cell receptor in positive and negative selection of developing cells. Science *248*:1335-1341.

Burman P, Mardh S, Norberg L, and Karlsson FA (1989) Parietal cell antibodies in pernicious anemia inhibit H, K-adenosine triphosphatase, the proton pump of the stomach. Gastroenterology 96: 1434-1438.

Callaghan JM, Toh BH, Pettitt JM, Humphris D C, and Gleeson P A (1990) Poly-*N*-acetylactosamine-specific tomato lectin interacts with gastric parietal cells. Identification of a tomato-lectin binding 60-90 x 10^3 M_r membrane glycoprotein of tubulovesicles. J. Cell Sci. *95*: 563-576.

Callaghan JM, Khan MA, Alderuccio F, van Driel IR, Gleeson PA, and Toh BH (1993) α and β subunits of the gastric H/K-ATPase are concordantly targeted by parietal cell autoantibodies associated with autoimmune gastritis. (submitted).

Gleeson PA and Toh BH (1991) Molecular targets in pernicious anaemia. Immunol. Today *12*: 233-238.

Hoedemaeker PJ and Ito S (1970) Ultrastructural localisation of gastric parietal cell antigen with peroxidase-coupled antibody. Lab. Invest. *22*: 184-188.

Jones CM, Toh BH, Pettitt JM, Martinelli TM, Humphris D, Callaghan J M, Goldkorn I, Mu F-T, and Gleeson P A (1991a) Monoclonal antibodies specific for the core protein of the β-subunit of the gastric proton pump. Eur. J. Biochem.*197*: 49-59.

Jones CM, Callaghan JM, Gleeson PA, Mori Y, Masuda T and Toh BH (1991b) The parietal cell autoantibodies recognised by neonatal thymectomy-induced murine gastritis are the α and β subunits of the gastric proton pump. Gastroenterology *101*: 287-294.

Karlsson FA, Burman P, Loof L, and Mardh S (1988) Major parietal cell antigen in autoimmune gastritis with pernicious anemia is the acid producing H+,K+-Adenosine Triphosphatase of the stomach. J. Clin. Invest..*81*:475-479

Kojima A and Prehn RT (1981) Genetic susceptibility to post-thymectomy autoimmune diseases in mice. Immunogenetics *14*:15-27.

Sakaguchi S and Sakaguchi N (1990) Thymus and autoimmunity: Capacity of the normal thymus to produce self-reactive T cells and conditions required for their induction of autoimmune disease. J. Exp. Med. *172*:537-545.

Strickland R and Mackay IR (1973) A reappraisal of the nature and significance of chronic atrophic gastritis. Am. J. Dig. Dis. *18*: 426-440

Smith H, Lou YH, Lacy P, and Tung SK (1992) Tolerance mechanisms in experimental ovarian and gastric autoimmune diseases. J. Immunol. *149*:2212-2218.

Toh BH, Gleeson PA, Simpson RJ *et al.* (1990) The 60-90 kDa parietal cell autoantigen associated with autoimmune gastritis is a β subunit of the gastric H+/K+-ATPase (proton pump). Proc. Nat. Acad. Sci. USA *87*: 6418-6422.

Toh BH, van Driel IR, and Gleeson PA (1992) Autoimmune gastritis: tolerance and autoimmunity to the gastric H+/K+ ATPase (proton pump). Autoimmunity *13*:165-172.

van Ewijk W, Ron Y, Monaco J, Kappler J, Marrack P, *et al.* (1988) Compartmentalization of MHC class II gene expression in transgenic mice. Cell *53*:357-370.

Acidic and non-acidic endosomes in kidney epithelial cells: their role in cell-specific membrane recycling processes

Dennis Brown and Ivan Sabolić
Renal Unit
Massachusetts General Hospital
Boston, MA 02114
U. S. A.

SUMMARY
Epithelial cells lining some kidney tubule segments have developed highly-specialized, yet independent systems for the rapid modulation their plasma membrane composition in response to different stimuli. In all cases, this process involves the movement of intracellular transporting vesicles that carry specific protein cargoes to and from the plasma membrane. These vesicles have transport features that differ among the various cell types, and this brief review will examine endosomes from the proximal tubule, and collecting duct intercalated principal cells, with respect to their content of two important proteins: water channels and proton pumps ($H^+ATPase$). The endosomal content of these two proteins varies in relation to the physiological role of the epithelial cell types. Apical endosomes from the proximal tubule contain both proteins, intercalated cell endosomes contain proton pumps but not water channels, and principal cell apical endosomes contain water channels but lack functional proton pumps. This endosomal division of labour along the urinary tubule is summarized in Figure 1 and has been revealed using a combination of morphological, immunocytochemical and functional techniques.

INTRODUCTION
The plasma membrane composition of virtually all eukaryotic cells is established, maintained and modified by the process of membrane recycling. Specific plasma membrane components are inserted by exocytosis of transport vesicles, and are removed by endocytosis of segments of the membrane in which particular proteins are concentrated. It is particularly important for kidney function that epithelial cells lining the urinary tubule can respond rapidly to changes in the composition of their environment; epithelial cells lining the

NATO ASI Series, Vol. H 89
Molecular and Cellular Mechanisms
of H+ Transport
Edited by Barry H. Hirst
© Springer-Verlag Berlin Heidelberg 1994

Figure 1

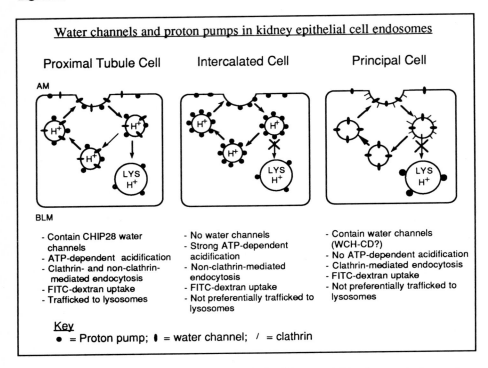

Water channels and proton pumps in kidney epithelial cell endosomes

Proximal Tubule Cell   Intercalated Cell   Principal Cell

AM

BLM

- Contain CHIP28 water
  channels
- ATP-dependent acidification
- Clathrin- and non-clathrin-
  mediated endocytosis
- FITC-dextran uptake
- Trafficked to lysosomes

- No water channels
- Strong ATP-dependent
  acidification
- Non-clathrin-mediated
  endocytosis
- FITC-dextran uptake
- Not preferentially trafficked to
  lysosomes

- Contain water channels
  (WCH-CD?)
- No ATP-dependent acidification
- Clathrin-mediated endocytosis
- FITC-dextran uptake
- Not preferentially trafficked to
  lysosomes

Key
● = Proton pump; ◖ = water channel; ╱ = clathrin

kidney collecting duct have developed highly-specialized, yet independent systems for the rapid modulation of the composition of their plasma membranes in response to different stimuli. Vasopressin induces the cycling of vesicles that carry water channels to and from the apical plasma membrane of principal cells, thus rapidly modulating the water permeability of this membrane (Brown, 1991, Harris, et al., 1991b, Verkman, 1989). In the intercalated cells of the collecting duct, hydrogen ion secretion is controlled by the recycling of vesicles carrying proton pumps to and from the plasma membrane (Brown, 1989, Schwartz and Al-Awqati, 1985). In both cell types, "coated" carrier vesicles are involved; clathrin-coated vesicles participate in water channel recycling, whereas the vesicles in intercalated cells are coated with the cytoplasmic domains of proton pumps. In proximal tubule epithelial cells, the apical plasma membrane has an extremely vigorous rate of constitutive endocytotic activity, and there is an extensive clathrin coat on the inter-microvillar domain of this membrane. Proximal tubule endosomes contain both water channels, and a vacuolar type $H^+ATPase$. In addition to these rapidly recycling membrane components, kidney epithelial cells are highly polarized with respect to many other membrane proteins that play an integral role in tubular function in all regions of the urinary tubule. These molecules have been localized using a variety of techniques from physiological measurements of tubular function to immunocytochemical detection at the light and/or electron microscopical level.

PROXIMAL TUBULE ENDOSOMES CONTAIN PROTON PUMPS AND WATER CHANNELS

The proximal tubule is a major site of reabsorption of fluid, electrolytes and other molecules such as amino acids, glucose and peptides that are present in the ultrafiltrate. The apical plasma membrane of proximal tubule epithelial cells demonstrates extensive membrane recycling, and an impressive endocytotic activity. This recycling appears to be constitutive, and so far there is little or no evidence for a stimulated recycling pathway in these cells.

Endosome acidification in the proximal tubule

A major function of the proximal tubule is bicarbonate reabsorption. This is an indirect result of active proton secretion by the proximal tubule epithelial cells. Both Na^+/H^+ exchange and a proton ATPase participate in proximal proton secretion (Aronson, 1983), and a vacuolar-type $H^+ATPase$ has been located on the apical plasma membrane and on apical endosomes in these cells both functionally, and by immunocytochemical studies using specific antibodies against subunits of the vacuolar $H^+ATPase$ (Brown, et al., 1988b, Sabolic, et al., 1985). Isolated cortical endosomes from kidney show a marked intraluminal acidification that is stimulated by ATP, and inhibited by n-ethylmaleimide (NEM) and bafilomycin. Material that is internalized by these endosomes is delivered to lysosomes, where it is degraded (Christensen, 1976). In contrast, proteins associated with the endosomal membrane, such as gp330, are returned to the apical membrane (Gutmann, et al., 1989, Kerjaschki, et al., 1984). Thus, proximal tubule endosomes represent an example of a "conventional" internalization pathway, which results in delivery of some internalized material to lysosomes, and recycling of membrane proteins back to the plasma membrane. The endocytotically active inter-microvillar region of the proximal tubule apical membrane has an extensive clathrin coat (Rodman, et al., 1984), which is presumably involved in the internalization of apical proteins and associated ligands. However, it is not known whether a second clathrin-independent pathway of internalization also exists at the apical plasma membrane of these cells.

Proximal tubule endosomes contain water channels

Up to 70% of the glomerular ultrafiltrate volume is reabsorbed in the proximal tubule. It has been demonstrated using isolated membrane fractions that plasma membranes of proximal tubule epithelial cells are highly-permeable to water due to the presence of a mercurial-sensitive water channel (Sabolic, et al., 1992a). Furthermore, the same proximal tubule endosomes contain both water channels and proton pumps (Ye, et al., 1989). It is now established that the proximal tubule water channel is identical to the erythrocyte water channel, CHIP28 (Preston and Agre, 1991, Zhang, et al., 1993). This protein has been localized to apical and basolateral plasma membranes in the proximal tubule by immunocytochemistry, and it is also found in endosomal membranes (Nielsen, et al., 1993, Sabolic, et al., 1992a). However, CHIP28 is not concentrated in clathrin coated pits on the apical membrane, implying that it may be internalized by a clathrin-independent mechanism in these cells. The nature of the vesicles involved in CHIP28 endocytosis in this tubule segment has not been established, although they clearly are in continuity with the acidic endosomal pathway. CHIP28 has not been detected in lysosomes - instead it appears to be recycled back to the apical membrane.

INTERCALATED CELL ENDOSOMES CONTAIN PROTON PUMPS BUT LACK WATER CHANNELS

Intercalated cells of the collecting duct are highly-specialized for the transport of protons across the collecting duct epithelium. This proton transport is mediated by a vacuolar type $H^+ATPase$ that can be inserted into and removed from the plasma membrane of these cells under different physiological conditions.

Endocytic vesicles in intercalated cells are acidic

Distinctive morphological alterations occur in intercalated cells during an increase in H^+ secretion by collecting ducts. Following induced acidosis in rats, an increase in the apical surface area of intercalated cells is accompanied by a decrease in the number of specialized "coated" cytoplasmic vesicles, known as tubulovesicles (Madsen, et al., 1991), as a result of their exocytotic fusion with the apical plasma membrane. Previous studies in turtle bladder and the collecting duct had shown that acidic vesicles containing proton pumps are inserted into the apical membrane of these carbonic anhydrase rich cells during acidosis (Gluck, et al., 1982, Schwartz and Al-Awqati, 1985). Tracer studies with horseradish peroxidase demonstrated that these specialized coated vesicles are also involved in endocytotic events (Brown, et al., 1987). Immunocytochemical studies showed that the vesicle coating material contained cytoplasmic subunits of the proton pump, including the 31, 56, and 70kD subunits (Brown, et al., 1987, Brown, et al., 1988b) The "coat" seen on the cytoplasmic side of the plasma membrane of some intercalated cells also contained the same pump subunits and the structure of the membrane coat was elucidated using the rapid-freeze, deep-etch technique. The membrane-coating material had a structure that was identical to that of immunoaffinity-purified proton pumps, incorporated into phospholipid liposomes (Brown, et al., 1987), thus confirming its identity as part of the proton pumping ATPase responsible for distal urinary acidification.

Opposite polarity of proton pumps in A and B intercalated cells

This and other work suggested that tubulovesicles fuse with the apical plasma membrane during acidosis and incorporate their membrane into this plasma membrane domain. However, in the cortical collecting duct (Madsen and Tisher, 1986) and in the turtle urinary bladder (Stetson and Steinmetz, 1985), two morphological variants of "carbonic anhydrase-rich" cell have been described, which are believed to be proton-secreting (A-cells) or bicarbonate-secreting (B-cells). Both types co-exist in kidney cortical collecting ducts, in accord with the ability of this tubule segment to secrete either net acid or net base under different physiological conditions of acidosis or alkalosis. Using anti-proton pump antibodies, we demonstrated that while all medullary intercalated cells have proton pumps associated with their apical plasma membrane, cortical intercalated cells had three patterns of labeling : apical, basolateral, and diffuse cytoplasmic or bipolar (Brown, et al., 1988a). This finding provided direct support for the idea that different intercalated cells in the cortex are responsible for either proton (apical pumps) or bicarbonate (basolateral pumps) secretion into the tubule lumen.

Schwartz et al. (1985) suggested that A and B cells may be interconvertible, and may change their functional polarity by inserting proton pumps into either apical or basolateral plasma membranes, depending on prevailing acid-base conditions of the animal. Although a definitive answer has not yet been achieved, more recent work suggests that A and B cells do not interconvert during alterations in acid-base status (Koichi, et al., 1992, Verlander, et al., 1988).

Intercalated cell endosomes do not contain water channels

The vesicles involved in proton pump recycling in intercalated cells are heavily-loaded with 6-carboxy fluorescein or FITC-dextran after in vitro infusion of these fluorescent probes (Lencer, et al., 1990a). Endosomes can be prepared from the inner stripe of the outer medulla, a region which contains many intercalated cells but also contains other cell types including principal cells and thick ascending limb (TAL) cells. It was shown that intercalated cell endosomes (and TAL endosomes) do not contain water channels by preparing vesicles from vasopressin-deficient Brattleboro rats, injected with either 6-CF alone, or 6-CF plus vasopressin. When 6-CF alone was injected, the resulting endosomes showed no rapid volume-dependent fluorescent quenching, indicating that they lack water channels. When vasopressin was injected simultaneously, a population of the harvested endosomes now contained water channels (Lencer, et al., 1990a). The simplest interpretation of these data is that the water-channel containing endosomes are derived from principal cells, where water channel recycling is activated only in the presence of vasopressin (see below), and that intercalated cell endosomes never contain water channels.

PRINCIPAL CELL ENDOSOMES HAVE WATER CHANNELS BUT LACK FUNCTIONAL PROTON PUMPS

By modulating the water permeability of their apical plasma membrane, principal cells of the collecting duct influence the degree of osmotic equilibration that can occur between the hypertonic interstitial space and the luminal fluid within the tubule. This process is under the control of vasopressin in mammals, and in the absence of this hormone, collecting ducts are relatively impermeable to water. It is believed that after vasopressin stimulation, proteinaceous water channels located in the limiting membranes of some cytoplasmic vesicles are delivered to the apical membrane by exocytosis. Once in the apical membrane, these channels appear to be responsible for the observed increase in membrane and, therefore, epithelial water permeability (Brown, 1991, Harris, et al., 1991b, Verkman, 1989). Upon vasopressin withdrawal, membrane patches containing the channels are removed from the apical membrane by endocytosis, and reside on vesicles in the apical cytoplasm. A CHIP28-related water channel, known as

WCH-CD, has recently been identified in collecting duct principal cells (Fushimi, et al., 1993)

Endocytosis of water channels is clathrin-mediated in principal cells

A series of studies using a combination of freeze-fracture electron microscopy, horseradish peroxidase (HRP) and fluorescent tracer studies, and functional assays revealed that the endocytotic phase of water channel recycling in the kidney collecting duct occurs via clathrin-coated pits and vesicles (Brown and Orci, 1983, Brown, et al., 1988, Verkman, et al., 1988, Verkman, et al., 1989). This process takes place at the apical plasma membrane, where the antidiuretic hormone, vasopressin, induces a 6 to 10-fold increase in HRP endocytosis coupled with a similar increase in the number of clathrin-coated pits on the apical membrane. Independent studies in perfused rabbit cortical collecting ducts confirmed this role of clathrin in vasopressin-induced recycling processes (Strange, et al., 1988). The results imply that in the presence of vasopressin, water channels are continually recycled between an intracellular compartment and the apical plasma membrane. Upon vasopressin withdrawal, the endocytotic pathway takes precedence, and a large increase in the internalization of apical fluid phase markers occurs.

Detection of water channels in endosomes

By infusing rats with 6-carboxyfluorescein or FITC-dextran, and using a fluorescence quenching assay to measure the water permeability of isolated microsomal vesicles, we showed that membranes of papillary endosomes that are induced to recycle by vasopressin-treatment of Brattleboro rats have a high water permeability, consistent with the presence of water channels in these membranes (Verkman, et al., 1988). In contrast, apical membrane segments that are recycled constitutively (in the absence of vasopressin) show only characteristics associated with diffusional water permeability, with no detectable water channels. Among the features of bulk water flow across membranes are a high Pf (membrane permeability coefficient), and low Ea (energy of activation). An Ea greater than 10 kcal mol^{-1} is associated with diffusive water movement across membranes, and Pf in lipid bilayers not containing water channels is around 0.0001 - 0.0005 cm sec^{-1} (79). The high Pf (0.03 cm s^{-1}) and low Ea (3.8 kcal mol^{-1}) of endosomes isolated from vasopressin-treated rats indicates the presence of water channels that are not detectable in endosomes that recycle constitutively in the absence of the hormone. In addition, the sensitivity of the rapid water flow to mercurial compounds is also a hallmark of bulk flow through water channels (Verkman, 1989).

Endosomes that internalize water channels are not acidic

The endosomes from rat renal papilla that show a high, $HgCl_2$-sensitive water permeability, consistent with the presence of vasopressin-sensitive water channels, do not exhibit ATP-dependent luminal acidification or NEM-sensitive ATPase activity (Lencer, et al., 1990b). Furthermore, immunoblotting and immunocytochemistry showed that they lack the 16kD, 31kD and 70kD subunits of the vacuolar proton pump (Sabolic, et al., 1992b). In contrast, the 56kD subunit of the proton pump was present in papillary endosomes, and was localized at the apical pole of principal cells by immunocytochemistry. Therefore, early endosomes derived from the apical plasma membrane of collecting duct principal cells fail to acidify because they lack functionally important subunits of a vacuolar-type proton pumping ATPase, including the 16kD transmembrane domain that serves as the proton-conducting channel, and the 70kD cytoplasmic subunit that contains the ATPase catalytic site. This specialized, non-acidic early endosomal compartment appears to be involved primarily in the hormonally-induced recycling of water channels to and from the apical plasma membrane of vasopressin-sensitive cells in the kidney collecting duct. A similar non-acidic compartment recycles water channels in another vasopressin-sensitive tissue, the toad urinary bladder (Harris, et al., 1991a, Shi, et al., 1990). This recycling mechanism is, therefore, quite distinct from the endocytotic pathway in proximal tubule epithelial cells in which acidic endosomes form part of a pathway that leads to lysosomal degradation of internalized material.

REFERENCES

Aronson PS (1983) Mechanisms of active H^+ secretion in the proximal tubule. Am. J. Physiol. 245: F647-F659

Brown D (1989) Membrane recycling and epithelial cell function. Am. J. Physiol. 256: F1-F12

Brown D (1991) Structural-functional features of vasopressin-induced water flow in the kidney collecting duct. Sem. Nephrol. 11: 478-501

Brown D, Gluck S, Hartwig J (1987) Structure of the novel membrane-coating material in proton-secreting epithelial cells and identification as an H^+ATPase. J. Cell Biol. 105: 1637-1648

Brown D, Hirsch S, Gluck S (1988a) An H^+ATPase is present in opposite plasma membrane domains in subpopulations of kidney epithelial cells. Nature 331: 622-624

Brown D, Hirsch S, Gluck S (1988b) Localization of a proton-pumping ATPase in rat kidney. J. Clin. Invest 82: 2114-2126

Brown D, Orci L (1983) Vasopressin stimulates the formation of coated pits in rat kidney collecting ducts. Nature 302: 253-255

Brown D, Weyer P, Orci L (1987) Non-clathrin coated vesicles are involved in endocytosis in kidney collecting duct intercalated cells. Anat. Rec. 218: 237-242

Brown D, Weyer P, Orci L (1988) Vasopressin stimulates endocytosis in kidney collecting duct epithelial cells. Eur. J. Cell Biol. 46: 336-340

Christensen EI (1976) Rapid protein uptake and digestion in proximal tubule lysosomes. Kidney Int 10: 301-310

Fushimi K, Uchida S, Hara Y, Hirata Y, Marumo F, Sasaki S (1993) Cloning and expression of apical membrane water channel of rat kidney collecting tubule. Nature 361: 549-552

Gluck S, Cannon C, Al-Awqati Q (1982) Exocytosis regulates urinary acidification in turtle bladder by rapid insertion of H^+ pumps into the luminal membrane. Proc. Natl Acad. Sci. U.S.A. 79: 4327-4331

Gutmann EJ, Niles JL, McCluskey RT, Brown D (1989) Colchicine-induced redistribution of an endogenous apical membrane glycoprotein (gp 330) in kidney proximal tubule epithelium. Am. J. Physiol. 257: C397-C407

Harris HW, Zeidel ML, Hosselet C (1991a) Quantitation and topography of membrane proteins in highly water-permeable vesicles from ADH-stimulated toad bladder. Am. J. Physiol. 261: C143-C153

Harris HWJ, Strange K, Zeidel ML (1991b) Current understanding of the cellular biology and molecular structure of the antidiuretic hormone-stimulated water transport pathway. J. Clin. Invest. 88: 1-8

Kerjaschki D, Noronha-Blob L, Sacktor B, Farquhar MG (1984) Microdomains of distinctive glycoprotein composition in the kidney proximal tubule brush border. J Cell Biol 98: 1505-1513

Koichi Y, Satlin LM, Schwartz GJ (1992) Adaptation of rabbit cortical collecting duct t0 in vitro acidification. Am. J. Physiol. 263: F749-F756

Lencer WI, Brown D, Ausiello DA, Verkman AS (1990a) Endocytosis of water channels in rat kidney: cell specificity and correlation with in vivo antidiuresis. Am. J. Physiol. 259: C920-C932

Lencer WI, Verkman AS, Arnaout A, Ausiello DA, Brown D (1990b) Endocytic vesicles from renal papilla which retrieve the vasopressin-sensitive water channel do not contain a functional H^+ATPase. J. Cell Biol. 111: 379-390

Madsen KM, Tisher CC (1986) Structure-function relationships along the distal nephron. Am. J. Physiol. 250: F1-F15

Madsen KM, Verlander JW, J. K, Tisher CC (1991) Morphological adaptation of the collecting duct to acid-base disturbances. Kidney Int. 40 (Suppl. 33): S57-S63

Nielsen S, Smith BL, Christensen EI, Knepper MA, Agre P (1993) CHIP28 water channels are localized in constitutively waterpermeable segments of the nephron. J. Cell Biol. 120: 371-383

Preston GM, Agre P (1991) Isolation of the cDNA for erythrocyte integral membrane protein of 28kilodaltons: member of an ancient channel family. Proc. Natl. Acad. Sci. 88: 11110-11114

Rodman JS, Kerjaschki D, Merisko E, Farquhar MG (1984) Presence of an extensive clathrin coat on the apical plasmalemma of the rat kidney proximal tubule cell. J. Cell Biol. 98: 1630-1636

Sabolic I, Haase W, Burkhardt G (1985) ATP-dependent H^+ pumps in membrane vesicles from rat kidney cortex. Am. J. Physiol. 248: F835-F844

Sabolic I, Valenti G, Verbavatz J-M, et al. (1992a) Localization of the CHIP28 water channel in rat kidney. Am. J. Physiol. 263: C1225-C1233

Sabolic I, Wuarin F, Shi L-B, et al. (1992b) Apical endosomes isolated from kidney collecting duct principal cells lack subunits of the proton pumping ATPase. J. Cell Biol. 119: 111-122

Schwartz GJ, Al-Awqati Q (1985) Carbon dioxide causes exocytosis of vesicles containing H^+ pumps in isolated perfused proximal and collecting tubules. J. Clin. Invest. 75: 1638-1644

Shi L-B, Brown D, Verkman AS (1990) Water, proton and urea transport in toad bladder endosomes that contain the vasopressin-sensitive water channel. J. Gen. Physiol. 95: 941-960

Stetson DA, Steinmetz PR (1985) A and B types of carbonic anhydrase-rich cells in turtle bladder. Am. J. Physiol. 249: F553-F565

Strange K, Willingham MC, Handler JS, Harris HW Jr. (1988) Apical membrane endocytosis via coated pits is stimulated by removal of antidiuretic hormone from isolated, perfused rabbit cortical collecting tubule. J. Membr. Biol. 103: 17-28

Verkman AS (1989) Mechanisms and regulation of water permeability in renal epithelia. Am. J. Physiol. 257: C837-C850

Verkman AS, Lencer WI, Brown D, Ausiello DA (1988) Endosomes from kidney collecting tubule cells contain the vasopressin -sensitive water channel. Nature 333: 268-269

Verkman AS, Weyer P, Brown D, Ausiello DA (1989) Functional water channels are present in clathrin coated vesicles from bovine kidney but not from brain. J. Biol. Chem. 264: 20608-20613

Verlander JW, Madsen KM, Tisher CC (1988) Effect of acute respiratory acidosis on two populations of intercalated cells in rat cortical collecting duct. Am. J. Physiol. 253: F1142-F1156

Ye R, Shi LB, Lencer WI, Verkman AS (1989) Functional colocalization of water channels and proton pumps in endosomes from kidney proximal tubule. J. Gen. Physiol. 93: 885-902

Zhang R, Skach W, Hasegawa H, Van Hoek AN, Verkman AS (1993) Cloning, functional analysis and cell localization of a kidney proximal tubule water transporter homologous to CHIP28. J. Cell Biol. 120: 359-369

Regulation and Properties of the Coated Vesicle Proton Pump

Yu Feng, Melanie Myers[1] and Michael Forgac[2]
Department of Cellular and Molecular Physiology
Tufts University School of Medicine
136 Harrison Ave.
Boston, MA 02111   U.S.A.

Introduction

Vacuolar acidification plays a crucial role in a number of basic cellular processes in eukaryotic cells (for review see Forgac, 1989). Thus,
following receptor–mediated endocytosis, exposure of ligand–receptor complexes to a low pH within the endosomal compartment activates ligand–receptor
dissociation, thus allowing receptor–recycling to occur. Receptor–recycling
in turn controls the rate of uptake into cells of macromolecules such as
LDL, transferrin and asialoglycoprotein. Receptor–recycling is also crucial
in controlling the sensitivity of cells to hormones and growth factors, such
as insulin and EGF. Exposure to a low pH within endosomes is also responsible for entry of the cytotoxic portions of many envelope viruses and
toxins into cells. In addition to its role in the endocytic pathway, vacuolar acidification is crucial for the correct intracellular targeting of
newly synthesized lysosomal enzymes from the Golgi to lysosomes, for the
processing and degradation of macromolecules in secretory and digestive
organelles, and for coupled transport processes in various vacuolar compartments.

Vacuolar acidification is carried out by a novel class of ATP–driven
proton pumps termed the vacuolar (H^+)–ATPases (or V–ATPases) (Forgac,
1989). V–ATPases have been identified in a variety of intracellular
compartments, including clathrin-coated vesicles, endosomes, lysosomes,
Golgi–derived vesicles, secretory vesicles and the vacuoles of fungi and
plants. V–ATPases are also present in the plasma membrane of certain
specialized cells, including intercalated cells of the kidney, osteoclasts,
neutrophils and macrophages, where they function either to alkalinize the

---

1. Current address: Ribozyme Pharmaceuticals, Inc., Boulder, CO 80301.
2. To whom correspondence should be addressed.

NATO ASI Series, Vol. H 89
Molecular and Cellular Mechanisms
of H^+ Transport
Edited by Barry H. Hirst
© Springer-Verlag Berlin Heidelberg 1994

cytoplasm or acidify the extracellular environment (see other chapters in this volume). The V-ATPases display common transport and inhibitor properties which indicate that they represent a family of (H)-ATPases which are distinct from both the P-type and F-type ATPases (Forgac, 1989).

Structure and Subunit Function of the Coated Vesicle Proton Pump

Our work has focused on the V-ATPase from clathrin-coated vesicles (Forgac, et al., 1983; Forgac, 1992a). We have demonstrated that this enzyme is composed of nine polypeptides of molecular weight 17,000-100,000 which are immunoprecipitated as a complex using monoclonal antibodies which recognize the native (H$^+$)-ATPase (Arai, et al., 1987a). We have shown that the 73 kDa A subunit possesses the catalytic nucleotide binding site (Arai, et al., 1987a) while the 17 kDa c subunit is a highly hydrophobic polypeptide which is responsible for the sensitivity of proton translocation to inhibition by the carboxyl reagent DCCD (Arai, et al., 1987b). Using quantitative amino acid analysis to determine subunit stoichiometry and labeling by both membrane impermeant and hydrophobic reagents to determine the topography of the various V-ATPase subunits, we were able to propose a structural model for the V-ATPases (Arai, et al., 1988). Using proteolysis of a sided, reconstituted preparation of the purified V-ATPase (Adachi, et al., 1990a) as well as dissociation of peripheral polypeptides, reversible covalent crosslinking and lectin binding studies (Adachi, et al., 1990b), we have further refined our structural model, which is shown in Fig.1.

Our data indicate that the V-ATPases are composed of two structural domains. The peripheral V_1 domain of molecular weight 500,000 has the structure $73_3 58_3 40_1 34_1 33_1$, is present exclusively on the cytoplas-mic side of the membrane and possesses all of the nucleotide binding sites of the complex. We have also recently demonstrated that the 50 kDa subunit of the adaptin complex AP-2 binds specifically and stoichiometrically to the V_1 domain (Myers and Forgac, 1993a) (see below). The 73 kDa A subunits possess the catalytic nucleotide binding sites while the 58 kDa B subunits possess additional, noncatalytic sites which are believed to play a regulatory role. The 40, 34 and 33 kDa subunits appear to link the V_1 domain to the integral V_0 sector. The integral V_0 domain of molecular weight 250,000 has the structure $100_1 38_1 19_1 17_6$ and is responsible for proton translocation across the membrane. The 17 and 19 kDa subunits are both extensively buried in the lipid bilayer while the 38 kDa subunit is

exposed on the cytosolic surface and the 100 kDa polypeptide is a transmembrane glycoprotein. One important conclusion which has emerged from
these studies is the close structural relationship between the V-ATPases and
the F-ATPase family, which normally function in proton-coupled ATP synthesis
rather than ATP-driven proton transport. The evolutionary relationship
implied by this structural similarity has been further demonstrated by the
observed amino acid sequence homology between the nucleotide binding subunits as well as the DCCD reactive c subunits of the V and F-ATPases (for
references see Forgac, 1989).

To further address the function of individual subunits in the V-ATPase
complex, we have established a procedure for dissociation of the peripheral
V_1 subunits from the integral V_0 domain followed by reassembly of these
dissociated subunits to give a functional V-ATPase complex (Puopolo and
Forgac, 1990). We have demonstrated that in the absence of the V_0 domain
the V_1 subunits are able to assemble into a V_1(-40 kDa) subcomplex and
that a functional V_1V_0 complex lacking the 40 kDa subunit can be formed

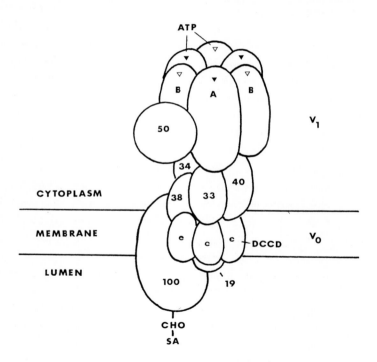

Fig.1 - Structural model of the coated vesicle V-ATPase (see text).

which possesses about 50% of the proton transport activity of the holo-
complex (Puopolo, et al., 1992b). The 40 kDa subunit does, however, appear
to play an important role in stabilizing the reassembled V-ATPase complex.
We have also investigated the structural and functional properties of the
V_O domain and have demonstrated that, unlike the homologous Fo domain of
the F-ATPases, the V_O domain does not function as a passive, DCCD-inhibit-
able proton channel following removal of V_1 (Zhang, et al., 1992).
Similarly, the reassembled V_1 domain, unlike F_1, is not a functional
ATPase (Puopolo, et al., 1992b).

We have begun to probe the structure of the nucleotide binding sites on
the V-ATPase complex and have identified Cys254 of the 73 kDa A subunit as
the residue responsible for the sensitivity of the V-ATPases to sulfhydryl
reagents, such as NEM (Feng and Forgac, 1992). Cys254 is located in the
consensus nucleotide binding sequence GXGKTV which forms part of a glycine
rich loop shared among many nucleotide binding proteins, including the
F-ATPase beta subunit and adenylate kinase. This residue is conserved as a
cysteine in all V-ATPase A subunit sequences (Puopolo, et al., 1991), from
bovine to yeast, but is a valine residue in all F-ATPase beta subunit
sequences, consistent with the respective sensitivities of these families to
inhibition by NEM. Based on these results, we have proposed a model for the
structure of the catalytic nucleotide binding site of the V-ATPases (Forgac,
1992b).

Regulation of Vacuolar Acidification

Immunocytochemical analysis of MDBK cells indicates that V-ATPases exist
in endocytic coated vesicles (Marquez-Sterling, et al., 1991). Nevertheless, several lines of evidence indicate that the earliest endocytic compartments, including endocytic coated vesicles, are near neutral pH (for review
see Forgac, 1992). These results suggest that the activity of the V-ATPases
is suppressed in the earliest endocytic compartments.

Recently, we have obtained evidence that Cys254 plays an important role
in regulation of V-ATPase activity (Feng and Forgac, 1992b). Thus,
disulfide bond formation between Cys254 and a second cysteine residue of the
A subunit results in a complete loss of V-ATPase activity which is reversed
upon reduction of the disulfide bond with DTT. Moreover, when coated
vesicles are isolated under conditions which prevent any change in the oxida-
tion state of Cys254, approximately 40% of the V-ATPase is found in an

inactive, disulfide bonded state. These results suggest that reversible disulfide bond formation at the catalytic site of the A subunit may play an important role in controlling V-ATPase activity in vivo (see Fig.2).

A second mechanism for regulation of vacuolar acidification involves control of assembly of the V-ATPase complex. We have demonstrated that assembled V_1 domains exist free in the cytoplasm of MDBK cells (Myers and Forgac, 1993b) and that free V_O domains exist in clathrin-coated vesicles (Zhang, et al., 1992). Moreover, the free V_O domains are silent with respect to passive proton translocation (Zhang, et al., 1992). Thus, activation of vacuolar acidification may involve the attachment of V_1 domains to preexisting V_O domains. Alternatively, control may be acheived by regulating the tightness of coupling of proton transport and ATP hydrolysis. We have shown that several conditions alter this coupling, including

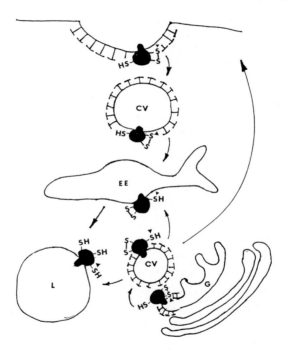

Fig.2 - Model for role of reversible disulfide bond formation in regulation of V-ATPase activity. Disulfide bond formation between Cys254 and a second A subunit cysteine residue maintains the V-ATPases in an inactive state in the plasma membrane and endocytic coated vesicles until delivery to early endosomes (EE) where disulfide exchange leads to activation.

high ATP concentrations (Arai, et al., 1989), mild proteolysis (Adachi, et al., 1990a) and membrane environment (Arai, et al., 1987b), suggesting that the V-ATPase is poised to exist in a partially coupled state.

A further potential mechanism for control of V-ATPase activity involves isoform substitution. In bovine tissues, the V-ATPase B subunit is coded for by multiple genes which are expressed in a tissue specific manner (Puopolo, et al., 1992a). The existence of multiple isoforms of the B subunit may also have important implications for targeting of V-ATPases in the cell.

A final mechanism for regulation of vacuolar acidification involves control of chloride conductance. Vacuolar acidification requires the activity of a chloride channel to dissipate the membrane potential generated during electrogenic proton transport (Arai, et al., 1989). We have demonstrated that both chloride channel activity and acidification in coated vesicles are modulated by protein kinase A-dependent phosphorylation of this chloride channel (Mulberg, et al., 1991).

Assembly and Targeting of V-ATPases

We have used antibodies against the coated vesicle V-ATPase to investigate the pathway for assembly of V-ATPases in vivo (Myers and Forgac, 1993b). Our results indicate that the peripheral V_1 subunits are able to assemble onto the V_0 sector in the endoplasmic reticulum, but that there is also a soluble pool of V_1 complexes which exists at steady state in parallel with those attached to the membrane. These results, together with the observation that intact coated vesicles possess a significant population of free V_0 domains (Zhang, et al., 1992), suggest the possibility that assembly of V_1 complexes onto preexisting V_0 domains may play a role in triggering vacuolar acidification.

Most recently, we have demonstrated that the 50 kDa polypeptide of the adaptin complex AP-2 specifically associates with the coated vesicle V-ATPase (Myers and Forgac, 1993a). The adaptins (AP-1 and AP-2) are multisubunit complexes which mediate the interaction between clathrin and receptors, with AP-2 localized to the plasma membrane and AP-1 to the trans-Golgi network (Pearse and Robinson, 1990). The receptor specificity is believed to reside on the 100 kDa adaptin subunits while the 50 kDa polypeptide, although posessing autokinase activity, is of unknown function. We have demonstrated that the 50 kDa adaptin protein both copurifies and coimmunoprecipitates with the coated vesicle V-ATPase using

monoclonal antibodies directed against the V–ATPase, and that the
stoichiometry of binding is one 50 kDa polypeptide per V–ATPase complex
(Myers and Forgac, 1993a). Incubation of the purified V–ATPase with
[^{32}P]–ATP results in labeling of both the 50 kDa polypeptide and a band
which comigrates with the V–ATPase B subunit. These results suggest a
possible mechanism by which V–ATPases are targeted to their correct
intracellular destinations (see Fig.3). Thus, V–ATPases in the plasma
membrane become concentrated in endocytic coated vesicles through their
interaction with the 50 kDa subunit of the AP–2 adaptin complex. By
contrast, V–ATPases in the trans–Golgi become concentrated in Golgi–derived
coated vesicles through their interaction with the corresponding 47 kDa
subunit of AP–1.

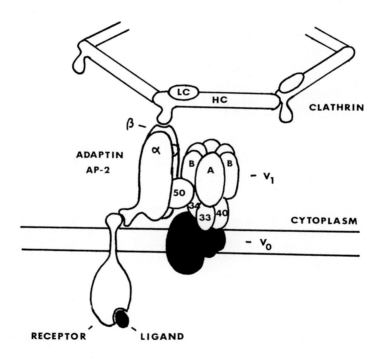

Fig.3 – Model for role of AP–2 in internalization of V–ATPases and cell
surface receptors in clathrin–coated pits (see text).

References

Adachi,I., Arai,H., Pimental,R. and Forgac,M. (1990a) "Proteolysis and Orientation on Reconstitution of the Coated Vesicle Proton Pump" J.Biol.Chem. 265, 960-966.

Adachi,I., Puopolo,K., Marquez-Sterling,N., Arai,H. and Forgac,M. (1990b) "Dissociation, Crosslinking and Glycosylation of the Coated Vesicle Proton Pump" J.Biol.Chem. 265, 967-973.

Arai,H., Berne,M., Terres,G., Terres,H., Puopolo,K. and Forgac,M. (1987a) "Subunit Composition and ATP-Site Labeling of the Coated Vesicle (H$^+$)-ATPase" Biochemistry 26, 6632-6638.

Arai,H., Berne,M. and Forgac,M. (1987b) "Inhibition of the Coated Vesicle Proton Pump and Labeling of a 17,000 Dalton Polypeptide by N,N'-Dicyclo hexylcarbodiimide" J.Biol.Chem. 262, 11006-11011.

Arai,H., Terres,G., Pink,S. and Forgac,M. (1988) "Topography and Subunit Stoi chiometry of the Coated Vesicle Proton Pump" J.Biol.Chem. 263, 8796-8802.

Arai,H., Pink,S. and Forgac,M. (1989) "Interaction of Anions and ATP with the Coated Vesicle Proton Pump" Biochemistry 28, 3075-3082.

Feng,Y. and Forgac,M. (1992a) "Cysteine 254 of the 73-kDa A Subunit is Responsible for Inhibition of the Coated Vesicle (H$^+$)-ATPase Upon Modification by Sulfhydryl Reagents" J.Biol.Chem. 267, 5817-5822.

Feng,Y. and Forgac,M. (1992b) "A Novel Mechanism for Regulation of Vacuolar Acidification" J.Biol.Chem. 267, 19769-19772.

Forgac,M. (1989) "Structure and Function of the Vacuolar Class of ATP-Driven Proton Pumps" Physiol.Rev. 69, 765-796.

Forgac,M. (1992) "Structure and Properties of the Coated Vesicle (H$^+$)-ATPase" J.Bioenerg.Biomemb. 24, 341-350.

Forgac,M. (1992) "Structure, Function and Regulation of the Coated Vesicle V-ATPase" J.Exp.Biol. 172, 155-169.

Forgac,M., Cantley,L., Wiedenmann,B., Altstiel,L. and Branton,D. (1983) "Clathrin-Coated Vesicles Contain an ATP-Dependent Proton Pump" Proc.Natl. Acad.Sci. 80, 1300-1303.

Marquez-Sterling,N., Herman,I.M., Pesecreta,T., Arai,H., Terres,G. and Forgac,M. (1991) "Immunolocalization of the Vacuolar-Type (H$^+$)-ATPase from Clathrin-Coated Vesicles" Eur.J.Cell Biol. 56, 19-33.

Mulberg,A.E., Tulk,B.M. and Forgac,M. (1991) "Modulation of Coated Vesicle Chloride Channel Activity and Acidification by Reversible Protein Kinase A-Dependent Phosphorylation" J.Biol.Chem. 266 20590-20593.

Myers,M. and Forgac,M. (1993a) "The Coated Vesicle Vacuolar (H$^+$)-ATPase Associates with and Is Phosphorylated by the 50 kDa Polypeptide of the Clathrin Assembly Protein AP-2" J.Biol.Chem. 268, 9184-9186.

Myers,M. and Forgac,M. (1993b) "Assembly of the Peripheral Domain of the Bovine Vacuolar H$^+$-ATPase" J.Cell.Physiol. 156, 35-42.

Pearse,B.M.F. and Robinson,M.S. (1990) Ann.Rev.Cell Biol. 6, 151-171.

Puopolo,K. and Forgac,M. (1990) "Functional Reassembly of the Coated Vesicle Proton Pump" J.Biol.Chem. 265, 14836-14841.

Puopolo,K., Kumamoto,C., Adachi,I. and Forgac,M. (1991) "A Single Gene Encodes the Catalytic "A" Subunit of the Bovine Vacuolar H$^+$-ATPase" J.Biol.Chem. 266 24564-24572.

Puopolo,K., Kumamoto,C., Adachi,I., Magner,R. and Forgac,M. (1992a) "Differential Expression of the "B" Subunit of the Vacuolar H$^+$-ATPase in Bovine Tissues" J.Biol.Chem. 267, 3696-3706.

Puopolo,K., Sczekan,M., Magner,R. and Forgac,M. (1992b) "The 40-kDa Subunit Enhances but Is Not Required for Activity of the Coated Vesicle Proton Pump" J.Biol.Chem. 267, 5171-5176.

Zhang,J., Myers,M. and Forgac,M. (1992) "Characterization of the V$_0$ Domain of the Coated Vesicle (H$^+$)-ATPase" J.Biol.Chem. 267, 9773-9778.

Characterization of the Osteoclast Vacuolar H+ ATPase : Biochemistry, Subunit Composition and Cloning of Catalytic Subunits.

Roland Baron, Marcjanna Bartkiewicz, Diptendu Chatterjee, Munmun Chakraborty, Pe'er David, Chris Fabricant and Natividad Hernando
Departments of Cell Biology and Orthopedics,
Yale University School of Medicine,
333 Cedar Street,
New Haven, CT 06510,
USA

Introduction

Bone resorption, a process necessary for bone growth and remodeling as well as bone repair, is performed by the multinucleated osteoclast. Abnormal regulation of bone resorption is the cause of several clinical entities. Osteoclastic bone resorption can be increased and lead to local or systemic bone loss (bone tumors, osteoporosis, osteoarthritis) or genetically deficient as in various types of osteopetrosis.

The osteoclast is a highly motile cell which attaches to, and moves along, the bone surface, mostly at the interface between bone and bone marrow (endosteum). The multinucleated osteoclast, formed by the asynchronous fusion of mononuclear precursors derived from the bone marrow within the monocyte-macrophage lineage, attaches to the mineralized bone matrix by forming a tight ring-like zone of adhesion, the sealing zone. This attachment involves the specific interaction between adhesion molecules of the integrin family in the cell's membrane and RGD-containing bone matrix proteins, some of which may even be secreted by the osteoclast to form an attachment substratum. The space contained inside this ring of attachment and between the osteoclast and the bone matrix constitutes the bone resorbing compartment. The osteoclast synthesizes several proteolytic enzymes, both lysosomal enzymes and metalloproteinases, which are then vectorially transported and secreted into this extracellular bone resorbing compartment.

NATO ASI Series, Vol. H 89
Molecular and Cellular Mechanisms
of H+ Transport
Edited by Barry H. Hirst
© Springer-Verlag Berlin Heidelberg 1994

Simultaneously, the osteoclast lowers the pH of this compartment by extruding protons across the membrane facing the bone matrix, i.e. the ruffled border membrane. The concerted action of the enzymes and the low pH in the bone resorbing compartment leads to the extracellular digestion of the mineral and organic phases of the bone matrix (Baron et al.1985; Blair et al.1989; Baron, 1989).

Although the H+ ATPase present in osteoclast membranes has been reported to be of the classical V-type, sensitive to NEM and Bafilomycin A1, and expressing the classical V_1 subunits A and B (Blair et al.1989; Bekker and Gay, 1990; Vaananen et al.1990), the possibility that subtle differences existed in the properties and structure of mammalian V-ATPases (Nelson and Taiz, 1989; Forgac, 1989; Nelson, 1991; Wang and Gluck, 1990) led us to investigate the possibility that variations in isoforms of the multisubunit vacuolar proton pump(s) may constitute the basis for the differential targeting, properties and regulation of H+-ATPases present in different organelles in the same cell, in different cells or in different organs (Nelson and Taiz, 1989; Forgac, 1989; Wang and Gluck, 1990).

The work performed in our laboratory during the last several years has been aimed at determining whether the H^+ ATPase present in osteoclast membranes and responsible for the acidification of the bone resorbing compartment differed from other V- ATPases such as those present in the kidney and in intracellular organelles. Although this project is still ongoing, we have identified pharmacological and, possibly, structural differences between these proton pumps as well as previously unrecognized biochemical mechanisms that may apply to other members of this family of ATPases.

Proton transport in membranes derived from osteoclasts reveals novel pharmacological properties

Osteoclasts isolated from calcium deficient laying hens (Zambonin Zallone et al.1982) were highly purified and the microsomal fraction was further fractionated, achieving a 10-fold enrichment in plasma membrane markers relative to the initial microsomal preparation . Most of the proton transport system(s) present in this microsomal fraction were derived from osteoclasts. ATP was found to be the only substrate for transport in the vesicles, confirming the negligible level of contamination with endocytic vesicles. High magnification electron microscopy on negatively stained preparations showed high densities of ball-and-stalk structures, similar to those observed in kidney tubule apical membranes which are enriched in V-ATPases (Brown et al.1987), in 30-40% of the osteoclast microsome membranes. These microsomal preparations were used to study the pharmacological properties of proton transport and for immunochemical characterization of the enzyme subunits.

The proton transport assay revealed a sharp and narrow pH dependance, with a pH optimum of 7.4. Analysis of H+ transport as a function of ATP concentration by the methods of Hill and Eadie-Hofstee was performed and demonstrated the presence of a single Km for ATP (800μM) with a Hill coefficient of 0.9, suggesting that only one type of H^+ ATPase was present and consistent with the reported Km for other proton ATPases.

Pharmacologically, inhibitors of V-ATPases inhibited 100 % of the acidification by osteoclast membrane vesicles. The IC_{50} for NEM, DCCD and Bafilomycin A1 were 0.1 μM, 35 μM and 6 nM respectively. The mitochondrial proton ATPase inhibitors, oligomycin, azide and fluoride had no effect. This data confirmed that, as reported by others (Blair et al.1989; Bekker and Gay, 1990; Vaananen et al.1990), the osteoclast proton pump exhibits properties of the vacuolar type, a finding in agreement with the high density of ball and stalk structures observed in electron microscopy on the osteoclast-derived microsomal vesicles used in this assay.

However, and, according to the literature, unlike any other V-ATPase, the OC-ATPase could also be inhibited by vanadate , which blocked 100% of the acidification at a concentration of 1mM, with an k1/2 of 100 μM (Chatterjee et al.1992) and by nitrate at similar concentrations (Chatterjee et al.1993).

Studies on the ATPase activity of the purified enzyme:

Purified osteoclast membranes were used for purification of NEM/vanadate sensitive H+-ATPase by both immunoaffinity column as described (Chatterjee et al.1993) and by glycerol gradient technique as described by Moriyama and Nelson (Moriyama and Nelson, 1987) for the purification of NEM-sensitive H+-ATPase from chromaffin granule membranes. By affinity purification, the osteoclast membrane H+-ATPase specific activity was enriched 85-90 fold compared to that in the total cell homogenate.

The H+-ATPase activity was assayed according to Xie X-S et al (Xie and Stone, 1989). The pH optima of the purified enzyme was found to be in between pH 5.8-6.0, with only one peak. Substrate specificity indicated that ATP was the best substrate for the osteoclast membrane H+-ATPase, and GTP, ITP, CTP or UTP could not significantly support the enzyme activity.

Substrate kinetics study demonstrated only one Km value of 0.75±.07 mM for ATP with a Vmax of 13.8±0.9 µmol/min/mg protein (Fig. 1C).

To further investigate whether NEM/vanadate sensitive H+-ATPase takes active part in osteoclast mediated bone acidification, we reconstituted the purified enzyme and found that H+ transport could be inhibited by 1mM NEM or 1mM sodium orthovanadate with an IC$_{50}$ of about 0.8 µM and 100 µM respectively.

Two dimensional polyacrylamide gel electrophoresis and phosphorylation of the purified enzyme and cloning of the catalytic domain :

Two-dimensional polyacrylamide gel electrophoresis of the purified osteoclast H+-ATPase revealed six major polypeptides of molecular weights 100kD, 63kD, 60kD, 42kD, 31kD and 16kD (Fig. 2A). There were some minor proteins also present in this preparation in the range of 115kD, 70kD and 60kD molecular weights which most probably came from the H+-ATPase of the contaminating cells or organelles.

Screening a chicken osteoclast cDNA library, we have recently isolated a clone encoding a B subunit isoform which had the same immunological properties as the B subunit expressed in bovine and human brain. Analysis of the deduced aminoacid sequence showed >90% identity with other B subunit sequences in the central region of the molecule but both the N- and C-terminal ends were markedly divergent from the *N. crassa* , bovine or human kidney isoform. In contrast, the osteoclast-derived B subunit is closely related to the recently reported human and bovine brain isoform of subunit B (Puopolo et al.1992; Nelson et al.1992). Regarding the A subunit, and despite the immunochemical differences that we previously reported (Chatterjee et al.1992), both PCR and screening chicken and a human osteoclast libraries generated several clones which were identical to the previously reported mammalian A subunits.

As the osteoclast H+-ATPase was found to have vanadate sensitivity, we then investigated whether the catalytic subunit could form a phosphorylated intermediate. Despite the apparent identity between the osteoclast A subunit and A subunits from other sources, our preliminary results indicate that, in osteoclasts but not in brain, the catalytic subunit gets rapidly phosphorylated in a vanadate-, hydroxylamine- and pH-sensitive way, and that this could be chased with cold ATP, all charasteristics of a transient phosphorylated intermediate.

Conclusion

We conclude from our studies that Vacuolar- H^+ ATPases differ in their pharmacology and in their primary structures. Although less sensitive than E1-E2 ATPases, the enzyme that we have purified and reconstituted from osteoclast membranes can be inhibited by 1mM vanadate or nitrate. The catalytic domain of the enzyme may be formed by a classical A subunit and a brain or brain-like B subunit. Unlike the brain enzyme, the A subunit of the osteoclast proton pump may be transiently phosphorylated during ATP hydrolysis and proton transport.

Acknowledgements: This work was supported by a grant from the NIH (AR-41339) to R.B.

References

Baron, R., Neff, L., Louvard, D., and Courtoy, P.J. (1985). Cell-mediated extracellular acidification and bone resorption: Evidence for a low pH in resorbing lacunae and localization of a 100 kD lysosomal membrane protein at the osteoclast ruffled border. J. Cell Biol. 101, 2210-2222.

Baron, R. (1989). Molecular mechanisms of bone resorption by the osteoclast. Anat. Rec. 224, 317-324.

Bekker, P.J. and Gay, C.V. (1990). Biochemical characterization of an electrogenic vacuolar proton pump in purified chicken osteoclast plasma membrane vesicles. J. Bone Min. Res. 5, 569-579.

Blair, H.C., Teitelbaum, S.L., Ghiselli, R., and Gluck, S. (1989). Osteoclastic Bone Resorption by a Polarized Vacuolar Proton Pump. Science 245, 855-857.

Brown, D., Gluck, S., and Hartwig, J. (1987). Structure of the Novel Membrane-coating Material in Proton-secreting Epithelial Cells and Identification as an H^+ATPase. J. Cell Biol. 105, 1637-1648.

Chatterjee, D., Chakraborty, M., Leit, M., Jamsa-Kellokumpu, S., Fuchs, R., and Baron, R. (1992). Sensitivity to vanadate and isoforms of subunits A and B distinguish the osteoclast proton pump from other vacuolar H+ ATPases. Proc. Natl. Acad. Sci. USA 89, 6257-6261.

Chatterjee, D., Neff, L., Chakraborty, M., Fabricant, C., and Baron, R. (1993). Sensitivity to nitrate and other oxyanions further distinguishes the vanadate-sensitive osteoclast proton pump from other vacuolar H+ ATPases. Biochemistry 32, 2808-2812.

Forgac, M. (1989). Structure and function of vacuolar class of ATP-driven proton pumps. Physiol. Rev. 69, 765-796.

Moriyama, Y. and Nelson, N. (1987). The purified ATPase from chromaffin granule membranes is an anion-dependent proton pump. J. Biol. Chem. 262, 9175-9180.

Nelson, N. and Taiz, L. (1989). The evolution of H^+-ATPases. Trends in Biochem. Sci. 14, 113-116.

Nelson, N. (1991). Structure and pharmacology of the proton-ATPases. Trends in Pharm. Sci. 12, 71-75.

Nelson, R.D., Guo, X.-L., Masood, K., Brown, D., Kalkbrenner, M., and Gluck, S. (1992). Selectively amplified expression of an isoform of the vacuolar H^+-ATPase 56-kilodalton subunit in renal intercalated cells. Proc. Natl. Acad. Sci. USA 89, 3541-3545.

Puopolo, K., Kumamoto, C., Adachi, I., Magner, R., and Forgac, M. (1992). Differential Expression of the "B" Subunit of the Vacuolar H^+-ATPase in Bovine Tissues. J. Biol. Chem. 267, 3696-3706.

Vaananen, H.K., Karhukorpi, E.K., Sundquist, K., Wallmark, B., Roininen, I., Hentunen, T., Tuukkanen, J., and Lakkakorpi, p. (1990). Evidence for the presence of a proton pump of the vacuolar H+-ATPase type in the ruffled border of osteoclasts. J. Cell Biol. 111, 1305-1311.

Wang, Z.-Q. and Gluck, S. (1990). Isolation and properties of bovine kidney brush border vacuolar H^+-ATPase. A proton pump with enzymatic and structural differences from kidney microsomal H^+-ATPase. J Biol. Chem. 265, 21957-21965.

Xie, X.S. and Stone, D.K. (1989). Isolation and reconstitution of the clathrin-coated vesicle proton translocating complex. J. Biol. Chem. 261, 2492-2495.

Zambonin Zallone, A., Teti, A., and Primavera, M.V. (1982). Isolated osteoclasts in primary culture: First observations on structure and survival in culture media. Anat. Embryol. 165, 405-413.

The Renal H-K-ATPase: Function and Expression

Charles S. Wingo[1,2], Xiaoming Zhou[2], Adam Smolka[3], Kirsten Madsen[1], C. Craig Tisher[1], Kevin A. Curran[1], W. Grady Campbell[4], Brian D. Cain[4]
Division of Nephrology,
  Hypertension and Transplantation[1]
University of Florida
Gainesville, Fl 32610-0224
USA

Evidence from several laboratories indicates that the collecting duct (CD) possesses a mechanism of active potassium reabsorption coupled to apical proton secretion. This process exhibits many of the characteristics of proton and potassium transport observed in the parietal cells of the gastric mucosa, and the accumulated evidence is consistent with the hypothesis that this process is due to a proton-potassium-activated adenosinetriphosphatase (an H-K-ATPase).

The direct and indirect evidence that supports the presence of an H-K-ATPase in the CD come from multiple lines of investigation. These include anatomic changes in the CD during K depletion (9,17,18). Such changes are particularly prominent in the intercalated cells of the outer medullary CD (OMCD) (9). Physiological evidence also indicates that the OMCD exhibits net potassium absorption during potassium restriction (1,10,20), and biochemical evidence demonstrates that the CD possesses a potassium-stimulated ATPase (K-ATPase) activity (3,7,8). This enzymatic activity exhibits pharmacologic characteristics of the gastric H-K-ATPase, and is stimulated several fold during potassium depletion (7). These facts suggest that K-ATPase represents the enzymatic activity of an H-K-ATPase. Immunohistochemical evidence further suggests that the protein involved in proton and potassium transport in the CD is similar to the H-K-ATPase of the parietal cells (23). Transport studies also demonstrate that proton and potassium transport in the CD are inhibited by several H-K-ATPase inhibitors (5,21,22,24-26). Specifically, four structurally dissimilar inhibitors of the gastric H-K-

---

[2]Department of Physiology, University of Florida, Gainesville, FL 32610-0274
[3]Division of Gastroenterology, Medical University of South Carolina, Charleston, SC 29425-2203
[4]Department of Biochemistry and Molecular Biology, University of Florida, Gainesville, FL 32610-0245

NATO ASI Series, Vol. H 89
Molecular and Cellular Mechanisms
of H+ Transport
Edited by Barry H. Hirst
© Springer-Verlag Berlin Heidelberg 1994

ATPase substantially decrease proton and K/Rb flux in the collecting duct or renal medullary membrane vesicles. Additional studies in individual intercalated cells demonstrate a potassium-dependent pH recovery that is inhibited by the gastric H-K-ATPase inhibitor SCH 28080 (16). Finally, studies utilizing the tools of molecular biology have reported the presence of mRNA for both α and β subunits of the gastric H-K-ATPase or the putative colonic H-K-ATPase in the kidney (4,6,15). Thus, evidence is beginning to emerge that more than one H-K-ATPase α subunit may be expressed in the kidney. In the present report we shall focus on three particular considerations. First, we shall address how alterations in *in vitro* ambient pCO_2 appear to be an important stimulus for activation of the renal H-K-ATPase and the mechanism by which this occurs. Second, we shall review recent immunohistochemistry using site-directed antibodies to the α subunit of hog gastric H-K-ATPase. Third, we shall review the molecular biological evidence which demonstrates the presence of a gastric-like H-K-ATPase β subunit in the kidney.

Regarding the physiological studies, an increase in the partial pressure of CO_2 (pCO_2) markedly stimulates Rb/K absorption by a related H-K-ATPase in the colonic epithelium (13). Moreover, McKinney and Davidson (12) demonstrated that increasing pCO_2 dramatically stimulates the rate of luminal acidification in the isolated perfused cortical collecting duct (CCD). In the CD and other tight epithelia, an increase in pCO_2 results in the exocytotic insertion of subapical vesicles into the luminal membrane(2,11). This process is dependent on changes in intracellular calcium concentration, requires the presence of functional calmodulin, and involves alteration in the cytoskeleton. Similarly, McKinney and Davidson demonstrated that the stimulation of acidification by pCO_2 was abolished by maneuvers that prevented rapid changes in intracellular calcium, inhibited calmodulin or inhibited tubulin polymerization (12). Because alterations in pCO_2 appeared to affect acidification in the CD and have clear implications to our understanding of renal acid-base and potassium homeostasis, we conducted a series of studies designed to examine whether the effect of alterations in ambient pCO_2 might be related to changes in the activity of the renal H-K-ATPase in the isolated perfused CCD of K-restricted rabbits. Potassium restriction was used to enhance the component of potassium absorption mediated by H-K-ATPase. All animals were conditioned to a low potassium diet, (Teklad #87433, Madison WI) for 96 hours prior to sacrifice. We addressed the following questions. Firstly, do alterations in ambient pCO_2 enhance potassium absorption? Isotopic ^{86}Rb efflux was used as a qualitative marker of unidirectional potassium lumen-to-bath flux. Secondly, if increasing pCO_2 enhances Rb absorption, is this enhanced Rb efflux attributable to an H-K-ATPase? Thirdly, if increasing pCO_2 enhances Rb efflux, is the effect of pCO_2 on Rb efflux dependent on either alterations in cytosolic calcium activity or the cytoskeleton.

Tubules were initially bathed with solutions that were equilibrated with 5% CO_2, (pH 7.4±0.03) and then were exposed to an identical solution equilibrated with 10% CO_2, (pH 7.1±0.04). Exposure to 10% CO_2 increased Rb efflux in the CCD of these K-restricted rabbits and the stimulation was apparent at the earliest time point measured, 30-60 minutes. To examine whether this enhanced Rb efflux was mediated by an H-K-ATPase, we repeated these studies in the presence of 10 μM luminal SCH 28080, a specific and potent gastric H-K-ATPase inhibitor. Under identical conditions to the previous study, SCH 28080 totally abolished the stimulatory effect of 10% CO_2 to enhance Rb efflux, indicating that the enhancement of Rb efflux induced by 10% CO_2 was dependent on an H-K-ATPase.

The effect of 10% CO_2 to stimulate proton secretion in both the turtle urinary bladder and the CD is dependent on alterations in intracellular calcium activity (2,12,19). Moreover, the effect of 10% CO_2 on acidification of the CD is dependent on calmodulin and a functional cytoskeleton. Thus, we examined the response of 10% CO_2 on Rb efflux in the presence of maneuvers that are known to attenuate changes in intracellular calcium, or inhibit calmodulin or tubulin polymerization.

When the tubules were perfused in the presence of the intracellular calcium buffer, 0.5 μM 1,2-bis-(2-amino-5-methyl-phenoxy)-ethane-N-N-N'-N'-tetraacetic acid tetraacetoxymethyl ester (MAPTAM, Sigma St. Louis, MO), the response to 10% CO_2 was abolished, suggesting that alterations in intracellular calcium were necessary for stimulation of H-K-ATPase-mediated Rb efflux. When tubules were exposure to 10% CO_2 solutions in the presence of 0.5 μM N-(6-aminohexyl)-5-chloro-1-naphthalene-sulfonamide (W7, Sigma, St. Louis, MO), an inhibitor of calmodulin, the stimulatory effect of 10% CO_2 on Rb efflux was also fully abolished. In fact, 10% CO_2 actually significantly inhibited Rb efflux. These facts suggest that the process of activation of H-K-ATPase-mediated Rb efflux depends on the presence of functional calmodulin.

In the CD the enhancement in acid secretion by 10% CO_2 can be abolished by agents which inhibit tubulin polymerization (19). To investigate whether the activation of H-K-ATPase mediated Rb efflux requires the presence of an intact cytoskeleton and a functional microtubule system, the CCD was exposed simultaneously to 10% CO_2 and colchicine, an inhibitor of microtubule polymerization. The stimulatory effect of 10% CO_2 on Rb efflux was abolished by the simultaneous addition of 0.5 mM colchicine. However, if Rb efflux was stimulated first by 10% CO_2, the subsequent addition of colchicine (0.5 mM) had no inhibitory effect. These data indicate that the failure to increase Rb efflux upon simultaneous exposure to 10% CO_2 and colchicine was not due to opposing effects of colchicine and 10% CO_2 on Rb efflux. The most consistent interpretation of these data is that 10% CO_2 results in a cascade of events initiated by alterations in intracellular pH which result in changes in intracellular calcium activity (Ca_i). These changes in Ca_i lead to an activation of calmodulin and perhaps other intracellular calcium-dependent events. These calcium-dependent events appear to involve changes in the cytoskeleton with polymerization of microtubules, a process which is inhibited by colchicine. This entire cascade is consistent with morphologic findings that acute acidosis results in the insertion of tubular-vesicular profiles from the subapical region into the luminal membrane by a process of exocytosis(2,11).

Although these studies indicate the functional importance of an H-K-ATPase on K/Rb absorption in the collecting duct, they do not establish whether the mechanism is mediated by a transport system that is molecularly related to the gastric H-K-ATPase. However, immunohistochemical studies demonstrate that intercalated cells of the CD possess an antigen that cross-reacts with a site-directed antibody against the amino terminus of the α subunit of the hog gastric H-K-ATPase. To determine whether the molecular structure of the renal H-K-ATPase exhibits homology to the gastric H-K-ATPase, we used a molecular biological approach.

The renal H-K-ATPase β subunit was selected as the primary target of our work because of the suggestion that β subunits were likely to serve regulatory roles in the P-type ATP-driven ion pumps. Furthermore, the H-K-ATPase β subunit has greater divergence from Na-K-ATPase β subunits (approximately 40% homology at the nucleotide level) than the α subunit (approximately 60% homology at the nucleotide level), suggesting that β subunit cDNA would make a more specific probe for long term studies, such as *in situ* hybridization. Na-K-ATPase is present in high abundance in the kidney and is a possible source of difficulty in all molecular studies of the renal H-K-ATPase.

The strategy involved polymerase chain reaction (PCR) amplification of an internal segment of the coding sequence of the ß subunit. Two separate sources were studied as templates for PCR: Total rabbit renal RNA and rabbit medullary cDNA library (Stratagene Cloning Systems). Using the cDNA template, a single 570 base pair (bp) fragment was amplified. Concerns about PCR artifacts led us to repeat the experiment using a separate set of reagents and pipettes from a separate laboratory with a fresh aliquot of template. The 570 bp product could be reproducibly amplified. This 570 bp PCR product was cloned into the plasmid TA 2000 (Invitrogen) and the nucleotide sequence determined. The nucleotide sequence was found to be identical to the target internal segment of the gastric H-K-ATPase ß subunit reported by Sachs and colleagues (14). We have since used the 570 bp product as a probe to screen a rabbit medullary cDNA library and a likely complete cDNA clone of renal H-K-ATPase ß subunit has been identified.

The 570 bp PCR product was also used as a probe to examine mRNA levels for the renal H-K-ATPase ß subunit. Because H-K-ATPase is important in potassium conservation and the activity of this enzyme increases with potassium depletion, we anticipated that mRNA levels for the ß subunit might increase with potassium depletion. For this reason, we examined the effect of a normal and a low-K diet on ß subunit steady state mRNA levels. Animals were adapted to a low potassium diet for seven days, at which time total RNA was prepared from stomach and kidney. The RNA was size-fractionated by argrose gel electrophoresis and transferred to hybridization membranes. The blots were then probed with the [^{32}P] 570 bp PCR product labelled by the random priming method.

Thirty μg of total RNA was run in each lane and loading checked by ethidium bromide staining. The initial experiments were very encouraging. In two rabbits, one adapted to a normal K diet and one adapted to a low-K diet, there appeared to be an increase in signal intensity from kidney in the animal adapted to the low-K diet (Figure 1). Equally important, the signal was clearly present under normal as well as low-K conditions. It should be emphasized that the electrophoretic mobility of the renal signal was identical to the gastric signal. The gastric signals were developed at 3 hours whereas the renal signals were developed at 96 hours.

However, when we repeated the experiment with three additional animals in each group, the results were not as clear cut (Figure 2). In fact, at present, we cannot conclude that a low-K diet increases H-K-ATPase ß subunit mRNA in the kidney. The surprise observation was the effect of potassium depletion on the stomach H-K-ATPase ß subunit mRNA (Figure 3). In all animals studied there was a clear and precipitous fall in stomach H-K-ATPase ß subunit steady-state mRNA levels in response to potassium depletion.

Thus, the present studies have identified a 570 bp cDNA from kidney that is identical to the ß subunit of the gastric H-K-ATPase. These observations constitute very strong evidence for a gastric-like H-K-ATPase in the kidney. Moreover, these studies suggest that this message is present in the kidney regardless of the dietary potassium content. Finally, these studies indicate that mRNA levels for the H-K-ATPase ß subunit in the stomach are sensitive to dietary potassium content. Clearly, the H-K-ATPase ß subunit in the kidney is not under the same regulatory control as that observed in the stomach.

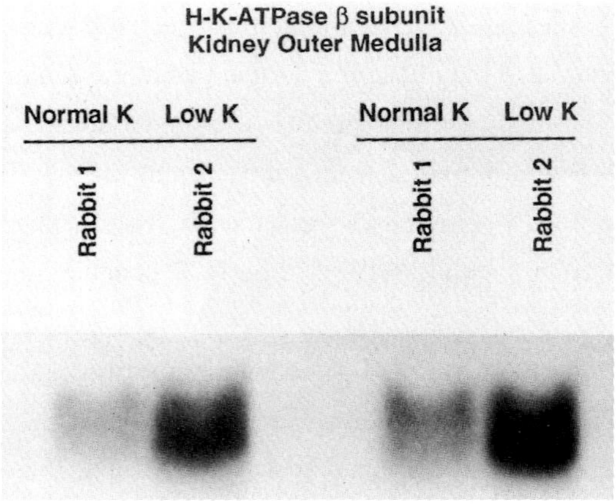

Figure 1. Northern analysis of H-K-ATPase ß subunit mRNA in kidney outer medulla using [^{32}P] labelled 570 bp PCR product. Analysis of the stomach demonstrated a signal of identical electrophoretic mobility (Figure 3). Autoradiogram exposure was 96 hours.

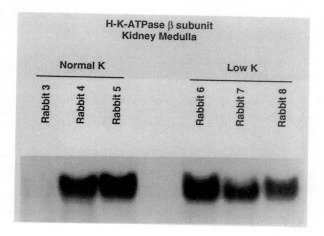

Figure 2. Northern analysis of H-K-ATPase ß subunit mRNA in kidney medulla using [^{32}P] labelled 570 bp PCR product. Autoradiogram exposure was 96 hours.

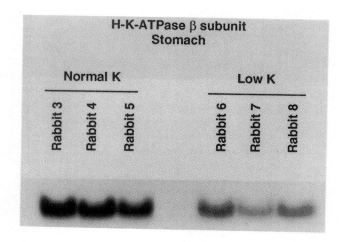

Figure 3. Northern analysis of H-K-ATPase ß subunit mRNA in stomach using [^{32}P] labelled 570 bp PCR product. Autoradiogram exposure was 3 hours.

References

Backman, K.A. and J.P. Hayslett. 1983. Role of the medullary collecting duct in potassium conservation. *Pflugers Arch.* 396:297-300.

Cannon, C., J. Van Adelsberg, S. Kelly, and Q. Al-Awqati. 1985. Carbon-dioxide-induced exocytotic insertion of H+ pumps in turtle-bladder luminal membrane: Role of cell pH and calcium. *Nature* 314:443-446.

Cheval, L., C. Barlet-Bas, C. Khadouri, E. Feraille, S. Marsy, and A. Doucet. 1991. K+-ATPase-mediated Rb+ transport in rat collecting tubule: modulation during K+ deprivation. *Am. J. Physiol. Renal,Fluid Electrolyte Physiol.* 29:F800-F805.

Crowson, M.S. and G.E. Shull. 1992. Isolation and characterization of a cDNA encoding the putative distal colon H+,K+-ATPase. Similarity of deduced amino acid sequence to gastric H+,K+-ATPase and Na+,K+-ATPase and mRNA expression in distal colon, kidney, and uterus. *J. Biol. Chem.* 267:13740-13748.

Curran, K.A., M.J. Hebert, B.D. Cain, and C.S. Wingo. 1992. Evidence for the presence of a K-dependent acidifying adenosine triphosphatase in the rabbit renal medulla. *Kidney Int.* 42:1093-1098.

Curran, K.A., I.K. Hornstra, R. Cruz, L.G. McMahon, C.S. Wingo, and B.D. Cain. 1992. Partial cDNA clone of an apparent renal H-K-ATPase ß subunit. *J. Am. Soc. Nephrol.* 3:775. Abstract.

Doucet, A. and S. Marsy. 1987. Characterization of K-ATPase activity in distal nephron: Stimulation by potassium depletion. *Am. J. Physiol. Renal,Fluid Electrolyte Physiol.* 253:F418-F423.

Garg, L.C. and N. Narang. 1988. Ouabain-insensitive K-adenosine triphosphatase in distal nephron segments of the rabbit. *J. Clin. Invest.* 81(4):1204-1208.

Hansen, G.T., C.C. Tisher, and R.R. Robinson. 1980. Response of the collecting duct to disturbances of acid base and potassium balance. *Kidney Int.* 17:326-337.

Linas, S.L., L.N. Peterson, R.J. Anderson, G.A. Aisenbery, F.R. Simon, and T. Berl. 1979. Mechanism of renal potassium conservation in the rat. *Kidney Int.* 15:601-611.

Madsen, K.M. and C.C. Tisher. 1983. Cellular response to acute respiratory acidosis in rat medullary collecting duct. *Am. J. Physiol. Renal,Fluid Electrolyte Physiol.* 245:F670-F679.

McKinney, T.D. and K.K. Davidson. 1988. Effects of respiratory acidosis on HCO_3^- transport by rabbit collecting tubules. *Am. J. Physiol. Renal,Fluid Electrolyte Physiol.* 255:F656-F665.

Perrone, R.D. and D.E. McBride. 1988. Aldosterone and pCO_2 enhance rubidium absorption in rat distal colon. *Am. J. Physiol. Gastrointest. Liver Physiol.* 17:G898-G906.

Reuben, M.A., L.S. Lasater, and G. Sachs. 1990. Characterization of a ß subunit of the gastric H+/K+-transporting ATPase. *Proc. Natl. Acad. Sci. U. S. A.* 87:6767-6771.

Reuben, M.A., F.L. Starr, S. Birmingham, G. Sachs, and J.A. Kraut. 1993. Characterization of a renal H,K ATPase alpha subunit mRNA found to be identical with gastric H,K ATPase mRNA. *Gastroenterology* 104:A177. Abstract.

Silver, R.B. and G. Frindt. 1993. Functional identification of H-K-ATPase in intercalated cells of cortical collecting tubule. *Am. J. Physiol. Renal,Fluid Electrolyte Physiol.* 264:F259-F266.

Spargo, B. 1954. Kidney changes in hypokalemic alkalosis in the rat. *J. Lab. Clin. Med.* 43:802-814.

Stetson, D.L., J.B. Wade, and G. Giebisch. 1980. Morphologic alterations in the rat medullary collecting duct following potassium depletion. *Kidney Int.* 17:45-56.

Van Adelsberg, J. and Q. Al-Awqati. 1986. Regulation of cell pH by Ca^{+2}-mediated exocytotic insertion of H+-ATPase. *J. Cell Biol.* 102:1638-1645.

Wingo, C.S. 1987. Potassium transport by the medullary collecting tubule of rabbit: effects of variation in K intake. *Am. J. Physiol. Renal,Fluid Electrolyte Physiol.* 22:F1136-F1141.

Wingo, C.S. 1989. Active proton secretion and potassium absorption in the rabbit outer medullary collecting duct - functional evidence for proton-potassium activated adenosine triphosphatase. *J. Clin. Invest.* 84:361-365.

Wingo, C.S. and F.E. Armitage. 1992. Rubidium absorption and proton secretion by rabbit outer medullary collecting duct via H-K-ATPase. *Am. J. Physiol. Renal,Fluid Electrolyte Physiol.* 263:F849-F857.

Wingo, C.S., K.M. Madsen, A. Smolka, and C.C. Tisher. 1990. H-K-ATPase immunoreactivity in cortical and outer medullary collecting duct. *Kidney Int.* 38:985-990.

Zhou, X. and C.S. Wingo. 1992. Mechanisms for enhancement of Rb efflux by 10% CO_2 in cortical collecting duct (CCD). *Clin. Research* 40:179A. Abstract.

Zhou, X. and C.S. Wingo. 1992. Mechanisms of rubidium permeation by rabbit cortical collecting duct during potassium restriction. *Am. J. Physiol. Renal,Fluid Electrolyte Physiol.* 263:F1134-F1141.

Zhou, X. and C.S. Wingo. 1992. H-K-ATPase enhancement of Rb efflux by cortical collecting duct. *Am. J. Physiol. Renal,Fluid Electrolyte Physiol.* 32:F43-F48.

Apical K$^+$-ATPase and Active Potassium Absorption in the Distal Colon

Henry J. Binder
Department of Internal Medicine
Yale University
333 Cedar Street
New Haven, CT   06510
U.S.A.

Na,K-ATPase and gastric parietal cell H,K-ATPase are two P-type ATPases that have been extensively studied.   P-type ATPases, dimers with distinct α and β subunits, are inhibited by ortho-vanadate.   During the past several years isoforms of both the α and β subunits of Na,K-ATPase and H,K-ATPase have been isolated and sequenced and, at least for the α subunit of Na,K-ATPase, there have been extensive structure-function relationship studies (Lingrel et al, 1990; Horisberger et al, 1991).   Na,K-ATPase is essential for both epithelial and non-epithelial cell function by maintaining a low intracellular [Na] and a high intracellular [K] as a result of extrusion of Na in exchange for K.   In most epithelial cells Na,K-ATPase is located on the basolateral membrane and until recently ouabain had been considered a specific inhibitor.   Na,K-ATPase is not inhibited by either omeprazole or SCH 28080, inhibitors of gastric acid secretion and gastric H,K-ATPase which, in contrast, is located on the apical membrane of parietal cells.   Gastric H,K-ATPase functions as a K-H exchange and is inhibited by omeprazole and SCH 28080, but not by ouabain (Wallmark et al, 1987). Despite these pharmacological differences, comparison of its amino acid sequence to that of the Na pump reveals partial (60%) homology (Shull & Lingrel, 1986).

Potassium absorption and proton secretion have been observed in the distal colon and distal cortical collecting ducts, and K-H exchanges have been proposed to explain these physiological events (Foster, et al, 1984; Suzuki & Kaneko, 1987; Wingo, 1989). Since in rabbit cortical collecting duct proton secretion, K

NATO ASI Series, Vol. H 89
Molecular and Cellular Mechanisms
of H$^+$ Transport
Edited by Barry H. Hirst
© Springer-Verlag Berlin Heidelberg 1994

absorption and K-activated ATPase activity are not affected by ouabain (Zhou & Wingo, 1992), but are inhibited by omeprazole and/or SCH 28080, it has been anticipated that the renal H,K-ATPase is closely related to the gastric H,K-ATPase. Further, polyclonal antibodies raised to unique peptide segments of the gastric H,K-ATPase amino acid sequence immunoblot with renal membranes suggesting at least partial homology between these two ATPases (Okusa et al, manuscript in preparation). To date, the renal H,K-ATPase has not been cloned so that its exact structure is not known. Potassium absorption, proton secretion and K-ATPase activity also occur in the mammalian distal (but not proximal) colon (Foster et al, 1984; Suzuki & Kaneko, 1987; Sweiry & Binder, 1990; del Castillo et al 1991), and this present chapter will summarize the recent information of the function and structure of the colonic K-ATPase/K absorption with particular emphasis on the studies performed in the rat distal colon.

Potassium transport has been extensively studied under voltage clamp conditions in the rat distal colon with ^{42}K and ^{86}Rb (Foster et al, 1984; Perrone & McBride, 1988; Sweiry & Binder, 1990; Pandiyan et al, 1992). The results of these experiments in the rat differ in part from those that have been performed in the rabbit distal colon (McCabe et al 1984; Halm & Frizzell, 1986), and although an overall model of active K absorption is not available for the mammalian colon, recent experiments do permit speculation that is consistent with these many observations. Active K absorption in the rat distal colon can be summarized as electroneutral, Na-independent and partially Cl-dependent (and partially Cl-independent) (Foster et al, 1984; Sweiry & Binder, 1990); these observations are consistent with a K-H exchange. Active K absorption is also inhibited by ortho-vanadate and mucosal ouabain, but not by omeprazole or SCH 28080, and K-activated ATPase activity is enriched in apical membranes of rat distal colon (del Castillo, et al 1991). This K-ATPase activity was inhibited by ortho-vanadate and partially by ouabain, but not by omeprazole and SCH 28080. These inhibitory studies provide strong evidence that active K absorption is energized by a novel apical membrane K-ATPase with properties that overlap both Na,K-ATPase and gastric H,K-ATPase. Such a possibility was reinforced by studies by Okusa et al (manuscript in preparation) in which one of the two polyclonal antibodies to portions of gastric H,K-ATPase identify a 95-100 kDa protein in apical membranes of rat distal colon.

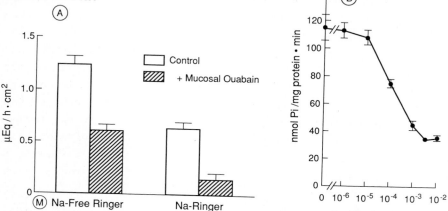

Fig. 1. A: Net Rb(K) absorption in rat distal colon determined under voltage clamp conditions in normal rats. Ouabain was present in serosal bath. One mM ouabain in mucosal bath completely inhibited Rb(K) absorption only in presence of mucosal Na. Therefore, mucosal Na-insensitive component is mucosal ouabain-sensitive and mucosal Na-sensitive component is mucosal ouabain-insensitive (from Pandiyan et al, 1992). B: K-ATPase activity in apical membranes prepared from rat distal colon. This study reveals a ouabain-sensitive and a ouabain-insensitive component (from del Castillo et al, 1991).

Several recent observations provide strong evidence, however, that there are **two** active K absorption/K-ATPase mechanisms in the rat distal colon. First, although K absorption is Na-independent, removal of mucosal Na results in an increase in active K absorption. As a result, K absorption has both mucosal Na-sensitive and mucosal Na-insensitive fractions (Pandiyan et al, 1992). Second, the effect of mucosal ouabain on K absorption is a function of mucosal Na. That is, mucosal ouabain completely inhibits K absorption in the presence of mucosal Na, i.e. ouabain inhibits the mucosal Na-insensitive fraction, but does not affect the mucosal Na-sensitive fraction (Fig 1A) (Pandiyan et al, 1992). Third, apical K-ATPase activity is partially inhibited by ouabain such that there are also ouabain-sensitive and ouabain-insensitive fractions of K-ATPase activity (Fig 1B) (del Castillo, 1991). Fourth, aldosterone induces active K secretion and inhibition of aldosterone-induced K secretion unmasks a marked enhancement of active K absorption. Thus, aldosterone stimulates both active K absorption and secretion. When aldosterone-stimulated active K absorption is analyzed with regard to the mucosal Na-sensitive and Na-insensitive components, aldosterone enhances the mucosal Na-sensitive component, not the mucosal Na-insensitive component (Fig 2) (Pandiyan et al, 1992). These experimental observations provide compelling evidence that there are *two* active K absorption/K-ATPase fractions, at least in the rat distal colon.

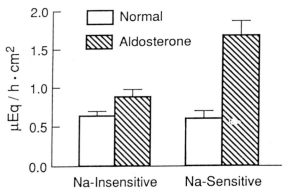

Fig. 2. Net Rb(K) absorption in rat distal colon determined under voltage clamp conditions in normal and aldosterone-treated rats. Ouabain was present in serosal bath; mucosal Na-insensitive Rb(K) absorption is that observed in presence of 125 mM Na, while mucosal Na-sensitive Rb(K) transport is the difference between that observed at 125 mM and 0 mM Na. (from Pandiyan et al, 1992).

Evidence of partial Cl dependence of active K absorption was also demonstrated in the transport studies. Since these results clearly provide demonstration of Cl-independence, we propose a model (Fig 3) in which Cl-dependency of K absorption represents the coupling of apical K-H exchange to an apical $Cl-HCO_3$ exchange or to a basolateral electroneutral K-Cl cotransport mechanism or both. Specific experiments are requried to test these possibilities. Although this model depicts a single K-H exchange, the experimental observations summarized above supports the presence of _two_ distinct K transport systems. It is not known whether both K transport processes are K-H exchanges.

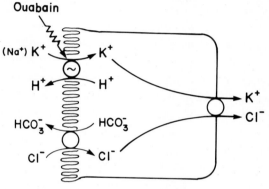

Fig. 3 - Model of active K absorption in rat distal colon. Model is based on recent studies (Foster et al, 1986; Sweiry & Binder, 1990; Pandiyan et al, 1992): ouabain inhibits K absorption only in presence of mucosal Na, and Na is a competitive inhibitor of K transport. Mechanism of Cl-dependence is speculative.

Potassium transport, proton secretion and K-activated ATPase activities have also been described in the distal colon of both rabbit and guinea pig. In the guinea pig Suzuki & Kaneko (1987, 1989) have studied mucosal acidification which is mucosal K-dependent, but not mucosal Na-dependent and is inhibited by mucosal ouabain. These results in the guinea pig are quite similar to those in the rat distal colon so that it is likely that the K-ATPase in the guinea pig is almost identical to the mucosal ouabain-sensitive form of the rat. In contrast, in the distal colon of the rabbit active K absorption is <u>not</u> inhibited by mucosal ouabain and in microsome-derived membranes K-dependent proton transport is inhibited by both vanadate and SCH 28080 (Wills & Biagi, 1982; Kaunitz & Sachs, 1986). An interesting speculation is that only the mucosal ouabain-insensitive fraction of the rat distal colon is expressed in the rabbit distal colon.

During the past year two different groups have reported the isolation and sequencing of a putative colonic K-ATPase from rat distal colon (Crowson & Shull, 1992; Jaisser et al, 1993). Crowson & Shull (1992) isolated a cDNA that encoded a polypeptide of 1036 amino acids with 63% homology to both Na,K-ATPase and gastric H,K-ATPase; 50% identity exists between these three ATPases. Northern blot analysis with the 3' untranslated fraction revealed message primarily in distal colon with lower levels in kidney and uterus, but not in the stomach. Using an entirely different cloning strategy Jaisser et al (1993) isolated a cDNA with an identical sequence. These investigators performed in situ hybridization localization and found this putative K-ATPase transcript restricted to the surface epithelial (and the upper 20% of the cells lining the colonic gland) cells of the distal colon. Three different approaches: Northern blot analysis, RNAse protection assay and in situ hybridization did <u>not</u> identify the colonic transcript in the kidney. Reasons for these differences are not evident. Based on the experimental data that two distinct K-ATPases are present in rat distal colon it is anticipated that a cDNA for a second isoform remains to be isolated and sequenced.

Aldosterone has several effects on ion transport in the rat distal colon (Binder & Rajendran, 1991): Besides induction of amiloride-sensitive electrogenic Na absorption and inhibition of electroneutral Na absorption, aldosterone stimulates net K secretion as a result of both enhanced potential-dependent K secretion and an induction of active K secretion via apical K channels. As noted above, inhibition of active K secretion (either with apical K channel blockers or serosal addition of bumetanide or ouabain) unmasks an accelerated rate of active K absorption (Sweiry & Binder, 1989). Aldosterone both quantitatively and qualitatively alters active K absorption as aldosterone both enhances the mucosal Na-sensitive component (Fig. 2) and alters the kinetics of Na inhibition of K absorption from competitive to non-competitive inhibition. This increase is accompanied by a 60% increase in apical K-ATPase activity (del Castillo & Binder, unpublished observations). Preliminary Northern blot analysis with the 3' portion of the putative K-ATPase cDNA revealed that aldosterone increased its expression (Lee & Binder, unpublished observations). Since aldosterone stimulated the mucosal Na-sensitive component and since we believe that component is also the mucosal ouabain-insensitive component, we speculate that the cDNA that has been cloned represents the mucosal ouabain-insensitive component and that the cDNA for mucosal ouabain-sensitive component remains to be isolated and sequenced.

What is the function of this apical K-ATPase and putative K-H exchange? It is highly likely that active K absorption is energized by this apical K-ATPase, but it is also possible that this K-ATPase has a role in the regulation of intracellular pH (pHi). Apical Na-H exchange in the rat distal colon appears to regulate pHi, at least in part. As aldosterone both enhances active K absorption and down regulates apical Na-H exchange, we speculate that aldosterone may convert an apical Na-dependent acid extrusion mechanism to a K-dependent process. It should be noted that recent studies in Caco-2 cells indicate that pHi is regulated by both Na-H and K-H exchanges (Abrahamse et al, 1992). Specific experiments of pHi are required to test this possibility.

Acknowledgements: These studies were supported in part by USPHS Research Grant DK 18777 from the National Institute of Diabetes, Digestive and Kidney Diseases.

References:
Abrahamse SL, Bindels RJM, vanOs CH (1992) The colon carcinoma cell line Caco-2 contains an H^+/K^+-ATPase that contributes to intracellular pH regulation. Pflugers Archiv 421:591-597
Binder HJ, Rajendran VM (1991) Aldosterone alters active Na, Cl and K transport in rat proximal and distal colon. In: Aldosterone: Fundamental Aspects. Ed. JP Bonvalet, et al Colloque INSERM John Libbey Eurotext Ltd., Paris pp 229-238
Crowson MS, Shull G (1992) Isolation and characterization of a cDNA encoding the putative distal colon H^+,K^+-ATPase. J Biol Chem 267:13740-13748
del Castillo JR, Rajendran VM, Binder HJ (1991) K^+-activated ATPase activities in rat distal colon: Apical membrane localization of ouabain-sensitive and ouabain-insensitive components. Am J Physiol 261:G1005-G1011
Foster ES, Hayslett JP, Binder HJ (1984) Mechanism of active potassium absorption and secretin in the rat colon. Am J Physiol 246:G611-G617
Halm DR, Frizzell RA (1986) Active K transport across rabbit distal colon: relation to Na absorption and Cl secretion. Am J Physiol 251:C252-C267
Horisberger J, Lemas D, Kraehenbuhl V, Rossier BC (1991) Structure-function of Na,K-ATPase. Ann Rev Physiol 53:565-584
Jaisser F, Coutry N, Farman N, Binder HJ, Rossier BC (1993) A putative H,K-ATPase is selectively expressed in surface epithelial cells of rat distal colon. Am J Physiol (in press)
Kaunitz JD, Sachs G (1986) Identification of a vanadate sensitive potassium-dependent proton pump from rabbit colon. J Biol Chem 261:14005-14010
Lingrel JB, Orlowski J, Shull MM, Price EM (1990) Molecular genetics of Na,K-ATPase. Progress in Nucleic Acid Research 38:37-89
McCabe RD, Smith PL, Sullivan LP (1984) Potassium transport by rabbit descending colon. Am J Physiol 251:C252-C267
Pandiyan V, Rajendran VM, Binder HJ (1992) Mucosal ouabain and Na^+ inhibit active $Rb^+(K^+)$ absorption in normal and sodium-depleted rat distal colon Gastroenterology 102:1846-1853
Perrone RD, McBride DE (1988) Aldosterone and Pco_2 enhance rubidium absorption in rat distal colon. Am J Physiol 254:G898-G906
Shull GE, Lingrel JB (1986) Molecular cloning of the rat stomach $(H^+ + K^+)$-ATPase. J Biol Chem 261:16788-16791

Suzuki Y, Kaneko K (1987) Acid secretion in isolated guinea pig colon. Am J Physiol 253:G155-G164

Suzuki Y, Kaneko K (1989) Ouabain-sensitive H^+-K^+ exchange mechanism in the apical membrane of guinea pig colon. Am J Physiol 256:G979-G988

Sweiry JH, Binder HJ (1989) Characterization of aldosterone-induced potassium secretion in rat distal colon. J Clin Invest 83:844-851

Sweiry JH, Binder HJ (1990) Active potassium absorption in rat distal colon. J Physiol (London) 423:155-170

Wallmark B, Briving C, Fryklund J, Munson K, Jackson R, Mendelein J, Rabon E, Sachs G (1987) Inhibition of gastric H^+-K^+-ATPase and acid secretion by SCH 28080, a substituted pyridyl $(1,2\alpha)$-imidazole. J Biol Chem 262:2077-2084

Wills NK, Biagi B (1982) Active potassium transport by rabbit descending colon epithelium. J Membr Biol 64:195-203

Wingo CS (1989) Active proton secretion and potassium absorption in the rabbit outer medullary collecting duct. J Clin Invest 84:361-365

Zhou X, Wingo CS (1992) H-K-ATPase enhancement of Rb efflux by cortical collecting duct. Am J Physiol 263:F43-F48

Amino Acid Residues Involved in Ouabain Sensitivity and Cation Binding

Jerry B Lingrel, James Van Huysse, William O'Brien, Elizabeth Jewell-Motz,
Patrick Schultheis
Department of Molecular Genetics, Biochemistry and Microbiology
University of Cincinnati College of Medicine
231 Bethesda Avenue
Cincinnati, Ohio 45267-0524

The Na,K-ATPase, which is found in the cells of all higher eukaryotes, utilizes ATP to transport Na^+ and K^+ across the cell membrane. For every three sodium ions transported out of the cell two potassium ions are transported in. The enzyme is composed of two subunits, a larger α subunit, which is thought to contain most of the catalytic sites, and a smaller α subunit which is required for the proper processing and maturation of the enzyme. Three isoforms exist for the α subunit ($\alpha1$, $\alpha2$ and $\alpha3$) and two isoforms exist for the β subunit ($\beta1$ and $\beta2$) in mammalian cells (Lingrel et al., 1990; Sweadner, 1989; Takeyasu et al, 1989). An additional isoform, $\beta3$, exists in Xenopous (Good et al., 1990). The $\alpha1$ isoform is found in all cells while $\alpha2$ is the primary isoform found in skeletal muscle, but is also present in the heart and nervous system. Expression of the $\alpha3$ isoform is limited to the heart and nervous system (Orlowski and Lingrel, 1988). The cDNAs and genes corresponding to these isoforms have been isolated and the mechanisms responsible for their differential expression are being investigated (Lingrel et al., 1990). The Na,K-ATPase is a member of the P-type family of ATPases, which also include the Ca-ATPases and the H,K-ATPases. These transport proteins are similar in structure and have in common an aspartyl phosphate intermediate during their catalytic cycle. Site-directed mutagenesis and expression (MacLennan, 1990; Clarke et al., 1990; Vilsen and Andersen, 1992; Andersen and Vilsen, 1992) studies have been carried out to define cation binding sites in the Ca-ATPase but, less information is available using this approach with the Na,K-ATPase. The Na,K- ATPase differs from the other P-type ATPases in that it is sensitive to cardiac glycosides, a class of drugs which are used in the treatment of congestive heart failure and certain arrhythmias. The sensitivity of the enzyme to these drugs provides a useful tool for investigating structure-function relationships. In this report we describe progress made toward identifying potential cation binding sites as well as amino acid residues that are involved in determining cardiac glycoside sensitivity.

NATO ASI Series, Vol. H 89
Molecular and Cellular Mechanisms
of H⁺ Transport
Edited by Barry H. Hirst
© Springer-Verlag Berlin Heidelberg 1994

Identification of Potential Cation Binding Sites

Because the Na,K-ATPase is required for growth and since this enzyme can be inhibited by cardiac glycosides, it is relatively simple to use selection with these drugs as a tool for determining which amino acids of the α subunit are required for function. Since the α subunit is responsible for cardiac glycoside sensitivity (Kent et al., 1987; Price and Lingrel, 1988) and since HeLa cells express a ouabain sensitive α1 isoform, these cells are killed by 1 µM ouabain (a cardiac glycoside used in the laboratory because of its inhibitory characteristics and its solublity in aqueous solutions). However, if an isoform naturally resistant to ouabain, such as rat α1, is expressed in HeLa cells, resistance to ouabain is conferred to these cells. This approach makes it possible to identify specific amino acid residues which are essential for enzyme activity. For example, if an amino acid substitution is made at a critical residue by mutating a resistant α isoform cDNA, and the cDNA encoding this substitution is expressed in HeLa cells, no colonies will appear in the presence of ouabain. Alternatively, if a substitution is introduced at a site which is not critical for Na,K-ATPase function, ouabain resistant colonies will appear following treatment with the drug.

EXTRACELLULAR FLUID

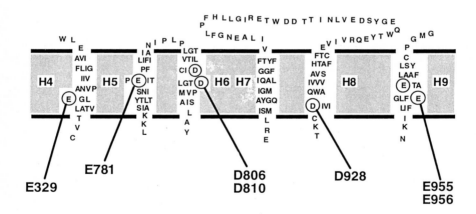

CYTOSOL

Figure 1. Location of negatively charged amino acids in transmembrane regions of the rat α1 isoform of Na,K-ATPase. Only those segments containing negatively charged amino acids are shown (transmembrane regions H4-H9).

Using this approach the functional role of each negatively charged amino acid residue localized in the transmembrane regions of the Na,K-ATPase is being investigated. Being negatively charged, these residues may be able to neutralize the positive charge of the cation and, as a result, are predicted to form at least part of the cation binding site(s) (Ovchinikov et al., 1987; Karlish et al., 1990; Pedemonte and Kaplan, 1990). The location of these amino acids are shown in Figure 1. Negatively charged amino acids occur at positions E329, E781, D806, D810, D928, E955 and E956 of the rat $\alpha 1$ isoform. In initial studies substitutions were made at glutamic acid residues 955 and 956. These two residues are highly conserved, and E955 was predicted by chemical labelling studies to be important in cation binding (Goldshleger et al., 1992). Both glutamines were converted to glutamines and aspartic acids (Van Huysse et al., 1993). None of the substitutions affected the ability of the Na,K-ATPase to confer resistance to HeLa cells. When enzymes carrying these substitutions were analyzed in microsomal membrane preparations, very little difference was observed in properties associated with the Na^+ or K^+ stimulation of Na,K-ATPase activity. Therefore, it is clear that the above amino acids are not absolutely required for ion transport and their role in cation binding, if any, is at most modest. Interestingly, glutamic acid residues 955 and 956 are not conserved in the Ca-ATPase. These studies indicate that the mutagenesis-expression-selection approach is feasible in rapidly screening for critical amino acids, and we are presently using it to analyze the effects of substitutions in the remaining negatively charged residues. It will be of interest to determine whether these residues are required for enzyme activity. Of the remaining five negatively-charged residues (i.e.: E329, E781, D806,D810 and D928), only one, D928, is not conserved in the Ca-ATPase. Depending on the results of these studies, insight may be gained into the basis for the differences in cation specificity in the Na,K-ATPase and the Ca- ATPase.

Determinants of Ouabain Sensitivity

Utilizing the selection approach described above it is also possible to identify amino acid residues responsible for ouabain sensitivity. In this case, however, site-directed mutagenesis is used to introduce amino acid replacements in the sensitive sheep $\alpha 1$ isoform. If a substitution prevents the drug from binding or reduces the affinity of the drug-enzyme interaction, then the altered enzyme will confer ouabain resistance when expressed in normally sensitive HeLa cells. The amino acids which are determinants of ouabain sensitivity identified using this approach are shown in Figure 2. Depicted in this figure is the ten transmembrane model of the sheep $\alpha 1$ subunit of the Na,K- ATPase. While the exact topology of this subunit has not been defined, this model is similar to that proposed for the Ca- ATPase (MacLennan et al., 1985). The amino acids that have been shown to date to influence the sensitivity of the enzyme to ouabain are indicated by residues in bold letters. Interestingly, many of the known determinants of ouabain affinity are located in the first extracellular region. Substitution of glutamine 111 and aspartic acid 122 with a variety of charged amino acids yields enzymes which confer resistance (Price et al., 1990). It is possible that these charged residues prevent the partial internalization of this region into the membrane when ouabain binds to the enzyme. In sensitive enzymes the naturally occurring amino acids at these two positions are uncharged. Substitution of aspartic acid residue 121 with asparagine also yields an enzyme able to confer resistance when expressed in sensitive cells (Price et al., 1989). It is possible that the negative charge of this amino acid contacts the lactone ring of ouabain. Two residues within the first transmembrane region also influence ouabain sensitivity (Schultheis and Lingrel, 1993; Canessa et al., 1992). These are amino acid residues cysteine 104 and tyrosine 108. When either residue is substituted with alanine, or when cysteine 104 is replaced with phenylalanine, ouabain resistant enzyme is produced. These studies suggest that the first transmembrane and first extracellular region play a critical role in determining ouabain sensitivity. Interestingly, the fourth extracellular loop is also involved in determining ouabain sensitivity (Schultheis et al., 1993). A DNA cassette encompassing this region was subjected to random chemical mutagenesis and inserted into the wild type sheep α1 cDNA. Expression of this cDNA in HeLa cells and selection in ouabain resulted in colony formation.

EXTRACELLULAR

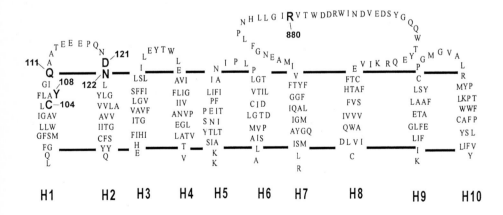

CYTOPLASMIC

Figure 2. Amino acid residues which alter sensitivity of the sheep α1 subunit to ouabain.

The transfected cDNA from one of these colonies was amplified by PCR, and sequence analysis of this DNA showed that arginine 880 was replaced by proline. Thus, arginine 880 appears to play a role in ouabain binding. At the present time it is not known whether these residues directly contact the cardiac glycoside or whether, in fact, the above substitutions allosterically alter the binding site. Much of the rest of the α subunit has not yet been explored and it is likely that amino acid residues will be found in other regions which alter sensitivity. Once a complete analysis has been carried out with respect to these amino acid residues, it should be possible to consider models of the actual binding site.

While it is unknown whether the amino acid determinants of ouabain affinity identified to date actually represent the binding site, an approach for identifying amino acids which interact with specific moieties of ouabain involves the use of structural analogs. Along this line, enzymes containing amino acid substitutions known to alter cardiac glycoside sensitivity were assayed for inhibition by both ouabain and ouabagenin (O'Brien et al., 1993). Ouabagenin has a structure similar to that of ouabain but lacks the sugar. While both compounds inhibit the enzyme, the I_{50} for ouabagenin is approximately 20 fold higher than that of ouabain. Therefore, the sugar portion of the drug contributes to the overall affinity of the interaction between the drug and the enzyme. Substitution of an amino acid residue which contacts the sugar group with an amino acid which does not interact with this moiety would be expected to yield an enzyme with similar I_{50} values for both ouabain and ouabagenin. Interestingly, a ouabagenin to ouabain I_{50} ratio of 20 was obtained for all the substitutions tested. It should be pointed out that the I_{50} for both drugs is displaced in each of the mutants depending on the level of resistance conferred by that mutant. Nevertheless, the I_{50} ratio of the drugs remains constant. Therefore, it is concluded that none of the amino acid residues tested to date actually contact or are affected by the sugar portion of ouabain. This implies that other amino acid residues must contact the sugar and a search is ongoing to identify such residues.

In summary, it is possible to use a mutagenesis-expression- selection approach to identify amino acid residues which alter ouabain sensitivity or to define residues involved in function such as cation-binding sites.

Acknowledgements

This work was supported by NIH Grants P01 HL 41496 and RO1 HL 28573.

References

Andersen, J.P. and Vilsen, B. (1992) Functional consequences of alterations to Glu309, Glu771, and Asp800 in the Ca^{2+}-ATPase of sarcoplasmic reticulum. J. Biol. Chem. 267:19383- 19387

Canessa, C.M., Horisberger, J.-D., Louvard, D. and Rossier, B.C. (1992) Mutation of a cysteine in the first transmembrane segment of Na,K-ATPase α subunit confers ouabain resistance. Embo J. 11:1681-1687

Goldshleger, R., Tal, D.M., Moorman, J., Stein, W.D. and Karlish, S.J.D. (1992) Chemical modification of Glu-953 of the α chain of Na^+,K^+-ATPase associated with inactivation of cation occlusion. Proc. Natl. Acad. Sci. USA 89:6911-6915

Good, P.J., Richter, K. and Dawid, I.B. (1990) A nervous system-specific isotype of the beta subunit of Na,K-ATPase expressed during early development of Xenopus laevis. Proc. Natl. Acad. Sci. USA 87:9088-9092

Karlish, S.J.D., Goldshleger,R. and Stein, W.D. (1990) A 19- kDA C-terminal tryptic fragment of the α chain of Na/K-ATPase is essential for occlusion and transport of cations. Proc. Natl. Acad. Sci. USA 87:4566-4570

Kent, R.B., Emanuel, J.R., Neriah, Y.B., Levenson, R. and Housman, D.E. (1987) Ouabain resistance conferred by expression of the cDNA for a murine Na^+,K^+-ATPase α subunit. Science 237:901-903

Lingrel, J.B, Orlowski, J., Shull, M.M., and Price, E.M. (1990) Molecular Genetics of Na,K-ATPase. In: Progress in Nucleic Acids Research and Molecular Biology, Academic Press, Inc., 38:37-89

MacLennan, D.H. Molecular tools to elucidate problems in excitraction coupling. (1990) Biophys. J. 58:1355-1365

MacLennan, D.H., C.J. Brandl, B. Korczak and N.M. Green. (1985) Amino-acid sequence of a Ca^{2+}, Mg^{2+}-dependent ATPase from rabbit muscle sarcoplasmic reticulum, deduced from its complementary DNA sequence. Nature 316:696-700

O'Brien, W.J., Wallick, E.T. and Lingrel, J.B. (1993) Amino acid residues of the Na,K-ATPase involved in ouabain sensitivity do not bind the sugar moiety of cardiac glycosides. J. Biol. Chem. 268:7707-7712

Orlowski, J. and Lingrel, J.B. (1988) Tissue-specific and developmental regulation of rat Na,K-ATPase catalytic α isoform and β subunit mRNAs. J. Biol. Chem. 263:10436-10442

Ovhinnikov, Y.A. (1987) Probing the folding of membrane proteins. Trends Biochem. Sci. 12:434-438

Pedemonte, C.H. and Kaplan, J.H. (1990) Chemical modification as an approach to elucidation of sodium pump structure-function relations. Am. J. Physiol. 258:C1-C23

Price, E.M. and Lingrel, J.B. (1988) Structure-Function Relationships in the Na,K-ATPase α Subunit: Site-Directed Mutagenesis of Glutamine-111 to Arginine and Asparagine-122 to Aspartic Acid Generates a Ouabain-Resistant Enzyme. Biochemistry 27:8400-8408

Price, E.M., Rice, D.A. and Lingrel, J.B. (1989) Site- directed mutagenesis of a conserved extracellular aspartic acid residue affects the ouabain sensitivity of sheep Na,K-ATPase. J. Biol. Chem. 264:21902-21906

Price, E.M., Rice, D.A. and Lingrel, J.B. (1990) Structure- function studies of Na,K-ATPase. J. Biol. Chem. 265:6638-6641

Schultheis, P.J. and Lingrel, J.B. (1993) Substitution of transmembrane residues with hydrogen-bonding potential in the α subunit of Na,K-ATPase reveals alterations in ouabain sensitivity. Biochemistry, 32:544-550

Schultheis, P.J., Wallick, E.T. and Lingrel, J.B. (1993) Kinetic analysis of ouabain binding to native and mutated forms of Na,K-ATPase and identification of a new region involved in cardiac glycoside interactions. J. Biol. Chem. in press.

Sweadner, K.J. (1989) Isozymes of the Na, K-ATPase. Biochem. Biophys. Acta., 988:185-220

Takeyasu, K., Renard, K.J., and. Aormino, J., Wolityky, B.A., Barnstein, A, Tamkum, M.M., and Fanbrough, D.M. (1989) Differential subunit and isoform expression involved in regulation of sodium pump in skeletal muscle. In Current Topics in Membranes and Transport. Academic Press, Inc. ,New York, 143- 165

Van Huysse, J.W., Jewell, E.A., and Lingrel J.B. (1993) Site Directed Mutagenesis of a Predicted Cation Binding Site of Na,K-ATPase. Biochemistry, 32:819-826

Vilsen, B. and Andersen, J.P. (1992) CrATP-induced Ca^{2+} occlusion in mutants of the Ca^{2+}-ATPase of sarcoplasmic reticulum. J. Biol. Chem. 267:25739-25743

The Use of Site-Directed Mutagenesis to Identify Functional Sites in the Ca^{2+} ATPase[*]

David H. MacLennan, Ilona S. Skerjanc [+], William J. Rice[+], Tip W. Loo and David M. Clarke[#]
Banting and Best Department of Medical Research
University of Toronto, Charles H. Best Institute, 112 College Street
Toronto, Ontario, M5G 1L6, Canada

INTRODUCTION

Our long term research objective has been to understand the mechanism of Ca^{2+} transport by the Ca^{2+} ATPase of the sarcoplasmic reticulum and our approach to the solution of this problem has been to investigate structure/function relationships in the enzyme. In recent years, we have used the powerful techniques of cloning, structural prediction and site-directed mutagenesis to gain insight into the roles of specific amino acids in the structure, function and assembly of the protein. These studies have allowed us to evaluate the functional roles of major regions of the Ca^{2+} ATPase and to define amino acids that are specifically involved in ATP-dependent Ca^{2+} translocation. We have been able to show that residues in the stalk sector are not involved in Ca^{2+} binding, but may be involved in regulation of Ca^{2+} transport, that specific residues within the transmembrane sector are involved with Ca^{2+} binding and Ca^{2+} translocation, that residues in the nucleotide binding and hinge domains are involved in ATP binding and in determination of apparent Ca^{2+} affinity, and that residues throughout the molecule are involved in conformational changes. From these data, we have developed a simple and easily understood hypothesis for the mechanism of ATP-dependent Ca^{2+} transport.

[*] Original research described in this review was supported by the National Institutes of Health, U.S.A. and the Medical Research Council of Canada.

[+] Recipient of postdoctoral or predoctoral fellowship support from the Medical Research Council of Canada.

[#] Recipient of a postdoctoral fellowship from the Heart and Stroke Foundation of Ontario.

NATO ASI Series, Vol. H 89
Molecular and Cellular Mechanisms
of H$^+$ Transport
Edited by Barry H. Hirst
© Springer-Verlag Berlin Heidelberg 1994

Structure of Ca^{2+} ATPases

The critical research goal for P type pumps is to integrate the structure of the protein with its function. In general, P type ATPases have similar structures, and deductions concerning the structure of one are applicable to the structures of others (Green, 1989; Green and MacLennan, 1989). Morphological studies have shown that the Ca^{2+} ATPase is comprised of cytoplasmic headpiece and stalk sectors and a transmembrane basepiece, making up a tripartite structure. Our structural model, based on primary sequence, predicted that the basepiece would consist of 10 transmembrane helices (M_1 to M_{10}), that the stalk sector would be made up of 5 α-helices contiguous with transmembrane helices M_1 through M_5, and that the headpiece would be made up of three globular domains, the first a seven membered ß-strand domain lying between stalk sequences 2 and 3 and the second and third occurring in alternating α-ß sequences lying between stalk sectors 4 and 5 (MacLennan *et al.*, 1985). The second and third domains, referred to as the phosphorylation domain and the nucleotide binding domain, would be rejoined by a long α-helix at the end of the third domain and this would form a hinge between them. Our predictions have been supported by images of tubular crystals of the Ca^{2+} ATPase, grown in EGTA and vanadate and embedded in amorphous ice, which were shown to diffract to 14 Å, allowing 3-dimensional structures to be reconstructed using helical symmetry (Toyoshima *et al.*, 1993). The molecule fits in a box 120 Å high and 50 Å x 85 A in diameter. Taylor and Green (1989) have analyzed the secondary structural predictions for several ion transport ATPases, noting highly conserved sequences that were predicted to be involved in nucleotide binding, and making use of earlier observations concerning chemical modification or crosslinking in formulating a model for the nucleotide binding fold in the Ca^{2+} ATPase.

Site-directed mutagenesis.

We initiated our studies of expression and site directed mutagenesis of the Ca^{2+} ATPase (Maruyama *et al.*, 1989) by asking the specific questions: Where does Ca^{2+} bind in the molecule?; Where does ATP bind in the molecule; and, Which residues are involved in conformational changes? Mutation of acidic residues in the stalk sector, even in clusters of 4, did not grossly affect Ca^{2+}

transport (Clarke *et al.*, 1989b). By contrast, mutation of 6 polar residues, Glu^{309}, Glu^{771}, Asn^{796}, Thr^{799}, Asp^{800} and Glu^{908}, in transmembrane sequences M_4, M_5, M_6 and M_8 abolished Ca^{2+} transport and Ca^{2+}-dependent phosphoenzyme formation from ATP (Clarke *et al.*, 1989a). The mutants were phosphorylated by P_i, but the reaction was no longer Ca^{2+}-inhibited, in line with loss of one or more Ca^{2+} binding sites. These residues have been mutated in different ways (Andersen and Vilsen, 1992; Clarke *et al.*, 1990a), showing that Ca^{2+} binding can often be preserved, but with altered affinity. Vilsen and Andersen (1992) have demonstrated that the Ca^{2+} binding mutants have lost their ability to occlude Ca^{2+}. We showed that only one of the two Ca^{2+} binding sites is lost for the Glu^{309} to Gln mutant (Skerjanc *et al.*, 1993). Mutation of several additional residues in transmembrane sequences also caused alterations of Ca^{2+} affinity (Andersen *et al.*, 1989; Clarke *et al.*, 1993). Thus, we have been able to position the Ca^{2+} binding and translocation sites in the center of the transmembrane domain of the Ca^{2+} ATPase.

We also used site-directed mutagenesis to define a series of residues in the nucleotide binding and hinge domains that affect ATP binding (Maruyama *et al.*, 1988; 1989b). Our studies of the phosphorylation site and the nucleotide binding domain have been fully consistent with, and have provided support for, the predictions concerning the residues involved in ATP binding (Taylor and Green, 1989). We have also found that mutation of another series of residues affects conformational changes. These mutations fall into two classes: those that block E_1P to E_2P conformational changes (Andersen *et al* 1989; Vilsen *et al.*, 1989, Clarke *et al.*, 1990a;1990b;1990c); and those that block the E_2P to E conformational changes (Andersen *et al.*, 1992; Clarke *et al.*, 1993). The E_1P to E_2P conformational change mutations are located in the head piece, ß-strand, stalk, and transmembrane domains (MacLennan *et al.*, 1992a), while the E_2P to E conformational change mutation are, to date, restricted to Gly^{310} (Andersen *et al.*, 1992) and Ala^{305} and Ala^{306} (Clarke *et al.*, 1993) in M_4. A summary of the locations of those mutations that destroy Ca^{2+} transport function is presented in Figure 1.

Our studies of mutagenesis have allowed us to propose a very simple and easily understood mechanism for Ca^{2+} transport (MacLennan, 1990; MacLennan *et al.*, 1992a; 1992b). Two high affinity Ca^{2+} binding and translocation sites are located in a channel created by M_4, M_5, M_6 and M_8. In the E_1 conformation, these sites are accessible to cytoplasmic Ca^{2+} and are stacked (Inesi, 1987), but they are not accessible to lumenal Ca^{2+}. At a distance of 40-60 Å away, in a cytoplasmic domain, an ATP hydrolytic or 'catalytic' site is created by closely apposed residues in loops between alternating α-helices and ß-strand structures that form the ATP binding cleft. Bound ATP is apposed to another exposed loop structure forming the complex phosphorylation site in the phosphorylation domain. The translocation site and the catalytic site, even though spatially very widely separated, mutually regulate each other through conformational changes that are propagated throughout the ATPase molecule and involve those residues in several domains that we have defined as 'conformational change' mutations. For example: the phosphoenzyme can only be formed from ATP when 2 Ca^{2+} ions are bound in the transmembrane domain; the presence of Ca^{2+} inhibits phosphorylation from P_i; and phosphorylation of the catalytic site by ATP is a prerequisite for Ca^{2+} translocation.

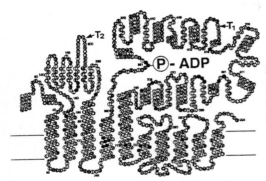

Fig 1. Structural model of the Ca^{2+} ATPase based on prediction and experimental evidence. Those residues which, when mutated, lead to complete loss of Ca^{2+} transport are darkened.

The conformational changes resulting from phosphoenzyme formation and decay induce conformational changes in the Ca^{2+} binding site. As a consequence, these Ca^{2+} binding sites lose access to the cytoplasm, gain access to the lumen, and change from narrow and ordered to wide and disordered (Hanel and Jencks, 1991; Orlowski and Champeil, 1991). These conformational changes, which probably cause rotation of helices relative to each other, result in disruption of the Ca^{2+} binding sites which were formed from 4 transmembrane helices. Ca^{2+} released simultaneously from the two low affinity sites now finds itself in the lumen, completing the process of Ca^{2+} transport. In this model, Ca^{2+} transport is much like passage through a turnstile. In order to pass the gate, a conformational change in the turnstile mechanism must be activated with a token. In the case of active transport, the token is hydrolysis of ATP.

Direct measurement of Ca^{2+} binding to the Glu309 to Gln mutant

The yield of mutant Ca^{2+}-ATPase protein from transient transfection systems is rather low, preventing us from carrying out direct measurement of Ca^{2+} and ATP binding to mutant proteins. Accordingly, in recent studies, we have tested the baculovirus expression system for its potential for synthesis of large amounts of Ca^{2+} ATPase protein (Skerjanc et al., 1993). The Ca^{2+} ATPase was expressed in infected SF9 cells at up to 3 mg protein/liter of cells. Vesicles made from Sf9 cells transported Ca^{2+} at rates 6 fold over background and with normal V_{max}, proving that the bulk of the protein expressed in Sf9 cells was functional. The mutant protein, E309Q, when expressed in these cells, exhibited only a background level of Ca^{2+} transport, as expected.

Because the yield of Ca^{2+} ATPase from Sf9 microsomes was still relatively small, we developed a single step method for isolation of the active protein. Total membranes containing about 1% Ca^{2+} ATPase protein were dissolved in a mixture of the detergent, $C_{12}E_8$, and Asolectin. The solubilized protein was then bound to the monoclonal antibody, A52, which was, in turn, bound to protein A sepharose beads. The Ca^{2+} ATPase protein was purified to greater than 90% homogeneity by these steps and was also highly active, but we could not elute it from the beads without loss of function using high or low pH or chaotropic salts. Nevertheless, with the protein bound to the beads, we could readily measure E_1P and E_2P formation for wild-type and E_2P formation for the E309Q mutant. We have used this material to measure Ca^{2+} binding to the mutant E309Q. In order to prove that the Ca^{2+} binding mutant, E309Q, was deficient in Ca^{2+} binding, we made direct measurements of Ca^{2+} binding to mutant and wild-type proteins expressed in the baculovirus system and immunoaffinity purified (Skerjanc et al., 1993). The proteins bound to A52-protein A Sepharose were incubated with different concentrations of Ca^{2+} and then filtered. Background counts representing radioactivity trapped in the filter and in the Sepharose were subtracted and counts above background were converted to bound Ca^{2+}.

We found that Ca^{2+} could be bound to the wild-type protein when Ca^{2+} was presented in different buffers favoring either the E_1 or the E_2 conformation of the protein. If Ca^{2+} were presented in a buffer favoring the E_1 conformation, half maximal Ca^{2+} binding occurred at about 0.3 µM, but, if presented in a buffer favoring the E_2 conformation, half maximal binding occurred at about 4 µM. The wild-type Ca^{2+} ATPase bound about 1.6 moles Ca^{2+}/mole protein in either the E_1 or the E_2 buffer. The mutant, however, bound 0.8 moles Ca^{2+}/mole protein in the E_2 buffer, but essentially no Ca^{2+} in the E_1 buffer. Our interpretation of these results is that the mutant retains one Ca^{2+} binding site, but access to the site is achieved only from the lumenal side and only when the protein is in the E_2 conformation (Fig. 2). Thus this normal Ca^{2+} binding site must be the more lumenal of the two 'stacked' sites, while the mutated site must be the more cytoplasmic one. Since Ca^{2+} binding was of relatively high affinity, we also propose that the presence of Ca^{2+} induced a transformation of the enzyme from a conformation with very low affinity for Ca^{2+} to one with relatively high Ca^{2+} affinity so that, at equilibrium, one mole of Ca^{2+} was bound per mole of enzyme with high affinity. This conformation may have been comparable to an E_1Ca^{2+} conformation. If the protein were maintained in the E_1 conformation, access to the Ca^{2+} binding site from the lumenal side would be blocked.

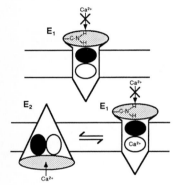

Fig 2. Model of the topology of Ca^{2+} binding sites based on studies of the E309Q mutant. In the E_1 conformation, in intact membranes, substitution of Gln for Glu[309] alters the affinity for Ca^{2+} of the cytoplasmic Ca^{2+} binding site and blocks Ca^{2+} access to the lumenal Ca^{2+} binding site. In the E_2 conformation, Ca^{2+} has access to both Ca^{2+} binding sites from the lumenal surface. Ca^{2+} can interact with the single unmutated site, driving the enzyme to an E_1 conformation with a single Ca^{2+} ion bound. Modified from Skerjanc et al (1993).

These ideas were strengthened by the observation that, whereas Ca^{2+}, even at 100 µM, could not inhibit E_2P formation in intact vesicles from the E309Q mutant, Ca^{2+} at 2 µM could completely inhibit E_2P formation in the solubilized protein (Skerjanc et al., 1993). We suggest that this occurs because Ca^{2+} gained access to the lumenal surface of the protein, the site of the unmutated Ca^{2+} binding site (Fig. 2). We also suggest that Ca^{2+} presented to the lumenal surface in the E_2 conformation, might form an E_1Ca^{2+} adduct that cannot be phosphorylated by P_i. It is now possible to ask whether similar topological experiments can be informative for every other Ca^{2+} binding and Ca^{2+} affinity mutant that we have detected. While the E309Q mutant in intact vesicles was indifferent to 100 µM Ca^{2+} (Clarke et al., 1989a) phosphorylation from Pi in the N796A Ca^{2+} binding mutant was, in fact, inhibited by high levels of medium Ca^{2+} in intact vesicles, but was unable to carry out the partial reaction of Ca^{2+} dependent phosphoenzyme formation from ATP, a reaction that requires that Ca^{2+} be bound to both sites (de Meis and Vianna, 1979). The same characteristics were found for the G801V mutant (Andersen *et al.*, 1992). Thus there is a good possibility that these mutations are affecting the more lumenal of the two stacked Ca^{2+} binding sites and are not affecting the more cytoplasmic site. Current studies will test these ideas.

Definition of a Ca^{2+} binding "face" in transmembrane sequence M_4

We have shown that mutation of Glu^{309} affected Ca^{2+} binding (Clarke *et al.*, 1989a), mutation of Gly^{310} blocked the E_2P to E_2 conformational change (Andersen *et al.*, 1992) and mutation of Pro^{312} blocked the E_1P to E_2P conformational change (Vilsen *et al.*, 1989). In an effort to determine whether additional interesting mutation sites were present in M_4, whether there was significance to the juxtaposition in 7 cases of large hydrophobic residues (Val, Leu, Ile) with small residues (Gly, Ala) in M_4 and whether large hydrophobic residues in M_4 were important, we mutagenized all remaining residues in this sequence by converting large hydrophobic residues (Leu, Ile, Val) to small residues (Gly, Ala) and *vice-versa* (Clarke *et al.*, 1993). We also mutated some hydrophobic residues to Ser and some to other large hydrophobic residues, and we changed Ala to Gly and *vice-versa*.

We found that mutation to Val of Ala^{305} or Ala^{306} gave rise to mutants blocked in the E_2P to E_2 conformational change and altered in Ca^{2+} affinity. Since these mutations altered residue size, we altered $Val^{304}Ala^{305}$ to AlaVal and $Ala^{306}Ile^{307}$ to IleAla to compensate for size changes. These mutants were also inactive and were found to be blocked at the E_2P to E_2 conformational transition, suggesting that not total bulk but precise positioning of these residues was critical to function. Mutation to Ala or Ser of the large hydrophobic residues led to losses of 40 to 85% of V_{max}, with no effect on K_m for Ca^{2+}.

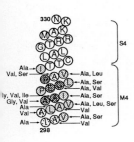

Fig 3. Evidence for a Ca^{2+} binding "patch" on part of the surface of M_4 in the Ca^{2+} ATPase. The mutation of Pro^{312}, Gly^{310}, Glu^{309}, Ala^{306}, or Ala^{305} leads to complete loss of Ca^{2+} transport ▨ , either through E_1P to E_2P or E_2P to E_2 conformational blocks, with accompanying alteration in Ca^{2+} affinity, or to loss of a Ca^{2+} binding site. Mutation of surrounding residues leads to reduced Ca^{2+} uptake and/or affinity ▨ ,or to no effect ▨. Reproduced from Clarke *et al* (1993).

An α-helical representation of M_4 (Fig. 3) shows that the critical residues Ala[305], Ala[306], Gly[309], Gly[310] and Pro[312] form a 'patch' on one surface of the M4 helix, which we think represents the active site of Ca^{2+} translocation. The opposite surface is predicted to be made up of a series of large hydrophobic residues, plus Ala[301] and Pro[308]. A small 'patch' near the lumenal surface made up of Ile[298], Ala[299] and Ala[301] and the single residue Ala[313] are unaffected by mutation. These residues could form the lower and upper ends of the 'patch' active in Ca^{2+} translocation. This dissection of an active 'patch' in M_4 raises the question of whether similar 'patches' can be found in M_5, M_6 and M_8, thereby aiding in our analysis of the active site of translocation, which is made up from these 4 transmembrane sequences (Clarke et al., 1993). The analysis of presently available mutants suggests that this will be the case.

References

Andersen J.P and Vilsen B (1992) Functional consequences of alterations to Glu[309], Glu[771], and Asp[800] in the Ca^{2+}-ATPase of Sarcoplasmic Reticulum. J. Biol. Chem. 267:19383-19387

Andersen JP, Vilsen B, Leberer E and MacLennan DH (1989) Functional consequences of mutations of the ß-strand sector of the Ca^{2+} ATPase of sarcoplasmic reticulum. J Biol Chem 264:21018-21023

Andersen JP, Vilsen B and MacLennan DH (1992) Functional consequences of alterations to Gly[310], Gly[770] and Gly[801] located in the transmembrane domain of the Ca^{2+} ATPase of sarcoplasmic reticulum. J Biol Chem 267:2767-2774

Clarke DM, Loo TW, Inesi G and MacLennan DH (1989a) Location of high affinity Ca^{2+} binding sites within the predicted transmembrane domain of the sarcopolasmic reticulum Ca^{2+} ATPase. Nature 339:476-478

Clarke DM, Maruyama K, Loo TW, Leberer E, Inesi, G and MacLennan DH (1989b) Functional consequences of glutamate, aspartate, glutamine and asparagine mutations in the stalk sector of the Ca^{2+} ATPase of sarcoplasmic reticulum. J Biol Chem 264:11246-11251

Clarke DM, Loo TW and MacLennan DH (1990a) Functional consequences of alterations to polar amino acids located in the transmembrane domain of the Ca^{2+} ATPase of sarcoplasmic reticulum. J Biol Chem 265:6262-6267

Clarke DM, Loo TW and MacLennan DH (1990b) Functional consequences of mutations of conserved amino acids in the ß-strand domain of the Ca^{2+}-ATPase of sarcoplasmic reticulum. J Biol Chem 265:14088-14092

Clarke DM, Loo TW and MacLennan DH (1990c) Functional consequences of alterations to amino acids located in the nucleotide-binding domain of the Ca^{2+}-ATPase of sarcoplasmic reticulum. J Biol Chem 265:22223-22227

Clarke DM, Loo TW, Rice W, Andersen JP, Vilsen B and MacLennan, DH (1993) Functional consequences of alterations to hydrophobic amino acids located in the M_4 trans-membrane sector of the Ca^{2+} ATPase of sarcoplasmic reticulum. J Biol Chem 268:18359-18364

de Meis L and Vianna AL (1979) Energy interconversion by the Ca^{2+}-dependent ATPase of the sarcoplasmic reticulum. Annu Rev Biochem 48:275-292

Green NM (1989) ATP-driven cation pumps: alignment of sequences. Biochem Soc Trans 17:970-972

Green NM and MacLennan DH (1989) ATP driven ion pumps: an evolutionary mosaic. Biochem Soc Trans 17:819-822

Hanel AM and Jencks WP (1991) Dissociation of calcium from the phosphorylated calcium-transporting adenosine triphosphatase of sarcoplasmic reticulum: Kinetic equivalence of the calcium ions bound to the phosphorylated enzyme. Biochemistry 50:11320-11330

Inesi G (1987) Sequential mechanism of calcium binding and translocation in sarcoplasmic reticulum adenosine triphosphatase. J Biol Chem 262:16338-16342

MacLennan DH (1990) Molecular tools to elucidate problems in excitation-contraction coupling. Biophys J 58:1355-1365

MacLennan DH, Brandl CJ, Korczak B and Green NM (1985) Sequence of a Ca^{2+} + Mg^{2+} dependent ATPase from rabbit muscle sarcoplasmic reticulum, deduced from its complementary DNA sequence. Nature 316:696-700

MacLennan DH, Clarke DM, Loo TW and Skerjanc I (1992a) Site-directed mutagenesis of the Ca^{2+} ATPase of sarcoplasmic reticulum. Acta Physiol Scand 146:141-150

MacLennan DH, Toyofuku T and Lytton J (1992b) Structure-function relationships in sarcoplasmic or endoplasmic reticulum type Ca^{2+} pumps. Ann NY Acad Sci 671:1-10

Maruyama K and MacLennan DH (1988) Mutation of aspartic acid-351, lysine-352 and lysine-515 alters the Ca^{2+} transport activity of the Ca^{2+} ATPase expressed in COS-1 cells. Proc Natl Acad Sci USA 85:3314-3318

Maruyama K, Clarke DM, Fujii J, Loo TW and MacLennan DH (1989a) Expression and mutation of Ca2+ ATPases of the sarcoplasmic reticulum. Cell Motil Cytoskel 14:26-34

Maruyama K, Clarke DM, Fujii J, Inesi G, Loo TW and MacLennan DH (1989b) Functional consequences of alterations to amino acids located in the catalytic center (isoleucine 348 to threonine 357) and nucleotide binding domain of the Ca^{2+} ATPase of sarcoplasmic reticulum. J Biol Chem 264:13038-13042

Orlowski S and Champeil P (1991) The two calcium ions initially bound to nonphosphorylated sarcoplasmic reticulum Ca^{2+}-ATPase can no longer be kinetically distinguished when they dissociate from phosphorylated ATPase toward the lumen. Biochemistry 30:11331-11342

Skerjanc IS, Toyofuku T, Richardson C and MacLennan DH (1993) Mutation of glutamate 309 to glutamine alters one Ca^{2+} binding site in the Ca^{2+} ATPase of sarcoplasmic reticulum expressed in Sf9 cells. J Biol Chem 268:15944-15950

Taylor WR and Green NM (1989) The homologous secondary structures of the nucleotide-binding sites of six cation-transporting ATPases lead to a probable tertiary fold. Eur J Biochem 179:241-248

Toyoshima C, Sasake H and Stokes DL (1993) Three-dimensional cryo-electron microscopy of the calcium ion pump in the sarcoplasmic reticulum membrane. Nature 362:469-471

Vilsen B and Andersen JP (1992) CrATP-induced Ca^{2+} Occlusion in mutants of the Ca^{2+}-ATPase of sarcoplasmic reticulum. J Biol Chem 267:25739-25743

Vilsen B Andersen JP Clarke DM and MacLennan DH (1989) Functional consequences of proline mutations in the cytoplasmic and transmembrane sectors of the Ca^{2+} ATPase of sarcoplasmic reticulum. J Biol Chem 264:21024-21030

Vilsen B Andersen J and MacLennan DH (1991) Functional consequences of alterations to amino acids located in the hinge domain of the Ca^{2+}-ATPase of sarcoplasmic reticulum. J Biol Chem 266:16157-16164

STRUCTURE-FUNCTION RELATIONSHIPS IN TRANSMEMBRANE SEGMENT IV OF THE YEAST PLASMA MEMBRANE [H+]-ATPASE

Anthony Ambesi and Carolyn W. Slayman
Departments of Genetics and Cellular & Molecular Physiology
Yale University School of Medicine
New Haven, CT 06510 USA

Introduction

The yeast plasma membrane [H+]-ATPase is essential for cell viability, pumping protons across the cell membrane to generate a large electrochemical gradient that provides the energy for nutrient uptake (reviewed by Goffeau & Slayman, 1981; Serrano, 1988; and Nakamoto & Slayman, 1989). The [H+]-ATPase is encoded by the PMA1 gene (Serrano et al., 1986) and belongs to a widely distributed family of transporters known as E_1E_2- or P-ATPases. Like other members of the group, it has a 100 kDa catalytic subunit that is firmly embedded in the lipid bilayer (Dufour & Goffeau, 1978) and alternates between two major conformational states (E_1 and E_2), hydrolyzing ATP by way of a covalent ß-aspartyl phosphate reaction intermediate (Dame & Scarborough, 1981; Amory & Goffeau, 1982).

Hydropathy analysis of the entire set of P-ATPase sequences points to a common topological plan for the 100 kDa subunit (reviewed by Nakamoto et al., 1989). In each case, a central hydrophilic domain containing well-conserved ATP-binding and phosphorylation sites is flanked by four hydrophobic segments towards the N-terminus and four to six hydrophobic segments towards the C-terminus. Although little direct structural information is available, it seems certain that the hydrophobic segments assemble in the membrane to form a channel- or pore-like structure through which cations are pumped.

Consistent with the fact that members of the P-ATPase family have different cation specificities, ranging from H+ to Na^+, K^+, Mg^{2+}, Ca^{2+}, Cu^{2+}, and Cd^{2+}, it is not surprising that there has been little sequence conservation within the membrane-spanning segments. The only exception is segment IV, which is illustrated in Fig. 1. Segment IV is unusually long (26 to 30 residues) and may exist in the membrane as a kinked α-helix, owing to the presence of a proline residue midway along the hydrophobic stretch. This proline forms part of a conserved motif (PVGL) in the fungal and plant [H+]-ATPases, which becomes PEGL in the animal cell [H+,K+]-, [Na+,K+]-, and [Ca2+]-ATPases. Of particular interest, recent site-directed mutagenesis studies on rabbit sarcoplasmic reticulum [Ca2+]-ATPase have identified the glutamate within the PEGL motif as one of six Ca^{2+} binding residues (Clarke et al., 1990), and

NATO ASI Series, Vol. H 89
Molecular and Cellular Mechanisms
of H+ Transport
Edited by Barry H. Hirst
© Springer-Verlag Berlin Heidelberg 1994

have demonstrated that several of the neighboring hydrophobic residues are also important for Ca^{2+} transport (Vilsen et al., 1991). Based on these observations, together with the fact that the strongly-conserved phosphorylation site (Asp378 in the yeast [H+]-ATPase) lies only a short distance downstream, it has been suggested that segment IV may play a role in coupling the energy from ATP hydrolysis to the translocation of cations across the membrane.

ATPase		Sequence	Reference
E. coli [K]	(254)	VLVALLVCLIPTTIGGLLSASAVAGMS	Hesse et al. (1984)
S. cerevisiae [H]	(325)	YTLGITIIGVPVGLPAVVTTTMAVGAAYLA	Serrano et al. (1986)
S. pombe [H]	(323)	YTLAITIIGVPVGLPAVVTTTMAVGAAYLA	Ghislain et al. (1987)
C. albicans [H]	(302)	YTLAITIIGVPVGLPAVVTTTMAVGAAYLA	Monk et al. (1991)
N. crassa [H]	(325)	FTLAITIIGVPVGLPAVVTTTMAVGAAYLA	Hager et al. (1986)
A. thaliana [H]	(277)	NLLVLLIGGIPIAMPTVLSVTMAIGSH	Harper et al. (1989)
Rat gastric [H, K]	(333)	FFMAIVVAYVPEGLLATVTVCLSLTA	Shull and Lingrel (1986)
Sheep kidney [Na, K]	(323)	FLIGIIVANVPEGLLATVTVCLTLTA	Shull et al. (1985)
Rabbit SR [Ca]	(298)	IAVALAVAAIPEGLPAVITTCLALGT	Brandl et al. (1986)
Rat PM [Ca]	(419)	FFIIGVTVLVVAVPEGLPLAVTISLAYSV	Shull and Greeb (1988)

Figure 1.   Sequence alignment of membrane-spanning segment IV among various P-ATPases.

To explore the functional role of membrane segment IV in the yeast [H+]-ATPase, we have undertaken a systematic study by site-directed mutagenesis, replacing each of the residues in turn by alanine. Mutant forms of the ATPase have been expressed in yeast secretory vesicles (Nakamoto et al., 1991) and characterized by measurements of ATP hydrolysis and ATP-dependent proton pumping. The results have identified two sites at which amino acid substitutions lead to a partial uncoupling of transport from hydrolysis.

Methods

Strain SY4 of *Saccharomyces cerevisiae* (MATα; ura3-52; leu2-3,112; his4-619; sec6-4; GAL; pma1::YIpGAL-PMA1) was used throughout this work. As described in detail by Nakamoto et al. (1991), the chromosomal PMA1 gene of SY4 has been placed under control of the GAL1 promoter by gene disruption. SY4 also contains a temperature-sensitive sec6-4 mutation that blocks the final step in plasma membrane biogenesis (Schekman & Novick, 1982). For expression studies, SY4 is transformed with a centromeric plasmid carrying a mutagenized copy of the PMA1 gene under the control of a heat-shock promoter (P$_{HSE}$-pma1; Nakamoto et al., 1991). Cells are grown to mid-exponential phase in supplemented minimal medium containing 2% galactose at 23°C (wild-type gene on; mutant gene off) and then shifted to medium containing 2% glucose at 37°C (wild-type gene off; mutant gene on). Following the shift, newly synthesized mutant ATPase is trapped in secretory vesicles, which are isolated by differential centrifugation and gel filtration (Walworth & Novick, 1987).

Oligonucleotide-directed mutagenesis was carried out on a 615bp BstEII-EcoRI restriction fragment of the PMA1 gene subcloned into a modified version of Bluescript (Stratagene, La Jolla, CA) according to the method of Taylor et al. (1985). After sequencing, the fragment was re-cloned into plasmid pPMA1.2 (Nakamoto et al., 1991) and moved into the vector YCp2HSE (Nakamoto et al., 1991), bringing the mutant allele under control of the heat-shock promoter.

Relative expression levels of the various mutant forms of the ATPase polypeptide were determined in isolated secretory vesicles by quantitative radioimmunoassay (Rao & Slayman, 1993). ATP hydrolysis was measured as previously described (Nakamoto et al., 1991). ATP-dependent proton pumping was monitored by fluorescence quenching of the pH-sensitive dye acridine orange. Briefly, secretory vesicles were suspended in 0.6 M sorbitol, 0.1 M KCl, 10 mM HEPES-KOH, pH 6.7, 2 μM acridine orange, and varying amounts of Na_2-ATP (0.5 to 5 mM). Proton pumping was initiated by the addition of $MgCl_2$ (to give a final concentration 5 mM in excess of the ATP concentration). Fluorescence quenching was monitored on a Hitachi F2000 fluorimeter (excitation, 430 nm; emission, 530 nm).

Results and Discussion

The first step in this study was to ask whether alanine replacements at each of the positions in segment IV allowed the ATPase polypeptide to be processed normally along the yeast secretory pathway. Isolated secretory vesicles were examined by quantitative radioimmunoassay using wild-type vesicles, in which the PMA1 ATPase constitutes ca. 8% of total vesicular protein (Nakamoto et al., 1991), as a standard. In 16 of the 18 cases studied to date, the mutant proteins were expressed at or near normal levels (65 to 100% of the wild-type control), while in two mutants (Gly333Ala, 20%; Gly349Ala, 25%), expression was substantially reduced. Even for the latter two cases, however, the measured rates of ATP hydrolysis and H^+ pumping fell well within the detectable range and, after correction for differences in protein expression, gave usable values for comparison with those from wild-type enzyme. Thus, alanine replacements along segment IV do not appear to cause any drastic perturbation in the folding or processing of the ATPase polypeptide.

When ATPase activity was assayed under optimal conditions (pH 6.7; 5 mM MgATP), the hydrolysis rate varied with the position of the mutation along the membrane segment (Fig. 2). Alanine substitutions towards the beginning of the segment (at Ile331, Ile332, Gly333,

Val334, Pro335, Leu338, Pro339) caused hydrolytic activity to fall to values less than 25% of the wild-type control, while substitutions further along the segment generally had little or no effect.

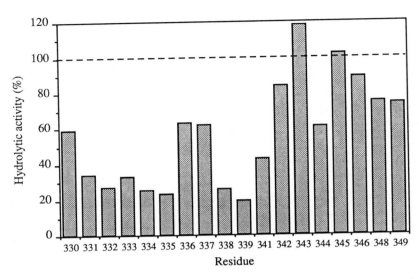

Figure 2. Vanadate-sensitive ATPase activity in secretory vesicles expressing mutant forms of the [H+]-ATPase. Mutant forms of the [H+]-ATPase were assayed for hydrolytic activity in the presence and absence of 100 μM vanadate as previously described (Nakamoto et al., 1991). Activities were corrected for expression and compared with that of wild-type control (n=1 to 2).

The final step was to assess the ability of the mutant enzymes to carry out ATP-dependent proton pumping. As described in the Methods section, this was done over a range of MgATP concentrations (typically, 0.5 to 5 mM), using acridine orange fluorescence quenching as a measure of proton transport. In parallel, ATP hydrolysis was assayed under the identical conditions (see Methods above), and ATP-dependent H+ pumping was then plotted as a function of the initial rate of ATP hydrolysis. When this was done for vesicles containing wild-type ATPase, the relationship was linear over most of the range, with an apparent saturation of H+ transport (probably due to the characteristics of the acridine orange assay) only at the highest ATP concentrations tested (Fig. 3A). All but two of the mutants so far examined gave data that fell along the same line (see Ile331Ala and Thr343Ala in Fig. 3A as examples). In two cases, however (Val336Glu and Gly337Ala), there was a clearcut partial uncoupling of proton transport from ATP hydrolysis (Fig. 3B). Thus, these two adjacent residues appear to define a region of segment IV that is critical for proper energy transduction. As direct structural data become available (see, for example, Toyoshima et al., 1993), it will be interesting to learn whether segment IV actually lines the transport pore and/or whether it participates in helix-helix interactions that are important for the overall structure of the membrane domain.

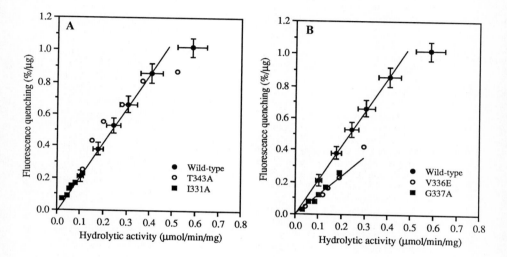

Figure 3.   **ATP-dependent proton transport measured as a function of ATP hydrolytic activity.** Fluorescence quenching of the pH-sensitive dye acridine orange was used to monitor proton pumping over a range of substrate concentrations (0.5 to 5.0 mM ATP) and compared directly with ATP hydrolytic activity assayed under identical conditions. *A.* Two mutants, Thr343Ala and Ile331Ala, differing in activity (see Fig. 2) but displaying normal coupling. *B.* Two mutants, Val336Glu and Gly337Ala, with partial uncoupling of transport from hydrolysis.

Acknowledgments

This work was supported by NIH research grant GM15761 and postdoctoral fellowship GM15827 to A. A. from the National Institute of General Medical Sciences. The authors are grateful to Kenneth E. Allen for expert technical assistance.

References

Amory A & Goffeau A (1982) Characterization of the ß-aspartyl phosphate intermediate formed by the H^+-translocating ATPase from the yeast *Schizosaccharomyces pombe*. J Biol Chem 257:4723-4730

Brandl CJ, Green NM, Korczak B, & MacLennan DH (1986) Two Ca^{2+}-ATPase genes: Homologies and mechanistic implications of deduced amino acid sequences. Cell 44:597-607

Clarke DM, Loo TW, & MacLennan DH (1990) Functional consequences of alterations to polar amino acids located in the transmembrane domain of the Ca^{2+}-ATPase of sarcoplasmic reticulum. J Biol Chem 265:6262-6267

Dame JB & Scarborough GA (1980) Identification of the phosphorylated intermediate of the *Neurospora* plasma membrane H^+-ATPase as a ß-aspartyl phosphate. J Biol Chem 256:10724-10730

Dufour JP & Goffeau A (1978) Solubilization by lysolecithin and purification of the plasma membrane ATPase of the yeast *Schizosaccharomyces pombe*. J Biol Chem 253:7026-7032

Ghislain M, Schlesser A, & Goffeau A (1897) Mutation of a conserved glycine residue modifies the vanadate sensititvity of the plasma membrane H^+-ATPase from *Schizosaccharomyces pombe*. J Biol Chem 262:17549-17555

Goffeau A & Slayman CW (1981) The proton-translocating ATPase of the fungal plasma membrane. Biochim Biophys Acta 639:197-223

Hager KM, Mandala SM, Davenport JW, Speicher DW, Benz EJ, & Slayman CW (1986) Amino acid sequence of the plasma membrane ATPase of *Neurospora crassa*: Deduction from genomic and cDNA sequences. Proc Natl Acad Sci USA 83:7693-7697

Harper JF, Surowy TK, & Sussman MR (1989) Molecular cloning and sequence of cDNA encoding the plasma membrane proton pump (H^+-ATPase) of *Arabidopsis thaliana*. Proc Natl Acad Sci USA 86:1234-1238

Hesse J, Wieczorek L, Altendorf K, Reicin AS, Dorus E, & Epstein W (1984) Sequence homology between two membrane transport ATPases, the Kdp-ATPase of *Escherichia coli* and the Ca^{2+}-ATPase of sarcoplasmic reticulum. Proc Natl Acad Sci USA 81:4746-4750

Monk BC, Kurtz MB, Marrinan JA, & Perlin DS (1991) Cloning and characterization of the plasma membrane H^+-ATPase from *Candida albicans*. J Bact 173:6826-6836

Nakamoto RK, Rao R, & Slayman CW (1989) Transmembrane segments of the P-type cation-transporting ATPases. Ann N Y Acad Sci 574:165-179

Nakamoto RK, Rao R, & Slayman CW (1991) Expression of the yeast plasma membrane $[H^+]$ATPase in secretory vesicles. A new strategy for directed mutagenesis. J Biol Chem 266:7940-7949

Nakamoto RK & Slayman CW (1989) Molecular properties of the fungal plasma-membrane $[H^+]$-ATPase. J Bioenergetics Biomembranes 21:621-632

Schekman R & Novick PJ (1982) in *The Molecular Biology of the Yeast Saccharomyces: Metabolism and Gene Expression* (Strathern JN, Jones EW, & Broach JR, eds.) pp. 361-398, Cold Spring Harbor Laboratory, Cold Spring Harbor, NY

Serrano R (1988) Structure and function of proton translocating ATPase in plasma membranes of plants and fungi. Biochim Biophys Acta 947:1-28

Serrano R, Kielland-Brandt MC, & Fink GR (1986) Yeast plasma membrane ATPase is essential for growth and has homology with the $(Na^+ + K^+)$, K^+-, and Ca^{2+}-ATPases. Nature 319:689-693

Shull GE & Greeb J (1988) Molecular cloning of two isoforms of the plasma membrane Ca^{2+} transporting ATPase from rat brain. J Biol Chem 263:8646-8657

Shull GE & Lingrel JB (1986) Molecular cloning of the rat stomach ($H^+ + K^+$)-ATPase. J Biol Chem 261:16788-16791

Shull GE, Schwartz A, & Lingrel JB (1985) Amino-acid sequence of the catalytic subunit of the ($Na^+ + K^+$) ATPase deduced from complementary DNA. Nature 316:691-695

Taylor JW, Ott J, & Eckstein F (1985) The rapid generation of oligonucleotide-directed mutations at high frequency using phosphorothioate-modified DNA. Nuc Acids Res 13:8765-8785

Toyoshima C, Sasabe H, & Stokes DL (1993) Three-dimensional cryo-electron microscopy of the calcium ion pump in the sarcoplasmic reticulum membrane. Nature 362:469-471

Vilsen B, Andersen JP, & MacLennan DH (1991) Functional consequences of alterations to hydrophobic amino acids located at the M_4S_4 boundary of the Ca^{2+}-ATPase of sarcoplasmic reticulum. J Biol Chem 266:18839-18845

Walworth NC & Novick PJ (1987) Purification and characterization of constitutive secretory vesicles from yeast. J Cell Biol 105:163-174

The role of two small proteolipids associated to the H$^+$-ATPase from yeast plasma membrane

Catherine Navarre, Serge Leterme, Michel Ghislain and André Goffeau.

Unité de Biochimie Physiologique,
Université Catholique de Louvain,
Place Coix du Sud, 2-20,
1348 louvain-la-Neuve,
Belgium.

Introduction

The plasma membrane of *Saccharomyces cerevisiae* contains a well-characterized ATPase that pumps protons out of the cell energizing the membrane for nutrient uptake. This H$^+$-ATPase is similar to the other members of the cation-translocating ATPase family and has a single catalytic subunit of about 100 kDa, which forms an aspartyl phosphate catalytic intermediate during ATP hydrolysis (see reviews by Serrano, 1988; Goffeau and Green, 1990).

However, upon electrophoretic conditions adapted to separating low-molecular-weight proteins (Schägger and von Jagow, 1987), this purified H$^+$-ATPase is contaminated by two small Coomassie-stained compounds of 7.5 and 4 kDa. Chloroform/methanol extraction, followed by ether precipitation of these two small compounds establish their proteolipidic nature (Navarre et al., 1992).

Sequence of two plasma membrane isopoproteolipids

Microsequencing of the proteolipidic fraction extracted from *S.cerevisiae* plasma membrane reveals two very similar 38-residue polypeptides with a calculated molecular weight of about 4250 Da. These two polypeptides, termed PMP1 and PMP2 (for Plasma Membrane Proteolipids), differ by only one residue and are two isoforms of the same proteolipid (Figure 1).

We have isolated and sequenced the two corresponding genes, referred to as PMP1 and

NATO ASI Series, Vol. H 89
Molecular and Cellular Mechanisms
of H$^+$ Transport
Edited by Barry H. Hirst
© Springer-Verlag Berlin Heidelberg 1994

PMP2, from a genomic DNA library of *S.cerevisae*, using an oligonucleotide probe derived from a portion of the amino acid sequence. The nucleotide sequences are 92% identical and confirm the amino acid sequence of the 2 proteins (Figure 1). The amino-terminal part of the PMP proteins, which was cleaved in the mature proteolipids, is 3 amino acids longer in PMP2 than in PMP1. The mature proteolipids PMP1 and PMP2 are otherwise identical, except for the residue 21, which is an alanine for PMP1 and a serine for PMP2.

The hydropathy analysis of these two isoproteolipids predicts amphipatic peptides with a single hydrophobic transmembrane domain (residues 1-22) and a highly basic C-terminal cytoplasmic tail (residues 23-38) (Figure 2). This bipartite structure is similar to that of phospholamban, a 52-residue proteolipid reported to regulate the Ca^{++}-ATPase of cardiac sarcoplasmic reticulum (see review by Tada et al., 1988).

```
PMP1 ATG --- --- --- ACT TTA CCA GGT GGT GTT ATT TTA GTT TTC
      M            T   L   P   G   G   V   I   L   V   F
           L   M   S
PMP2 ATG TTG ATG AGC ACG TTA CCA GGT GGT GTT ATC TTA GTT TTT


ATT TTG GTC GGT TTG GCT TGT ATT GCC ATT ATT GCT ACC ATT ATC
                                          A
 I   L   V   G   L   A   C   I   A   I   I   T   I   I
                                          S
ATT CTA GTC GGT TTG GCT TGT ATC GCC ATC ATT TCT ACC ATT ATC


TAC AGA AAA TGG CAA GCT AGA CAA AGA GGA TTG CAA AGA TTC TAA
 Y   R   K   W   Q   A   R   Q   R   G   L   Q   R   F   -
TAC AGA AAA TGG CAA GCT AGA CAA AGA GGT TTA CAA AGA TTC TAA
```

Figure 1. Nucleotide and deduced amino acids sequences of the PMP1 and PMP2 genes.
The amino acids are shown in standard one letter code. The residues cleaved during protein processing are underlined. The nucleotides which are different between the two isogenes are in bold.

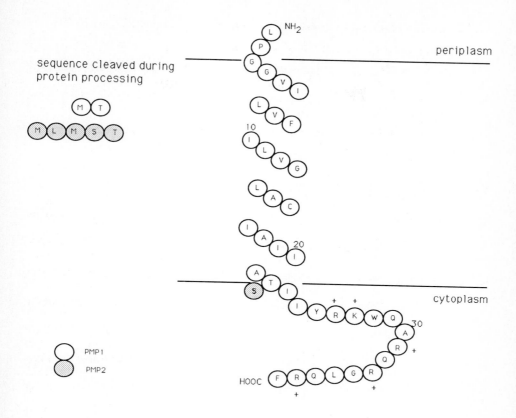

Figure 2. Cartoon model of the proteolipids catography.

Phenotype associated with the deletion of PMP genes

We have constructed single and double null mutants of PMP genes from the *S. cerevisiae* haploid strain W 303-1B (MATα leu2 his3 ade2 trp1 ura3); the PMP1 and PMP2 genes were entirely deleted and replaced by the URA3 and HIS3 markers, respectively.

Plasma membranes were purified from the *S.cerevisiae* strains W303-1B α, W303-1B Δ PMP1 α, W303-1B Δ PMP2 α and W303-1B Δ PMP1 Δ PMP2 α. The 7.5-kDa band corresponding to the dimeric proteolipid is present in the wild type strain and in the strains deleted in one of the two PMP genes (Figure 3A). In contrast, this 7.5-kDa band is not detected in the double PMP null mutant. To confirm the absence of the isoproteolipids in the plasma membrane of the strain deleted in the two PMP genes, the proteolipidic fraction was extracted. As expected, this fraction does not show any proteic bands (Figure 3B) and NH_2-terminal sequencing of this extract does not reveal any proteolipid sequence.

The plasma membrane H^+-ATPase activity was modified by the deletion of the PMP genes. When the cells were incubated in glucose before the purification of the membranes, the Vmax of the H^+-ATPase activity of the doubly PMP deleted strain was less than 50% of that of the wild type (Table 1). If only one PMP gene (PMP1 or PMP2) was deleted, the Vmax of the enzyme was intermediate between that of the wild type and that of the double null mutant (Table 1). However, the strong decrease of the ATPase Vmax value in the null mutant was not associated with a modification of the growth rate in normal conditions (data not shown).

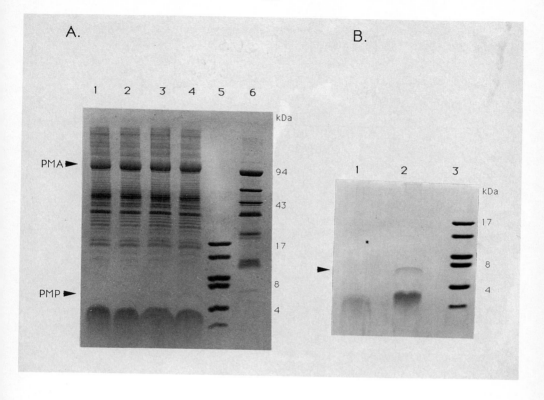

Figure 3. Coomassie Blue-stained Tricine SDS-PAGE analysis of plasma membranes proteins from PMP deleted strains.

A. Lanes 1, W303-1BΔPMP1ΔPMP2 plasma membranes (50 μg). lane 2, W303-1BΔPMP2 plasma membranes (50 μg). Lane 3, W303-1BΔPMP1 plasma membranes (50 μg). Lane 4, W303-1B plasma membranes (50 μg). Lanes 5 and 6, light and heavy molecular weight markers.

B. Lane 1, W 303-1BΔPMP1ΔPMP2 proteolipidic fraction. Lane 2, W 303-1B proteolipidic fraction. Lane 3, light molecular weight markers.

Table 1. Effects of deletion of PMP genes on plasma membrane H$^+$-ATPase activity from glucose-activated yeast strains

Strain	Vmax (μmole Pi/min/mg)		Km (mM MgATP)
W303-1B	10.1	(100%)	1.1
W303-1BΔPMP1	6.5	(64%)	1.1
W303-1BΔPMP2	7.4	(74%)	0.9
W303-1BΔPMP1ΔPMP2	3.9	(39%)	0.7

Conclusions

We have purified and sequenced two small proteolipids of 38-residues contaminating the preparations of plasma membrane H$^+$-ATPase of *Saccharomyces cerevisiae*. The two corresponding genes have been isolated and disrupted. Preliminary evidences of a possible regulation of the plasma membrane H$^+$-ATPase Vmax activity by these two small proteolipids have been presented.

References

Goffeau A. and Green M. (1990) Monovalent cations in biological systems. Paternak C. and Phill D eds. CRC Press, Inc, Boca raton, FL. 155-169.

Navarre C., Ghislain M., Leterme S., Ferroud C., Dufour J-P. and Goffeau A (1992) Purification and complete sequence of a small proteolipid associated with the plasma membrane H$^+$-ATPase of *Saccharomyces cerevisae*. J. Biol. Chem. 267: 6425-6428.

Schägger H. and von Jagow G. (1987) Tricine-sodium dodecyl-sulfate polyacrylamide gel electrophoresis for the separation of proteins in the range from 1 to 100 kDa. Anal. Biochem. 166: 368-379.

Serrano R. (1988) Structure and function of proton translocating ATPase in plasma membranes of plant and fungi. Biochem. Biophys. Acta 947: 1-28.

Tada M., Kadoma M., Inui M. and Fujii J. (1988) Regulation of Ca^{++}-Pump from cardiac sarcoplasmic reticulum. Methods Enzymol. 157: 107-154.

Ion Translocation Stoichiometries of Two Endomembrane H^+-Pumps Studied by Patch Clamp

Julia M. Davies, Ian Hunt & Dale Sanders
Biology Department
University of York
York YO1 5DD
United Kingdom

The perception of the plant cell vacuole as primarily a storage organelle has become increasingly out-moded in recent years. That plant biologists can now describe the vacuole as the cell's "micro-kidney" (Taiz, 1992) without blushing, testifies to the strength of opinion that this is indeed a dynamic organelle, critical to metabolic regulation and the cell cycle.

Certainly, occupation of around 90% of cell volume in the typical mature plant cell is good qualification for a role in storage. Inorganic (K^+, Cl^-, NO_3^-) and organic solutes (sucrose, amino acids, proteins) are known to be partitioned (Taiz, 1992) and potentially toxic solutes may be sequestered (e.g. Martinoia et al., 1993). In part as a consequence of its storage facility, the vacuole contributes to the turgor potential which is vital to cell expansion. In addition, the presence of acid hydrolases (the pH of the vacuolar lumen may be two units or more below that of the cytoplasm) allows the vacuole to function as a lytic compartment. A role for the vacuole in intracellular signalling is indicated by the presence of $InsP_3$- and voltage-gated Ca^{2+} channels in the vacuolar membrane (tonoplast), each of which is capable of mobilising lumenal Ca^{2+} (Schumaker and Sze, 1987; Johannes et al., 1992) which is stored at mM concentrations.

Two Pumps - One Membrane

Energisation of the tonoplast and the concomitant acidification of the lumen are critical to vacuolar function and are facilitated by two electrogenic H^+-pumps - a V-type H^+-ATPase and an inorganic pyrophosphatase, H^+-PPase (Rea and Sanders, 1987). The H^+ electrochemical gradient established across the tonoplast drives H^+-

NATO ASI Series, Vol. H 89
Molecular and Cellular Mechanisms
of H+ Transport
Edited by Barry H. Hirst
© Springer-Verlag Berlin Heidelberg 1994

coupled secondary transport systems; the membrane potential ($\Delta\Psi$) component may regulate channel activity (Johannes et al., 1992) while the acidified lumen facilitates optimal hydrolytic activity. The reason for the co-residence of the two H^+-pumps in the tonoplast has been the subject of intense speculation. Studies with intact vacuoles showed that both were capable of generating a $\Delta\bar{\mu}_{H+}$ under putative *in vivo* conditions (Johannes and Felle, 1990). As the ATPase shows the greater activity in mature cells one possibility is that the H^+-PPase is an auxilliary pump which regulates cytosolic PPi (since soluble pyrophosphatases are restricted to plastids). Thus hydrolytic free energy is put to useful purpose while PPi-producing reactions are pulled to completion by virtue of control of cytoplasmic PPi levels (Rea and Sanders, 1987).

A clearer understanding of the physiological role of the H^+-PPase has come from a consideration of the energetics of cellular K^+ partitioning and the obligatory dependence of the enzyme on K^+ for transport activity. In most cells K^+ is the predominant vacuolar cation. With vacuolar K^+ concentrations typically around 200 mM (and as high as 500 mM in guard cells), cytoplasmic K^+ homeostatically maintained around 100 mM and an *in vivo* cytoplasm-negative tonoplast membrane potential of -20 mV to -50 mV, K^+ transport into the lumen would be against a gradient of 3 to 7 kJ.mol^{-1}. Although the tonoplast is known to contain a K^+-H^+ antiport which could in theory facilitate K^+ accumulation, its activity would be severely inhibited at typical cytoplasmic K^+ concentrations (Cooper et al., 1991). This raises the possibility that the H^+-PPase is not only stimulated by K^+ but also translocates it into the vacuole. A test of this hypothesis is not feasible with methodology based on radiometric assays in tonoplast vesicles, since K^+ leakiness in this experimental system precludes meaningful measurement of PPi-driven K^+ fluxes. However, higher plant vacuoles are, among endomembranes, uniquely amenable to use of the patch clamp technique. This permits the high resolution assay of voltage-dependent transport activity in defined solutions. We have therefore patch clamped intact vacuoles from *Beta vulgaris* root storage tissue using a configuration analogous to the "whole cell" mode which allows the macroscopic current generated by a whole population of membrane pumps to be measured.

The Case for a K⁺-H⁺-PPase

In a series of vectorial substitution experiments, it was shown that the H^+-PPase requires K^+ specifically at the cytoplasmic face of the enzyme to operate hydrolytically (forward mode) and pump positive charge out of the cytosol (Davies et al., 1991). Conversely, the H^+-PPase can only be reversed by orthophosphate (Pi) and translocate positive charge into the cytosol when K^+ is present solely at the lumenal enzyme face (Davies et al., 1992). Such results are consistent with PPi-driven K^+-H^+ symport but do not substantiate the hypothesis, since kinetic activation of the pump by K^+ is not distinguished from K^+ transport. Direct evidence for PPi-driven K^+ transport was gained from a thermodynamic study of pump action. If the PPase were to translocate K^+ according to the relationship:

$$n[H^+]_c + m[K^+]_c + PPi \Leftrightarrow n[H^+]_v + m[K^+]_v + 2Pi \qquad \text{Eq.1}$$

then the reversal potential (E_{rev}) of the enzyme-generated current (the point at which the enzyme is at equilibrium and the membrane potential at which no net current passes) becomes:

$$E_{rev} = \{RT/(n+m)F\}\ln\{[PPi]^2.[H^+]_v^n.[K^+]_v^m/K_{PPi}.[PPi].[H^+]_c^n.[K^+]_c^m\} \qquad \text{Eq.2}$$

where [] signifies chemical activity, subscripts c and v refer to cytoplasmic and vacuolar compartments respectively, n and m are, respectively, the coupling ratios of H^+ and K^+ translocated per PPi, K_{PPi} is the equilibrium constant for PPi hydrolysis and R, T, F have their usual meanings. K_{PPi} may be computed for given conditions of pH, K^+, Mg^{2+}, PPi and Pi (Davies et al., 1992, 1993). The prevailing K^+ and H^+ gradients and the respective translocation stoichiometries therefore determine the value of E_{rev} in an unequivocal fashion.

Absolute values of E_{rev} and shifts in E_{rev} as trans-tonoplast gradients were altered were found to depend on both H^+ *and* K^+ (Davies et al., 1992). For example, as cytosolic K^+ was increased from 30 to 100 mM (with a fixed pH gradient) E_{rev} became more postive, as predicted by Eq. 2. Direct K^+ translocation by the PPase is the simplest explanation of such results.

Physiological action of the K^+-H^+-PPase

Quantitative comparison of experimental and predicted values of E_{rev} for a variety of possible values of n and m revealed that the enzyme is likely to pump three positive charges per cycle. However, our data are not consistent with integer values for n and m individually; rather, the respective coupling ratios are 1.3 H^+ and 1.7 K^+ for the range of conditions employed in these studies (Fig. 1).

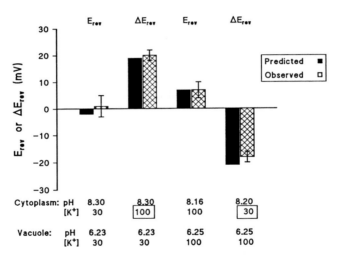

Figure 1. Comparison of observed with predicted E_{rev} values for shifts in $[K^+]_c$ and a coupling ratio of 1.3 H^+:1.7 K^+ per PPi.

It may well be that of the three cation transport sites on the enzyme, at least one translocates K^+ or H^+, with competition between the two ions leading to the phenomenological non-integer coupling ratios. A preliminary test of this hypothesis is the examination of the relationship between the K_m for K^+ stimulation of hydrolysis and H^+ concentration. Data from tonoplast vesicles show a clear linear relationship between the two parameters (Fig.2).

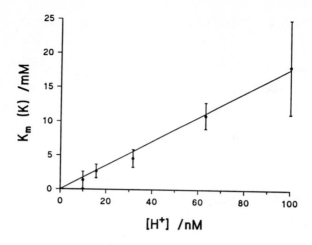

Figure 2. Variation with H^+ concentration of the K_m for K^+ stimulation of H^+-PPase hydrolysis in *Beta vulgaris* tonoplast vesicles. (Unpublished data; I. Williamson, University of York).

For typical cytosolic conditions (Davies et al., 1993), vacuolar K^+ could reach 900 mM before its concentration brings the pump to equilibrium. This capacity would be of special significance for guard cells in which vacuolar K^+ accumulation results in osmotically-induced swelling and stomatal opening. Thus as the pump responsible for vacuolar K^+ accumulation, the H^+-PPase has a clear physiological purpose.

An additional possible role for the H^+-PPase comes from consideration of the hydrolytic free energy available from PPi during anoxia. Calculation of K_{PPi} values for simulated *in vivo* conditions of anaerobiosis has suggested that the free energy of hydrolysis (ΔG) for PPi comes to within 4 to 5 kJ.mol^{-1} that of ATP under the condition of lowered cytosolic pH likely to prevail in the anaerobic plant cell. As PPi levels are known to remain constant during prolonged anoxia while those of ATP fall (Dancer and ap Rees, 1988), not only may the relative increase in available energy from PPi help sustain cellular reactions but the H^+-PPase could remain active in vacuolar H^+ sequestration thus preventing toxic reduction of cytosolic pH and enabling the plant cell to endure anoxia.

Additional Binding Sites for the H^+-PPase

In vitro the PPase requires Mg^{2+}, K^+ and PPi for activity. Its response to changes in total Mg^{2+} and PPi are complex, regardless of the source tissue, and the presence of metal ion and protonated complexes of PPi makes reaction analyses especially tortuous. Reaction kinetic modelling of *in vitro* hydrolysis by oat root H^+-PPase has been used to assess the importance of various complexes in PPase activity (Leigh et al., 1992). This has revealed that, in contrast to many soluble PPases, the substrate may be Mg_2PPi (rather than MgPPi) but that this complex can also act as a non-competitive inhibitor, requiring a separate binding site. Free PPi may act as a competitive inhibitor of hydrolysis (acting at the substrate site) and free Mg^+ activates the enzyme, again requiring a discrete regulatory binding site. Under *in vivo* conditions, the enzyme is likely to operate at around 70% of its maximum *in vitro* activity.

Unlike the vacuolar H^+-ATPase, the H^+-PPase is strongly inhibited by free Ca^{2+} (inhibition is apparent even at sub-micromolar free Ca^{2+}), a characteristic which could afford independent regulation of the two H^+ pumps by changes in cytoplasmic Ca^{2+}. Inhibition is not competitive with free Mg^{2+} and kinetic modelling suggests the presence of a discrete Ca^{2+} regulatory site on the enzyme (Rea et al., 1992). As such, Ca^{2+} signalling could be involved in cytoplasmic PPi homeostasis (PPi levels are known to be stable at around 0.25 mM). This scheme is supported by evidence which indicates that sucrose synthesis is inhibited by suppression of PPi turnover (Brauer et al., 1990) in conditions where free cytoplasmic Ca^{2+} is also elevated (Gilroy et al., 1987).

Variable Coupling Ratio of the H^+-ATPase

Kinetic estimates of the H^+-ATPase nH^+:ATP ratio suggest a value of two (e.g., Schmidt and Briskin, 1993). However, this ratio would constrain the pH gradient attainable across the tonoplast; neither is it consistent with the strong homology of the enzyme with the F_oF_1 H^+-ATPase. Moreover, kinetic estimates are subject to intrinsic experimental errors which tend to lead to underestimation of n. Arguably the most severe of these are unquantified pump uncoupling and H^+ leakage, to which vesicles are especially prone. Patch clamp may be applied in the measurement of H^+-ATPase stoichiometry, as it was with the H^+-PPase, where:

$$E_{rev} = \{RT/nF\} \ln \{[ADP].[Pi].[H^+]_v^n/K_{ATP}.[ATP].[H^+]_c^n\}$$ Eq.3

Using bafilomycin A_1 as a specific inhibitor of V-ATPases (Bowman et al., 1988) to identify the H^+-ATPase-generated current in *Beta vulgaris* vacuoles (and hence the E_{rev} for a specific trans-tonoplast H^+ gradient) we have found that n varies critically with cytoplasmic pH (Davies, Hunt & Sanders; unpublished data). At pH_c 7, n is 3 but drops to below 2 as pH_c rises to 8 (bafilomycin inhibition of H^+-ATPase hydrolytic action was unaffected by this pH change). Moreover, contrary to previous speculation (e.g., Moriyama and Nelson, 1988) membrane potential does not induce this uncoupling. *In vivo* predictions of pump action suggest that this intrinsic uncoupling would allow the H^+-ATPase to acidify the lumen to below pH 3. It is envisaged that there are three (possibly more) H^+ binding sites and that the enzyme can operate with one or more sites unoccupied. This hitherto unobserved uncoupling capability would allow the H^+-ATPase to play a more critical role in cytosolic pH regulation than had been previously thought and has strong evolutionary implications when viewed in the context of its homology with the F_oF_1 H^+-ATPase.

References

Bowman EJ, Siebers A, Altendorf K (1988) Bafilomycins: A class of inhibitors of membrane ATPases from microrganisms, animal cells and plant cells. PNAS USA 85:7972-76

Brauer M, Sanders D, Stitt M (1990) Regulation of photosynthetic sucrose synthesis; a role for calcium? Planta 182:236-43

Cooper S, Lerner HR, Reinhold L (1991) Evidence for a highly specific K^+/H^+

antiporter in membrane vesicles from oil-seed rape hypocotyls. Plant Physiol 97:1212-20

Dancer JE, ap Rees T (1989) Effects of 2,4-dintrophenol and anoxia on the inorganic-pyrophosphate content of the spadix of *Arum maculatum* and the root apices of *Pisum sativum*. Planta 178:421-24

Davies JM, Rea PA, Sanders D (1991) Vacuolar proton-pumping pyrophosphatase in *Beta vulagris* shows vectorial activation by potassium. FEBS Lett 278:66-68

Davies JM, Poole RJ, Rea PA, Sanders D (1992) Potassium transport into plant vacuoles energized directly by a proton-pumping inorganic pyrophosphatase PNAS USA 89:11701-05

Davies JM, Poole RJ, Sanders D (1993) The computed free energy change of hydrolysis of inorganic pyrophosphate and ATP: apparent significance for inorganic-pyrophosphate-driven reactions of intermediary metabolism . Biochim Biophys Acta 1141:29-36

Gilroy S, Hughes WA, Trewavas AJ (1987) Calmodulin antagonists increase free cytosolic calcium levels in plant protoplasts *in vivo*. FEBS Lett 212:133-37

Johannes E, Felle HH (1990) Proton gradient across the tonoplast of *Riccia fluitans* as a result of the joint action of two electroenzymes. Plant Physiol 93:412-17

Johannes E, Brosnan JM, Sanders D (1992) Parallel pathways for intracellular Ca^{2+} release from the vacuole of higher plants. The Plant Journal 2:97-102

Leigh RA, Pope AJ, Jennings IR, Sanders D (1992) Kinetics of the vacuolar H^+-pyrophosphatse. Plant Physiol 100:1698-1705

Martinoia E, Grill E, Tommasini R, Kreuz K, Amrhein N (1993) ATP-dependent glutathione S-conjugate 'export' pump in the vacuolar membrane of plants. Nature 364:247-249

Moriyama Y, Nelson N (1988) The vacuolar H^+-ATPase, a proton pump controlled by a slip. In "The ion pumps: Structure, function and regulation" pp. 387-394. Ed. WD Stein. Pub. Alan Liss Inc.

Randall SK, Sze,H (1986) Properties of the partially purified tonoplast H^+-pumping ATPase from oat roots. J Biol Chem 261:1364-71

Rea PA, Sanders D (1987) Tonoplast energization: two H^+ pumps, one membrane. Physiol Plant 71:131-41

Rea PA, Britten CJ, Jennings IR, Calvert CM, Skiera LA, Leigh RA, Sanders D (1992) Regulation of vacuolar H^+-pyrophosphatase by free calcium. Plant Physiol 100:1706-15

Schmidt AL, Briskin DP (1993) Energy transduction in tonoplast vesicles from red beet (*Beta vulgaris*) storage tissue: H^+/substrate stoichiometries for the H^+-ATPase and H^+-PPase. Arch Biochem Biophys 301:165-173

Schumaker KS, Sze H (1987) Inositol-1,4,5-trisphosphate releases Ca^{2+} from vacuolar membrane vesicles of oat roots. J Biol Chem 262:3944-46

Taiz L (1992) The plant vacuole. J Exp Biol 172:113-22

Molecular Dissection of Vacuolar H+-Pyrophosphatase

Philip A. Rea, Eugene J. Kim, Yongcheol Kim and Rui-Guang Zhen
Plant Science Institute
Department of Biology
University of Pennsylvania
Philadelphia
PA 19104-6018, U.S.A.

1. Introduction

It is now established that inorganic pyrophosphate (PPi) is a major energy source for electrogenic H+-translocation across the vacuolar membrane (tonoplast) of plant cells (Rea *et al* 1992a; Rea and Poole 1993). The enzyme responsible, the vacuolar H+-translocating pyrophosphatase (V-PPase; EC 3.6.1.1), which is an abundant membrane constituent, comprising 1-10% of total vacuolar membrane protein, exclusively utilizes PPi as substrate (Rea and Poole 1986) and, in parallel with the V-ATPase (EC 3.6.1.3), generates an inside-acid, inside-positive transtonoplast H+-electrochemical potential difference. The H+ gradient so generated serves to energize a wide range of ΔpH and/or $\Delta\psi$-coupled transport processes, including the vacuolar accumulation of low molecular weight solutes (sugars, amino acids, carboxylic acids, mineral ions) and the lumenal localization of storage proteins and lysosomal-type hydrolases. Because the plant vacuole is the largest intracellular organelle known, frequently accounting for 40-99% of total intracellular volume, and directly participates in many fundamental physiological processes such as cytosolic pH stasis, "excretion" of metabolically perturbing agents (e.g. xenobiotics and alkaloids), sequestration of regulatory Ca^{2+}, turgor regulation and nutrient storage and retrieval, any meaningful account of plant cell metabolism is contingent on an understanding of the functional characteristics and organization of the primary energizers responsible for the establishment of the requisite transmembrane ion gradients: the V-PPase and V-ATPase.

In this chapter we deal with four facets of the V-PPase: its molecular identity, the use of 1,1-diphosphonates as type-specific inhibitors, biochemical studies directed at defining the organization of the enzyme in the membrane and the results of expression studies which demonstrate that the "substrate-binding" subunit, alone, is sufficient for all of the known functions of the pump. Our treatment is, by necessity, brief: those readers requiring a more comprehensive and physiologically-oriented review of the literature are directed to Rea and Poole (1993).

2. Molecular Identity

The V-PPase has a rudimentary subunit composition. Whereas all characterized V-ATPases are "F-like" in structure and consist of 7-10 distinct polypeptide species organized into discrete heteromultimeric, nucleotide-binding, peripheral (V_1) and H+-translocating, integral (V_0) sectors (Nelson 1992; Sze *et al* 1992), only one polypeptide species with an apparent M_r of 64,500-66,800 has been shown unequivocally to be associated with the V-PPase (Britten *et al*

NATO ASI Series, Vol. H 89
Molecular and Cellular Mechanisms
of H+ Transport
Edited by Barry H. Hirst
© Springer-Verlag Berlin Heidelberg 1994

1989; Maeshima and Yoshida 1989; Rea *et al* 1992a). Detergent-solubilization and chromatography of tonoplast vesicles yields a single M_r 64,500-66,800 polypeptide which strictly copurifies with K^+-activated PPi hydrolytic activity (Britten *et al* 1989; Maeshima and Yoshida 1989; Rea *et al* 1992b; Sarafian and Poole 1989).

Direct participation of the M_r 64,500-66,800 polypeptide in substrate (Mg_2PPi) binding is demonstrated by its kinetics of labeling by the sulfhydryl reagent [^{14}C]-*N*-ethylmaleimide (NEM) (Britten *et al* 1989; Rea *et al* 1992b). The V-PPase activity of native tonoplast vesicles is subject to ligand-modified, irreversible inhibition by NEM. Inhibition is pseudo-first order and quantitative protection is conferred by substrate (Mg^{2+} + PPi) whereas free PPi (PPi minus divalent cation) increases the potency of NEM by a factor of two versus control membranes incubated with NEM in the absence of ligands. And, in accord with the purification data, treatment of isolated tonoplast vesicles with [^{14}C]-NEM after pretreatment with [^{12}C]-NEM in the presence of Mg^{2+} + PPi - to block nonprotectable, nonessential NEM-reactive groups and protect, protectable groups - generates a single differentially ^{14}C-labeled polypeptide of M_r 64,500-66,800 (Britten *et al* 1989; Rea *et al* 1992b). Since labeling is abolished by Mg^{2+} + PPi and potentiated by free PPi, in exact correspondence with the kinetics of inactivation of the enzyme, and the same polypeptide is found in purified preparations, the M_r 64,500-66,800 polypeptide is deduced to contain the NEM-reactive, substrate-binding ("catalytic") site whose alkylation by NEM is responsible for irreversible inhibition of the V-PPase (also see *Section 5*).

3. A New Category of Ion Translocase

Three near-identical cDNAs encoding the V-PPase from *Arabidopsis thaliana* and *Beta vulgaris* have recently been isolated and characterized (Sarafian *et al* 1992; Kim, Y., Kim, E.J. and Rea, P.A., unpublished). All three clones, designated AVP, BVP-1 and BVP-2, for *Arabidopsis* **V**acuolar **P**yrophosphatase and isoforms 1 and 2 of the enzyme from **B**eta, encode a 770 amino acid polypeptide with a predicted mass of 81 kDa.

Computer-assisted hydropathy plots establish that AVP, BVP-1 and BVP-2 encode an extremely hydrophobic integral membrane protein containing 13-16 transmembrane spans. In addition, several of the putative hydrophilic domains contain clusters of charged residues which may participate in cation (Mg^{2+}, Ca^{2+}, K^+) or anion (PPi^{4-}, $MgPPi^{2-}$) binding. The topological model proposed on the basis of the sequence data (Figure 1) is necessarily speculative, and its validity will depend on the results of direct biochemical investigations (see *Section 5*), but the inferred overall orientation of the 81 kDa polypeptide is in accord with the "positive-inside" rule wherein, for most polytopic membrane proteins, the majority of the positively charged amino acid residues are disposed to the cytoplasmic face of the membrane.

Notwithstanding the need for detailed protein chemical and/or mutagenic studies to validate, refute or refine such sequence-based models, a fundamental conclusion to come from the molecular cloning of the V-PPase is recognition of its novelty. The V-PPase is representative of a new class of primary ion translocase. All previously defined phosphoanhydride-energized H⁺ pumps fall into one of three major categories - F, P and V - based on their organization, mechanism and inhibitor sensitivity (Pedersen and Carafoli 1987). Thus, F-, P- and V-type ATPases are subject to inhibition by azide, orthovanadate and bafilomycin, respectively, and the enzymes within each category show significant sequence indentities, indicative of a common ancestry (Nelson and Taiz 1989; Pedersen and Carafoli 1987). The V-PPase, on the other hand, is not inhibited by any of these type-specific inhibitors and computer searches of the nucleotide sequences of AVP, BVP-1 and BVP-2, and the deduced sequences of their protein products, reveal no detectable homology between this pump and any other sequenced ion translocase (Sarafian *et al* 1992; Rea *et al* 1992a).

4. Type-Specific Inhibitors

Because of its novelty there has been demand for type-specific inhibitors of the V-PPase. However, other than through the measurement of azide-, orthovanadate- and bafilomycin-insensitive, K⁺-activated PPi-dependent H⁺-translocation (Rea and Turner 1990), strict criteria for the identification of V-PPase activity in uncharacterized membrane fractions have been lacking. In an attempt to remedy this deficiency we have screened a range of PPi analogs, primarily 1,1-diphosphonates, for activity against the V-PPase and other phosphohydrolases (Zhen *et al* 1994).

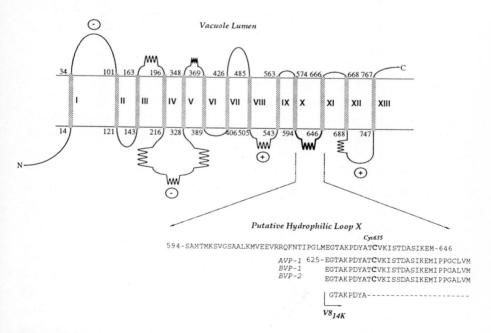

Figure 1. Tentative topological model of V-PPase showing location of NEM-reactive cysteine residue (Cys 635) in putative hydrophilic loop X, based on the alignment of the N-terminal sequence of the [¹⁴C]-NEM-labeled 14 kDa peptide fragment (V8₁₄K) generated by digestion of the purified enzyme with V8 protease.

Two strategically important findings have resulted from these investigations: (i) 1,1-diphosphonates containing a heteroatom (NH_2 or OH) on the bridge carbon are exquisitely potent V-PPase inhibitors. When tested at a substrate concentration corresponding to the K_m of the enzyme (30 μM), the inhibitory potency (apparent inhibition constants, Ki_{app} values, μM, in parenthesis) of the PPi analogs follows the sequence: aminomethylenediphosphonate, O_3-P-CH(NH_2)-P-O_3 (1.8) > hydroxymethylenediphosphonate, O_3-P-CH(OH)-P-O_3 (5.7) ≅ ethane-1-hydroxy-1,1-diphosphonate, O_3-P-C(CH_3)(OH)-P-O_3 (6.5) > imidodiphosphate, O_3-P-NH-P-O_3 (12) > methylenediphosphonate, O_3-P-CH_2-P-O_3 (68) >> dichloromethylenediphosphonate, O_3-P-C(Cl_2)-P-O_3 (>500). (ii) When examined at a PPi concentration (300 μM) corresponding to ten times the K_m of the V-PPase for substrate - the usual concentration employed to assay the enzyme *in vitro* (Rea and Turner 1990) and the prevailing cytosolic concentration in plants (Weiner *et al* 1987; Takeshige and Tazawa 1989) - the most efficacious inhibitor, aminomethylenediphosphonate, is not only potent but also type-specific. It is 6- to 38-fold more active as an inhibitor than the two PPi analogs most commonly employed for investigations of the V-PPase, methylenediphosphonate and imidodiphosphate (e.g. Chanson and Pilet 1987), and under appropriate assay conditions exerts no inhibitory effect against other plant ion translocases (V-ATPase and plasma membrane H^+-ATPase) and soluble PPi hydrolases (soluble PPase, alkaline phosphatase and non-specific phosphomonoesterase). Hence, aminomethylenediphosphonate is an inhibitor of unsurpassed potency which should prove invaluable as a diagnostic probe for the preliminary identification of V-PPase activity and membrane fractions with which the enzyme is associated.

5. Alkylation of Cys635 is Responsible for Inhibition by NEM

The results of studies directed at defining the location of the residue(s) involved in enzyme inactivation by NEM provide the first direct test of the organization of the V-PPase in the membrane and refute earlier speculations concerning the identity of putative catalytic motifs associated with substrate binding. X-ray crystallographic and site-directed mutagenesis studies of the *soluble* PPase from *Saccharomyces* have disclosed 17 residues which are thought to directly participate in catalysis (Cooperman *et al* 1992). 11-16 of these residues (depending on alignment procedure) are conserved in all sequenced soluble PPases. It has therefore been considered to be of potential functional significance that eight of the active site residues of soluble PPases fall into two configurations - E(X)$_7$KXE and D(X)$_4$DXK(X)$_4$D - beginning at positions 48 and 146, respectively, in the sequence of the enzyme from *Saccharomyces* and that variants of these motifs - D(X)$_7$KXE and E(X)$_4$DXK(X)4D - with an equivalent spacing and alternation of acidic (Asp and Glu) and basic (Lys) residues are found at positions 257 and 119, respectively, in the deduced sequence of the V-PPase (Rea *et al* 1992a) (Figure 1). Although the V-PPase and soluble PPases appear to be remote evolutionarily, it has been proposed that they may share convergent motifs related to the need for both classes of enzyme to interact with the same substrates, inhibitors and activators (MgPPi, Mg_2PPi, Ca^{2+}, Mg^{2+}) (Rea *et al* 1992a). Moreover, because both of the V-PPase motifs are tentatively assigned a cytosolic orientation on the basis of hydropathy analyses and the motif starting at position 119 contains a Cys residue (Cys128) and is immediately flanked by another, it has been speculated that the sensitivity of the enzyme to inhibition and covalent modification by NEM and protection by substrate at the cytosolic face of the membrane is due to alkylation of one or both of these cysteine residues. It is now apparent, however, from peptide mapping experiments that this notion, at least in its most restrictive form, is not correct (Figure 1).

Differential labeling of tonoplast vesicles with [^{14}C]-NEM, purification of the M_r 66,000 subunit and its digestion with V8 protease yields a single M_r 14,000, ^{14}C-labeled peptide (V8$_{14K}$) after electrophoresis on Tricine gels and fluorography. The N-terminal sequence of this fragment - GTAKPDYA - unambiguously aligns the peptide to the C-terminal segment of the V-PPase, starting with the scissile Glu at position 625 (Figure 1). Since V8$_{14K}$ encompasses only one Cys residue (Cys 635) which is conserved between AVP, BVP-1 and BVP-2, and only one other (non-conserved) Cys residue is to be found on the C-terminal side of Glu125 (in AVP), Cys635 is inferred to be the residue which undergoes covalent modification by NEM to render the V-PPase inactive. It therefore appears that the substrate-protectable, NEM-reactive segment of the V-PPase corresponds to putative hydrophilic loop X (residues 594-646), not loop II as speculated previously. These results do not automatically preclude the participation of loops II and IV in substrate binding and/or turnover but the labeling data cannot be reconciled with direct involvement of the "soluble PPase-like" motifs (specifically that containing Cys128) in irreversible inhibition of the enzyme by NEM.

6. Sufficiency of Polypeptide Encoded by AVP for V-PPase Function

It has been argued that because cDNAs encoding the substrate-binding subunit of the V-PPase specify a transmembrane polypeptide satisfying the minimum structural requirements of a PPi-dependent H$^+$-translocase - direct interaction with substrate and continuity across the phospholipid bilayer - additional subunits are not required (Rea et al 1992a). Unfortunately, attempts to test this assertion biochemically through reconstitution of the pump have been largely unproductive. For instance, Britten et al (1992) detail a procedure for reconstitution of the transport function of the V-PPase purified from Vigna to produce proteoliposomes exhibiting high rates of K$^+$-activated, PPi-dependent H$^+$-translocation but SDS-PAGE of the final preparation reveals the coinsertion of two polypeptide species of M_r 20,000 and 21,000 in addition to the M_r 66,000 substrate-binding subunit. Furthermore, since the results of experiments aimed at determining whether the M_r 20,000 and/or 21,000 polypeptides participate in PPi-dependent H$^+$-translocation are self-contradictory, neither their identity nor function is known. Separation of the M_r 66,000 from the M_r 20,000 and 21,000 polypeptides before reconstitution by ion-exchange chromatography generates proteoliposomes active in PPi hydrolysis but deficient in H$^+$-translocation, whereas the complementary experiment - addition of the M_r 20,000 and 21,000 polypeptides back to the M_r 66,000 subunit before reconstitution does not restore transport (Britten et al 1992). Consequently, while the high transport activity of the reconstituted enzyme may facilitate future functional studies, reconstitution of the pump has not resolved the specific question of whether the M_r 66,000 polypeptide, alone, is sufficient for V-PPase-catalyzed transport. It was for this reason that the independent but complementary strategy of heterologously expressing cDNAs encoding the substrate-binding subunit and screening for PPi-dependent H+-translocation was adopted. If, and only if, the M_r 66,000 subunit is the sole constituent of the transport-competent enzyme should PPi-energized H+-translocation be demonstrable upon heterologous expression of AVP, BVP-1 or BVP-2, alone.

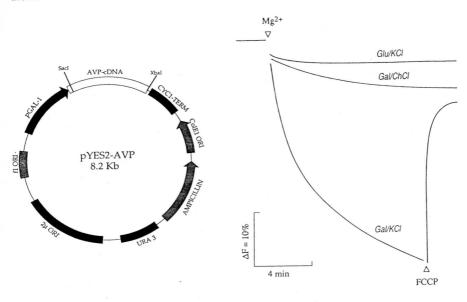

Figure 2. Map of the construct pYES2-AVP and characteristics of PPi-dependent H+-translocation by microsomes isolated from *Saccharomyces cerevisiae* (strain AACY1) after transformation with pYES2-AVP. Glu/KCl, membranes from glucose-grown transformants assayed for H+-translocation in the presence of 50 mM KCl; Gal/ChCl, membranes from galactose-induced cells assayed in the presence of 50 mM choline chloride; Gal/KCl, membranes from galactose-induced cells assayed in the presence of 50 mM KCl. Intravesicular acidification was measured with the fluorescent ΔpH indicator, acridine orange, after the addition of 1.3 mM Mg^{2+} to media containing 1 mM Tris-PPi.

The yeast, *Saccharomyces cerevisiae*, was the organism of choice for these studies. It lacks endogenous V-PPase (Rea *et al* 1992a; Manolson, M.F., *pers comms*) but possesses vacuoles and is amenable to rigorous genetic manipulation. Heterologous expression in this system was therefore thought to have the potential of fulfilling two major experimental objectives: (i) The provision of a near unequivocal test of the sufficiency of the M_r 66,000 subunit of the V-PPase for PPi-dependent H+-translocation. (ii) The development of an expression system for the analysis of *in vitro* mutagenized enzyme. Thus, the entire open reading frame of AVP was amplified with VENT polymerase using primers engineered to contain *Sna*BI and *Xba*I restriction sites at their termini to facilitate directional cloning. The resulting 2.3 kb, PCR-amplified fragment was cloned into the multiple cloning site of pYES2 expression vector to generate a plasmid containing the AVP insert with its translation initiation site located within 60 bp of the transcription start site. Most of the amplified DNA was then replaced with the original cDNA to ensure sequence fidelity and generate a construct in which AVP, lacking only the 5'-untranslated sequence, was inserted directionally between the GAL1 promoter and the CYC1 termination sequences of the plasmid (Figure 2).

Several ura⁻ strains of *Saccharomyces* were transformed with the (URA3-containing) construct, designated pYES2-AVP, by the LiOAc/PEG method, and selected for uracil prototrophy. The resulting transformants were grown on minimal selection medium containing 2% galactose and 0.2% fructose and aliquots of the cells from these, and untransformed, lines were harvested and lysed. In all of the transformants examined, but none of the untransformed strains, immunoreactive V-PPase M_r 66,000 polypeptide was detectable after SDS-PAGE.

One of the transformed strains, AACY1, was examined further and by all criteria - Northern and Western analysis, hydrolytic and transport assays - was found to contain membrane-localized, functionally integral *Arabidopsis* V-PPase. Moreover, the preparation of membranes enriched for the V-PPase from vesiculated spheroplasts yielded membrane vesicles at the 10/30% interface of discontinuous sucrose density gradients active in PPi-dependent H⁺-translocation. Untransformed yeast lack this activity and the activity seen in the pYES2-AVP transformants is found only after galactose induction before membrane extraction (Figure 2). Since the V-PPase activity obtained is specifically activated by K⁺ and inhibited by Ca²⁺ and aminomethylenediphosphonate, as is the endogenous plant enzyme (Rea and Poole 1993; Zhen *et al* 1994), all of the functional characteristics of the V-PPase are conserved in the heterologously expressed enzyme, showing that the polypeptide encoded by AVP is sufficient for the assembly of fully functional, transport-competent pump. Thus, the direct participation of other polypeptides, such as the M_r 20,000 and 21,000 species detected in reconstituted preparations of the V-PPase, need not be invoked to account for pump function.

7. Conclusions

The V-PPase is a novel, abundant, plant-ubiquitous H⁺ pump in which all of the known catalytic functions - PPi hydrolysis, H⁺-translocation, K⁺-activation and Ca²⁺ regulation - map to the same hydrophobic subunit. Thus, the V-PPase is both worthy of and amenable to detailed molecular dissection. The ready availability of cDNAs encoding the enzyme, the recent identification of aminomethylenediphosphonate as a type-specific V-PPase inhibitor and the development of *Saccharomyces* as a versatile system for heterologous expression of transport-competent pump will enable the parallel application of *in vivo* expression, site-directed mutagenic and protein chemical techniques to analyses of its cellular function and molecular organization. The V-PPase is the first PPi-dependent H⁺-translocase to have been characterized molecularly. This, in conjunction with its unique origins and recognition of PPi (the limiting case of a high energy phosphate) as a key metabolite in plants, will make future studies of this pump of broad biological significance.

8. Literature Cited

Britten CJ, Turner JC, Rea, PA (1989) Identification and purification of substrate-binding subunit of higher plant H+-translocating inorganic pyrophosphatase. FEBS Lett 256: 200-206

Britten CJ, Zhen R-G, Kim EJ, Rea PA (1992) Reconstitution of transport function of vacuolar H+-translocating inorganic pyrophosphatase. J Biol Chem 267: 21850-21855

Chanson A, Pilet P-E (1987) Characterization of the pyrophosphate-dependent proton transport in microsomal membranes from maize roots. Physiol Plant 74: 643-650

Cooperman BS, Baykov AA, Lahti R (1992) Evolutionary conservation of the active site of soluble inorganic pyrophosphatase. Trends Biochem Sci 17: 262-266

Nelson N. (1992) Evolution of organellar proton-ATPases. Biochim Biophys Acta Bio-Energetics 1100: 109-124

Nelson N. and Taiz L. (1989) The evolution of H+-ATPases. Trends Biochem Sci 14: 113-116

Maeshima M, Yoshida Y (1989) Purification and properties of vacuolar membrane proton-translocating inorganic pyrophosphatase from mung bean. J Biol Chem 264: 20068-20073

Parry RV, Turner JC, Rea PA (1989) High purity preparations of higher plant vacuolar H+-ATPase reveal additional subunits. Revised subunit composition. J Biol Chem 264: 20025-20032

Pedersen PL, Carafoli E (1987) Ion motive ATPases. I. Ubiquity, properties and significance to cell function. Trends Biochem Sci 12: 146-150

Rea PA, Kim Y, Sarafian V, Poole RJ, Davies JM, Sanders D (1992a) Vacuolar H+-translocating pyrophosphatases: a new category of ion translocase. Trends Biochem Sci 17: 348-353

Rea PA, Britten CJ, Sarafian V (1992b) Common identity of substrate-binding subunit of vacuolar H+-translocating inorganic pyrophosphatase of higher plant cells. Plant Physiol 100: 723-732

Rea PA, Poole RJ (1986) Chromatographic resolution of vacuolar H+-translocating pyrophosphatase from H+-translocating ATPase of higher plant tonoplast. Plant Physiol 81: 126-129

Rea PA, Poole RJ (1993) Vacuolar H+-translocating pyrophosphatase. Annu Rev Plant Physiol Plant Mol Biol 44: 157-180

Rea PA, Turner JC (1990) Tonoplast adenosine triphosphatase and inorganic pyrophosphatase. Methods Plant Biochem 3: 385-405

Sarafian V, Poole RJ (1989) Purification of an H+-translocating inorganic pyrophosphatase from vacuole membranes of red beet. Plant Physiol 91: 34-38

Sarafian V, Kim Y, Poole RJ, Rea PA (1992) Molecular cloning and sequence of cDNA encoding the pyrophosphate-energized vacuolar membrane proton pump of *Arabidopsis thaliana* 89: 1775-1779

Sze H, Ward JM, Lai S (1992) Vacuolar H+-translocating ATPases from plants: structure, function and isoforms. J Bioenerg Biomembr 24: 371-381

Takeshige K, Tazawa M (1989) Determination of the inorganic pyrophosphate level and its subcellular localization in *Chara corallina*. J Biol Chem 264: 3262-3266

Weiner H, Stitt M, Heldt HW (1987) Subcellular compartmentation of pyrophosphate and alkaline phosphatase in leaves. Biochim Biophys Acta 893: 13-21

Zhen R-G, Baykov AA, Bakuleva NP, Rea PA (1994) Aminomethylenediphosphonate: a potent type-specific inhibitor of both plant and phototrophic bacterial H+-pyrophosphatases. Plant Physiol 102: in press

Acknowledgments: This work was supported by grants from the Department of Energy (Grant № DE-FG02-91ER20055) and the National Science Foundation (Grant № DCB-9005330).

F-type H+-ATPase: Catalysis and Proton Transport

Atsuko Iwamoto, Hiroshi Omote, Robert K. Nakamoto,
Masatomo Maeda and Masamitsu Futai
Department of Organic Chemistry and Biochemistry
Institute of Scientific and Industrial Research,
Osaka University
Ibaraki, Osaka 567, Japan

Introduction

F_0F_1 H+ATPase (or F-type ATPase) catalyzes ATP synthesis or hydrolysis coupling with proton translocation (for reviews, see Futai *et al.*, 1989; Senior, 1990; Fillingame, 1990). The F-type ATPase of *Escherichia coli* is similar to those found in inner mitochondrial or chloroplast thylakoid membranes, and has contributed greatly to the understanding of this complicated enzyme. The catalytic site of the enzyme is in the β subunit or at the interface between the α and β subunits of the membrane extrinsic F_1 sector. The proton pathway is formed from the *a, b,* and *c* subunits of the membrane intrinsic F_0 sector. The γ, δ, and ε subunits of F_1 are required functionally and structurally to connect the catalytic subunits to the F_0 sector. The mechanism of ATP hydrolysis can be studied using purified F_1 (F_1 - ATPase).

Mutational analysis of the *E. coli* enzyme defined, at least partly, the catalytic site in the β subunit (Futai *et al.*, 1992ab; Iwamoto *et al.*, 1993). Furthermore, the active role(s) of the γ subunit in energy coupling between the chemical (ATP synthesis/hydrolysis) and osmotic reaction (proton translocation) has been recognized by similar analysis. In this article, we report our recent results indicating that the catalytic site is located near the ATP γ phosphate and discuss the role of the γ subunit in energy coupling.

Catalytic Site Defined by Biochemical and Genetic Studies
βLys-155 and βThr-156 residues

The βLys-155 and βThr-156 of the β subunit are in the Gly-Gly-Ala-Gly-Val-Gly-Lys-Thr sequence (residues 149 - 156 of the *E. coli* β subunit). A similar sequence, Gly-X-X-X-X-Gly-Lys-Thr/Ser, is conserved in ATP or GTP binding proteins and is known as a glycine-rich sequence or P-loop (Walker *et al.*, 1984). βLys-155 has been

NATO ASI Series, Vol. H 89
Molecular and Cellular Mechanisms
of H+ Transport
Edited by Barry H. Hirst
© Springer-Verlag Berlin Heidelberg 1994

shown to be near the γ phosphate moiety of ATP by affinity labeling using adenosine triphosphopyridoxal (Ida *et al.*, 1991), and the importance of βThr-156 has been suggested from studies on a defective mutant enzyme with an insertion of a glycine residue between βLys-155 and βThr-156 (Takeyama *et al.*, 1990).

For examination of the roles of the two residues, a series of mutants were constructed: βLys-155 → Ala, Ser, or Thr; βThr-156 → Ala, Cys, or Asp; βLys-155/βThr-156 → βThr-155/βLys-156; and βThr-156/βVal-157 → βAla-156/βThr-157 (Iwamoto *et al.*, 1991; Omote *et al.*, 1992). All mutants were defective in ATP synthesis and had very low (< 0.1% of wild type) membrane ATPase activities. Membrane vesicles prepared from these mutants did not show ATP-dependent proton translocation. Purified βAla-155, βSer-155, βAla-156, and βCys-156 mutant enzymes showed low rates of multisite (steady state) and unisite (single site) catalysis (< 0.02% and < 1.5%, respectively, of the wild-type rate). The rates of ATP binding (k_1) in unisite catalysis of mutant enzymes were much lower than that of the wild type: not detectable with the βAla-156 and βCys-156 enzymes and 10^2 fold lower than the wild type with the βAla-155 and βSer-155 enzymes. These results suggest that βLys-155 and βThr-156 are essential for catalysis (Iwamoto *et al.*, 1991; Omote *et al.*, 1992). Results on the βLys-155 residue are consistent with recent reports by Senior and coworkers (Senior & Al-Shawi, 1992; Senior *et al.*, 1993).

On the other hand, the βThr-156 → Ser mutant was active in ATP synthesis and had about 1.5-fold higher membrane ATPase activity than the wild-type. The H^+ translocation by mutant membranes was essentially the same as that by wild-type membranes. This result is consistent with the finding that the glycine-rich sequence of the β subunit can be replaced by that of the p21 *ras* protein (Gly-Ala-Gly-Gly-Val-Gly-Lys-Ser, residues 10 - 17) (Takeyama *et al.*, 1990): βThr-156 corresponds to Ser-17 in the glycine-rich sequence of the *ras* protein. Thus, the hydroxyl moiety of the βThr-156 residue is essential for catalysis. We used a similar approach to study the region between βGlu-161 and βLys-201, and found that βGlu-181 and βArg-182 are essential residues for catalysis (Park *et al.*, in preparation). A glycine-rich sequence is also found in the α subunit of the F-type ATPase (Gly-Asp-Arg-Gln-Thr-Gly-Lys-Thr, residues 169-176). Mutational studies indicated that αLys-175 and αThr-176, corresponding to βLys-155 and βThr-156, respectively, are not absolutely essential for catalysis, but that the location of these residues, or possibly the entire conserved sequence, is in the domain required for the subunit/subunit interactions (Jounouchi *et al.*, 1993). Such interactions are essential for enzyme stability and multisite catalysis.

Residues near the glycine-rich sequence

Residues located near the glycine-rich sequence or functionally related to this sequence can be mapped genetically by isolating second-site mutants: pseudo-revertants of mutants of the glycine-rich sequence may indicate two functionally/structurally related amino acid residues (from the first and the second-site mutations).

This approach has been fruitful in identifying residues near βGly-149. We found that the effect of the βSer-174 → Phe mutation was suppressed by the βGly-149 → Ser mutation (Iwamoto *et al.*, 1991). Other replacements (βAla-149 or βCys-149) at position 149 could also suppress the βPhe-174 mutation, whereas the βGly-149 → Thr or βGly-150 → Ser mutation could not. These results strongly suggest that βGly-149 and βSer-174 are located close together near the γ phosphate moiety of ATP. A single βCys-149 mutant enzyme was defective, indicating that the combination of the two deleterious mutations (βCys-149 and βPhe-174) gave an active enzyme. Results on the defective βCys-149 mutant prompted us to isolate other pseudo-revertants: we found that βGly-172 → Glu, βSer-174 → Phe, βGlu-192 → Val, or βVal-198 → Ala replacement suppressed

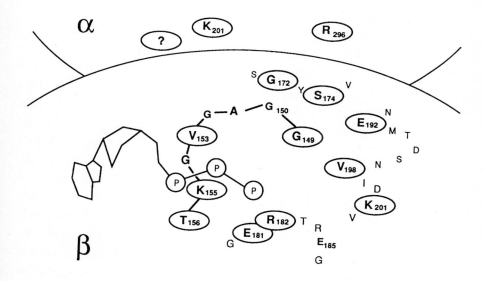

Fig. 1 **Model of the catalytic site of the F-type ATPase.** A model of the β subunit near the γ phosphate moiety of ATP is shown. The combined approach of affinity labeling with AP3-PL and analysis of random and directed mutants and pseudo-revertants suggests the amino acid residues near the γ phosphate moiety of ATP. See text for details. Modified from Futai *et al.* (1992ab) and Iwamoto *et al.* (1993).

the βCys-149 mutation (Iwamoto *et al.*, 1993). These results suggest that βGly-149, βGly-172, βSer-174, βGlu-192, and βVal-198 are located close together in the catalytic site. We proposed a model of the catalytic site (Futai *et al.*, 1992ab; Iwamoto *et al.*, 1993) (Fig. 1). βGlu-192 and βLys-155 are binding residues for DCCD (dicyclohexylcarbodiimide) (Yoshida *et al.*, 1982) and AP3-PL (adenosine triphosphopyridoxal) (Ida *et al.*, 1991), respectively. A part of AP3-PL also binds to βLys-201 (Ida *et al.*, 1991) Thus, we thought that double mutants might become less sensitive to these reagents if βGly-149, βGly-172, βSer-174, and βVal-198 are near βGlu-192 and βLys-155 (or βLys-201). As expected, F1-ATPases with the double mutations βCys-149/βGlu-172, βCys-149/βSer-174, βCys-149/βVal-192 and βCys-149/βAla-198 were less sensitive than the wild-type enzyme to DCCD and AP3-PL (Iwamoto *et al.*, 1993). These results support the proposed model of the catalytic site (Fig. 1).

Subunit/subunit interactions during catalysis can be studied by a similar approach. Recently, we found that the defect of the βSer-174 → Phe mutant was suppressed by an αArg-296 → Cys mutation, suggesting that close interaction between the α and β subunits is necessary for enzyme catalysis and that the βSer-174 residue in the catalytic site is possibly near αArg-296 of the α subunit (Omote *et al.*, in preparation).

Role of the γ Subunit in Energy Coupling

Minor Subunits (γ, δ, and ε) of F1

Although the α and β subunits have ATP binding sites, the minimum complex requied for ATPase activity is a combination of the α, β and γ subunits (Futai, 1977). Addition of the δ and ε subunits to the αβγ complex resulted in formation of an active F1 (αβγδε) complex capable of binding to Fo (Dunn & Futai, 1980). From these findings, it seemed likely that the minor subunits (γ, δ and/or ε) have roles in coupling the catalytic reaction carried out by the α and β subunits to proton transport through the Fo sector. If this is so, mutations in one of the minor subunits might cause defective energy coupling. Mutagenesis studies on the δ subunit suggest that this subunit is essential for binding F1 to Fo, but that individual residues may not have strict functional roles (Jounouchi *et al.*, 1992a). Similar results were obtained with the ε subunit by analyses of a series of mutants (Jounouchi *et al.*, 1992b). However, a mutant lacking 16 residues from the amino terminus gave interesting results. Membrane vesicles from this mutant had about 1/3 of the wild-type ATPase activity, but this ATPase was less effectively coupled with ATP hydrolysis than reconstituted membranes of the wild type having the same ATPase activity. Such membranes with defective coupling were not obtained from other ε mutants including those with missense mutations of conserved amino acid residues. Thus it seems reasonable to conclude that the amino-terminal deletion mutant membranes had low efficiency in forming an electrochemical proton gradient because the subunit mutation induced a structural alteration of the ε subunit or of the complex of the minor subunits and the α and β subunits.

Role of the γ Subunit in Regulation of H⁺ Transport

Extensive mutagenesis experiments suggested that the conserved carboxyl-terminal region is important for catalysis and energy coupling. The alignment of known γ subunits clearly indicated conservation of the amino-terminal region near γMet-23 and the carboxyl-terminal region. Initial mutagenesis studies on the carboxyl-terminal region suggested that this region is important for catalysis and energy coupling: γAla-283 → end and γThr-277 → end mutants could grow by oxidative phosphorylation and had 63 % and 14 %, respectively, of the wild-type membrane ATPase activity, whereas a γGln-269 → end mutant was defective in oxidative phosphorylation and had only very low membrane ATPase activity (Iwamoto et $al.$, 1990). Mutations between γThr-277 and γGln-269 suggested the importance of this region in catalysis and energy coupling. γGln-269 → Glu or Leu, γThr-273 → Val or Gly and γGlu-275 → Lys or Gln showed decreased ATPase activity and growth by oxidative phosphorylation. Interestingly, different mutants with similar membrane ATPase activities showed different proton-pumping abilities: γGln-269 → Leu and γThr-277→end mutants, and a frameshift mutant had lower proton pumping activity than γGlu-275 → Lys although the four mutants had similar membrane ATPase activities (about 15 % of the wild-type) (Iwamoto et $al.$, 1990).

We also introduced mutations to examine the role of the conserved amino-terminal region (between positions 19 and 33, 83 and 87, and at 165) of the γ subunit. Most of the mutations had no effects on oxidative phosphorylation. Exciting exceptions were γMet-23 → Lys and γMet-23 → Arg mutations (Shin et $al.$, 1992): although these mutants grew only very slowly by oxidative phosphorylation, they had high membrane ATPase activities (65 and 100 %, respectively, of the wild-type). These membranes formed only a low proton gradient. In contrast, the γMet-23 → Asp, γMet-23 → Glu, and γMet-23 → Leu mutants formed as high proton gradients as the wild type and grew well by oxidative phosphorylation. These results indicate that substitutions of γMet-23 by residues with positively charged side chains perturbed energy coupling, and suggest the role of the γ subunit in energy coupling.

The role of the γ subunit in energy coupling was confirmed by isolation of second-site mutations that could be mapped in the same subunit (Nakamoto *et al.*, 1993). We introduced random mutations in the γ subunit gene with a γLys-23 mutation and screened for pseudo-revertants capable of growth by oxidative phosphorylation. Strains carrying γArg-242 → Cys, γGln-269 → Arg, γAla-270 → Val, γIle-272 → Thr, Thr-273 → Ser, γGlu-278→ Gly, γIle-279 → Thr and γVal-280 → Ala substitutions in combination with the γLys-23 mutation were capable of growth by oxidative phosphorylation. γMet-23 → Lys, γGln-269 → Arg and Thr-273 → Ser as single mutations showed low growth yields at 37° by oxidative phosphorylation and positive growth at 30°. The double mutants γMet-23 → Lys/γGln-269 → Arg and γMet-23 → Lys/γThr-273 → Ser could grow at both temperatures. These results suggest that γMet-23, γArg-242, and the region between γGln-269 and γVal-280 are close to each other and interact for efficient energy coupling.

Summary

The molecular biological approach to the understanding of the structure and function of H$^+$ATPase is discussed. Results showed that βLys-155 and βThr-156 in the glycine-rich sequence (Gly-Gly-Ala-Gly-Val-Gly-Lys-Thr, residues 149-156) are essential for catalysis, and that βGly-149 is close to βGly-172, βSer-174, βGlu-192 and βVal-198. From results on mutants, a model of the catalytic site is proposed that is consistent with the inhibitor sensitivities of the mutant enzymes. Genetic studies suggested that the γ subunit plays a role in regulating energy coupling (coupling between catalysis and proton transport). For the regulation, the interaction of the amino-terminal with carboxyl-terminal region of the γ subunit was found to be essential.

This study was supported in part by research grants from the Ministry of Education, Science and Culture of Japan and the Human Frontier Science Program.

References

Dunn, S. D., Futai, M. (1980) Reconstitution of a functional coupling factor from the isolated subunits of *Escherichia coli* F1ATPase. J. Biol. Chem. 255: 113-118.

Fillingame, R. H. (1990) Molecular mechanism of ATP synthesis by F1Fo-type H+transporting ATP synthases. The Bacteria XII (Krulwich, T. A., ed) pp. 345 - 391, Academic Press, New York.

Futai, M. (1977) Reconstitution of ATPase activity from the isolated α, β, and γ subunits of the coupling factor, F1, of *Escherichia coli*. Biochem. Biophys. Res. Commun. 79: 1231-1237.

Futai, M., Noumi, T., Maeda, M. (1989) ATP synthase (H+-ATPase): Results by combined biochemical and molecular biological approaches. Annu. Rev. Biochem. 58: 111-136.

Futai, M., Iwamoto, A., Omote, H., Maeda, M. (1992a) A glycine-rich sequence in the catalytic site of F-type ATPase. J. Bioenerg. Biomemb. 24: 463-467.

Futai, M., Iwamoto, A., Omote, H., Orita, Y., Shin, K., Nakamoto, R.K., Maeda, M. (1992b) *Escherichia coli* ATP synthase (F-ATPase): Catalytic site and regulation of H+ translocation. J. Exp.Biol. 172: 443-449.

Ida, K., Noumi, T., Maeda, M., Fukui, T., Futai, M. (1991) Catalytic site of F1-ATPase of *Escherichia coli* Lys-155 and Lys-201 of the β subunit are located near the γ-phosphate group of ATP in the presence of Mg2+. J. Biol. Chem. 266: 5424-5429.

Iwamoto, A., Miki, J., Maeda, M., Futai, M. (1990) H+-ATPase γ subunit of *Escherichia coli*.: Role of the conserved carboxyl-terminal region. J. Biol. Chem. 265: 5043-5048.

Iwamoto, A., Omote, H., Hanada, H., Tomioka N., Itai, A, Maeda, M, Futai, M. (1991) Mutations in Ser-174 and the glycine-rich sequence (Gly-149, Gly-150, and Thr-156) in the β subunit of *Escherichia coli* H+-ATPase. J. Biol. Chem. 266: 16350-16355.

Iwamoto, A., Park, M.-Y., Maeda, M., Futai, M. (1993) Domains near ATP γ phosphate in the catalytic site of H+-ATPase. Model proposed from mutagenesis and inhibitor studies. J. Biol. Chem. 268: 3156-3160.

Jounouchi, M., Takeyama, M., Chaiprasert, P., Noumi, T., Moriyama, Y., Maeda, M., Futai, M. (1992a) *Escherichia coli* H+-ATPase : Role of the δ subunit in binding F1 to the Fo sector. Arch. Biochem. Biophys. 292: 376-381.

Jounouchi, M., Takeyama, M., Noumi, T., Moriyama, Y., Maeda, M., Futai, M. (1992b) Role of the amino terminal region of the ε subunit of *Escherichia coli* H+-ATPase (FoF1) Arch. Biochem. Biophys. 292: 87-94.

Jounouchi, M., Maeda, M., Futai, M. (1993) The α subunit of ATP synthase (FoF1): The Lys-175 and the Thr-176 residues in the conserved sequence (Gly-X-X-X-X-Gly-Lys-Thr/Ser) are located in the domain required for stable subunit-subunit interaction. J. Biochem. 114: 171-176.

Nakamoto, R. K., Shin, K., Iwamoto, A., Omote, H., Maeda, M., Futai, M. (1992) *Escherichia coli* FoF1-ATPase: residues involved in catalysis and coupling. Ann. N.Y. Acad. Sci. 671: 335-344.

Nakamoto, R. K., Maeda, M., Futai, M. (1993) The γ subunit of the *Escherichia coli* ATP synthase. Mutations in the carboxyl-terminal region restore energy coupling to the amino-terminal mutant γMet-23 → Lys. J. Biol. Chem. 268: 867-872.

Omote, H., Maeda, M., Futai, M. (1992) Effects of mutations of conserved Lys-155 and Thr-156 residues in the phosphate-binding glycine-rich sequence of the F1-ATPase β subunit of *Escherichia coli*. J. Biol. Chem. 267: 20571-20576.

Omote, H., Maeda, M., Futai, M. in preparation.

Park, M.-Y., Omote, H., Iwamoto, A., Maeda, M., Futai, M. in preparation.

Senior, A. E. (1990) The proton-translocating ATPase of *Escherichia coli*. Annu. Rev. Biophys. Biophys. Chem. 19: 7-41.

Senior, A. E., Al-Shawi, M. K. (1992) Further examination of seventeen mutations in *Escherichia coli* F1-ATPase β-subunit. J. Biol. Chem. 267: 21471-21478.

Senior, A. E., Wilke-Mounts, S., and Al-Shawi, M. K. (1993) Lysine 155 in β-subunit is a catalytic residue of *Escherichia coli* F1 ATPase. J. Biol. Chem. 268: 6989-6994.

Shin, K., Nakamoto, R. K., Maeda, M., Futai, M. (1992) F₀F1-ATPase γ subunit mutations perturb the coupling between catalysis and transport. J. Biol. Chem. 267: 20835-20839.

Takeyama, M., Ihara, K., Moriyama, Y., Noumi, T., Ida, K., Tomioka, N., Itai, A., Maeda, M., Futai, M. (1990) The glycine-rich sequence of the β subunit of *Escherichia coli* H+-ATPase is important for activity. J. Biol. Chem. 265: 21279-21284.

Walker, J. E., Saraste, M., Gay, N. J. (1984) The *unc* operon: Nucleotide sequence, regulation and structure of ATP-synthase. Biochim. Biophys. Acta 768: 164-200.

Yoshida, M., Allison, W. S., Esch, F. S., Futai, M. (1982) The specificity of carboxyl group modification during the inactivation of the *Escherichia coli* F1-ATPase with dicyclohexyl (¹⁴C)carbodiimide. J. Biol. Chem. 257: 10033-10037.

Contribution of infrared spectroscopy to the understanding of the structure of the *Neurospora crassa* plasma membrane H$^+$-ATPase

Erik Goormaghtigh, Laurence Vigneron and Jean-Marie Ruysschaert
Free University of Brussels
Campus Plaine CP206/2
Bld du Triomphe
B1050 Brussels
Belgium

The E$_1$-E$_2$ ATPases form a class of rather simple ATPases since in most cases a single polypeptide seems to be responsible for the ATP hydrolysis and ion transport (although the possible role of a short peptide is still considered). Among the E$_1$-E$_2$ ATPases, the *Neurospora crassa* plasma membrane H$^+$-ATPase is well characterised (see Slayman 1987) and was shown to be active as a monomer when purified and inserted into a lipid membrane (Goormaghtigh et al., 1986). Modelling of its structure should then be as simple as possible.

Even though the chemiosmotic theory of Mitchell (1961) satisfactorily describes the thermodynamics of ion transport and ATPase activity of the E$_1$-E$_2$ ATPases, the mechanism of the coupling between the hydrolysis of ATP and the active transport of ions through the membrane is still to be understood. This understanding requires a molecular description of the enzyme during its catalytic cycle which involves the orientation of the protein with respect to the lipid membrane. A number of experimental techniques including various labelling of the protein, protein chemistry, protein engineering, kinetic measurements, fluorescence spectroscopy, circular dichroism, ESR,... have brought important contributions to the knowledge of some structural details of the E$_1$-E$_2$ ATPases in their catalytic cycle. More recently, the primary sequence availability for a large number of related ATPases has prompted investigators to resort to prediction methods to obtain 1°) the localisation of membrane embedded segments and 2°) the secondary structure of membrane proteins. The main problem for evaluating the potentialities of the predictive methods for membrane proteins is the scarcity of experimental data. High resolution structures have been obtained for only three unrelated membrane proteins: the photosynthetic complex of *Rhodopseudomonas viridis* (Michel 1982), bacteriorhodopsin (Henderson et al., 1990) and bacterial porin (Weiss et al., 1991). Localisation of transmembrane segments is predicted with a reasonably good score for the two first proteins but fails for porin. The large proportion of polar and charged amino acids and even of

NATO ASI Series, Vol. H 89
Molecular and Cellular Mechanisms
of H$^+$ Transport
Edited by Barry H. Hirst
© Springer-Verlag Berlin Heidelberg 1994

prolines in transmembrane helices, especially in transport proteins (Brandl and Deber 1986), is expected to make the prediction of transmembrane segments a complex task in the general case. Prediction of secondary structure for soluble proteins is not yet very efficient as reviewed by Nishikawa and Nogushi (1991) and was found to be worst than a random choice for membrane proteins by Wallace and al. (1986), probably because the folding in a membrane environment can not be described by the rules elaborated for the soluble proteins.

In the present paper we propose to use infrared spectroscopy to gain experimental structure information on the *Neurospora crassa* plasma membrane H^+-ATPase reconstituted in model membranes or dispersed in a detergent.

Materials and Methods

The *Neurospora crassa* plasma membrane H^+-ATPase was purified as described by Smith and Scarborough (1984). Reconstitution into asolectin vesicles was obtained by removal of deoxycholate by gel filtration (Goormaghtigh et al., 1988) with protein/lipid ratio's (w/w) ranging from 1/100 to 1/10. Elimination of unincorporated ATPase and partial elimination of empty liposomes was achieved by ultracentrifugation on glycerol density gradient. This enrichment in the protein content is useful for the spectroscopic investigations. Activity reaches 23 μmol/min/mg protein. Alternatively, the ATPase was placed in a Mes-Tris buffer pH 6.8 containing 2.5 μg/ml of lysophosphatidylglycerol (lysoPG) by gel filtration. LysoPG is able to completely activate the ATPase (Hennessey and Scarborough 1988) and the low detergent/protein ratios makes possible some specific spectroscopic investigations. The hexameric assembly found for the purified ATPase (Chadwick et al., 1987) is not destroyed in these conditions.

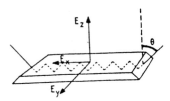

Figure 1: Germanium internal reflection element used for ATR-FTIR. The incident light angle $\theta = 45°$. The infrared light pathway within the germanium internal reflection crystal is indicated by the dotted line. Ex, Ey and Ez are the three components of the electric field of the light.

Infrared spectra were recorded by attenuated total reflection infrared spectroscopy of films deposited on germanium crystal internal reflection elements (figure 1). Films of ATPase-lysoPG complex or oriented multilayers were obtained as described elsewhere (Cabiaux et al., 1989) by slowly evaporating 40-100 µl of the sample on one side of the ATR plate under N_2 flow at room temperature. The presence of trehalose in the buffer allows the sample to be transiently dried without loss of activity (Crowe et al., 1987). The ATR plate was then sealed in an universal sample holder (Perkin-Elmer 186-0354) and hydrated by flushing D_2O-saturated N_2 (room temperature) for 90 min. Upon deuteration the "random structure" shifts from about 1655 cm^{-1} to about 1640 cm^{-1} upon H/D exchange allowing to differentiate the alpha-helix from the "random" structure. Protons belonging to the peptide bonds of ordered structures as alpha-helices, and especially transmembrane alpha-helices are not expected to exchange significantly in our experimental conditions. Details about the application of infrared attenuated total reflection (ATR) spectroscopy to membrane research were reviewed elsewhere (Goormaghtigh and Ruysschaert 1990). Attenuated total reflection infrared spectra were obtained on a Perkin Elmer 1720X FTIR spectrophotometer equipped with a liquid nitrogen cooled MCT detector at a resolution of 4 cm^{-1}. 256 scans were averaged. Every 4 scans, reference spectra of a clean Germanium plate were automatically recorded by a sample shuttle accessory and ratioed against the recently run sample spectra. The spectrophotometer was continuously purged with dry air. The internal reflection element was a germanium ATR plate (50x20x2 mm, Harrick EJ2121) with an aperture angle of 45° yielding 25 internal reflections (figure 1).

Secondary structure determination has been carried out as previously described (Goormaghtigh et al., 1990) from the analysis of the shape of amide I'. Briefly, in a first step we use Fourier self-deconvolution using a Lorentzian line shape (full width at half height FWHH=30 cm^{-1}) and a Gaussian line shape for the apodization. The spectrum arising from the lipid part of the system was found to be completely flat between 1700 and 1600 cm^{-1} and was therefore not subtracted. A first least square iterative curve fitting is performed with Lorentzian bands. The input parameters for this first curve fitting are chosen by our program as described (Goormaghtigh et al., 1990). This first curve fitting is carried out on spectra deconvoluted with K=2. In such conditions, the amide I' band presents well defined maxima and the fitting is carried out. In a next step we use the bands resulting from this curve fitting as input parameters for a fitting on the spectrum deconvoluted with K=1. The resulting fitting is analysed as follows: each band is assigned to a secondary structure according to the frequency of its maximum (see figure 2). The areas of all bands assigned to a given secondary structure are then summed up and divided by the total area. The number so obtained is taken to be the proportion of the polypeptide chain in that conformation. Potentialities and limitations of this methods have been discussed before (Goormaghtigh et al., 1990).

Hydrogen/deuterium (H/D) exchange kinetics were started by flushing the cell containing the ATPase film with D_2O-saturated N_2. A home made software allows the recording of spectra at logarithm spaced time intervals, starting with 0.25 min intervals.

Results

Spectrum of the *Neurospora crassa* plasma membrane H^+-ATPase reconstituted in asolectin oriented multibilayers appears in figure 2 as well as the final result of the Fourier self-deconvolution - curve fitting procedure designed to obtain the secondary structure of the protein.

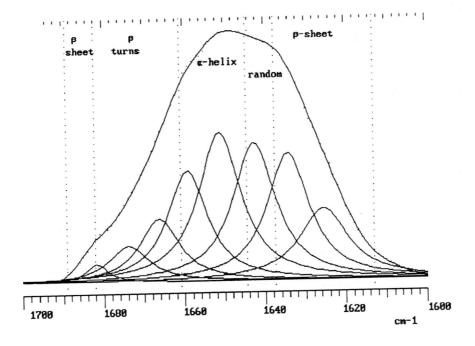

Figure 2: FTIR-ATR spectrum of the ATPase reconstituted in asolectin oriented multibilayers in the amide I region. Sample has been deuterated for 90 min. Decomposition of the spectrum into Lorentzian components as described in Materials and Methods.

Investigation of the secondary structure of the protein in different conformations can be carried out since it has been shown that this enzyme exists in different conformations in the presence of various ligands (Addison and Scarborough 1982, Mandala and Slayman 1988). The similarity of the spectra recorded in the presence of different ligands (not shown) demonstrates that the secondary structure is not significantly affected by the ligands. Results of the Fourier self-deconvolution and

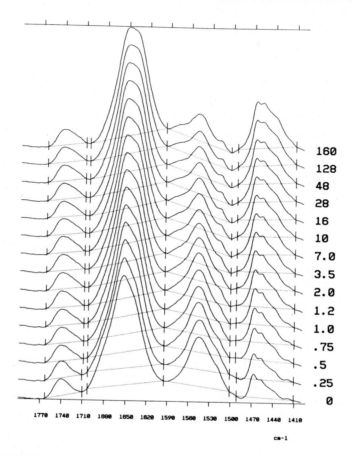

160
128
48
28
16
10
7.0
3.5
2.0
1.2
1.0
.75
.5
.25
0

1770 1740 1710 1680 1650 1620 1590 1560 1530 1500 1470 1440 1410

cm-1

Figure 3: ATR-FTIR spectra of the ATPase in the 1800-1400 cm^{-1} frequency range recorded as a function of the deuteration time which appears in the right hand side of the figure (in min). A selection of deuteration times appears on this figure which shows the decrease of the amide II (amide N-H deformation) band area (1600-1500 cm^{-1}) and the increase of the amide II' (amide N-D deformation) band area (1500-1400 cm^{-1}). The Amide I (amide C=O stretching) band (1700-1600 cm^{-1}) experiences simultaneously a small shift towards lower frequencies.

curve fitting procedures confirm that no significant conformational change at the level of the secondary structure can be put forward by the present approach: conditions tested are 100 µM vanadate, 10 mM Mg^{++}, 100 µM vanadate + 10 mM Mg^{++}, 2 mM MgATP + 100 µM vanadate, 2 mM MgADP. The α-helix content is found to be 38% and the β-sheet content 29% with a standard deviation of 1.2% for the helix and 3.5% for the β-sheet, i.e. not significantly different for the different ligands according the precision of the method (Goormaghtigh et al., 1990).

Since the secondary structure of the protein is very similar for the different conformations detected by limited proteolysis, it seems that the conformational difference concerns essentially the tertiary structure. We suggest here that one potential way to investigate the tertiary structural change is to measure H/D exchange kinetics. The rate of H/D exchange is indeed a measure of the accessibility of the amide groups to the solvent provided that other parameters such as the pH and the secondary structure remain constant. Figure 3 illustrates how the H/D exchange can be followed as a function of the time by FTIR-ATR.

Plotting the area of amide II or amide II' reported to the area of amide I for normalisation of the areas to a same protein amount in all experiments as a function of the time of deuteration allows to estimate the number of amide group exchanged per time unit. Figure 4 reports an example of such kinetic data obtained in the absence of ligands. Comparison with the kinetics obtained in the presence of ligands (not shown) indicates that the exchange rate is slower in the presence of ligands: Mg^{++} is similar to the control experiment but all experiments containing vanadate, Mg-vanadate, MgATP-vanadate or MgADP display a slowing down of the H/D exchange. In fact 10-15% of the amino acid residues seems to resist exchange in the presence of these ligands from the beginning of the deuteration and this value is constant up to the longer deuteration time carried out (160 min). Control experiments indicate that neither the ligands by themselves when tested on an unrelated protein nor the thickness of the film play any role on the changes observed.

In conclusion, it appears that the ligands which are known from limited proteolysis experiments to lock the Neurospora ATPase in different conformations do not significantly modify the secondary structure of the enzyme but do modify the surface of the protein accessible to the solvent. H/D exchange measurements can quantitatively evaluate this surface modification. This can be of significance for the understanding of the structure of the protein during the catalytic cycle.

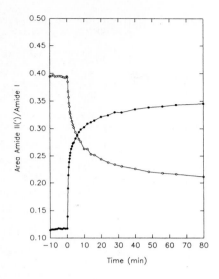
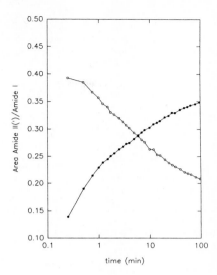

Figure 4: rate of H/D exchange in the absence of ligands. Open symbols indicate the decrease of the amide II area and filled symbols indicate the increase of the amide II' area. Data are reported on a linear (left side of the figure) and on a logarithm (right side of the figure) time scale.

References

Addison R. and Scarborough G.A. (1982) Conformational change of the Neurospora plasma membrane H^+-ATPase during its catalytic cycle. J. Biol. Chem. 257, 10421-10426.

Brandl C.J. and Deber C.M. (1986) Hypothesis about the membrane-buried proline residues in transport proteins. Proc. Natl. Acad. Sci. U.S.A. 83, 917-921.

Cabiaux V., Brasseur R., Wattiez R., Falmagne P., Ruysschaert J.M. and Goormaghtigh E. (1989) Secondary structure of diphtheria toxin and its fragments interacting with acidic liposomes studied by polarized infrared spectroscopy. J. Biol. Chem.,264, 4928-4938.

Chadwick C.C., Goormaghtigh E. and Scarborough G.A. (1987) A hexameric form of the Neurospora crassa plasma membrane H^+-ATPase. Arch. Biochem. Biophys. 252, 348-356.

Crowe J.H., Spargo B.J. and Crowe L.M. (1987). Preservation of dry liposomes does not require retention of residual water. Proc. Natl. Acad. Sci. USA 84, 1537-1540

Goormaghtigh E. and Ruysschaert J.M. (1990). Polarized attenuated total reflection infrared spectroscopy as a tool to investigate the conformation and orientation of membrane components. in "Molecular description of biological membrane components by computer aided conformational analysis", CRC, vol I pp 285-329.

Goormaghtigh E., Cabiaux V. and Ruysschaert J.M. (1990). Secondary structure and dosage of soluble and membrane proteins by attenuated total reflection FTIR spectroscopy on hydrated films. Eur. J. Biochem 193, 409-420.

Goormaghtigh E., Chadwick C. and Scarborough G.A. (1986). Monomers of the Neurospora plasma membrane H^+-ATPAse catalyze efficient proton translocation. J. Biol. Chemistry 261, 7466-7471.

Goormaghtigh E., Ruysschaert J.M. and G.A. Scarborough (1988) High yield incorporation of the Neurospora plasma membrane H^+-ATPase into proteoliposomes. Prog. Clin. Biol. Res. 273, 51-56.

Henderson R., Baldwin J.M., Ceska T.A., Zemlin F., Beckmann E. and Downing K.H. 1990 Model for the structure of bacteriorhodopsin based on high-resolution electron cryo-microscopy. J. Mol. Biol. 213, 899-929.

Hennessey J.P. and Scarborough G.A. (1988). Secondary structure of the *Neurospora crassa* plasma membrane H^+ ATPase as estimated by circular dichroïsm. J. Biol. Chem. 263, 3123-3130.

Mandala S.M. and Slayman C. (1988) Identification of tryptic cleavage sites for two conformational states of the Neurospora plasma membrane H^+-ATPase. J. Biol. Chem. 263, 15122-15128.

Michel H. 1982 Three-dimensional crystals of a membrane protein complex. The photosynthetic reaction centre from Rhodopseudomonas viridis. J. Mol. Biol. 158, 567-572.

Mitchell P. (1961), Coupling of phosphorylation to electron and hydrogen transfer by a chemi-osmotic type of mechanism. Nature 191, 144-148.

Nishikawa K. and Nogushi T. 1991 Predicting protein secondary structure based on amino acid sequence. Methods Enzymol. 202, 31-44.

Slayman C.L. (1987). The plasma membrane ATPase of Neurospora: a proton-pumping electroenzyme. J. Bioenerg. Biomembr. 19, 1-20.

Smith R. and Scarborough G.A. (1984), Large scale isolation of the Neurospora plasma membrane H^+-ATPase. Anal. Biochem. 138, 156-163.

Wallace B.A., Cascio M. and Mielke D.L. (1986). Evaluation of methods for the prediction of membrane protein secondary structures. Proc. Natl. Acad. Sci. (USA) 83, 9423-9427

Weiss M.S., Kreusch A., Schiltz E., Nestel U., Welte W., Weckesser J, and Schulz G.E. 1991 The structure of porin from Rhodobacter capsulatus at 1.8 Å resolution. FEBS Lett. 280, 379-382.

PARTIAL REACTION CHEMISTRY AND CHARGE DISPLACEMENT BY THE FUNGAL PLASMA-MEMBRANE H+-ATPASE

Clifford L. Slayman, Adam Bertl*, Michael R. Blatt#

Dept. of Cellular and Molecular Physiology
School of Medicine, Yale University
New Haven, CT 06510, USA

INTRODUCTION

Over the past five years, progress in understanding primary charge-displacement reactions in electrogenic ion pumps has come largely from studies on the sodium pump enzyme, the so-called Na^+,K^+-ATPase. This multiple-ion transporter/exchanger has been shown to occlude both sodium and potassium ions during the transport cycle, and to effect the major charge displacement either during sodium deocclusion or in a debinding step immediately following deocclusion. For kinetic reasons (Nakao & Gadsby, 1986; Stürmer et al., 1991; Gadsby et al., 1993) the sodium pump is now viewed as an ion-activating mechanism operating in series with a high-resistance transmembrane channel, such that the juncture between the two would serve as an ion well, exactly in the sense propounded by Mitchell (1969) for the F-type ATPases, to convert energy of electric fields into energy of chemical concentrations.

Both Na^+ deocclusion and charge displacement seem to be slow reactions, with rate constants of 20-180 sec^{-1} at normal temperatures (Forbush, 1984; Nakao & Gadsby, 1986; Läuger, 1991). Independent data on the voltage-dependent reaction step(s) in another P-type ATPase, the fungal plasma-membrane H^+-ATPase, show it to be nearly two orders of magnitude faster than charge displacement in the Na^+,K^+-ATPase. This result prompts a somewhat different view of the energetics and structure of P-type ATPases, including a simple device to obviate the high-resistance channel, explicitly for the *proton*-pumping enzyme.

H+-PUMP REACTION SCHEME

A general reaction diagram for the fungal H^+-ATPase is shown in Fig. 1-left. The vertical portion of the diagram, from H^+E_1ATP clockwise to E_1, has been drawn from the concensus reaction cycle for the Na^+,K^+-ATPase by substituting one H^+ for 3 Na^+ and leaving out all steps which would involve counterfluxing ions (i.e., K^+). Identification of the phospho-protein intermediate $[(H^+)E_1\sim P]$ came by isolation of a stable phosphoprotein from ATPase-enriched plasmalemma vesicles of *Neurospora* (Dame & Scarborough, 1980), and evidence for the precursor $[H^+E_1ATP]$ and product $[H^+E_2-P]$ states has come from reactivity studies on the *Saccharomyces* and *Schizosaccharomyces* enzymes (Amory et al., 1982; Wach & Gräber, 1991). Existence of the E_2-P state was inferred from $^{32}PO_4$-binding studies, but the

---

*Pflanzenphysiologisches Institut, Universität Göttingen, 3400 Göttingen, Germany
#Biochemistry & Biological Sciences, Wye College, Univ. of London, Wye TN25 5AH, UK

NATO ASI Series, Vol. H 89
Molecular and Cellular Mechanisms
of H+ Transport
Edited by Barry H. Hirst
© Springer-Verlag Berlin Heidelberg 1994

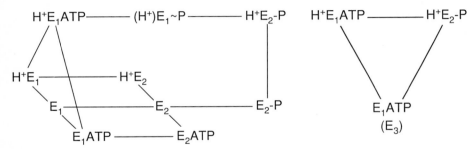

Figure 1. Reaction diagrams for the fungal plasma-membrane H+-ATPase. **Left:** "Realistic" model based on known partial-reaction chemistry (see text). **Right:** Lumped 3-state model for the purpose of electrical-kinetic analysis. Top line in both represents the likely voltage-dependent steps.

K_m for reaction is ~200 mM (Amory et al., 1982; cf. 5 mM for the Na+,K+-ATPase). Existence of an E_2-state is implied by the enzyme's vanadate reactivity; and the horizontal portion of the diagram, reaction of E_2 with either H+ or ATP and conversion to the corresponding E_1 substates, is supported by the finding on *Neurospora* enzyme that steady-state levels of the vanadate-reactive state are reduced either by elevating ATP at high pH or by lowering pH at zero ATP (Nagel et al, 1992). For energetic reasons (see below), outward proton transport must be associated with the top line in this scheme (H+E_1ATP→→H+E_2-P).

PUMP BLOCKADE BY ATP WITHDRAWAL

The principal strategy for obtaining electrical kinetic data on transport systems within intact biological membranes is to measure differences of membrane current with a specific transporter functioning, then blocked. Patch-clamp measurements of ion channels are intrinsically differential, and though the same might also be true for ion pumps, their much smaller individual currents (~$5 \cdot 10^{-17}$ A) at present demand synchronized mass blockade of pumps via chemical inhibitors. No fast-acting and specific inhibitor has yet been found for the fungal plasma-membrane H+-ATPase, and since perfusion of fungal membranes via patch pipettes has only recently become feasible (A. Bertl, unpub. expts.), we used general metabolic inhibitors to block the pump by ATP withdrawal.

The simplest agent for this purpose is cyanide (ca. 1 mM; pH 5.8), which blocks >98% of respiration in *Neurospora*, resulting in depletion of cytoplasmic ATP with a time-constant of a few seconds. Several other effects of sudden cyanide blockade are shown in Fig. 2; in this experiment, the plasma membrane depolarized by ~80 mV within seconds, repolarized transiently, then depolarized again to a quasi-steady level near -130 mV (Fig. 2A, top trace). The cytoplasmic ATP concentration, [ATP]$_i$, measured at discrete intervals in parallel experiments, declined rapidly to 10-15% of the control level (Fig. 2B, ●); and cytoplasmic proton concentration, [H+]$_i$, rose (more slowly) to about 0.4 μM (Fig. 2B, ○; pH$_i$ ~7.0 → ~6.4), resulting in only a small shift (20-25 mV) of the pump's equilibrium voltage (E_P). All changes were reversed, albeit with associated transients, by washout of cyanide (13 min. in Fig. 2).

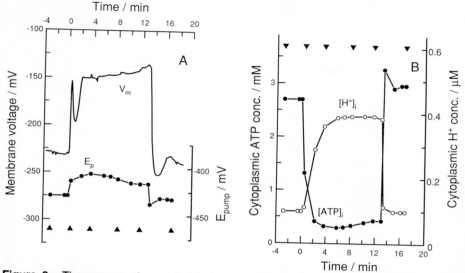

Figure 2. Time-course of cyanide blockade in *Neurospora* spherocytes. **A**: Membrane voltage (V_m) and H+-pump reversal voltage (E_P); **B**: Intracellular concentrations of H+ and ATP (parallel experiments). Cells grown overnight in (EG) ethylene glycol-containing medium, then preincubated in EG-free, glucose-free medium as previously described (Blatt et al., 1987). Carets designate the times at which I-V scans were taken (six from a total of 14).

EVOLUTION OF MEMBRANE CURRENTS

Comparison of the curve for V_m in Fig. 2A with that for E_p indicates that the kinetic effect of ATP withdrawal, rather than a change in driving force, is mainly responsible for the observed voltage changes. Quantitative analysis of the situation requires detailed knowledge of the relationship between ion flow through the H+-pump and membrane voltage (the so-called I-V relationship of the pump). Suitable data were obtained via a voltage-clamp procedure (Fig. 3, legend), imposed in 8-sec. scans where designated by the carets in Fig. 2. The six corresponding membrane I-V plots are shown in Fig. 3A, in the order (top-down): #'s 1(●), 14(□), 3(▼), 5(Δ), 10(○), and 8(▲) from the set of 14 scans. Even though no data were obtained during the transient *re*polarization, clearly the major *de*polarizing effect of cyanide was rapid, occurring at least within the time of ATP decay. More importantly, the I-V plots changed *shape* as well as scale, during sustained inhibition (cf., plots 8 and 3 with plot 1); and both were restored rapidly after washout of the inhibitor (cf. plot 14 with plot 1).

QUANTITATIVE FORMULATION

Analysis of these membrane I-V plots requires separate formulations for the proton pump and the background non-pump currents. *Neurospora* spherocytes (see legend to Fig. 2), when studied during pump blockade, show outward rectification, with negative and positive current asymptotes which extrapolate through the origin (V,I=0)--behavior which mimics

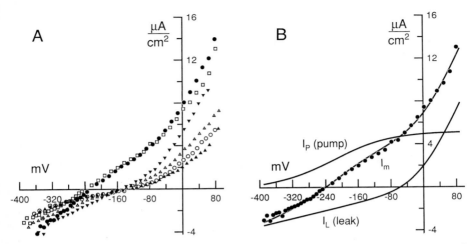

Figure 3. Current-voltage curves from the plasmalemma of *Neurospora* spherocytes. **A:** Six membrane I-V plots, corresponding to the 6 carets in Fig. 2. **B:** Replot of #14 (fully recovered from cyanide blockade), along with the constituent pump (I_P) and background leak (I_L) curves, drawn via text Eqs. 1-4. Data were obtained via a voltage-clamp procedure using a double-barreled microelectrode: one barrel to measure V_m, and the other to deliver the clamping current (I_m). A computer-generated double staircase of 100-msec pulses was imposed, scanning the voltage range from about. -350 mV to +100 mV, and the corresponding staircase of steady currents was recorded. Each such scan occupied about 8 sec.

a simple constant-field regime (Blatt et al., 1987). A generalized description of such current (I_L) can be written as follows:

$$I_L = P_X X_o F U \frac{e^{(U-U_L)} - 1}{e^U - 1} \quad , \tag{1}$$

in which X_o is the extracellular concentration of a presumed major permeant ion (X^+), U represents the reduced membrane voltage (FV_m/RT; Hansen et al., 1981), U_L is the reduced equilibrium voltage for X^+ (i.e., $\ln[X_o/X_i]$), and P_X is membrane permeability to X^+. This equation, used to fit plots of I_L versus V_m for all systems except the H^+ pump, contains two adjustable parameters: U_L and the product $P_X X_o$, which are related to the limiting slope conductance (G_L) at large positive membrane voltages by $G_{L\infty} = (P_X X_o F^2/RT) \exp(-U_L)$.

A general and relatively simple equation for the pump current can be derived by lumping together the likely voltage-dependent steps (upper line of Fig. 1-left), and separately lumping the voltage-independent steps into two groups: one for ATP binding, and one for H^+ binding. This leads to a condensed 3-state model for the pump, shown in Fig. 1-right. To simplify nomenclature, let the lumped reaction states H^+E_1ATP, H^+E_2-P, and E_1ATP be represented by E_1, E_2, and E_3, respectively, with the corresponding transition "rate" constants k_{12}, k_{23}, and k_{31} in the forward (clockwise) direction, and k_{13}, k_{32}, and k_{21} in the reverse direction. Because of the lumping procedure, these *operational* rate constants are complicated and need to be somewhat expanded for present purposes:

$$k_{12} = k_{12}^0 e^{(1-\delta)U}, \quad k_{21} = k_{21}^0 e^{-\delta U}, \quad k_{23} = k_{23}^0 [ATP], \quad k_{31} = k_{31}^0 [H^+]_i , \quad \text{(2a-d)}$$

where 0-superscripted rate constants designate standard conditions (i.e., $V_m = 0$ mV, $[ATP]_i$ = 1 M, and $[H^+]_i$ = 1 M), and voltage-dependence is assumed to arise from an Eyring barrier positioned a fraction (δ) of the way through the membrane dielectric (Hansen et al, 1981). Finally, pump current (I_P) is given by

$$I_p = FN \frac{k_{23}k_{31}k_{12} - k_{13}k_{32}k_{21}}{(k_{31}+k_{23}+k_{32})k_{12} + (k_{32}+k_{13}+k_{31})k_{21} + (k_{23}+k_{32})k_{13} + k_{31}k_{23}} , \quad \text{(3)}$$

where N is the density of transporter molecules in the membrane (i.e., sites/area). Then total membrane current becomes $I_m = I_P + I_L$ (Eq. 4), which must be fitted to the whole ensemble of membrane I-V curves.

A single fitted curve (#14) is shown in Fig. 3B (center), along with its decomposition into I_P (upper) and I_L (lower). The full sets of pump and leak curves are displayed separately in Fig. 4A and 4B. Several qualitative conclusions can be reached by inspection of Fig. 4 and comparison with Fig. 2. The saturating pump current (approx. at +80 mV) was roughly proportional to $[ATP]_i$. The failure of membrane voltage to track I_P, after the initial sudden decline of $[ATP]_i$ and/or pH_i, owed to a progressive decline of leak conductance (G_L: curve slopes in Fig. 4B). But changes in G_L lagged behind the changes in pump current, thus in a general way accounting for voltage transients. [Compare pairs of curves in Fig. 4A with corresponding curves in Fig. 4B: #'s 1 & 3, and #'s 8 & 14.]

ANALYTIC DETAILS

Since there is not sufficient information in the membrane I-V curves per se to extract all of the parameters for Eqs. 1-4, we have made several simplifications: i) Fixing E_P, which is a function of the six *standard* rate constants, at the values calculated in Fig. 2A. ii) Assigning a specific value to N: ~3100 sites/μm^2, estimated from freeze-fracture data (Slayman et al., 1990). Some such assignment is necessary because only relative values of the k's can be determined without a value for N. iii) For the pump, forcing k^0_{12} to the smallest allowable value. This was done to *minimize quantitative differences* between the computed rates for the charge transitions in the H^+-ATPase, versus the Na^+,K^+-ATPase. And finally, iv) forcing as many parameters as possible to common values for all 14 fitted curves, including the 6 in Fig. 2A. Satisfactory fits could be obtained only when at least one one parameter in I_L and one in I_P were adjusted from plot to plot. For the leak, adjustment was required in P_XX_0. For the pump, fits were obtained most easily by adjusting k^0_{23}; but to avoid changing *reaction energy*, fitting was carried out by allowing both k^0_{23} and k_{32} to vary, *in a fixed ratio*. A synopsis of the final optimized parameters is given in Fig. 4, legend.

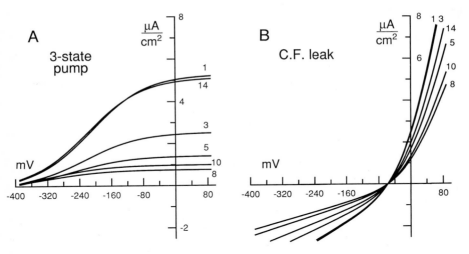

Figure 4. Synopsis of constituent fitted pump (**A**) and leak (**B**) I-V curves, for the six membrane I-V plots of Fig. 3A. All curves were obtained from a single joint fit of the full set (14) of data plots, with two parameters (k^0_{23} and P_XX_0) allowed to vary. Optimized parameters for these curves are as follows:

Parameters found separately for each curve			Parameters held common
Symbol (Fig. 3A)	Cv#	k^0_{23} (10^4 sec^{-1}M^{-1})	$G_{L\infty}$ (μS·cm^{-2})
●	1	3.90	105
▼	3	3.65	115
△	5	8.80	76.5
▲	8	4.32	47.8
○	10	4.75	57.4
▢	14	3.38	86.0

Parameters held common:

$k^0_{12} = 3.36 \times 10^3$ (sec^{-1})
$k^0_{21} = 3.92 \times 10^{-4}$ "
$k_{13} = 1.77 \times 10^4$ "
$k^0_{31} = 3.3 \times 10^{12}$ (sec^{-1}M^{-1})
$\delta = 0.625$
$k^0_{23}/k_{32} = 48.7$ (M^{-1})
$E_L = -57.5$ (mV)

Although the major effect of the cyanide-induced decline of cytoplasmic ATP concentration is a decline of pump current (Fig. 4A), the time-dependent variations of k^0_{23} and G_L indicate further complexities. Leak conductance declines when pH$_i$ falls, as might be expected, e.g., with pH-dependent K$^+$ channels (observed in *Saccharomyces*; A. Bertl, unpub. expts.); but *pump* reactivity actually seems to *increase*. Under most conditions, rate-limitation in enzyme forward cycling is associated with the ATP reaction, since k^0_{23}[ATP]$_i$ is never much larger than 100 sec^{-1}, while $k^0_{12} \geq 3360$ sec^{-1} and k^0_{31}[H$^+$]$_i$ approximates 3 x 10^5 sec^{-1}. The ratio of the two voltage-dependent parameters, $k^0_{12}/k^0_{21} = 8.6 \times 10^6$, corresponds to about 400 mV (= (RT/F)ln(k^0_{12}/k^0_{21})), localizing about 80% of the free energy shift from ATP hydrolysis (~500 mV; Warncke & Slayman, 1980) to voltage-dependent steps within the pump. Finally, the absolute magnitude of k^0_{12}, ≥ 3360 sec^{-1}, contrasts strongly with the pulse-relaxation data and Na$^+$-deocclusion results on the Na$^+$, K$^+$-ATPase.

SPECULATIVE RECONCILIATION

The concept of a high-resistance access channel between the membrane exterior and the catalytically reactive sites of the Na+,K+-ATPase has the virture of describing a wide range of kinetic data. But a physical consequence of that arrangement is intrinsic inefficiency: a major fraction of the energy of ATP hydrolysis must be dissipated (I^2R) across the internal resistance, and so is unavailable to form *external* gradients, either electrical or chemical. Such a pump would be a poor electric generator. In addition, a high-resistance access channel could screen the primary chemical reactions--and charge displacements within them--from easy measurement with external electrodes.

Thus, an interesting way in which to reconcile the apparent slowness (~100 sec^{-1}) of the major voltage-dependent reaction in the Na+,K+-ATPase with the speed ($\geq$ 3360 sec^{-1}) of the lumped voltage-dependent step(s) in the fungal H+-ATPase would be to suppose that the latter reflects the fundamental chemical reactions whereas the former reflects mainly a filtering process. The basic chemical reactions--Na+-catalyzed- versus H+$_3$O-catalyzed protein phosphorylation and rearrangement--could be the same, but with the latter feeding into a *low-resistance access path*. This should have structural implications.

It has been pointed out many times (see e.g., Nagle & Nagle, 1983) that protons are special ions, which could *in principle* transfer across biological membranes via ice-like structures, by flipping hydrogen bonds, rather than by the literal transport required for ordinary inorganic ions and radicals. Such a "proton wire" need not involve the protein conformational changes presumed to underlie gating and/or transit of sodium ions, for example, through membrane channels. And without either the literal transfer of H+$_3$O or protein conformational changes, a proton wire could be expected to operate rapidly, in a low-resistance mode rather than a high-resistance mode. There is evidence that K+ and Na+ bind to the (E_2-form of) the fungal proton ATPase at low [H+]$_i$, but cannot replace H+ in the catalytic cycle (G. Nagel & E. Bashi, unpub. expts.). This might happen either because Na+ and K+ cannot catalyze formation of the phosphoprotein (corresponding to (H+)E_1~P in Fig. 1-left), or because the transmembrane passageway admits only protons. A proton wire should be *very* selective.

There are, of course, other possible ways to resolve the large difference in apparent rate constants for charge-displacement steps in the fungal H+-ATPase and the animal Na+,K+-ATPase. Assuming the rate calculations to be correct for both sets of measurements, it may be that the fundamental reactions in the H+-ATPase really do occur more rapidly than in the Na+,K+-ATPase. Or perhaps charge-displacement occurs at different steps of the chemical reaction for the two enzymes. Or perhaps the idea of a high-resistance access channel into the Na+,K+-ATPase needs modification, as is suggested by the I^2R problem (see above). Nevertheless, it is an attractive notion that the important difference between these two P-type ATPases lies not in the ion-activation process per se, but in the mode by which activated ions are carried through the membrane segments of the enzyme.

For the H+-ATPase, perhaps hydronium ions are activated, but protons are transported.

ACKNOWLEDGEMENTS

Supported by Research Grants GM-15858 from the National Institute of General Medical Sciences, and ER-13359 from the U.S. Department of Energy, and by Forschungs-Stipendium Be 1181/2-1 from the Deutsche Forschungsgemeinschaft (to AB). The authors are indebted to Dr. U.-P. Hansen, to Mr. John Rose, and to Mr. Duncan Wong for computer programming and assistance with data analysis.

REFERENCES

Amory A, Goffeau A, McIntosh DB, Boyer PD (1982) Exchange of oxygen between phosphate and water catalyzed by the plasma membrane ATPase from the yeast *Schizosaccharomyces pombe*. J Biol Chem 257:12509-12516.

Blatt MR, Rodriguez-Navarro A, Slayman CL (1987) Potassium-proton symport in *Neurospora*: Kinetic control by pH and membrane potential. J Membr Biol 98:169-189.

Bowman BJ, Mainzer SE, Allen KE, Slayman CW (1978) Effects of inhibitors on the plasma membrane and mitochondrial adenosine triphosphatases of *Neurospora crassa*. Biochim Biophys Acta 512:13-28.

Dame JB, Scarborough GA (1980) Identification of the hydrolytic moiety of the *Neurospora* plasma membrane H+-ATPase and demonstration of a phosphoryl-enzyme intermediate in its catalytic mechanism. Biochem 19:2931-2937.

Forbush B (1984) Na+ movement in a single turnover of the Na pump. Proc Natl Acad Sci USA 81:5310-5314.

Gadsby DC, Rakowski RF, DeWeer P (1993) Extracellular access to the Na,K pump: Pathway similar to ion channel. Science 260:100-103.

Hansen U-P, Gradmann D, Sanders D, Slayman CL (1981) Interpretation of current-voltage relationships for "active" ion transport systems: I. Steady-state reaction-kinetic analysis of Class-I mechanisms. J Membr Biol. 63:165-190.

Läuger P (1991) Electrogenic Ion Pumps. Sinauer Assoc., New York.

Mitchell P (1969) Chemiosmotic coupling and energy transduction. *In* Cole A, ed., Experimental Biophysics. Marcel Dekker, New York. pp. 159-216.

Nagel G, Bashi E, Firouznia F, Slayman CL (1992) Probing the molecular mechanism of a P-type H+-ATPase by means of fluorescent dyes. *In* Quagliariello E, Palmieri F, eds., Molecular Mechanisms of Transport. Elsevier, Amsterdam. pp. 33-40.

Nagle JF, Tristram-Nagle S (1983) Hydrogen bonded chain mechanisms for proton conduction and proton pumping. J Membr Biol 74:1-14.

Nakao M, Gadsby DC (1986) Voltage dependence of Na translocation by the Na/K pump. Nature 323:628-630.

Slayman CL, Kaminski P, Stetson D (1990) Structure and function of fungal plasma-membrane ATPases. *In* Kuhn PJ, Trinci APJ, Jung MJ, Goosey MW, Copping LG, eds., Biochemistry of Cell Walls and Membranes in Fungi. Springer-Verlag, Berlin. pp. 299-316.

Stürmer W, Bühler R, Apell H-J, Läuger P (1991) Charge translocation by the Na,K-pump: II. Ion binding and release at the extracellular face. J Membr Biol 121:163-176.

Wach A, Gräber P (1991) The plasma membrane H+-ATPase from yeast. Effects of pH, vanadate and erythrosine B on ATP hydrolysis and binding. Eur J Biochem 201: 91-97.

Warncke J, Slayman CL (1980) Metabolic modulation of stoichiometry in a proton pump. Biochim Biophys Acta 591:224-233.

MOLECULAR PROPERTIES, KINETICS AND REGULATION OF MAMMALIAN Na$^+$/H$^+$ EXCHANGERS

Chung-Ming Tse, Susan A. Levine, C.H. Chris Yun, Steven R.Brant, Samir Nath, Jacques Pouysségur*, and Mark Donowitz.

Departments of Medicine and Physiology, GI Unit, The Johns Hopkins University School of Medicine and Dept of Biochemistry, University of Nice, France

INTRODUCTION

Na$^+$/H$^+$ exchange was first described by Murer, Hopfer and Kinne [19] in renal brush border membrane vesicles. This process is mediated by Na$^+$/H$^+$ exchangers which catalyze the exchange of extracellular Na$^+$ for intracellular H$^+$ with a stoichiometry of 1:1. Na$^+$/H$^+$ exchangers have multiple functions, including pH homeostasis, volume regulation, cell proliferation, and transcellular Na$^+$ absorption [reviewed in 12]. In no cell is it the only mechanism for any one of these functions. For instance, multiple mechanisms of pH homeostasis are present in most eukaryotic cells including a Cl$^-$/HCO$_3^-$ exchanger, a NaHCO$_3^-$ co-transporter, a Na$^+$-dependent Cl$^-$/HCO$_3^-$ exchanger and multiple mechanisms of H$^+$ extrusion [reviewed in 15], including the H-K-ATPase pump. In this review, we will focus on recent advances in identification and understanding of the structure/function relationships and acute protein kinase regulation of members of the mammalian Na$^+$/H$^+$ exchanger gene family.

MOLECULAR IDENTIFICATION OF Na$^+$/H$^+$ EXCHANGER GENE FAMILY.

Molecular identification of the mammalian Na$^+$/H$^+$ exchanger was pioneered by Pouyssegur *et al* who used genetic complementation [6,12] with fibroblast cell lines that they had selected to lack all endogenous Na$^+$/H$^+$ exchangers (the Chinese hamster lung fibroblast derived cell line PS120 and the mouse fibroblast derived cell line LAP1 [10,22]). Since then, three additional members of this gene family have been identified by our group and by Orlowski and Shull. We have named them in

NATO ASI Series, Vol. H 89
Molecular and Cellular Mechanisms
of H$^+$ Transport
Edited by Barry H. Hirst
© Springer-Verlag Berlin Heidelberg 1994

the order of their molecular identification as NHE1(standing for N̲a̲⁺/H̲⁺ E̲xchanger), NHE2, etc. NHE1 is the Na⁺/H⁺ exchanger isoform initially cloned by Pouyssegur et al [27]. It has been cloned from human, rabbit, rat, pig (LLC-PK₁ cells) and Chinese hamster; and contains 815-820 amino acids (species variation). NHE2 has been cloned from rabbit and rat and has 809 and 813 amino acids, respectively [35,38]. NHE3 has been cloned from rat, rabbit and human and has 831, 832 and 832 amino acids, respectively [21,30, Brant, Tse and Donowitz, unpublished results]. NHE4 has been cloned from rat and has 717 amino acids [21]. The corresponding predicted sizes of NHE1, NHE2, NHE3 and NHE4 based on amino acid composition as predicted from cDNAs without considering glycosylation are ~91 kD, ~91 kD, ~93 kD and ~81 kD, respectively. All NHEs exhibit about 50 % amino acid identity with respect to each other. NHE2 and NHE4 are most closely related with 60 % amino acid identity and NHE3 is the least related isoform [32].

As predicted from NHE primary structure, with differences being present throughout the entire amino acid sequence, NHEs are separate gene products. NHE1 gene has been localized to human chromosome 1 p35-p36.1 by in situ hybridization [18] and the NHE3 gene has been physically mapped to the distal portion of chromosome 5p 15.3 [4].

TISSUE AND CELLULAR DISTRIBUTION OF Na⁺/H⁺ EXCHANGER MESSAGE AND PROTEIN

Based on Northern analysis and ribonuclease protection assays, NHE1 message is present in nearly all mammalian cells [21,36]. The only mammalian cells studied in which NHE1 message was not identified are the OK cells (Opossum renal proximal tubule cell line) and rat proximal tubule cortical segments S₁ and S₂ [16]. All of these cells are known to lack functional basolateral membrane Na⁺/H⁺ exchangers. Using NHE1 antibody, NHE1 has been found in plasma membrane of fibroblasts and A431 epidermoid cells [35]. In the rabbit ileum it is restricted to the basolateral membrane of both the villus epithelial cell and the crypt epithelial cell, but appears to be diffusely present in the plasma membrane of goblet cells [36]. In addition, it is restricted to the basolateral membrane of the Cl⁻ secretory human colon cancer cell line, Caco-2 [39] and to the basolateral membrane of the porcine renal epithelial cell line, LLC-PK1 [24]. In rabbit kidney, NHE1 is on the basolateral membrane of proximal tubule cells, distal convoluted tubules, thick ascending limb of Henle's loop, and the collecting duct but is absent from glomeruli, the thin descending limb of Henle's loop, intercalated cells of the connecting tubules and collecting ducts and non-epithelial cells of the renal cortex and medulla [2].

NHE2, NHE3 and NHE4 are more restricted in message distribution and are predominantly expressed in stomach, intestine and kidney and thus are epithelial cell isoform Na^+/H^+ exchangers [21,30,35]. NHE2 message is present in large amounts in stomach, uterus, kidney, intestine, adrenal gland and much less in trachea and skeletal muscle [30,35]. The message is most expressed in the kidney medulla exceeding that in the kidney cortex. In the gastrointestinal tract the ascending colon has most message followed by jejunum > ileum > duodenum > descending colon.

NHE3 message is found exclusively in kidney, intestine, and stomach [21,30] Most message is present in the kidney cortex, exceeding the medulla. The area of second most message is rabbit ascending colon which is approximately equal to ileum > jejunum. NHE3 message is not present in the duodenum or descending colon. Thus, the presence of NHE3 message in rabbit gastrointestinal tract co-localizes with the expression of the neutral NaCl absorption process, supporting that NHE3 is involved in neutral NaCl absorption. Using a NHE3 antibody, Aronson et al showed that NHE3 is restricted to the brush border of renal proximal tubules cells (Aronson, P.S., unpublished results).

NHE4 message is present in largest amount in the rat stomach (maximum in gastric antrum) > proximal small intestine = cecum = proximal colon with much smaller amounts in the uterus, brain, kidney, and skeletal muscle [21]

TOPOLOGY OF THE MAMMALIAN NHE GENE FAMILY: TWO FUNCTIONAL DOMAINS

The primary structure (amino acids) and the predicted secondary structure (hydrophobicity profile) of all four identified mammalian Na^+/H^+ exchanger gene family members are similar [34]. Based on the hydrophobicity plots using the method of Engelman or Kyte and Doolittle, they are all predicted to have 10-12 membrane spanning domains and a long cytoplasmic C-terminus. Antibody studies confirm the NHE membrane topology that the C-terminus for NHE1 and NHE2 is intracellular [25,31], based on the requirement for membrane permeabilization to visualize the epitope.

The most highly conserved portions of the molecule among the identified isoforms are the membrane spanning domains and of these, membrane spanning domains 5A and 5B are the most conserved. Each protein is predicted to contain a signal peptide at the N-terminus and the first membrane spanning domain appears to be cleaved off in the intact protein. NHE1 is N-glycosylated on extracytoplasmic loop a. The areas least related among the Na^+/H^+ exchanger isoforms include: (a) the first membrane spanning domain (the signal peptide) and extracellular loop a; and (b) the intracytoplasmic C-terminal domain. The C-terminus for all isoforms contains multiple putative protein kinase consensus sequences which are extremely variable among members of the gene family.

Structure-function studies using NHE1 and NHE3 have suggested each NHE molecule can be divided into two functional domains, i.e., the transporter domain and the regulatory domain. The transporter domain is the N-terminal membrane spanning domain, which in NHE1 consists of amino acids from 1-515 [37] and in NHE3 consists of amino acids from 1-475 [34]. This N-terminal membrane spanning domain retains the ability to (a) be inserted into the plasma membrane; (b) catalyze amiloride-sensitive Na^+/H^+ exchange activity, although at a much slower rate than the wild type exchangers; (c) be activated by intracellular H^+ i.e. with an H^+-modifier site. The regulatory domain is the long cytoplasmic C-terminus which mediates growth factor regulation of NHEs. NHE1 is stimulated by growth factors *via* a change in the affinity of the exchanger for intracellular H^+ [17,20]. Wakabayashi *et al* [37] constructed a set of deletion mutants within the cytoplasmic carboxyl-terminus of human NHE1 and stably expressed the truncated cDNAs in PS120 fibroblasts. It was found that deletion of the cytoplasmic C-terminus of NHE1, in particular the region from amino acids 567-635, completely abolished the growth factor activation. On the other hand, NHE3 can be up and down regulated, in particular, serum and fibroblast growth factor (FGF) stiumulate NHE3 while phorbol myristate acetate (PMA) inhibits it [17,33]. In contrast to NHE1, NHE3 is regulated by growth factors *via* a V_{max} change of Na^+/H^+ exchange. Deletion of the cytoplasmic C-terminus of NHE3 abolished the regulation by PMA and FGF but not the stimulation of NHE3 by serum. Based on a preliminary study of NHE3 C-terminus deletion mutants, separate domains responsible for up and down regulation of NHE3 have been identified [34].

AMILORIDE BINDING DOMAIN

Clark and Limbird [6] reviewed the amilorde sensitivity of Na^+/H^+ exchangers in different tissues and cells. Table 1 summarizes the amiloride sensitivity of the recently cloned Na^+/H^+ exchangers (NHE1, NHE2 and NHE3) expressed either in PS120 cells or CHO cells. For all three expressed exchangers, Na^+/H^+ exchange [17,20] is entirely inhibited by amiloride and its 5-amino substituted amiloride analogues. NHE1 and NHE2 are equally sensitive to amiloride [35,41], while NHE3 is the amiloride resistant isoform. NHE1 is sensitive to ethylisopropylamiloride while NHE2 and NHE3 are resistant, with NHE3 more resistant than NHE2.

The putative fourth membrane spanning domain has been implicated in amiloride binding [32]. Counillon and Pouyssegur made [7] a series of site-directed mutants of NHE1 in the fourth transmembrane domain. Their rational was based on the identification of a point-mutation in AR300 cells which contain NHE1 which is resistant to amiloride and EIPA. These cells contain a mutated NHE1 which differs in only one amino acid from wild type NHE1 (L^{167}-> F) [7]; and point mutations were made in wild type NHE1 to mimick NHE2, NHE3, NHE4 and AR300 since NHE2, NHE3 and NHE4 differ is this area as well. They showed that the NHE1 mutants, F165Y (mimicking NHE4) and L167F (mimicking AR300), are 30 fold and 5 fold more resistant to amiloride, respectively, when compared with the wild-type NHE1; and suggested that Phe 165 and Leu 167 are important for amiloride inhibition of Na^+/H^+ exchange.

Although NHE2 has the same amiloride sensitivity as NHE1, it is 20-50 fold more resistant to EIPA than NHE1 [35]. NHE1 has a Phe 168 in the fourth transmembrane domain, which corresponds to Tyr 144 in NHE2. Therefore, we made a site-directed mutant of NHE2, the mutant Y144F which mimicks NHE1 [41]. Interestingly, this mutant behaved as the wild type NHE2, being resistant to EIPA. Conversely, it was found that the NHE1 mutant, F168Y which mimicks NHE2, had the same sensitivity to EIPA as the wild-type NHE1 [7]. Thus, we additionally conclude that there are additional amiloride binding domains elsewhere in the NHE or other parts of the exchanger which affect amiloride binding. The affinity for external Na$^+$ is not significantly changed by any of these mutants in the fourth transmembrane domain [41]. This supports that the amiloride binding site and the external Na$^+$ binding site are not identical, as was previously suggested based on functional studies of native fibroblast Na$^+$/H$^+$ exchangers [9].

KINETICS

The kinetics of Na$^+$/H$^+$ exchange in cells and tissues has been previously described [12]. We will focus here on the kinetics of the cloned mamalian Na$^+$/H$^+$ exchangers expressed in Na$^+$/H$^+$ exchanger deficient cells.

Na KINETICS

NHE1, NHE2, and NHE3 have been studied when stably expressed in the PS120 fibroblast cell line and in CHO cells [17,20,38]. The cloned exchangers show similar kinetic characteristics when undergoing Na$^+$-dependent pH recovery following an acid load [17]. The kinetics for external Na$^+$ follows a classical Michaelis-Menten model with a K$_m$ of 4.7-18 mM and a Hill coefficient of 1 [17], suggesting that there is a single binding site for external Na$^+$. When rabbit NHEs were expressed in PS120 cells, we found that all isoforms had similar K$_m$(Na$^+$) (15-18 mM) and there was no significant difference among the isoforms. However, when rat NHEs were expressed in CHO cells [20], Orlowski found that K$_m$(Na$^+$) of NHE1 was 2-fold lower than that of NHE2 and NHE3. The reason for the discrepancy that different K$_m$(Na$^+$) values were observed among rat NHE isoforms and the same K$_m$(Na$^+$) was found in all rabbit NHE isoforms is not known.

H KINETICS

The kinetics with respect to internal H$^+$ are also similar for the three exchangers with all three deviating from the hyperbolic response expected with Michaelis-Menten kinetics and all having a Hill coefficient of 2-3 [17]. The data describing the effect of intracellular H$^+$ best fits an allosteric model with at least 2 independent binding sites for H$^+$ [1]. In addition to the internal H$^+$ transport site, there is thought

TABLE 1.   Effect of amiloride and its analogues on NHE1, NHE2 and NHE3. Values are taken from references shown in superscripts.

Inhibition constants (K_i) (μM)

Inhibitor	NHE1	NHE2	NHE3
Amiloride	1 [41] 3 [7] 1.6 [20]	1 [41]	39 [41] 100 [20]
Ethylisopropyl-amiloride	0.02 [20,7,41] 0.05 [29]	1 [41]	8 [41] 2.4 [20]
Methylpropyl-amiloride	0.05 [7]		
Dimethyl-amiloride	0.02 [20]		14 [20]
Benzamil	120 [20]		100 [20]

TABLE 2.   Effect of various agents on rate of Na+/H+ exchange in the three transfected PS120 cell lines, with kinetic parameters affected shown in parentheses [17].

Agent	NHE1	NHE2	NHE3
FBS (10%)	stimulation (K')	stimulation (V_{max})	stimulation (V_{max})
FGF (10 ng/ml)	stimulation	stimulation (V_{max})	stimulation (V_{max})
thrombin (1 U/ml)	not done	stimulation (V_{max})	stimulation (V_{max})
PMA (1 μM)	stimulation (K')	stimulation (V_{max})	inhibition (V_{max})
8-Br-cAMP (0.5 mM)	no effect	no effect	no effect
8-Br-cGMP (0.5 mM)	no effect	no effect	no effect

to be an internal modifier site for intracellular H^+, which can regulate the activity of the exchanger [1]. The pK values for intracellular $[H^+]$ (pK$[H^+]$) are similar among different isoforms of NHE when measured by the spectroflourometric method using the pH sensitive dye, BCECF, under conditions in which cells expressing NHE1 are not serum deprived.

REGULATION OF CLONED Na$^+$/H$^+$ EXCHANGERS BY GROWTH FACTORS / PROTEIN KINASE IN PS120 FIBROBLASTS

When the three cloned Na$^+$/H$^+$ exchangers (NHE1, NHE2 and NHE3) were stably expressed in the same cell line (PS120), they differed in their response to growth factors [17,33]. Thus, there are intrinsic differences among NHE proteins that allows them to respond differently to growth factors. Table 2 summarizes the effects of various agents on the rate of Na$^+$/H$^+$ exchange by NHE1, NHE2 and NHE3 stably expressed in PS120 cells. Fetal bovine serum and FGF stimulate all three cloned exchangers. Of interest, PMA stimulates NHE1 and NHE2 but inhibits NHE3. cAMP does not have any effect on the cloned exchangers. Of note, Only β-NHE1 (cloned from trout red blood cells) is regulated by cAMP when expressed in PS120 cells [4]. This shows that the PS120 cell model is adequate to demonstrate a cAMP effect on Na$^+$/H$^+$ exchange.

The biochemical mechanism by which protein kinases regulate Na$^+$/H$^+$ exchangers has been studied most in detail by Pouyssegur et al with NHE1 [25,26]. They demonstrated that NHE1 is stimulated by thrombin and EGF, with thrombin acting via phosphatidylinositol turnover, while EGF acts by affecting tyrosine phosphorylation [26]. NHE1 is a phosphoprotein [25,26] and the amount of phosphorylation on serine of the same set of specific phosphopeptides in NHE1 changes with exposure to EGF and thrombin in parallel with the changes in intracellular pH, likely due to a change in Na$^+$/H$^+$ exchange rate. Also the phosphatase 1 and 2A inhibitor okadaic acid, which would be expected to increase phosphorylation on the assumption that phosphorylation is present under basal conditions, increases basal phosphorylation of NHE1 and also increases the basal rate of Na$^+$/H$^+$ exchange. Okadaic acid adds to the stimulatory effects of EGF and thrombin on both phosphorylation of NHE1 and rate of Na$^+$/H$^+$ exchange [26]. Nonetheless, it is unknown if changes in the phosphorylation of the Na$^+$/H$^+$

exchanger or in an associated protein lead to changes in the Na$^+$/H$^+$ exchange rate.

Phosphopeptide mapping demonstrated that NHE1 has changes in phosphorylation in response to both thrombin and EGF but that these changes occur only on serine residues with no tyrosine phosphorylation identified [25], and the same set of phosphopeptides are changed by both agents. This indicates that an intermediate kinase is involved in EGF regulation. This has been postulated to involve MAP kinase (mitogen activated protein kinase)[25]. In addition, it has been postulated that NHE1 may be more directly regulated by a kinase associated with the exchanger (NHE1 kinase) [26], perhaps similar to the regulation of the ß-adrenergic receptor by a receptor related kinase.

Concerning mechanisms of regulation, the C-terminus of NHEs is required for growth factor/protein kinase regulation. For NHE1, Pouyssegur *et al* proposed that phosphorylation of the C-terminus is required to allow interaction of the C-terminus with the H$^+$-modifier site since depleting intracellular ATP alters the H$^+$ modifier site functionally [37]. However, the domain on the C-terminus of NHE1 which is phosphorylated by protein kinase and growth factors is different from the domain that interacts with the H$^+$-modifier site [Pouyssegur, J, unpublished results].

REGULATION OF CLONED Na$^+$/H$^+$ EXCHANGERS BY ATP DEPLETION

Another way to study the role of phosphorylation in regulation of Na$^+$/H$^+$ exchange is via studying the effects of ATP depletion. In all three cloned exchangers, cellular ATP depletion eliminates regulation of Na$^+$/H$^+$ exchange by growth factors [17]. Interestingly, cellular ATP depletion also changes the kinetics of Na$^+$/H$^+$ exchange. ATP depletion reduces the co-operative nature of Na$^+$/H$^+$ exchange with Hill coefficient decreased from 2 to 1.2, which is not significantly different from 1 and fits Michaelis-Menten kinetics [30]. This observation suggests that: (a) the hydrogen modifer site of the Na$^+$/H$^+$ exchangers requires phosphorylation or another ATP dependent process to function; (b) thus, under ATP depleted conditions, the affinity of the H$^+$ substrate site can be distinguished from the H$^+$ modifer site since only one H$^+$ binding site is acting (Hill coefficient = 1) and that must be the substrate transport site since Na$^+$/H$^+$ exchange still occurs. Therefore, it can be concluded that NHE1 has lower affinity for intracellular H$^+$ at the substrate site than NHE2 and NHE3 which have similar affinity. Further, ATP depletion decreases the V_{max} of all three cloned exchangers suggesting that basal phosphorylation of the NHE and/or regulatory protein(s) associated with NHEs is required for exchange activity.

REGULATION OF Na^+/H^+ EXCHANGE BY CELL SHRINKAGE

Osmotic shrinkage activates Na^+/H^+ exchange in a variety of cell types such as lymphocytes [11] and red blood cells [5]. Mechanistically, cell shrinkage activates Na^+/H^+ exchange by increasing its sensitivity to intracellular H^+, an effect similar to that of growth factor activation on NHE1 [11]. Recently, Grinstein et al showed that NHE1 is involved in cellular volume regulation [13]. They showed that NHE1, stably transfected in PS120 cells, responds to hyperosmolarity and causes cellular alkalinization. Although this process is ATP dependent, the hyperosmolar activation of NHE1 changes neither the amount of phosphorylation of NHE1 nor the pattern of the NHE1 phosphopeptide map. Thus, they proposed that osmotic activation of NHE1 does not require phosphorylation of the Na^+/H^+ exchanger and that this mechanism of regulation of NHE1 is distinct from the growth factor regulation which is phosphorylation dependent [13].

EFFECT OF CELL TYPE ON THE NHE PLASMA MEMBRANE LOCATION AND PROTEIN KINASE REGULATION

There is evidence that the cell type and membrane location (apical vs basolateral membrane in epithelial cells) can influence second messenger regulation of Na^+/H^+ exchange in addition to the nature of the Na^+/H^+ exchanger isoform. PMA stimulates cloned NHE1 expressed in PS120 cells [17,36], but has no effect on the cloned NHE1 expressed in LAP cells [28]. PMA inhibits endogenous NHE1 in human A431 cells [28] but has no effect on endogenous NHE1 in granulocytic HL-60 cells [23]. In polarized cells, Caco-2 cells have NHE1 on its basolateral membrane but this NHE1 is not regulated acutely by serum, phorbol esters or growth factors [16]; while NHE1 in the basolateral membrane of other epithelial cells is regulated by second messengers [for review, see 32]. Further, OK cells have an endogenous amiloride resistant Na^+/H^+ exchanger on the apical membrane but lack any basolateral membrane Na^+/H^+ exchanger. When cloned NHE1 is stably expressed in OK cells, it is expressed partially in the basolateral membrane. The transfected NHE1 on the basolateral membrane of OK cells is inhibited by protein kinase C but is stimulated by protein kinase A [14]. In PS120 cells transfected with NHE1, NHE1 is stimulated by protein kinase C and protein kinase A has no effect. Also of note is that NHE1 does not have any cAMP dependent protein kinase consensus sequence. These studies raise the possibility that in addition to the nature of the NHE isoforms, NHEs are closely associated with other "NHE specific kinase(s) or regulatory protein(s)" which adds to the diversity in regulation of Na^+/H^+ exchangers in different cell types and even in different cellular domains in the same cells (polarized cells).

NHE1 IS THE HOUSEKEEPING ISOFORM Na^+/H^+ EXCHANGER AND NHE3 IS THE BRUSH BORDER Na^+/H^+ EXCHANGER OF INTESTINAL AND RENAL EPITHELIAL CELLS

By Northern blot analysis and by immunocytochemistry, NHE1 has been found ubiquitously in nearly all cells [2,36]. It is sensitive to amiloride inhibition, activated by cell shrinkage [13] and stimulated by growth factors and protein kinases, in particular protein kinase C. These characteristics support that NHE1 is the "housekeeping isoform" Na^+/H^+ exchanger.

NHE3 is resistant to amiloride and its message is restricted to intestine and kidney [30,33]. The expression of NHE3 message in rabbit gastrointestinal tract co-localizes with the expression of the neutral NaCl absorptive process. When expressed in PS120 cells, NHE3 is the only isoform exchanger that is inhibited by protein kinase C. Glucocorticoid stimulates ileal brush border Na^+/H^+ exchange. Only NHE3 message, but not NHE1 and NHE2 message, increases with glucocorticoid stimulation of Na^+/H^+ exchange [39]. Further, Aronson et al recently successfully raised an antibody against NHE3 and showed that NHE3 is the brush border Na^+/H^+ exchanger of kidney proximal tubule cells. These pharmacological, functional and immunolocalization studies suggest that NHE3 is a brush-border Na^+/H^+ exchanger of intestinal and renal epithelial cells.

CONCLUSIONS

With the identification of the Na^+/H^+ exchanger gene family, there has been an advance in understanding of the molecular properties, physiology and structure-function relationships of Na^+/H^+ exchangers. Four NHEs have so far been identified. However, additional members of the mammalian gene family almost certainly remain to be identified.

Why are there so many Na^+/H^+ exchanger isoforms? As discussed in the Introduction, it is predicted that each Na^+/H^+ exchanger assumes a specialized function. NHE1 has been shown to be involved in regulation of cellular volume and cellular proliferation, in addition to pH homeostasis. NHE3 is a brush border Na^+/H^+ exchanger in intestinal and renal epithelial cells and as such is probably fine-tuned for Na^+ absorption. The physiological roles of NHE2 and NHE4 in addition to being a Na^+/H^+ exchanger to regulate intracellular pH are not clear. Defining how many mammalian Na^+/H^+ exchangers exist and their relative physiologic roles is an exciting challenge that will be accomplished in the next several years.

REFERENCES:

1.   Aronson, P.S., Nee, J., Suhm, N.A. 1982. Modifer role of internal H⁺ inactivating the Na⁺/H⁺ exchange in renal microvillus membrane vesicles. *Nature.* **299**:161-163.

2.   Biemesderfer, D., Reilly, R., Exner, M., Igarashi, P., Aronson, P. 1992. Immunocytochemical characterization of Na⁺/H⁺ exchanger isoform NHE1 in rabbit kidney. *Amer. J. Physiol.* **263**:F833-F840.

3.   Borgese, F., Sardet, C., Cappadoro, M., Pouyssegur, J., Motais, R. 1992. Cloning and expressing a cAMP-activated Na⁺/H⁺ exchanger: evidence that the cytoplasmic domain mediates hormonal regulation. *Proc. Natl. Acad. Sci. USA.* **89**:6768-6769.

4.   Brant, S.R., Bernstein, M., Wasmuth, J.J., Taylor, E.W., McPherson, J.D., Li, X., Walker, S.A., Pouyssegur, J., Donowitz, M., Tse, C.M., Jabs, E.W. 1993. Physical and genetic mapping of a human apical epithelial Na⁺/H⁺ exchanger (NHE-3) isoform to chromosome 5p15.3. *Genomics* **15**:668-672.

5.   Cala P.M. 1986 Volume sensitive alkali metal-H transport in Amphiuma red blood cells. *Current Topics in Membr. Transport* 26, 79-100.26. Sardet, C., Fafournox, P., Pouyssegur, J. 1991. Thrombin, epidermal growth factor, and okadaic acid activate the Na⁺/H exchanger, NHE1, by phosphorylating a set of common sites. *J. Biol. Chem.* **266**:19166-19171.

6.   Clark, J.D., Limbird, L.L. 1991. Na⁺/H⁺ exchanger subtypes: a predictive review. *Am. J. Physiol.* **261**:G945-G953.

7.   Counillon, L., Franchi, A., Pouyssegur, J. 1993. A point mutation of the Na⁺/H⁺ exchanger gene (NHE-1) and amplification of the mutated allele confer amiloride-resistance upon chronic acidosis. *Proc. Natl. Acad. Sci. USA.,* 90, 4508-4512.

8.   Counillon, L., and Pouyssegur, J. 1993. Nucleotide sequence of the Chinese hamster Na⁺/H⁺ exchanger NHE1. *Biochim. Biophys. Acta* 1172, 343-345.

9.   Franchi, A., Cragoe, E., Pouyssegur, J. 1986. Isolation and properties of fibroblast mutants overexpressing an altered Na⁺/H⁺ antiporter. *J. Biol. Chem.* **261**:14614-14620.

10.   Franchi, A., Perucca-Lostanten, D., Pouyssegur, J. 1986. Functional expression of a human Na⁺/H⁺ antiporter gene transfected into antiporter-deficient mouse L cells. *Proc. Natl. Acad. Sci. USA.* **83**:9388-9392.

11.   Grinstein, S., Cohen, S., Goetz, J.D., Rothstein, A. Mellors, A. and Gelfand, E.W. 1986. Activation of the Na⁺/H⁺ antiport by changes in cell

volume and by phorbol ester; possible role of protein kinase. *Current Topics in Membr. Transport* 26, 115-136.

12. Grinstein, S., Rotin, D., Marson, M.J. 1989. Na^+/H^+ exchanger and growth factor-induced cytosolic pH change. Role in cellular proliferation. *Biochim. Biophy. Acta.* **988**:73-91.

13. Grinstein, S., Woodside, M., Sardet, C., Pouyssegur, J. and Rotin, D. 1992. Activation of the Na^+/H^+ antiporter during cell volume regulation. Evidence for a phosphorylation-independent mechanism. *J. Biol. Chem.* 267, 23823-23828.

14. Helmle-Kolb, C., Counillon, L., Roux, D., Pouyssegur, J., Mrkic, B. and Murer, H., 1993. Na^+/H^+ exchange activates in NHE1-transfected OK-cells: cell polarity and regulation. *Pflugers Arh.* In press.

15. Krapf, R., Alpern, R.J. 1993. Cell pH and transepithelial H/HCO_3 transport in renal proximal tubule. *J. Membr. Biol.* **131**:1-10.

16. Krapf, G., Solioz, M. 1991. Na^+/H^+ antiporter in RNA expression in single nephron segments of rat kidney cortex. *J. Clin. Invest.* **88**:783-788.

17. Levine, S., Montrose, M., Tse, C.M., Donowitz, M. 1993. Kinetic and regulation of three cloned mammalian Na^+/H^+ exchangers stably expressed in a fibroblast cell line. *J. Biol. Chem.* in press.

18. Mattei, M.G., Sardet, C., Franchi, A., Pouyssegur, J. 1988. The human amiloride-sensitive Na^+/H^+ antiporters: localization to chromosome 1 by in situ hybridization. *Cytogenet. Cell. Censt.* **48**:6-8.

19. Murer, H., Hopfer, U., Kinne, R. 1976. Sodium, proton antiport in brush border membranes isolated from rat small intestine and kidney. *Biochem. J.***154**:597-602.

20. Orlowski, J. 1993 Heterologous expression and functional properties of amiloride high affinity (NHE1) and low affinity (NHE3) isoforms of the rat Na^+/H^+ exchanger. *J. Biol. Chem.* 268, 16369-16377.

21. Orlowski, J., Kandasamy, R.A., Shull, G.E. 1992. Molecular cloning of putative members of the Na^+/H^+ exchanger gene family. *J. Biol. Chem.* **267**:9332-9339.

22. Pouyssegur, J., Sardet, C., Franchi, A., L'Allemain, G., Paris, S, 1984. A specific mutation abolishing Na^+/H^+ antiport activity in hamster fibroblasts precludes growth at neutral and acidic pH. *Proc. Natl. Acad. Sci. USA.* **81**:4833-4837.

23. Rao, G.N., Sardet, C., Pouyssegur, J. and Berk, B.C. 1993. Phosphorylation of Na^+/H^+ antiporter is not stimulated by phorbol ester and acidification in granulocytic HL-60 cells. *Am. J. Physiol.* 264, C1278-C1284.

24. Reilly, R.F., Hildebrandt, F., Biemesderfer, D., Sardet, C., Pouyssegur, J., Aronson, P.S., Slayman, C.N., Igarashi, P. 1991. cDNA cloning and immunolocalization of a Na⁺/H⁺ exchange in LLC-PK₁ renal epithelial cells. *Am. J. Physiol.* **261**:F1088-F1094.

25. Sardet, C., Counillon, L., Franchi, A., Pouyssegur, J. 1990. Growth factors induce phosphorylation of the Na⁺/H⁺ antiporter, a glycoprotein of 110 kD. *Science.* **247**:723-726.

26. Sardet, C., Fafournox, P., Pouyssegur, J. 1991. Thrombin, epidermal growth factor, and okadaic acid activate the Na⁺/H exchanger, NHE1, by phosphorylating a set of common sites, *J. Biol. Chem.* **266**:19166-19171.

27. Sardet, C., Franchi, A., Pouyssegur, J. 1989. Molecular cloning, primary structure and expression of the human growth factor - activatable Na⁺/H⁺ antiporter. *Cell.* **56**:271-280.

28. Takaichi, K., Balkovetz, D.F., Meir, E. V., and Warnock, D. G. 1993. Cytosolic pH sensitivity of an expressed human NHE1 Na⁺/H⁺ exchanger. *Am. J. Physiol.* 264, C944-C950.

29. Takaichi, K., Wang, D., Blakovets, D.F., Warnock, D.G. 1992. Cloning, sequencing, and expression of Na⁺/H⁺ antiporter cDNAs from human tissues. *Am. J. Physiol.* **262**:C1069-C1076.

30. Tse, C.M., Brant, S.R., Walker, S., Pouyssegur, J., Donowitz, M. 1992. Cloning and sequencing a rabbit cDNA encoding an intestinal and kidney-specific Na⁺/H⁺ exchanger isoform (NHE-3). *J. Biol. Chem.* **267**:9340-9346.

31. Tse, C.M., Levine, S., Khurana, S. Montgomery, J.M., Pouyssegur, J. and Donowitz, M. 1993. Characterization of a polyclonal antibody against Na⁺/H⁺ exchanger-2 (NHE2): Evidence that NHE2 is not N-glycosylated. *XIIth International Congress of Nephrology* Abstract 122.

32. Tse, C.M., Levine, S., Yun, C., Brant, S., Counillon, L., Pouyssegur, J., and Donowitz, M. 1993. Structure/function studies of epithelial isoforms of the mammalian Na⁺/H⁺ exchanger gene family. 1993. *J. Membr. Biol.* **135**:93-108.

33. Tse, C.M., Levine, S.A., Yun, C.H., Brant, S.R., Pouyssegur, J., Montrose. M.H., Donowitz, M. 1993. Functional characteristics of a cloned epithelial Na⁺/H⁺ exchanger (NHE3): resistant to amiloride and inhibition of protein kinase C. *Proc. Natl. Acad. Sci. USA* in press.

34. Tse, C.M., Levine, S., Yun., C., Montrose, M. and Donowitz, M. 1993 The cytoplasmic domain of the epithelial specific Na⁺/H⁺ exchanger isoform (NHE3) mediates inhibition by phorbol myristate acetate (PMA) but not stimulation by serum. *Gastroenterology* 104, A285.

35. Tse, C.M., Levine, S.A., Yun, C.H.C., Montrose, M.H., Little, P.J., Pouyssegur, J., Donowitz. M. 1993. Cloning and expression of a rabbit

cDNA encoding a serum-activated ethylisopropyl amiloride resistant epithelial Na$^+$/H$^+$ exchanger isoform (NHE-2). *J. Biol. Chem.* **268**, 11917-11924.

36.   Tse, C.M., Ma, A.I., Yang, V.W., Watson, A.J.M, Potter, J., Sardet, C., Pouyssegur, J., Donowitz, M. 1991. Molecular cloning of cDNA encoding the rabbit ileal villus cell basolateral membrane Na$^+$/H$^+$ exchanger. *EMBO. J.* **10**:1957-1967.

37.   Wakabayashi, S., Fafournoux, P., Sardet, C., Pouyssegur, J. 1992. The Na$^+$/H$^+$ antiporter cytoplasmic domain mediates growth factor signals and controls H$^+$ - sensing. *Proc. Natl. Acad. Sci. USA.* **89**:2424-2428.

38.   Wang, Z., Orlowski, J and Shull, G.E. 1993. Primary structure and functional expression of a novel gastrointestinal isoform of the rat Na$^+$/H$^+$ exchanger. *J. Biol. Chem.* **268**, 11925-11928.

39.   Watson, A.J.M., Levine, S., Donowitz, M., Montrose, M.H. 1991. Kinetics and regulation of polarized Na$^+$/H$^+$ exchanger from Caco-2 cells, a human intestinal cell line. *Am. J. Physiol.* **261**:G229-G238.

40.   Yun, C.H., Gurubhagavatula, S., Levine, S.A., Montgomery, J.M., Brant, S.R., Cohen, M.E., Pouyssegur, J., Tse, C.M., Donowitz, M. 1993. Glucocorticoid stimulation of ileal Na$^+$ absorptive cell brush border Na$^+$/H$^+$ exchange and association with an increase in message for NHE-3, an epithelial isoform Na$^+$/H$^+$ exchanger. *J. Biol. Chem.* **268**:206-211.

41.   Yun. C.H.C., Little, P.J., Nath, S.K., Levine, S.A., Pouyssegur, J., Tse, C.M. and Donowitz, M. 1993. Leu143 in the putative fourth membrane spanning domain is critical for amiloride inhibition of an epithelial Na$^+$/H$^+$ exchanger isoform (NHE2) *Biochem. Biophys. Res. Comm.* **193**, 532-539.

Polarized Na/H exchangers in colonic epithelium: a problem of balance

Marshall H. Montrose
Department of Medicine
Johns Hopkins University
720 Rutland Avenue
Baltimore, MD 21205
USA

In the colon, bacterial fermentation of non-absorbed polysaccharides and protein produces high concentrations of short chain fatty acids (SCFAs) (Bugaut, 1987). As a result, SCFAs (such as acetate, propionate and butyrate) are the predominant osmolytes and anions in the unique environment of the colonic lumen (Cummings *et al*, 1987). Some mechanisms by which SCFAs benefit the colon are well documented. Some SCFAs are the favored metabolic substrates of colonocytes (Bugaut, 1987;Roediger, 1982), and SCFAs stimulate electroneutral Na^+ absorption by the colon (Bugaut, 1987; Roediger and Moore, 1981; Binder and Mehta, 1989; Gabel *et al*, 1991).

The ability of SCFAs to stimulate Na^+ absorption is not mediated primarily by metabolism of SCFAs, but rather by rapid nonionic diffusion of SCFAs into colonocytes and acidification of the cytoplasm (Horvath *et al*, 1986). Acidification is a consequence of the transmembane equilibration of the nonionized form of these weak acids, and subsequent deprotonation in the cytosol (Roos and Boron, 1981). As shown in Figure 1, this acidification activates amiloride-sensitive apical Na/H exchange to mediate electroneutral Na^+ uptake from the colonic lumen (Binder and Mehta, 1989; Gabel *et al*, 1991; Sellin and DeSoigne, 1990). These basic features of the model of SCFA-stimulated Na^+ absorption have been verified in several laboratories.

Two observations are difficult to resolve with this model. The first observation is that rat colonocytes have both apical (Foster *et al*, 1989; Binder *et al*, 1989) and basolateral (Dudeja *et al*, 1989; Rajendran *et al*, 1991) Na/H exchange. If cellular acidification can activate Na/H exchange at both membranes, how can SCFAs efficiently stimulate net Na^+

NATO ASI Series, Vol. H 89
Molecular and Cellular Mechanisms
of H⁺ Transport
Edited by Barry H. Hirst
© Springer-Verlag Berlin Heidelberg 1994

APICAL   BASOLATERAL

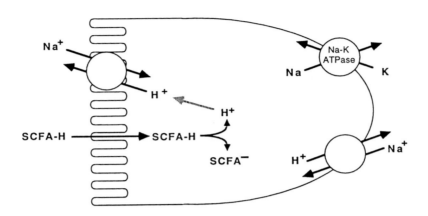

Figure 1. Model of SCFA-stimulated Na+ absorption. Nonionic diffusive uptake of SCFA leads to cellular acidification which activates apical Na/H exchange.

absorption?   Since rates of apical and basolateral Na/H exchange are comparable in isolated membranes from rat colon (Binder *et al*, 1989; Rajendran *et al*, 1991), activation of both exchangers would imply a significant expenditure of cellular energy in a futile cycle which would dissipate vectorial Na^+ transport.   In addition, there is no evidence that basolateral Na/H exchange is activated by apical SCFA since serosal-to-mucosal Na^+ fluxes in the colon are unchanged in this condition (Binder and Mehta, 1989; Gabel *et al*, 1991; Peterson *et al*, 1981).   A second alternative is that only the apical exchanger is sensitive to acidification. However, we and others have shown that basolateral Na/H exchange is activated by intracellular acidification at the level of whole cells (Watson *et al*, 1991) and isolated basolateral membrane vesicles from colon (Dudeja *et al*, 1989; Rajendran *et al*, 1991). It remains unresolved how efficient Na^+ absorption can be mediated in tissues which express both apical and basolateral Na/H exchange.

The second observation which stretches the credibility of the model is that apical SCFAs are required to efficiently stimulate Na^+ absorption. SCFAs are normally only present at the apical surface of colonocytes, but some experiments have tested the effects of adding SCFAs at different membrane surfaces. In gallbladder and sheep rumen, apical SCFAs strongly

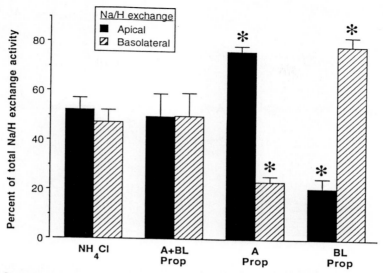

Figure 2. Transepithelial gradients of propionate modulate the extent of activation of Na/H exchange at the opposing membrane surfaces of HT29-C1 cells. Monolayers of HT29-C1 on permeant filter supports were loaded with the pH-sensitive dye BCECF and measured by digital imaging. Na-dependent pH recovery was quantified as described in the text.

stimulate net Na^+ absorption, but addition of basolateral SCFAs does not significantly increase either mucosal-to-serosal Na^+ fluxes or net Na^+ absorption (Gabel *et al*, 1991; Peterson *et al*, 1981). In rat, 25mM SCFA had to be added to the basolateral compartment to stimulate Na^+ absorption to a level which was only 34% percent of the response to 5 mM apical SCFA (Binder and Mehta, 1989). Assuming that SCFAs can acidify colonocytes via nonionic diffusion across the basolateral membrane (a function related only to the lipid solubility of the uncharged SCFA), SCFA added at either membrane domain should effectively activate apical Na/H exchange.

Our experiments have attempted to address these issues and discover mechanisms which would allow colonocytes to solve the problem of balancing the activity of exchangers in both membrane domains.

Using monolayers of BCECF-loaded HT29-C1 cells grown on a permeant filter support, we measured intracellular pH (pH_i) of single cells in a digital imaging microscope as described previously (32,59). We first compared the rate of pH_i recovery when cells were acidified either by NH_4Cl-prepulse (14,32), or by exposure to an isosmotic medium containing

APICAL

BASOLATERAL

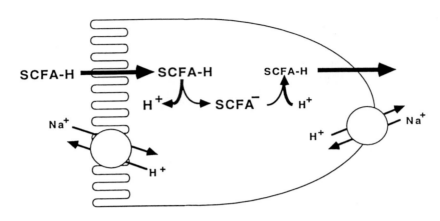

Figure 3. Model explaining generation of transcellular pH gradients by transepithelial SCFA gradients. Under physiologic conditions with high concentrations of apical SCFAs, nonionic diffusion of SCFA into the cell at the apical pole will lead to an acidification, which will be diminished by the net efflux of SCFA at the basolateral pole. In the continuing presence of a buffer gradient, this may lead to a standing pH_i gradient.

130 mM Na-propionate (substituted for 130 mM NaCl in our standard Ringer's). In both cases, cells acidified and Na^+ was required to initiate pH recovery. Addition of Na^+ to either the apical or basolateral medium was used to demonstrate the presence of polarized exchangers in each membrane domain (Montrose *et al*, 1987). We calculated the fraction of Na/H exchange activated at each surface versus the total Na/H exchange observed in the presence of apical and basolateral Na^+. When we exposed cells to acidifying conditions which were symmetric across the monolayer (NH_4Cl-prepulse or propionate in both apical and basolateral perfusates), we activated equivalent amounts of Na/H exchange at each surface (Figure 2, *two conditions on left*). This comparison suggests that propionate is not likely to activate Na/H exchangers via mechanisms other than acidification.

In contrast, transepithelial gradients of propionate (directed either apical to basolateral, or basolateral to apical) preferentially stimulated Na^+-dependent pHi recovery at the membrane domain exposed to the SCFA (Figure 2, *two conditions on right*). Other experiments showed that even under these conditions of transepithelial SCFA gradients, all Na^+-dependent pHi recovery was sensitive to 3 μM ethyl-isopropyl amiloride (EIPA). Our results therefore show that transepithelial gradients of SCFAs can selectively activate apical Na/H exchange.

The results suggest that transepithelial SCFA gradients may be required to allow a high efficiency of net Na^+ absorption in colonic tissues (such as rat colon) which express both apical and basolateral Na/H exchange. Assuming that propionate only activates Na/H exchange via pH_i, the most likely explanation is that transepithelial gradients of propionate generate transcellular pH gradients. In the simplest working model (Figure 3), acidification caused by nonionic diffusion of SCFA into the cell at one surface can be partially (or completely) offset by the net efflux of nonionized SCFA at the opposite surface.

Using the HT29-C1 tissue culture model (a cell line derived from a human colon carcinoma; Huet *et al*, 1987), we have also been able to duplicate the observation that apical SCFA is required for efficient activation of apical Na/H exchange, although cells acidify when exposed to either apical or basolateral SCFA. Results further suggest that it is the transepithelial gradient of SCFA which is responsible for the selective activation of either apical or basolateral Na/H exchange. This strongly suggests that we have discovered a fundamental limitation in the currently accepted model describing how SCFAs stimulate Na^+ absorption. Independent of exact mechanism, our work suggests that we have observed a new form of regulation in which the transepithelial gradient of a luminal nutrient regulates Na/H exchange.

Literature

Binder, H.J., Stange, G., Murer, H., Stieger, B. and H-P. Hauri. Sodium-proton exchange in colon brush-border membranes. Am. J. Physiol. 251:G382-G390, 1986.

Binder, H.J., and P. Mehta. Short-chain fatty acids stimulate active sodium and chloride absorption in vitro in the rat distal colon. Gastroenterology 96:989-996, 1989.

Bugaut, M. Occurrence, absorption and metabolism of short chain fatty acids in the digestive tract of mammals. Comp. Biochem. Physiol. 86B:439-472, 1987.

Cummings, J.H., Pomare, E.W., Branch, W.J., Naylor, C.P.E. and MacFarlane, G.T. Short chain fatty acids in human large intestine, portal, hepatic and venous blood. Gut 28:1221-1227, 1987.

Dudeja, R.K., Foster, E.S. and T.A. Brasitus. Na-H antioporter of rat colonic basolateral membrane vesicles. Am. J. Physiol. 257:G624-G632, 1989.

Foster, E.S., Dudeja, P.K. and T. A. Brasitus. Na+-H+ exchange in rat colonic brush-border membrane vesicles. Am. J. Physiol. 250:G781-G787, 1986.

Gabel, G., Vogler, S. and H. Martens. Short-chain fatty acids and CO2 as regulators of Na+ and Cl- absorption in isolated sheep rumen mucosa. J Comp Physiol B 161:419-426, 1991.

Horvath, P.J., Weiser, M.M. and Duffey, M.E. Propionate alters ion transport by rabbitdistal colon. Fed. Proc. 45:509, 1986.

Huet, C., Sahuquillo-Merino, C., Coudrier, E. and Louvard, D. Multipotent intestinalcell line (HT-29) provide new models for cell polarity and terminal differentiation. J. Cell Biol. 105:345-357, 1987.

Montrose, M.H., Friedrich, T. and H. Murer. Measurements of intracellular pH in single LLC-PK$_1$ cells: recovery from an acid load via basolateral Na$^+$/H$^+$ exchange. J. Membr. Biol. 97:63-78, 1987.

Petersen, K-U., Wood, J.R., Schulze, G. and K. Heintze. Stimulation of gallbladder fluid and electrolyte absorption by butyrate. J. Membrane Biol. 62:183-193, 1981.

Rajendran, V.M., M. Oesterlin, and H.J. Binder. Sodium uptake across basolateral membrane of rat distal colon. J. Clin. Invest. 88:1379-1385, 1991.

Roediger, W.E.W. Utilization of nutrients by isolated epithelial cells of the ratcolon. Gastroenterology 83:424-429, 1982.

Roediger, W.E.W. and A. Moore. Effect of short-chain fatty acid on sodium absorption in isolated human colon perfused through vascular bed. Dig. Dis. & Sci. 26(2):100-106, 1981.

Roos, A. and W.F. Boron. Intracellular pH. Physiol. Rev. 61: 296-434, 1981.

Sellin, J.H., and R. DeSoignie. Short-chain fatty acid absorption in rabbit colon in vitro. Gastroenterology 99:676-683, 1990.

Watson, A.J.M., Levine, S., Donowitz, M. and M.H. Montrose. Kinetics and regulation of a polarized Na$^+$/H$^+$ exchanger from a human intestinal cell line. Am. J. Physiol. 261:G229-G238, 1991.

The AE gene family of band 3-related anion exchangers

Seth L. Alper

Molecular Medicine and Renal Units, Beth Israel Hospital
Dept. of Cell Biology, Harvard Medical School
Boston, MA 02215

Introduction

Chloride/bicarbonate exchange is nearly ubiquitous among vertebrate cells (Alper, 1991). In erythrocytes, it serves to increase the CO_2 carrying capacity of the blood so as to increase delivery to the lungs for expiration of the CO_2 produced by cellular respiration (Tanner, 1993). In other cells, chloride/bicarbonate exchange participates in the maintenance of intracellular pH, volume, and chloride concentration. Chloride/bicarbonate exchange acid-loads cells in response to intracellular alkalinization, or participates with sodium/proton exchange in regulatory volume increase. Polarized epithelial cells restrict some of their chloride/bicarbonate exchange functions to apical or basolateral membranes in order to mediate vectorial transepithelial transport of bicarbonate and/or chloride. The only chloride/bicarbonate exchangers of known identity are the AE gene products.

AE genes and mRNAs

The band 3-related AE genes are at least three in number. The AE1 gene was first cloned and characterized from mouse (Kopito, et al., 1987), then from chicken (Kim et al., 1989), rat (Kudrycki and Shull, 1993), and human (Tanner, 1993). The exon-intron splice sites of the 20 exons of the murine gene are conserved in the human gene. The murine and rat genes are transcribed from a single erythroid cell promoter which uses several closely spaced inititation sites defining a staggered start to exon 1 and results in translational initiation in exon 2. Chicken erythroid cells use dual promoters which display equal potency and demonstrate no difference in developmental regulation. The two resultant AE1 translation products differ by 33 residues at the amino terminus. Erythroid AE1 cDNAs have been cloned from mouse (Kopito and Lodish, 1985), human (Lux et al., 1989), chicken (Kim, et al. 1988), rat (Kudrycki and Shull, 1993), and trout (Hubner et al., 1992).

The major site of nonerythroid AE1 expression is the kidney, where immunocytochemical studies have localized AE1 polypeptide to the basolateral surface of the

NATO ASI Series, Vol. H 89
Molecular and Cellular Mechanisms
of H⁺ Transport
Edited by Barry H. Hirst
© Springer-Verlag Berlin Heidelberg 1994

Type A acid-secreting intercalated cell. The predominant renal transcripts from mouse (Brosius et al., 1989), rat (Kudrycki and Shull, 1989), and human (Wagner et al., 1987) encode a protein identical to erythroid AE1 except for their translational initiation at the murine equivalent of Met 80 within exon 5. The renal promoter drives transcriptional initiation from several, clustered start sites within intron 3, and produces (at least in rat) two transcripts which vary in the inclusion or exclusion of a portion of intron 3, but share common sequences in exon 3 and downstream (Kudrycki and Shull, 1993).

The AE2 gene has been cloned from rat and human genomes (G. Shull, pers. comm.). AE2 cDNA has been cloned from mouse, human, rat, and rabbit (Alper et al., 1988; Gehrig et al., 1992; Kudrycki et al., 1990; Chow et al., 1992). Northern blot data suggest at least two AE2 transcripts which vary only in the 5' untranslated region. AE2 mRNAs are found widely in epithelial and mesenchymal tissues. Deduced amino acid sequences of AE2 proteins show greater than 90% identity across species, a degree of conservation greater than shown by AE1 proteins. The AE2 hydrophobic domain shares 67% amino acid identity with that of AE1, but only 33% identity with the aligning portion of the N-terminal cytoplasmic domain. The far N-terminal 250 residues of AE2 have no corresponding domain in the AE1 protein.

The rat AE3 gene encodes at least two RNA transcripts and two protein isoforms (Linn et al., 1992). The "brain" isoform cDNA (bAE3) has been cloned from mouse (Kopito et al., 1989), rat (Kudrycki et al., 1990), and human (Yannoukakos et al, submitted). bAE3 has also been cloned from stomach. The "cardiac" isoform cDNA (cAE3) cloned from rat and human comprises exon C1 within intron 6, followed by exons 7-23. cAE3 mRNA also has been detected at lower levels in skeletal and vascular smooth muscle. The bAE3 amino terminal residues 1-270 encoded by exons 1-6, and the cAE3 amino terminal residues 1-73 encoded by exon C1, are of distinct sequence, but the downstream amino acid sequences of the two isoforms are identical. Though only cAE3 was detected in adult mouse heart, adult heart of rat, rabbit, chick, and human heart express similar levels of bAE3 and cAE3 transcripts. AE3 amino acid sequences are highly conserved across species. Murine AE3 shares with AE2 69% amino acid identity across the hydrophobic membrane-spanning domain, 60% identity across the mid-portion of the polypeptide, but only 28% across the N-terminal 250 amino acids which are not alignable with AE1. The C1 exon amino acid sequence has no homologous domains in AE1 or AE2, and is itself less well conserved across species than is the remainder of the AE3 sequence.

The topography of the AE2 and AE3 proteins is modeled on that of AE1, commonly modeled with 14 transmembrane spans. The two termini of AE1 are known from proteolytic and antibody studies to be cytoplasmic. The disposition of the first eight transmembrane spans have also been defined with reasonable likelihood. There follows in the AE1 sequence one hydrophobic region which must cross the bilayer either one or three times. The last two hydrophobic regions of the protein each suffice in length to traverse the bilayer twice.

Molecular genetics of the AE genes

The AE genes are unlinked. The human AE1 gene is found in chromosome 17, between band q21 and the terminus of the long arm (Showe et al., 1987). The human AE2 gene maps to chromosome 7q35-36 (Palumbo et al., 1986). Localization of the human AE3 gene is awaited. Murine genes encoding AE1, AE2 and AE3 have been mapped, respectively, to murine chromosomes 11 (Love et al., 1990), 5, and 1 (White et al., submitted).

The recent description of human mutations in AE1 associated with heritable diseases of red blood cells has initiated the rapidly developing molecular human genetics of the AE gene family. The best studied mutation is that of Southeast Asian Ovalocytosis (SAO). SAO erythrocytes are heterozygous, with 50% wild-type AE1 protein and 50% AE1 protein carrying the Δ400-408 deletion at the putative initial insertion of the polypeptide into the lipid bilayer (Jarolim et al., 1991; Schofield et al., 1992). The mutation is caused by an in-frame deletion of 27 nucleotides. No individuals homozygous for the mutation have been identified, suggesting embryonic lethality of the homozygous mutation. SAO AE1 protein is present in the red cell membrane, where it appears to be nonfunctional and unable to bind DIDS or ^{3}H-H$_2$DIDS (Schofield, et al., 1992). The SAO membrane shows increased mechanical rigidity, increased resistance to plasmodial invasion, and a decreased mobile fraction of AE1 consistent with ankyrin-binding of increased affinity (Liu et al., 1990). The purified protein is preferentially tetrameric and has altered denaturation properties (Moriyama et al., 1992).

A reduced mass ratio of AE1 to spectrin polypeptide in the membrane of red cells from patients with hereditary spherocytosis led to the identification of Band 3 Prague (Jarolim et al, submitted). This AE1 mutation is caused by a genomic in-frame insertion of 10 nucleotides which leads to replacement of the last two putative transmembrane domains of AE1 with a neosequence of 70 amino acids of increased hydrophilicity. As was true for SAO, no individuals with the homozygous mutation have yet been found, suggesting embryonic lethality of the homozygous genotype. Heterozygous red cells show reduced AE1 function.

Two AE1 mutations have been associated with red cell membrane deficiencies of the AE1-binding protein 4.2. Band 3 Tuscaloosa is characterized by the P327R substitution (Jarolim, et al., 1992). This proline is conserved in mammalian but not trout AE1, and is absent in AE2 and AE3. Band 3 Montefiore displays the substitution E40K (Rybicki, et al., 1993). This glutamate is conserved in mammalian AE1 and AE2, but not in AE3 or in trout

AE1. The associated hemolytic anemia is recessively inherited, and also associated with a mild deficiency of red cell glyceraldehyde-phosphate dehydrogenase.

Functional insights from heterologous expression of recombinant AE proteins.

Functional demonstrations of anion exchange have been achieved for each AE cDNA, using at least three heterologous expression systems. The Xenopus oocyte system has documented bidirectional chloride/chloride and chloride/anion exchange mediated by AE1, AE2, and AE3. Transiently transfected, BCECF-loaded mammalian cells have provided evidence for chloride/bicarbonate exchange mediated by AE2 and AE3. Sf9 insect cells infected with recombinant baculovirus have also exhibited AE2-mediated chloride/bicarbonate exchange. Lastly, isotopic fluxes consistent with sulfate/chloride exchange mediated by AE1 and AE2 have been measured in membrane vesicles prepared from transiently transfected mammalian cells.

All AE protein-mediated bidirectional anion transport appears to be sodium-independent, to require a permeant trans-anion, and not to require the cytoplasmic domains (Kopito et al., 1989; Lindsey et al., 1990; Ruetz et al., 1993). Overexpression of AE2 or of AE3 has produced no change in resting pH_i (Lee et al., 1991). AE1 and AE2 have revealed two distinctions. The first is the reduced pharmacological sensitivity of AE2 to chloride transport inhibitors active against AE1. The second is the pH_i-dependence of AE2-mediated chloride/base exchange (He et al., 1993; Lee et al, 1992).

Initial attempts to assign functional significance to individual amino acid residues or domains have focused on the interaction of transport inhibitors with the exterior surface of the protein. The results indicated that several surface lysines important for covalent binding of impermeant inhibitors are not required for anion transport, but may contribute subtly to reversible inhibition or to anion transport (Bartel et al., 1989; Garcia and Lodish, 1989; Passow et al., 1992; Wood et al., 1992). Additional directed AE1 mutations have documented combinations of putatively endoplasmic residues which when mutated result in loss of functional expression in the Xenopus oocyte, but which singly are without effect. None of these mutations have yet been reported as to their effects on AE1 surface expression.

Intracellular trafficking of AE proteins

The only cell culture systems in which endogenous AE protein biosynthesis has been studied with pulse-chase kinetics are erythroid. Early studies with erythroleukemia lines agreed that spectrin and ankyrin biosynthesis preceeded that of AE1. These studies led to the proposal that stable incorporation of AE1 into the plasma membrane, if not into the endoplasmic reticulum, required a preassembled cytoskeleton (Lazarides and Woods, 1989;

Lehnert and Lodish, 1988), but more recent immunocytochemical studies of nontransformed erythroid precursors differentiating in culture have questioned the universality of this conclusion (Nehls, et al., 1993). Such data have fueled speculation about the identity of putative chaperonins for AE1, and oocyte expression experiments have suggested glycophorin A as a plausible candidate (Groves and Tanner, 1992). The normal levels of AE1 protein in red cells genetically deficient in glycophorin A (Tanner, 1993) undermine this otherwise attractive hypothesis.

AE1 membrane domain is dimeric in detergent solution, whereas detergent-solubilized AE1 holoprotein consists of stable dimers and tetramers. The tetrameric interaction in the red cell membrane seems to require the cytoplasmic domain (Reithmeier and Casey, 1992), and only the tetramer binds ankyrin, aldolase and other glycolytic enzymes, and hemoglobin (Schubert et al., 1992). The oligomeric state of AE1 freshly solubilized from the red cell membrane is thought to reflect its recent state in the membrane. Attempts to locate restricted binding sites or domains in the AE1 cytoplasmic domain for ankyrin and for protein 4.1 have suggested different sites according to the techniques used and to the reporting laboratory. Though recombinant AE2 is not co-immunoprecipitated with the repeat domain of erythroid ankyrin, recombinant AE1 and AE3 can be so precipitated (Morgans and Kopito, 1993).

All overexpression systems have been unable to process most heterologous mammalian AE protein beyond the host cis-Golgi stacks out to the plasma membrane. In the case of mammalian AE1, no evidence has yet been obtained for surface expression of the heterologous protein in host mammalian cells. However, chicken AE1 was stably expressed in chick embryo fibroblasts and quail tumor cells (Fuerstenberg et al., 1990).

Chloride/bicarbonate exchange activity is present variably in both apical and basolateral membranes of polarized epithelial cells in intestine and kidney. Although immunostaining of endogenous AE proteins to date has been restricted to basolateral surfaces of epithelia in situ, at least two systems have provided immunoblot evidence for the presence in fractions enriched in apical plasma membrane of AE1 (van Adelsberg et al., 1993) and AE2 (Chow et al., 1992).

References

Alper SL (1991) The band 3-related anion exchanger (AE) gene family. Annu Rev Physiol 53:549-64

Alper SL, Kopito RR, Libresco SM, Lodish HF (1988) Cloning and characterization of a murine band 3-related cDNA from kidney and from a lymphoid cell line. J Biol Chem 263:17092-9

Bartel D, Lepke S, Layh SG, Legrum B, Passow H (1989) Anion transport in oocytes of Xenopus laevis induced by expression of mouse erythroid band 3 protein--encoding cRNA and of a cRNA derivative obtained by site-directed mutagenesis at the stilbene disulfonate binding site. Embo J 8:3601-9

Brosius FC, Alper SL, Garcia AM, Lodish HF (1989) The major kidney band 3 gene transcript predicts an amino-terminal truncated band 3 polypeptide. J Biol Chem 264:7784-7

Chow A, Dobbins JW, Aronson PS, Igarashi P (1992) cDNA cloning and localization of a band 3-related protein from ileum. Am J Physiol G345-52

Fuerstenberg S, Beug H, Introna M, Khazaie K, Munoz A, Ness S, et al. (1990) Ectopic expression of the erythrocyte band 3 anion exchange protein, using a new avian retrovirus vector. J Virol 64:5891-902

Garcia AM, Lodish HF (1989) Lysine 539 of human band 3 is not essential for ion transport or inhibition by stilbene disulfonates. J Biol Chem 264:19607-13

Gehrig H, Muller W, Appelhans H (1992) Complete nucleotide sequence of band 3 related anion transport protein AE2 from human kidney. Biochim Biophys Acta 1130:326-8

Groves JD, Tanner MJ (1992) Glycophorin A facilitates the expression of human band 3-mediated anion transport in Xenopus oocytes. J Biol Chem 267:22163-70

He X, Wu X, Knauf PA, Tabak LA, Melvin JE (1993) Functional expression of the rat anion exchanger AE2 in insect cells by a recombinant baculovirus. Am J Physiol C1075-9

Hubner S, Michel F, Rudloff V, Appelhans H (1992) Amino acid sequence of band-3 protein from rainbow trout erythrocytes derived from cDNA. Biochem J 17-23

Jarolim P, Palek J, Amato D, Hassan K, Sapak P, Nurse GT, et al. (1991) Deletion in erythrocyte band 3 gene in malaria-resistant Southeast Asian ovalocytosis. Proc Natl Acad Sci U S A 88:11022-6

Jarolim P, Palek J, Rubin HL, Prchal JT, Korsgren C, Cohen CM (1992) Band 3 Tuscaloosa: Pro327-Arg327 substitution in the cytoplasmic domain of erythrocyte band 3 protein associated with spherocytic hemolytic anemia and partial deficiency of protein 4.2. Blood 80:523-9

Kim HR, Kennedy BS, Engel JD (1989) Two chicken erythrocyte band 3 mRNAs are generated by alternative transcriptional initiation and differential RNA splicing. Mol Cell Biol 9:5198-206

Kim HR, Yew NS, Ansorge W, Voss H, Schwager C, Vennstrom B, et al. (1988) Two different mRNAs are transcribed from a single genomic locus encoding the chicken erythrocyte anion transport proteins (band 3). Mol Cell Biol 8:4416-24

Kopito RR, Andersson M, Lodish HF (1987) Structure and organization of the murine band 3 gene. J Biol Chem 262:8035-40

Kopito RR, Lee BS, Simmons DM, Lindsey AE, Morgans CW, Schneider K (1989) Regulation of intracellular pH by a neuronal homolog of the erythrocyte anion exchanger. Cell 59:927-37

Kopito RR, Lodish HF (1985) Primary structure and transmembrane orientation of the murine anion exchange protein. Nature 316:234-8

Kudrycki KE, Newman PR, Shull GE (1990) cDNA cloning and tissue distribution of mRNAs for two proteins that are related to the band 3 Cl-/HCO3- exchanger. J Biol Chem 265:462-71

Kudrycki KE, Shull GE (1989) Primary structure of the rat kidney band 3 anion exchange protein deduced from a cDNA. J Biol Chem 264:8185-92

Kudrycki KE, Shull GE (1993) Rat kidney band 3 Cl-/HCO3- exchanger mRNA is transcribed from an alternative promoter. Am J Physiol F540-7

Lazarides E, Woods C (1989) Biogenesis of the red blood cell membrane-skeleton and the control of erythroid morphogenesis. Annu Rev Cell Biol 5:427-52

Lee BS, Gunn RB, Kopito RR (1991) Functional differences among nonerythroid anion exchangers expressed in a transfected human cell line. J Biol Chem 266:11448-54

Lehnert ME, Lodish HF (1988) Unequal synthesis and differential degradation of alpha and beta spectrin during murine erythroid differentiation. J Cell Biol 107:413-26

Lindsey AE, Schneider K, Simmons DM, Baron R, Lee BS, Kopito RR (1990) Functional expression and subcellular localization of an anion exchanger cloned from choroid plexus. Proc Natl Acad Sci U S A 87:5278-82

Linn SC, Kudrycki KE, Shull GE (1992) The predicted translation product of a cardiac AE3 mRNA contains an N terminus distinct from that of the brain AE3 Cl-/HCO3-

exchanger. Cloning of a cardiac AE3 cDNA, organization of the AE3 gene, and identification of an alternative transcription initiation site. J Biol Chem 267:7927-35

Liu SC, Zhai S, Palek J, Golan DE, Amato D, Hassan K, et al. (1990) Molecular defect of the band 3 protein in southeast Asian ovalocytosis. N Engl J Med 323:1530-8

Love JM, Knight AM, McAleer MA, Todd JA (1990) Towards construction of a high resolution map of the mouse genome using PCR-analysed microsatellites. Nucleic Acids Res 18:4123-30

Lux SE, John KM, Kopito RR, Lodish HF (1989) Cloning and characterization of band 3, the human erythrocyte anion- exchange protein (AE1). Proc Natl Acad Sci U S A 86:9089-93

Morgans CW, Kopito RR (1993) Association of the brain anion exchanger, AE3, with the repeat domain of ankyrin. J. Cell. Sci. 105:1137-1142

Moriyama R, Ideguchi H, Lombardo CR, Van DH, Low PS (1992) Structural and functional characterization of band 3 from Southeast Asian ovalocytes. J Biol Chem 267:25792-7

Nehls V, Zeitler ZP, Drenckhahn D (1993) Different sequences of expression of band 3, spectrin, and ankyrin during normal erythropoiesis and erythroleukemia. Am J Pathol 142:1565-73

Palumbo AP, Isobe M, Huebner K, Shane S, Rovera G, Demuth D, et al. (1986) Chromosomal localization of a human band 3-like gene to region 7q35--- -7q36. Am J Hum Genet 39:307-16

Passow H, Lepke S, Wood PG (1992) Exploration of the mechanism of mouse erythroid band 3-mediated anion exchange by site-directed mutagenesis. Prog. Cell. Res. 2:85-101

Raley-Susman KM, Sapolsky RM, Kopito RR (1993) Cl-/HCO3- exchange function differs in adult and fetal rat hippocampal neurons. Brain Res. 614:308-314

Reithmeier RAF, Casey JR (1992) Oligomeric structure of the human erythrocyte band 3 anion transport protein. Prog. Cell. Res. 2:181-190

Ruetz S, Lindsey AE, Ward CL, Kopito RR (1993) Functional activation of plasma membrane anion exchangers occurs in a pre-Golgi compartment. J Cell Biol 121:37-48

Rybicki AC, Qiu JJ, Musto S, Rosen NL, Nagel RL, Schwartz RS (1993) Human erythrocyte protein 4.2 deficiency associated with hemolytic anemia and a homozygous 40glutamic acid-->lysine substitution in the cytoplasmic domain of band 3 (band 3Montefiore). Blood 81:2155-65

Schofield AE, Reardon DM, Tanner MJ (1992) Defective anion transport activity of the abnormal band 3 in hereditary ovalocytic red blood cells. Nature 355:836-8

Schubert D, Huber E, Lindenthal S, Mulzer K, Schuck P (1992) The relationships between the oligomeric structure and the functions of human erythrocyte band 3 protein: the functional unit for the binding of ankyrin, hemoglobin and aldolase and for anion transport. Prog. Cell. Res. 2:209-217

Showe LC, Ballantine M, Huebner K (1987) Localization of the gene for the erythroid anion exchange protein, band 3 (EMPB3), to human chromosome 17. Genomics 1:71-6

Tanner MJ (1993) Molecular and cellular biology of the erythrocyte anion exchanger (AE1). Semin Hematol 30:34-57

van Adelsberg JS, Edwards JC, al Awqati Q (1993) The apical Cl/HCO3 exchanger of beta intercalated cells. J Biol Chem 268:11283-9

Wagner S, Vogel R, Lietzke R, Koob R, Drenckhahn D (1987) Immunochemical characterization of a band 3-like anion exchanger in collecting duct of human kidney. Am J Physiol F213-21

Wood PG, Muller H, Sovak M, Passow H (1992) Role of Lys 558 and Lys 869 in substrate and inhibitor binding to the murine band 3 protein: a study of the effects of site-directed mutagenesis of the band 3 protein expressed in the oocytes of Xenopus laevis. J Membr Biol 127:139-48

Proton-coupled peptide transport in the small intestine and kidney

Vadivel Ganapathy, Matthias Brandsch, Frederick H. Leibach
Department of Biochemistry and Molecular Biology, Medical
College of Georgia, Augusta, GA 30912, USA

Introduction

Carrier-mediated transport of small peptides is a well
established phenomenon in the small intestine and kidney
(Ganapathy & Leibach, 1986, 1991; Ganapathy et al., 1991;
Matthews, 1991; Silbernagl, 1988). In the small intestine,
absorption of intact peptides into the enterocytes across the
brush border membrane via the peptide transport system plays an
important role in the assimilation of dietary proteins. The
substrates for the peptide transport system are generated by
concerted actions of gastric and pancreatic proteases and brush
border peptidases. Of these enzymes, dipeptidylpeptidase IV
(DPP IV) and angiotensin converting enzyme (ACE), both of which
are associated with the brush border membrane, are of
significant importance because they directly generate dipeptides
from the amino terminus and the carboxy terminus respectively,
of large polypeptides and proteins by sequential action. The
enterocytes, in addition, possess highly active cytosolic
peptidases which hydrolyze the peptides entering the cell via
the brush border membrane peptide transport system. Thus, even
though protein digestion products are absorbed into the
enterocytes to an appreciable extent in the form of intact
peptides, what enters the circulation is predominantly in the
form of free amino acids. However, peptides which are resistant
to hydrolysis by cytosolic peptidases may enter the circulation
in intact form because there is evidence for the presence of a
peptide transport system in the basolateral membrane of normal
enterocytes (Dyer et al., 1990) as well as cultured intestinal
cell lines (Thwaites et al., 1993). The physiological functions
of the peptide transport system in the kidney are not as obvious

NATO ASI Series, Vol. H 89
Molecular and Cellular Mechanisms
of H+ Transport
Edited by Barry H. Hirst
© Springer-Verlag Berlin Heidelberg 1994

as in the small intestine. One of the most likely functions is in the conservation of amino nitrogen which is present in the glomerular filtrate in the form of peptides and proteins. The brush border membrane of renal tubular cells possesses a number of peptidases which generate free amino acids as well as small peptides from large polypeptides and proteins. In particular, the activities of DPP IV and ACE, whose hydrolytic products consist of dipeptides, are very high in this membrane. The peptide transport system present in the brush border membrane of the tubular cells serves to transport these peptides from the tubular lumen into the cells where these peptides undergo further hydrolysis by cytosolic peptidases. Peptides which are hydrolysis-resistant may leave the cell in intact form to enter the blood via the basolateral membrane peptide transport system (Barfuss et al., 1988).

Characteristics of the peptide transport system

Even though it is not known at this time whether the carrier proteins responsible for the transport of peptides in the small intestine and kidney are the same or different, the general characteristics of the transport process are very similar in both tissues. The intestinal as well as the renal peptide transport systems accept di- and tripeptides, consisting of L-amino acids, as substrates. Blocking of the amino group, carboxyl group or both in the peptide substrate drastically reduces its affinity for the transport system. Free amino acids are excluded by the system. Histidyl residues play an essential role in the catalytic function of the peptide transport system in both tissues.

Transport of peptides in the small intestine and kidney is an active process. Unlike most active transport systems available for organic solutes in these tissues, the peptide transport system does not depend on a Na^+ gradient for its driving force. The system is energized by an electrochemical H^+ gradient (Ganapathy & Leibach, 1991; Ganapathy et al., 1987; Hoshi, 1985). This is accomplished by the ability of the transport system to catalyze cotransport of H^+ with the peptide substrates (Thwaites et al., 1993). The transfer process is electrogenic (Ganapathy & Leibach, 1983; Ganapathy et al., 1984; Abe et al., 1987), and therefore the chemical component as well as the electrical component of the proton motive force participate in the energization of the transport system. There is ample evidence for the existence of a H^+ gradient across the brush border membrane of the intestinal and renal tubular cells (Ganapathy & Leibach, 1985; Ganapathy et al., 1987). This gradient is generated by the Na^+-H^+ exchanger which converts a Na^+ gradient into a H^+ gradient via electroneutral coupling between Na^+ influx and H^+ efflux. In addition to this exchanger, a V-type H^+-ATPase may also contribute to generation of the H^+ gradient. The presence of the H^+-ATPase has been demonstrated in the brush border membrane of renal tubular cells (Simon & Burckhardt, 1989), but it is not known at this time whether a similar mechanism is also operative in the small intestine.

Functions of the peptide transport system

In addition to its role in the transport of small peptides, the peptide transport system in the small intestine and kidney may have other functions. Recently, it has been shown that Zn^{2+}, in the form of peptide-Zn^{2+} chelate, can be transported into intestinal brush border membrane vesicles via the peptide transport system in a H^+ gradient-dependent manner (Tacnet et al., 1993). This suggests that the transport system may play a role in the intestinal absorption of not only Zn^{2+} but also other trace elements such as Cu^{2+} and Ni^{2+} which readily chelate with small peptides. Furthermore, many pharmacologically active compounds whose chemical structures are related to peptides have been shown to be substrates for the peptide transport system. This includes aminocephalosporin antibiotics (Inui et al., 1984; Tsuji, 1987) and the anticancer agent bestatin (Inui et al., 1992). The therapeutic efficacy of these drugs is likely to be determined by the rate of their absorption from the small intestine via the peptide transport system. Similarly, since these agents are excreted by the kidney, the renal peptide transport system may function in their reabsorption and thereby influence their half-life in the circulation. The transport system can also be exploited for efficient intestinal absorption of nonpeptide-like drugs if these drugs can be suitably modified to yield prodrugs which structurally resemble the peptide substrates. Once the prodrug is transported into the enterocyte via the peptide transport system, the parent drug can be released by the action of cytosolic peptidases (Bai et al., 1992).

The intestinal and renal peptide transport systems also have potential clinical applications. Absorption of amino nitrogen from the intestine in the form of small peptides rather than free amino acids appears to have nutritional advantages. This has clinical relevance to enteral nutrition. Even though most of the currently available enteral solutions are elemental diets containing amino nitrogen in the form of free amino acids, the possible advantages of enteral diets consisting of small peptides instead of free amino acids are receiving increasing attention in recent years. Such peptide-based formulas will reduce the osmolality of the enteral solution and also will enable inclusion, in the form of peptides, of amino acids such as tyrosine, glutamine, and cysteine which are generally omitted in elemental diets owing to their insolubility and/or instability. Use of peptide-based formulas instead of currently available amino acid-based formulas for parenteral nutrition also will have similar practical advantages. The presence of an efficient peptide transport mechanism and highly active peptidases in the kidney strongly suggests the feasibility of such an approach.

Tripeptides as substrates for the peptide transport system

Until recently, there has been no published report in which transport of a tripeptide via the peptide transport system has been directly demonstrated. The conclusion that the transport system accepts tripeptides as substrates was based upon competition experiments in which tripeptides were shown to inhibit dipeptide transport. Nonavailability of radiolabeled tripeptides and hydrolysis of tripeptides by brush border peptidases were the primary factors responsible in the past for the lack of direct studies on tripeptide transport. Recently, identification of a rat strain which does not express DPP IV activity has made it possible to investigate directly the characteristics of the transport of intact tripeptides. Fischer 344 rats obtained from commercial sources from Japan exhibit a total lack of DPP IV activity while Fischer 344 rats obtained within the USA have normal levels of this enzyme activity (Tiruppathi et al., 1990a). Due to the absence of this enzyme, brush border membrane vesicles isolated from the kidneys of the Japanese Fischer 344 rats are unable to hydrolyze tripeptides of X-Pro-Y type. These DPP IV-negative membrane vesicles have proved to be an excellent experimental system to elucidate the mechanism of transport of the tripeptide Phe-Pro-Ala (Tiruppathi et al., 1990b). This tripeptide accumulates inside the vesicles in intact form against a concentration gradient when an inwardly directed H^+ gradient is imposed across the membrane. The transport process is Na^+-independent. The H^+-dependent transport of the tripeptide is electrogenic, being stimulated by an inside-negative membrane potential and is inhibitable by several tripeptides as well as dipeptides. Detailed kinetic analysis with this experimental system has unequivocally demonstrated that dipeptides and tripeptides share a common transport system (Tiruppathi et al., 1990c).

Peptide transport in cultured cell lines of intestinal and renal origin

Availability of cell lines which express the peptide transport system will facilitate studies on the regulatory aspects of the transport system. Among various cell lines screened, only the intestinal cell line Caco-2 and the renal cell line MDCK have been shown to possess peptide transport activity (Brandsch M, Ganapathy V, Leibach FH, manuscript in preparation). In these two cell lines, transport of the dipeptide glycylsarcosine occurs via an active transport system that is specific for di- and tripeptides and the transport process is Na^+-independent and is energized by a H^+ gradient. Available data on the peptide transport system expressed in these cells strongly suggest that the activity of the system is under the regulation of protein kinase C. The function of the transport system is inhibited by treatment of the cells with activators of protein kinase C and this inhibition is blocked if the treatment is done in the presence of protein kinase C inhibitors. The protein kinase C-induced inhibition of the transporter function is primarily due to a decrease in the maximal velocity, the affinity of the system towards its substrates remaining unaffected. Future experiments are needed to determine the exact mechanism underlying the protein kinase C-mediated change in the peptide transporter activity.

Conclusion

The peptide transport system expressed in the mammalian intestine and kidney is unique because of the involvement of an electrochemical H^+ gradient as the driving force. H^+-Coupled solute transport systems are very common in bacteria, but are rare in animals. Most solute transport systems in animals are energized by an electrochemical Na^+ gradient. Future studies aimed at elucidation of the structural features of the peptide-H^+ cotransport system are expected to generate valuable information which will help us understand the molecular basis for the preference of this transport system for H^+ rather than Na^+ as the coupling ion.

Acknowledgements

This work was supported by the National Institutes of Health Grants DK 28389 and DK 42069. The authors thank Ida O. Thomas for excellent secretarial assistance.

References

Abe M, Hoshi T, Tajima A (1987) Characteristics of transmural potential changes associated with the proton-peptide co-transport in toad small intestine. J Physiol 394:481-499

Bai JPF, Hu M, Subramanian P, Mosberg HI, Amidon GL (1992) Utilization of peptide carrier system to improve intestinal absorption: targeting prolidase as a prodrug-converting enzyme. J Pharm Sci 81:113-116

Barfuss DW, Ganapathy V, Leibach FH (1988) Evidence for active dipeptide transport in the isolated proximal straight tubule. Am J Physiol 255:F177-F181

Dyer J, Beechey RB, Gorvel JP, Smith RT, Wootton R, Shirazi-Beechey SP (1990) Glycyl-L-proline transport in rabbit enterocyte basolateral membrane vesicles. Biochem J 269:565-571

Ganapathy V, Burckhardt G, Leibach FH (1984) Characteristics of glycylsarcosine transport in rabbit intestinal brush border membrane vesicles. J Biol Chem 259:8954-8959

Ganapathy V, Leibach FH (1983) Role of pH gradient and membrane potential in dipeptide transport in intestinal and renal brush border membrane vesicles. Studies with glycyl-L-proline and L-carnosine. J Biol Chem 258:14189-14192

Ganapathy V, Leibach FH (1985) Is intestinal peptide transport energized by a proton gradient? Am J Physiol 249:G153-G160

Ganapathy V., Leibach FH (1986) Carrier mediated reabsorption of small peptides in renal proximal tubule. Am J Physiol 251:F945-F953

Ganapathy V, Leibach FH (1991) Proton-coupled solute transport in the animal cell plasma membrane. Curr Opin Cell Biol 3:695-701

Ganapathy V, Miyamoto Y, Leibach FH (1987) Driving force for peptide transport in mammalian kidney and intestine. Contrib Infusion Ther Clin Nutr 17:54-68

Ganapathy V, Miyamoto Y, Tiruppathi C, Leibach FH (1991) Peptide transport across the animal cell plasma membrane: recent developments. Indian J Biochem Biophys 28:317-323

Hoshi T (1985) Proton-coupled transport of organic solutes in animal cell membranes and its relation to Na^+ transport. Jpn J Physiol 35:179-191

Inui KI, Okano T, Takano M, Saito H, Hori R (1984) Carrier-mediated transport of cephalexin via the dipeptide transport system in rat renal brush border membrane vesicles. Biochim Biophys Acta 769:449-454

Inui KI, Tomita Y, Katsura T, Okano T, Takano M, Hori R (1992) H^+ coupled active transport of bestatin via the dipeptide transport system in rabbit intestinal brush border membranes. J Pharmacol Exp Ther 260:482-486

Matthews DM (1991) Protein absorption. Development and present state of the subject. Wiley-Liss, New York, NY

Silbernagl S (1988) The renal handling of amino acids and oligopeptides. Physiol Rev 68:911-1007

Simon BJ, Burckhardt G (1989) Characterization of inside-out oriented H^+-ATPases in cholate-pretreated renal brush-border membrane vesicles. J Membr Biol 117:141-151

Tacnet F, Lauthier F, Ripoche P (1993) Mechanisms of zinc transport into pig small intestine brush-border membrane vesicles. J Physiol 465:57-72

Thwaites DT, Brown CDA, Hirst BH, Simmons NL (1993) Transepithelial glycylsarcosine transport in intestinal Caco-2 cells mediated by expression of H+-coupled carriers at both apical and basal membranes. J Biol Chem 268:7640-7642

Tiruppathi C, Miyamoto Y, Ganapathy V, Roesel RA, Whitford GM, Leibach FH (1990a) Hydrolysis and transport of proline-containing peptides in renal brush border membrane vesicles from dipeptidyl peptidase IV-positive and dipeptidyl peptidase IV-negative rat strains. J Biol Chem 265:1476-1483

Tiruppathi C, Ganapathy V, Leibach FH (1990b) Evidence for tripeptide-proton symport in renal brush border membrane vesicles. Studies in a novel rat strain with a genetic absence of dipeptidyl peptidase IV. J Biol Chem 265:2048-2053

Tiruppathi C, Ganapathy V, Leibach FH (1990c) Kinetic evidence for a common transporter for glycylsarcosine and phenylalanylprolylalanine in renal brush border membrane vesicles. J Biol Chem 265:14870-14874

Tsuji A (1987) Intestinal uptake of β-lactam antibiotics and its relationship to peptide transport. Adv Biosci 65:125-131

H$^+$-coupled solute transport in cultured intestinal epithelia

David T. Thwaites, Barry H. Hirst and Nicholas L. Simmons

Department of Physiological Sciences, The Medical School, University of Newcastle-upon-Tyne, Newcastle-upon-Tyne, NE2 4HH, UK.

Introduction

As described in the chapter by Ganapathy, Brandsch and Leibach, absorption of intact small peptides (di/tripeptides) via the intestinal di/tripeptide transporter plays an important role in nutrient absorption. Studies by these authors and others using intestinal brush-border membrane vesicles (BBMV) demonstrate that unlike many other intestinal transport mechanisms it is the electrochemical gradient for H$^+$, rather than Na$^+$, that plays a major role in di/tripeptide absorption across the apical membrane of the intestinal enterocyte (Ganapathy and Leibach, 1983; Ganapathy et al., 1984; Ganapathy and Leibach, 1985; Thwaites et al., 1993g). The electrogenic (and Na$^+$-independent) nature of this transport mechanism has been demonstrated in both intact tissues (Boyd and Ward, 1982) and BBMV (Ganapathy et al., 1984). However, despite such strong evidence, dipeptide accumulation by intestinal BBMV is not marked and only indirect demonstrations of H$^+$-coupled dipeptide transport are available in mammalian tissues (Ganapathy and Leibach, 1985). The exact mechanism(s) involved in di/tripeptide exit across the basolateral membrane are unclear, although a single report using rabbit intestinal basolateral membrane vesicles describes pH-dependent acceleration of dipeptide (Gly-Pro) uptake (Dyer et al., 1990). The driving force for this pH-dependent route of absorption is thought to be the acid microclimate (an area of low pH adjacent to the apical membrane) which has been demonstrated in the human small intestine both in vivo (Rawlings et al., 1987) and in vitro (Lucas et al., 1978). Recently there has been a great deal of interest from the pharmaceutical industry in this H$^+$-coupled transport route as it is believed to be responsible for the oral absorption of certain groups of drugs including some aminocephalosporin antibiotics (Okano et al., 1986) and some angiotensin-converting enzyme (ACE) inhibitors (Friedman and Amidon, 1989).

Using three separate but complementary techniques (radiolabel fluxes, intracellular

NATO ASI Series, Vol. H 89
Molecular and Cellular Mechanisms
of H$^+$ Transport
Edited by Barry H. Hirst
© Springer-Verlag Berlin Heidelberg 1994

pH measurements and short-circuit current (I_{sc}) determinations) we have examined transepithelial intestinal transport of the biologically stable dipeptide glycylsarcosine [Gly-Sar (Addison *et al.*, 1972)]. To accomplish this we have utilized the human intestinal epithelial cell line Caco-2 reconstituted as epithelial monolayers on permeable filters. The Caco-2 system has been identified as a useful model for the study of intestinal epithelial permeability (Hidalgo *et al.*, 1989). This cell system expresses a number of solute transporters, including those for sugars (Blais *et al.*, 1987), amino acids (Hu and Borchardt, 1992), bile acids (Hidalgo and Borchardt, 1990) and aminocephalosporins (Dantzig and Bergin, 1990; Inui *et al.*, 1992). Gly-Sar transport together with cellular accumulation were measured across both apical and basolateral cell surfaces of Caco-2 cell monolayers (Thwaites *et al.*, 1993a; 1993b). Direct measurement of H^+-coupled dipeptide transport across both apical and basal membranes was determined in Caco-2 cells loaded with the pH-sensitive fluorescent dye BCECF [2',7'-bis(2-carboxyethyl)-5(6)-carboxyfluorescein] (Thwaites *et al.*, 1993a; 1993b; 1993c). Furthermore, the electrogenic nature of this H^+-coupled dipeptide transport process was identified by I_{sc} measurements using voltage clamped Caco-2 cell monolayers (Thwaites *et al.*, 1993f). Finally, this combination of techniques has also been used to examine other substrates that may undergo H^+-coupled transport including aminocephalosporin antibiotics (Thwaites, 1993; Thwaites *et al.*, 1993b), ACE inhibitors, and interestingly a range of amino acids (Thwaites *et al.*, 1993d; 1993e; 1994).

Results

pH-dependent dipeptide transport

Results in Fig. 1 indicate that acidification of the apical environment stimulates both transepithelial apical-to-basal (J_{a-b}) dipeptide transport (Fig. 1a) and intracellular accumulation (Fig. 1b), with maximal stimulation between apical pH 6.0-6.5. This pH-stimulated intracellular accumulation across the apical membrane represents an approximate 12-fold accumulation above medium levels (Fig. 1b). Transepithelial dipeptide transport in the basal-to-apical direction (J_{b-a}) was stimulated by basolateral acidity, maximal transport (Fig. 1a) and intracellular accumulation (Fig. 1b) were detected at basolateral pH 5.5 (3-fold accumulation above medium levels). Clearly transepithelial dipeptide transport can occur in both absorptive (J_{a-b}) and secretory (J_{b-a}) directions. Fig. 2 indicates that in the absence of a transepithelial pH gradient (apical pH 7.4, basolateral pH 7.4) net dipeptide transport is in

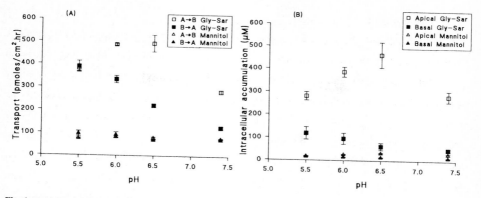

Fig. 1 (a) Absorptive (open symbols; apical pH 5.5-7.4, basolateral pH 7.4) and secretory (closed symbols; basolateral pH 5.5-7.4, apical pH 7.4) transport of [^{14}C]Gly-Sar (squares) and [^{3}H]mannitol (triangles) across Caco-2 cell monolayers as a function of extracellular pH. (b) Intracellular accumulation across either the apical (open symbols) or basolateral (closed symbols) membranes as a function of extracellular pH. Other details as part (a). Results are illustrated as mean ± 1 SEM, n=7-8. (Adapted from Thwaites *et al.*, 1993a).

the absorptive direction. Intracellular accumulation and net absorptive transport were both increased when the apical pH was lowered to pH 6.0 (Fig. 2). A net secretory state can also be imposed on this cell system when the basolateral pH is 6.0 (apical pH 7.4).

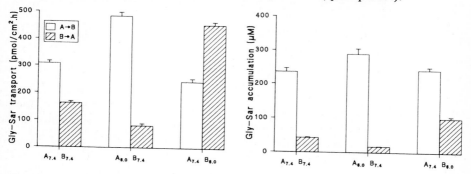

Fig. 2 (a) Transepithelial transport of [^{14}C]Gly-Sar across Caco-2 cell monolayers [J_{a-b} (open columns) and J_{b-a} (striped columns)]. Apical (A) and basal (B) bathing solutions were as indicated and contained 36μM Gly-Sar. Paired monolayers of the same cell batch were used to measure the bidirectional fluxes. Data are the mean ± SEM (n=4-6) for each condition. (b) Cellular accumulation of Gly-Sar from apical (open columns) or basal (striped columns) bathing solutions at the end of the flux period. Other details as for Fig. 2(a). (Adapted from Thwaites *et al.*, 1993b).

H⁺-coupled dipeptide transport

Results from the experiments described above indicate that extracellular acidity is associated with an increase in dipeptide transport. However, whether this effect represents a pH-stimulation of the carrier or H⁺-coupled substrate transport cannot be distinguished by radiolabel flux experiments alone. Therefore, an alternative experimental approach (involving Caco-2 cell monolayers loaded with the pH-sensitive fluorescent dye BCECF and continuous monitoring of pH$_i$) was devised. Addition of 20mM Gly-Sar (either at pH 7.4 or 6.0) to the apical surface of BCECF-loaded Caco-2 cell monolayers was associated with intracellular acidification (Fig. 3a). Perfusion of Gly-Sar at the basolateral membrane had a different effect on intracellular pH. In contrast to apical addition, addition of Gly-Sar (20mM) to the basolateral membrane resulted in cytosolic acidification at pH 6.0 only (Fig. 3b). Under optimal conditions for dipeptide transport, Gly-Sar flux is, therefore, associated with H⁺ flow across both apical and basolateral membranes.

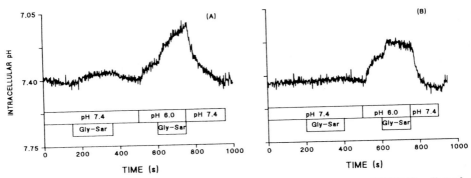

Fig. 3 Intracellular pH in monolayers of Caco-2 cells loaded with the pH sensitive fluorochrome, BCECF. The effects of altering the pH of the bathing solution and/or addition of the dipeptide Gly-Sar (20mM) to either (a) the apical surface and (b) the basolateral surface of Caco-2 cell monolayers. (With permission Thwaites *et al.*, 1993a).

This technique can also be used to distinguish between transport of an intact dipeptide (valylvaline or Val-Val) and its constituent amino acid (valine). Fig. 4 demonstrates that perfusion of Val-Val across the apical surface of Caco-2 cells is associated with intracellular acidification whereas valine is without effect. This observation complements results in Fig. 5 which indicate that Val-Val is a potent inhibitor of pH-stimulated apical-to-basal Gly-Sar transport whereas valine has no inhibitory effect.

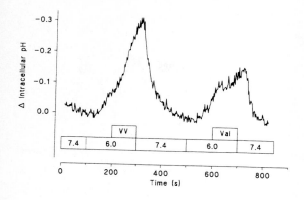

Fig. 4 A comparison of the effects of 10mM Val-Val (VV) and 20mM valine (V) on intracellular pH when perfused at the apical surface at pH 6.0 (basolateral pH 7.4). A single trace representative of three others. (With permission Thwaites *et al.*, 1993c).

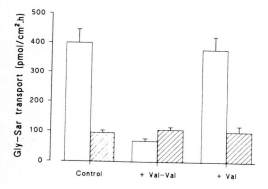

Fig. 5 Apical-to-basal transport of [^{14}C]Gly-Sar (open columns) and [^{3}H]mannitol (hatched columns) across Caco-2 cell monolayers in the presence of a transepithelial pH gradient (apical pH 6.0, basolateral pH 7.4). Transport was determined under control conditions (in the absence of unlabelled dipeptide or amino acid) and in the presence of 20mM Val-Val or valine (Val). Results are the mean±SEM, n=3-4. (Adapted from Thwaites *et al.*, 1993c).

As indicated in Fig. 2 another advantage of using intestinal epithelia grown on permeable supports is the relatively easy access to the basolateral surface. A comparison of the relative abilities of a dipeptide (Gly-Sar) and an aminocephalosporin antibiotic (cephalexin) to stimulate H$^+$-flow into the intestinal Caco-2 cells is indicated in Fig. 6. Cephalexin and Gly-Sar show a similar pattern of effects when perfused across the apical surface, both substrates causing intracellular acidification at pH 7.4 and to a greater extent at pH 6.0 (Fig. 6). However, when perfused at the basolateral surface at pH 7.4 cephalexin induces a large intracellular acidification whereas Gly-Sar is without effect.

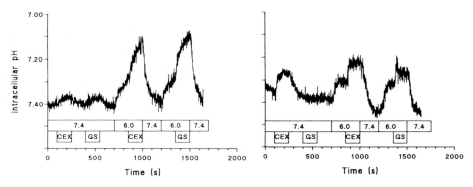

Fig. 6 Intracellular pH measured in BCECF-loaded Caco-2 monolayers. The effect of 20mM Gly-Sar (GS) and 20mM cephalexin (CEX) on [pH]_i after exposure to either the apical (left panel) or basolateral (right panel) membrane. Measurements were made at both pH 7.4 and 6.0. This experiment is representative of 4 others. (With permission Thwaites *et al.*, 1993b).

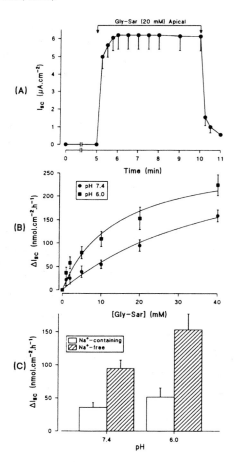

Fig. 7 Characteristics of the Gly-Sar stimulated I_{sc}.

(A) The response of I_{sc} (positive I_{sc} is an inward current) to apical Gly-Sar (20mM) in Na^+-free choline media (apical pH 6.0, basolateral pH 7.4). Data are the mean $\pm$ SEM of 9 epithelial layers. (B) Concentration dependence of the increment in I_{sc} to Gly-Sar at apical pH 6.0 (■) and 7.4 (●). Basolateral pH was pH 7.4. Data are the mean $\pm$ SEM of 6 determinations per concentration from 6 separate epithelial layers. The solid lines are the least-squares Michaelis-Menton fits of the data. (C) The effect of Na^+ and of apical pH on the increment in I_{sc} to 20mM Gly-Sar. (With permission Thwaites *et al.*, 1993f).

Electrogenic dipeptide transport

Over the pH range used in this study the dipeptide Gly-Sar will predominately be in the form of a zwitterion. It is likely, therefore, that H^+-coupled dipeptide transport is electrogenic. To test this hypothesis experiments were performed using voltage-clamped Caco-2 cell epithelial monolayers. Fig. 7 indicates that the dipeptide Gly-Sar can stimulate reversible inward I_{sc} in these intestinal epithelial monolayers (Fig. 7a), that is pH-dependent and saturable (Fig. 7b), and independent of Na^+ (Fig. 7c).

pH_i recovery after dipeptide-induced acid load

Fig. 8 demonstrates that the dipeptide (Val-Val)-induced intracellular acidification occurs in both Na^+-containing and Na^+-free conditions. The pH_i recovery from this acid load is dependent on the presence of Na^+ at the apical surface.

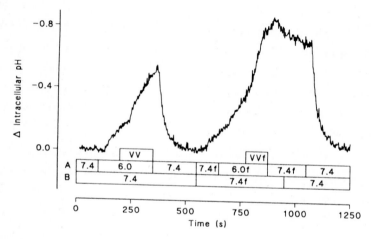

Fig. 8 Dipeptide (Val-Val)-induced intracellular acidification in the presence and absence of extracellular Na^+ in both apical (A) and basolateral (B) chambers, f denotes Na^+ free conditions. Note that Na^+ is reintroduced sequentially to the basolateral then apical surface. This experiment is representative of three others. (Adapted from Thwaites *et al.*, 1993c).

Discussion

(a) Dipeptide transport in Caco-2 epithelium

The main conclusions from the experiments described in the results section (and their significance) are enumerated below.

(i) It is apparent, from experiments with intestinal BBMV, that dipeptide uptake across the intestinal apical membrane is enhanced by an inwardly-directed pH gradient and is electrogenic (Ganapathy and Leibach, 1985). However, direct H^+-coupling or intravesicular accumulation have not been convincingly demonstrated in mammalian intestine. Our results indicate the presence of a functional dipeptide carrier in the apical membrane of human intestinal Caco-2 cells that is stimulated by apical acidity and is involved in both transepithelial transport and, perhaps most significantly, intracellular accumulation of the dipeptide Gly-Sar (Fig. 1; Thwaites *et al.*, 1993a). These results are in agreement with studies of cephalosporin uptake (Dantzig and Bergin, 1990) and transport (Inui *et al.*, 1992).

(ii) For transepithelial transport to occur, exit from the cytoplasm across the basolateral membrane must also be facilitated. Clearly, the basolateral membrane of Caco-2 cells also possesses a pH-stimulated dipeptide transport mechanism (Fig. 1; Thwaites *et al.*, 1993a). This observation is consistent with previous studies in rabbit enterocyte basolateral-membrane vesicles (Dyer *et al.*, 1990). Furthermore, efflux of cephradine from Caco-2 cells is also via a carrier-mediated process (Inui *et al.*, 1992). The relatively low intracellular accumulation of Gly-Sar across the basolateral membrane (Fig. 1-2), coupled to the lower transport seen over the pH range 5.5-7.4 suggest that the properties of H^+-coupled transport across the apical and basolateral membranes are different. Since intracellular Gly-Sar accumulation will depend on the relative rates of transfer across both membranes it appears that tranport across the apical membrane must be greater and that the basolateral transport step is the rate-limiting step in transepithelial transcellular dipeptide transport (Thwaites *et al.*, 1993b).

(iii) Clearly both peptide absorption and secretion can take place (Fig. 2; Thwaites *et al.*, 1993b). The physiological function of peptide secretion is uncertain although this transport mechanism could function to clear dipeptides released by hydolysis of bioactive peptides directed at the basolateral surface of intestinal cells *in vivo* (Thwaites *et al.*, 1993b).

(iv) Fig. 3 shows the first direct demonstration of dipeptide-associated H^+ flow in any mammalian intestinal experimental preparation (Thwaites *et al.*, 1993a). H^+-coupled dipeptide carriers are present at both apical and basolateral membranes of this intestinal epithelia.

(v) This novel method (pH_i measurement) for studying peptide transport distinguishes between transport of the intact substrate (dipeptide) and products of degradation (amino acids) (Fig. 4; Thwaites *et al.*, 1993c).

(vi) pH-stimulated dipeptide transport and intracellular accumulation in both absorptive and secretory directions are inhibited by cephalexin (Thwaites *et al.*, 1993b). However, although the cytosolic acidification induced by Gly-Sar and cephalexin show similarities at the apical surface, responses at the basolateral membrane (at pH 7.4) are clearly different (Fig. 6). Other differences in aminocephalosporin (cephradine) and dipeptide (Gly-Sar) transport have been noted in efflux studies. When Caco-2 cells are loaded with cephradine or Gly-Sar the majority of cephradine exits across the basolateral membrane (Inui *et al.*, 1992) whereas most of the Gly-Sar is recycled across the apical surface (Thwaites *et al.*, 1993b). Thus although dipeptides and aminocephalosporins may share a common transport mechanism at the apical surface their exit across the basolateral membrane may be via distinct routes (Thwaites *et al.*, 1993b). Since both Gly-Sar and cephalexin are in the zwitterionic form at pH 6.0, but 25% of cephalexin is in the form of an anion at pH 7.4 (Purich *et al.*, 1973), these alternative routes for cephalexin transport across the basolateral membrane could include H^+/cephalexin anion symport (Thwaites *et al.*, 1993b) or cephalexin anion/OH^- exchange.

(vii) Fig. 7 describes the first demonstration of the electrogenic nature of dipeptide transport in an intact mammalian cell system (Thwaites *et al.*, 1993f). The transport process is pH-dependent, Na^+-independent, reversible and saturable.

(viii) The recovery of pH_i after dipeptide-induced cytosolic acidification is dependent on the presence of apical Na^+ (Fig. 8; Thwaites *et al.*, 1993c) which suggests the involvement of an apically-localised Na^+/H^+ exchanger. This contrasts with previous reports confining Na^+/H^+ exchange activity to the basolateral membrane of Caco-2 cell monolayers (Watson *et al.*, 1991). The slower rate of recovery observed in the absence of extracellular Na^+ (Fig. 8) could be due to the H^+,K^+-ATPase identified recently in Caco-2 cells (Abrahamse *et al.*, 1993).

(ix) In conclusion dipeptide transport in the human intestinal epithelial cell line Caco-2 is mediated via a pH-dependent, Na^+-independent, H^+-coupled, electrogenic carrier mechanism. Whether one or many transporters exist at the apical surface of the intestinal enterocyte requires further study. However, the results described here suggest that the Caco-2 cell model and the three techniques used in this study will prove useful tools in the identification and comprehensive characterisation of substrates that undergo H^+-coupled transport across the intestinal epithelium.

(b) Other substrates/other transporters

(i) Aminocephalosporin antibiotics

When used in conjunction, these techniques (pH$_i$ measurement and radiolabel fluxes) prove to be useful predictors of the oral avilability of some drugs. The two orally available aminocephalosporin antibiotics cephalexin and cefadroxil are potent inhibitors of Gly-Sar transport in the Caco-2 cells and also stimulate intracellular acidification when perfused across the apical membrane of Caco-2 cell monolayers. In contrast, the non-orally available cephalosporin cefazolin neither inhibits dipeptide transport nor induces H$^+$ flow into the cells (Thwaites, Hirst and Simmons, manuscript in preparation). These observations complement and extend studies with aminocephalosporins in rabbit intestinal BBMV (Okano *et al.*, 1986).

(ii) Angiotensin-converting enzyme (ACE) inhibitors

The two orally-active ACE inhibitors enalapril maleate and captopril (Bai and Amidon, 1992) both inhibit Gly-Sar transport in the Caco-2 cells and also induce intracellular acidification. However, a third ACE inhibitor lisinopril [which is also orally-absorbed but with a lower permeability than enalapril or captopril (Bai and Amidon, 1992)] failed to inhibit Gly-Sar transport or induce intracellular acidification when perfused at the apical surface of BCECF-loaded Caco-2 cell monolayers. [^{14}C]Lisinopril transport across Caco-2 cell monolayers was lower than transport of the paracellular marker mannitol and was not stimulated by acidification of the apical medium. These results suggest that lisinopril absorption is mediated by the paracellular (passive) pathway (Thwaites, Hirst and Simmons, manuscript in preparation).

(iii) Amino acids

Recently we have identified a novel H$^+$-coupled amino acid transport system localised to the apical surface of Caco-2 cell monolayers. This transport system appears to be involved in the absorptive transport of a number of neutral amino acids (and analogues) including ß-alanine (Thwaites *et al.*, 1993d), L-proline (Thwaites *et al.*, 1993e) and α-methylaminoisobutyric acid [MeAIB (Thwaites *et al.*, 1994)]. The apical-to-basal transport of each of these amino acids is stimulated by lowering the apical pH to pH 6.0, and intracellular accumulation of the amino acid is independent of extracellular Na$^+$. All three amino acids stimulate intracellular acidification when perfused at the apical surface of

BCECF-loaded Caco-2 cell monolayers. The three substrates also increase inward I_{sc} in Na^+-free conditions.

(c) Overview

The ability of human intestinal epithelial layers to generate significant net transepithelial transport of substrates including dipeptides, certain antibiotics and amino acids even in Na^+-free conditions suggest that the H^+ electrochemical gradient may play a much wider role in nutrient absorption than recognised previously. The functional expression of the intestinal dipeptide transporter in these cells will allow study of the regulation of this carrier (see chapter by Ganapathy, Brandsch and Leibach). The utility of the Caco-2 cell line is highlighted by the identification of pH_i homeostatic mechanisms (Na^+/H^+ exchange, H^+,K^+-ATPase) in these cells. Furthermore, the expression of these transport mechanisms in a single epithelial cell line will allow the exploration of the interrelationship between dissipative solute-dependent H^+ transport and pH_i homeostasis in human intestinal epithelia.

Acknowledgements

Charlotte Ward provided excellent technical assistance. We thank Dr. M.J. Humphrey, Pfizer Central Research for the supply of $[^{14}C]$Gly-Sar. This study was supported by the Wellcome Trust (to DTT, grant 038748/Z/93) and the LINK Programme in Selective Drug Delivery & Targeting [funded by SERC/MRC/DTI and Industry (SERC grant GR/F 09747)].

References

Abrahamse, S.L., Bindels, R.J.M and van Os, C.H. (1993) The colon carcinoma cell line Caco-2 contains an H^+,K^+-ATPase that contributes to intracellular pH regulation. *Pflügers Arch.* 421: 591-597.

Addison, J.M., Burston, D. and Matthews, D.M. (1972) Evidence for active transport of the dipeptide glycylsarcosine by hamster jejunum *in vitro*. *Clin. Sci.* 43: 907-911.

Bai, J.P.F. and Amidon, G.L. (1992) Structural specificity of mucosal-cell transport and metabolism of peptide drugs: implication for oral peptide drug delivery. *Pharm. Res.* 9: 969-978.

Blais, A., Bissonnette, P. and Berteloot, A. (1987) Common characteristics for Na^+-

dependent sugar transport in Caco-2 cells and human fetal colon. *J. Membr. Biol.* 99: 113-125.

Boyd, C.A.R. and Ward, M.R. (1982) A micro-electrode study of oligopeptide absorption by the small intestinal epithelium of *Necturus maculosus*. *J. Physiol.* 324: 411-428.

Dantzig, A.H. and Bergin, L. (1990) Uptake of the cephalosporin, cephalexin, by a dipeptide transport carrier in the human intestinal cell line, Caco-2. *Biochim. Biophys. Acta* 1027: 211-217.

Dyer, J., Beechey, R.B., Gorvel, J.P., Smith, R.T., Wootton, R. and Shirazi-Beechey, S.P. (1990) Glycyl-L-proline transport in rabbit enterocyte basolateral-membrane vesicles. *Biochem. J.* 269: 565-571.

Friedman, D.I. and Amidon, G.L. (1989) Passive and carrier-mediated intestinal absorption components of two angiotensin converting enzyme (ACE) inhibitor prodrugs in rats: enalapril and fosinopril. *Pharm. Res.* 6: 1043-1047.

Ganapathy, V., Burckhardt, G. and Leibach, F.H. (1984) Characteristics of glycylsarcosine transport in rabbit intestinal brush-border membrane vesicles. *J. Biol. Chem.* 259: 8954-8959.

Ganapathy, V. and Leibach, F.H. (1983) Role of pH gradient and membrane potential in dipeptide transport in intestinal and renal brush-border membrane vesicles from the rabbit. *J. Biol. Chem.* 258: 14189-14192.

Ganapathy, V. and Leibach, F.H. (1985) Is intestinal peptide transport energized by a proton gradient? *Am. J. Physiol.* 249: G153-G160.

Hidalgo, I.J. and Borchardt, R.T. (1990) Transport of bile acids in the human intestinal epithelial cell line, Caco-2. *Biochim. Biophys. Acta* 1035: 97-103.

Hidalgo, I.J., Raub, T.J. and Borchardt, R.T. (1989) Characterization of the human colon carcinoma cell line (Caco-2) as a model system for intestinal epithelial permeability. *Gastroenterology* 96: 736-749.

Hu, M. and Borchardt, R.T. (1992) Transport of a large neutral amino acid in a human intestinal epithelial cell line (Caco-2): uptake and efflux of phenylalanine. *Biochim. Biophys. Acta* 1135: 233-244.

Inui, K.I., Yamamoto, M. and Saito, H. (1992) Transepithelial transport of oral cephalosporins by monolayers of intestinal epithelial cell line Caco-2: specific transport systems in apical and basolateral membranes. *J. Pharmacol. Exp. Ther.* 261:

195-201.

Lucas, M.L., Cooper, B.T., Lei, F.H., Johnson, I.T., Holmes, G.K.T., Blair, J.A. and Cooke, W.T. (1978) Acid microclimate in coeliac and Crohn's disease: a model for folate malabsorption. *Gut* 19: 735-742.

Okano, T., Inui, K.I., Maegawa, H., Takano, M. and Hori, R. (1986) H^+-coupled uphill transport of aminocephalosporins via the dipeptide transport system in rabbit intestinal brush-border membranes. *J. Biol. Chem.* 261: 14130-14134.

Purich, E.D., Colaizzi, J.L. and Poust, R.I. (1973) pH-partition behaviour of amino acid-like ß-lactam antibiotics. *J. Pharm. Sci.* 62: 545-549.

Rawlings, J.M., Lucas, M.L. and Russel, R.I. (1987) Measurement of jejunal surface pH *in situ* by plastic pH electrode in patients with coeliac disease. *Scand. J. Gastroenterol.* 22: 377-384.

Thwaites, D.T. (1993) Direct measurement of aminocephalosporin-induced H^+ transport in BCECF-loaded Caco-2 cells. *J. Physiol.* 467: 344P.

Thwaites, D.T., Brown, C.D.A., Hirst, B.H. and Simmons, N.L. (1993a) Transepithelial glycylsarcosine transport in intestinal Caco-2 cells mediated by expression of H^+-coupled carriers at both apical and basal membranes. *J. Biol. Chem.* 268: 7640-7642.

Thwaites, D.T., Brown, C.D.A., Hirst, B.H. and Simmons, N.L. (1993b) H^+-coupled dipeptide (glycylsarcosine) transport across apical and basal borders of human intestinal Caco-2 cell monolayers display distinctive characteristics. *Biochim. Biophys. Acta* 1151: 237-245.

Thwaites, D.T., Hirst, B.H. and Simmons, N.L. (1993c) Direct assessment of dipeptide/H^+ symport in intact human intestinal (Caco-2) epithelium: a novel method utilising continuous intracellular pH measurement. *Biochem. Biophys. Res. Comm.* 194: 432-438.

Thwaites, D.T., McEwan, G.T.A., Brown, C.D.A., Hirst, B.H. and Simmons, N.L. (1993d) Na^+-independent, H^+-coupled transepithelial ß-alanine absorption by human intestinal Caco-2 cell monolayers. *J. Biol. Chem.* 268: 18438-18441.

Thwaites, D.T., McEwan, G.T.A., Cook, M.J., Hirst, B.H. and Simmons, N.L. (1993e) H^+-coupled (Na^+-independent) proline transport in human intestinal (Caco-2) epithelial cell monlayers. *FEBS Lett.* 333: 78-82.

Thwaites, D.T., McEwan, G.T.A., Hirst B.H. and Simmons, N.L. (1993f) Transepithelial dipeptide (glycylsarcosine) transport across epithelial monolayers of human Caco-2 cells is rheogenic. *Pflügers Arch.* 425: 178-180.

Thwaites, D.T., Simmons, N.L. and Hirst, B.H. (1993g) Thyrotropin-releasing hormone (TRH) uptake in intestinal brush-border membrane vesicles: comparison with proton-coupled dipeptide and Na^+-coupled glucose transport. *Pharm. Res.* 10: 667-673.

Thwaites, D.T., McEwan, G.T.A., Hirst, B.H. and Simmons, N.L. (1994) Proton gradient driven α-methylaminoisobutyric acid transport in human intestinal Caco-2 cells. *J. Physiol.* (in press).

Watson, A.J.M., Levine, S., Donowitz, M. and Montrose, M.H. (1991) Kinetics and regulation of a polarized Na^+/H^+ exchanger from Caco-2 cells, a human intestinal cell line. *Am. J. Physiol.* 261: G229-G238.

Measurement of H^+ and Ca^{2+} extrusion from single isolated cells.

A.V. Tepikin, O.H. Petersen
Physiological Laboratory
University of Liverpool
PO Box 147
Crown Street
Liverpool L69 3BX

Two techniques for measurement of H^+ and Ca^{2+} extrusion from single isolated cells have been recently developed - the droplet technique and the technique of ionic clamp.

Ionic clamp.

The technique of ionic clamp allows fast and controlled changes of ion concentration inside isolated cells (Belan, Kostyuk, Snitsarev, Tepikin,1993). The simplified diagram of the technique is shown on Figure1.Cells loaded by an appropriate indicator (Fura-2 for intracellular calcium clamp on snail neurons, BCECF for intracellular pH clamp on mouse lacrimal acinar cells) were penetrated by a microelectrode allowing iontophoretic injection of Ca^{2+} or H^+ into the cells. To clamp intracellular ion concentration the cell fluorescence signal was compared with external command signal which determines the clamp level (circuit 1, Fig1). If there is a difference between the signal from the cell and the external signal this produces iontophoretic current and flow of ions into the cell (circuit 2, Fig1). The process continues until the signal of fluorescence reaches the command level. After that the system operates to maintain the command level of ion concentration in the cell. The iontophoretic current (clamping current) was registered using opto-isolated circuit (circuit3, Fig1). The values of iontophoretic current can be used to estimate the buffering capacity of the cytoplasm and the intensity of ion extrusion from large cells.

Droplet technique.

The droplet technique allows direct measurement of ion extrusion from single isolated cells (Tepikin, Kostyuk, Snitsarev, Belan, 1991; Tepikin, Gallacher, Voronina, Petersen, 1992). The main idea behind the droplet technique is to maintain cell in

NATO ASI Series, Vol. H 89
Molecular and Cellular Mechanisms
of H^+ Transport
Edited by Barry H. Hirst
© Springer-Verlag Berlin Heidelberg 1994

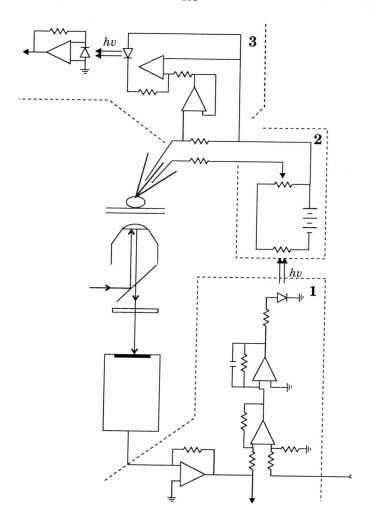

Fig.1. Simplified diagram of ionic clamp experiment.

a small droplet of extracellular solution (the volume of the droplet is 10-100 times larger than volume of the cell) in which the cell can produce measurable changes of extracellular ion concentration. A diagram of the droplet technique is shown in figure2.

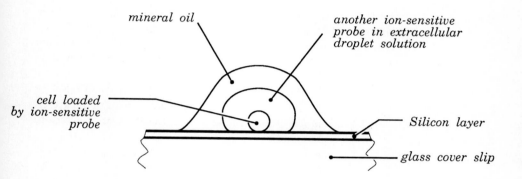

Fig.2. Schematic digram of droplet technique.

The cell and extracellular droplet solution contain indicators.The difference in the optical properties of these probes allow us to monitor simultaneously intracellular and extracellular ion concentrations. Pairs of indicators that have been used in our experiments are presented in the table below.

Combinations of intracellular and extracellular probes used in the droplet technique experiments:

Intracellular indicator.	Indicator in extracellular droplet solution.
Fura-2	Fluo-3
Fura-2	Calcium-Green-5N
Fura-2	Antipyrylazo-3
Fura-2	Ca-ion-selective microelectrode
Fura-2	BCECF
Fluo-3	Mag-Fura-2

Cell-containing droplets where constructed on the surface of silinised glass coverslips and covered by mineral oil to prevent evaporation. Stimulation of the cell by agonist was performed by injection from a microelectrode inserted into droplet solution or by diffusion from micropipette.

The droplet technique can be used to investigate ion extrusion during complex changes of intracellular ion concentration like calcium oscillations.

References.

Belan P.V., Kostyuk P.G., Snitsarev V.A., Tepikin A.V. (1993). Calcium clamp in isolated neurons of the snail Helix Pomatia. Journal of Physiology, 462, pp.47-58.

Tepikin A.V., Kostyuk P.G., Snitsarev V.A., Belan P.V. (1991). Extrusion of calcium from a single isolated neurons of snail Helix Pomatia. J. Membrane Biol.,123, pp.43-47.

Tepikin A.V., Voronina S.G., Gallacher D.V., Petersen O.H. (1992). Acetylcholine-evoked increase in the cytoplasmic Ca^{2+} concentration and Ca2+ extrusion measured simultaneously in single mouse pancreatic acinar cells. J.Biol. Chem., 267, 6, pp.3569-3572.

Tepikin A.V., Voronina S.G., Gallacher D.V., Petersen O.H. (1992). Pulsatile Ca^{2+} extrusion from single pancreatic acinar cells during receptor-activated cytosolic Ca^{2+} spiking. J. Biol. Chem., 262, 20, pp.14073-14076.

CFTR Chloride Channels and Pancreatic Ductal Bicarbonate Secretion

Barry E. Argent, John P. Winpenny and Michael A. Gray.
Department of Physiological Sciences
University Medical School
Framlington Place
Newcastle upon Tyne
NE2 4HH
U.K.

The main function of pancreatic duct cells is to secrete a HCO_3^--rich isotonic fluid (Fig. 1). This secretion flushes digestive enzymes, secreted by acinar cells located at the termini of the smallest ducts, along the ductal tree and into the duodenum (Fig. 1). It also neutralises acid chyme entering the duodenum from the stomach and thus helps to create the correct luminal environment for the digestion of food in the gut (Case & Argent, 1993; Argent & Case, 1994).

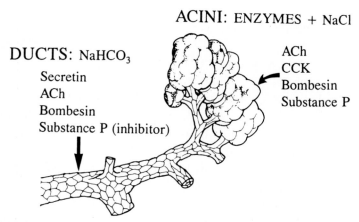

ACINI: ENZYMES + NaCl

DUCTS: NaHCO₃

Secretin
ACh
Bombesin
Substance P (inhibitor)

ACh
CCK
Bombesin
Substance P

Fig. 1. Structure of the exocrine pancreas and regulation of ductal secretion. Largely, based on data obtained from micropuncture experiments on isolated rat ducts. There are wide species variations in the actions and/or efficacy of the various hormones and neurotransmitters. Moreover, the HCO_3^- concentration in secretin-stimulated pancreatic juice varies markedly with species. Maximum values are 140-150 mM in cat, dog, pig, guinea-pig, hamsters, and humans; 120 mM in the rabbit, and 80 mM in the rat. Secretin uses cyclic AMP as an intracellular messenger whereas acetylcholine increases intracellular calcium concentration. The calcium pathway is particularly well developed in the duct cells of rats and mice. (Based on a diagram in Takahashi, 1984; with permission).

NATO ASI Series, Vol. H 89
Molecular and Cellular Mechanisms
of H⁺ Transport
Edited by Barry H. Hirst
© Springer-Verlag Berlin Heidelberg 1994

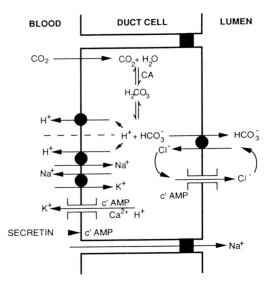

Fig. 2. Cellular model for pancreatic HCO_3^- secretion evoked by secretin. Largely based on morphological, fluorescence and electrophysiological studies on small ducts isolated from rat and pig pancreas. For references see Case & Argent (1993); Argent & Case (1994). CA, carbonic anhydrase.

Figure 2 shows the current model for ductal HCO_3^- secretion evoked by a rise in intracellular cyclic AMP. The initial step is diffusion of CO_2 into the duct cell, and its hydration by carbonic anhydrase (CA) to carbonic acid. This dissociates to form H^+ and HCO_3^-, and the proton is translocated back across the basolateral membrane either by an electrogenic H^+-ATPase or a Na^+/H^+ exchanger. Effectively, this is the active transport step for HCO_3^- and it leads to the accumulation of HCO_3^- inside the duct cell. HCO_3^- ions are then thought to exit across the apical membrane on a Cl^-/HCO_3^- exchanger. The rate at which this exchanger cycles will depend on the availability of luminal Cl^-, and on the rate at which intracellular chloride (accumulated by the exchanger) leaks from the cell. Both of these processes are controlled by an apical Cl^- channel which is activated following secretin-stimulation of the duct cell. Since HCO_3^- exit at the apical membrane generates a current, there must be equal current flow across the basolateral membrane during secretion. Some of this current is accounted for by K^+ efflux through a K^+ channel, and the remainder by cycling of the electrogenic pumps, namely H^+-ATPase and Na^+, K^+-ATPase. Finally, the negative transepithelial potential, generated by activation of the apical Cl⁻ conductance, draws Na⁺ and a small amount of K⁺ into the lumen via a cation-selective paracellular pathway. At the moment, the secretory model is largely based on the spatial distribution of transport elements and how the conductance pathways in the basolateral and apical membranes respond to stimulation. To confirm that the model actually works will require measurement of the electrochemical driving forces acting on Cl⁻ and HCO_3^- ions.

The chloride channel located on the apical membrane of the duct cell is cystic fibrosis transmembrane conductance regulator (CFTR), the protein that is encoded by the CF gene (Fig. 3). CFTR is expressed in ductal, but not acinar cells, of the human pancreas (Marino *et al.* 1991), and when it is either absent or dysfunctional pancreatic HCO_3^- and fluid secretion is markedly reduced (Durie & Forstner, 1989). This causes proteinaceous acinar secretions to become concentrated in the duct lumen, where they eventually precipitate, causing blockage of the small ducts and eventual destruction of the gland.

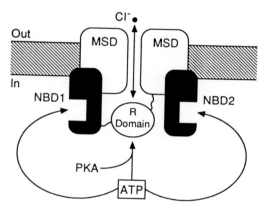

Fig. 3. Predicted structure of the cystic fibrosis transmembrane conductance regulator (CFTR) chloride channel. CFTR consists of 1480 amino acids and contains twelve hydrophobic sequences, which are taken to represent membrane spanning domains (MSD), plus two nucleotide binding domains (NBD) and a regulatory (R) domain. The R domain may act as a plug which closes the channel by blocking the pore formed by the twelve transmembrane spans. Unplugging the pore, and thus opening of the channel, requires phosphorylation of the R-domain which is catalysed by cyclic AMP-dependent protein kinase. Phosphorylation should increase the net negative charge on the R domain and, it has been speculated, lead to electrostatic repulsion of the domain from the internal face of the plasma membrane. In addition, it appears that ATP binding and perhaps hydrolysis at the nucleotide binding domains is also necessary to keep the channel open. PKA = protein kinase A. (From Anderson *et al.* 1991; with permission).

Figures 4 and 5 show that secretin and cyclic AMP increase the activity of CFTR chloride channels, and whole cell chloride currents, in pancreatic duct cells. The biophysical properties of the CFTR channel can be summarised as follows: (i) low conductance (human, 6-7 pS; rat, 3-4 pS); (ii) open and closed times of the order of tens to hundreds of milliseconds (Fig. 4); (iii) a linear or slightly rectifying (outward) current-voltage plot; and (iv) no marked voltage-dependence of opening (Gray *et al.* 1988; Gray *et al.* 1989; Gray *et al.* 1990). The CFTR channel is highly selective for Cl^- over Na^+ and K^+, and has a HCO_3^-/Cl^- permeability ratio of about 0.2, making it very unlikely that significant amounts of HCO_3^- could be secreted directly via the channel (Gray *et al.* 1990). Extracellular 4,4'-diisothiocyanostilbene-2,2'-disulphonic acid (DIDS) has no effect on the channel; however, 5-nitro-2-(3-phenylpropylamino)-benzoic acid (NPPB) inhibits CFTR channel activity (Gray *et al.* 1990). The chance of finding CFTR channels on stimulated duct cells is quite low; less than 1 in 5 patches have channel activity (Gray *et al.* 1988). However, the channels usually occur in a cluster; most often with two or three, but sometimes with up to 15 channels in each membrane patch (Gray *et al.* 1988).

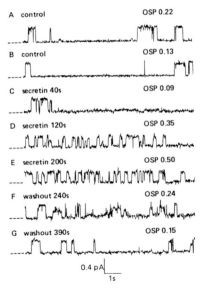

Fig. 4. Regulation of CFTR chloride channel activity by secretin. Cell-attached patch on the apical membrane of a human pancreatic duct cell. Dashed lines indicate the current level when all channels are closed. Each trace illustrates 10 sec. of data and OSP is the open state probability of the channel. (From Gray *et al.* 1989; with permission).

Only about 70% of duct cells have detectable CFTR chloride currents (Fig. 5); however, the cells are electrically coupled, and the epithelium probably functions as a syncitium *in vivo* (Gray *et al.* 1993). Using data from single channel and whole cell recordings it is possible to estimate that there are about 1000 active CFTR channels in the apical membrane of an unstimulated duct cell and that this number increases to about 4000 following exposure to cyclic AMP (Gray *et al.* 1993). Whether this reflects the insertion of additional channels into the plasma membrane or activation of quiescent channels that are already present is unclear at the moment. Whatever the explanation, it is likely that CFTR channels are not evenly distributed over the apical plasma membrane, but are clustered near the junctional complex (Gray *et al.* 1993).

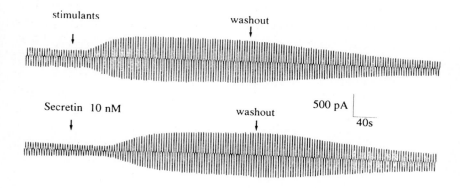

Fig. 5. Continuous whole-cell recording showing the effects of cyclic AMP and secretin on CFTR chloride currents in a cluster of 5 rat pancreatic duct cells. The pipette (intracellular) solution contained caesium ions to block basolateral potassium channels (see Fig. 2). The membrane potential was held at 0 mV and then alternatively clamped at $\pm$ 60 mV for 1 sec. periods. When normalised to input capacitance (a measure of cell surface area), the CFTR current density in stimulated cells is about 100 pA/pF at a membrane potential of 60 mV. From such data it can be calculated that the conductance of the apical plasma membrane is 14.6 mS.cm^{-2}, a value which is consistent with that obtained by circuit analysis of the epithelium (11.4 mS.cm^{-2}; Novak & Greger, 1991). Thus CFTR chloride channels form the major conductance pathway in the apical membrane of pancreatic duct cells that have been exposed to secretin or cyclic AMP. (From Gray *et al.* 1993; with permission).

About 70% of patients with CF are homozygous for the deletion of a phenylalanine at position 508 (delta F508), which is located within the first nucleotide binding domain (NBD1) of CFTR (Fig. 3). In terms of pancreatic function, delta F508 is a severe mutation and homozygous patients have pancreatic insufficiency (Kristidis *et al.* 1992). However, well over two hundred different CF mutations have now been described (many of them patient-specific), and some are not associated with overt pancreatic disease. Possible strategies for treating the pancreatic defect in CF are systemic gene therapy and/or the use of other chloride channels already present in the apical plasma membrane to bypass defective CFTR.

Recently, three groups have produced transgenic mice which are homozygous for a null mutation of the CF gene (Dorin *et al.* 1992; Snouweart *et al.* 1992; Ratcliff *et al.* 1993). These animals have either no detectable CFTR mRNA or very reduced levels of the message. Rather surprisingly, these CF mice do not exhibit marked pancreatic abnormalities, despite evidence of histological changes in their intestinal and respiratory tracts (Dorin *et al.* 1992; Snouweart *et al.* 1992; Ratcliff *et al.* 1993), together with defective cyclic AMP-mediated chloride ion transport in these tissues (Clarke *et al.* 1992; Dorin *et al.* 1992; Ratcliff *et al.* 1993). In contrast to CF mice, CF patients with null mutations, have severe pancreatic disease (Kristidis *et al.* 1992). We believe this difference is explained by the fact that mouse pancreatic duct cells have an alternate fluid secretory pathway which is activated by a rise in intracellular calcium concentration ($[Ca^{2+}]_i$). In support of this idea we have recently shown that an increase in $[Ca^{2+}]_i$, evoked by either ionomycin or acetylcholine, activates chloride channels in rat duct cells (Plant *et al.* 1993), and stimulates fluid secretion from isolated rat pancreatic ducts (Ashton *et al.* 1993) (Fig. 1).

Calcium-activated chloride currents can also be detected in mouse pancreatic duct cells (Fig. 6), and these currents have quite different biophysical properties to those mediated by CFTR (Gray *et al.* 1994). In particular, their current/voltage plot displays marked outward rectification, and the currents activate at depolarizing potentials and inactivate at hyperpolarizing potentials. This strongly suggests that the currents are carried by an ion channel that is distinct from CFTR. These calcium-activated currents are of similar magnitude in wild-type, heterozygote and homozygous cf/cf animals, and about 11-fold larger than the CFTR currents in the wild-type group. Fifty-four percent of wild-type cells, and 71% of homozygous cf/cf cells exhibited the calcium-activated currents. If the EGTA concentration in the recording pipette was increased from 0.2 to 2 mM (increasing the intracellular buffering capacity), the proportion of cells displaying currents was reduced to 10%, confirming that the currents are activated by an increase in $[Ca^{2+}]_i$ (Gray *et al.* 1994).

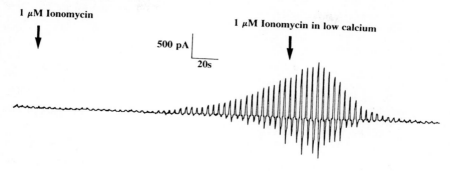

Fig. 6. Calcium-activated chloride currents evoked by ionomycin in a mouse pancreatic duct cell. Lowering the bath $[Ca^{2+}]$ from 2 mM to 0.1 μM completely reversed the currents suggesting that they are maintained by calcium influx. (From Gray *et al.* 1994; with permission).

Although the CF mouse is not a good model for the human pancreatic disease that occurs in CF, it does show that other chloride channels can effectively substitute for CFTR. This suggests that the alternate channel approach to therapy might have some chance of success. The obvious question is whether similar calcium-activated chloride channels are present in human pancreatic duct cells? One would expect the answer to be no, otherwise why would the human pancreas be affected by the disease? In fact, ionomycin and acetylcholine both increase ^{125}I efflux from human ductal adenocarcinoma cells (Lingard *et al.* 1994), and recently calcium-activated chloride currents have been described in CFPAC cells, a pancreatic adenocarcinoma cell line isolated from a CF patient with the delta F508 mutation (Warth & Greger, 1993). The fact that structural damage occurs to the pancreas of CF patients implies that this calcium pathway is either quantitatively less important than the cyclic AMP-activated (CFTR) pathway, or is not physiologically regulated in man.

Funded by grants from the Cystic Fibrosis Trust and the Medical Research Council (U.K.).

REFERENCES

Anderson MP, Berger HA, Rich DP, et al. (1991). Nucleoside triphosphates are required to open the CFTR chloride channel. Cell 67: 775-784.

Argent BE, Case RM. (1994). Pancreatic ducts: cellular mechanism and control of bicarbonate secretion. In: Johnson, LR, ed. Physiology of the Gastrointestinal Tract. Third Edition, New York: Raven Press. (in press).

Ashton NA, Evans RL, Elliott AC, et al. (1993). Regulation of fluid secretion and intracellular messengers in isolated rat pancreatic ducts by acetylcholine. J. Physiol. 471: 549-562.

Case RM, Argent BE. (1993). Pancreatic duct cell secretion. Control and mechanisms of transport. In: Go VLW, Di Magno E, Gardner J, Lebenthal E, Reber H, Scheele G, eds. The Pancreas: Biology, Pathobiology and Disease, 2nd Edition, New York: Raven Press, 301-350.

Clarke LL, Grubb BR, Gabriel SE, et al. (1992). Defective epithelial chloride transport in a gene-targeted mouse model of cystic fibrosis. Science 257: 1125-1128.

Dorin JR, Dickinson P, Alton EWFW, et al. (1992). Cystic fibrosis in the mouse by targeted insertional mutagenesis. Nature 359: 211-215.

Durie PR, Forstner GG. (1989). Pathophysiology of the exocrine pancreas in cystic fibrosis. J. Roy. Soc. Med. 82 (Suppl 16): 2-10.

Gray MA, Greenwell JR, Argent BE. (1988). Secretin-regulated chloride channel on the apical plasma membrane of pancreatic duct cells. J. Membrane Biol. 105: 131-142.

Gray MA, Harris A, Coleman L, et al. (1989). Two types of chloride channel on duct cells cultured from human fetal pancreas. Am. J. Physiol. 257: C240-C251.

Gray MA, Pollard CE, Harris A, et al. (1990). Anion selectivity and block of the small conductance chloride channel on pancreatic duct cells. Am. J. Physiol. 259: C752-C761.

Gray MA, Plant S, Argent BE. (1993). cAMP-regulated whole cell chloride currents in pancreatic duct cells. Am. J. Physiol. 264: C591-C602.

Gray MA, Winpenny JP, Porteous DJ, et al. (1994). CFTR and calcium-activated chloride currents in pancreatic duct cells of a transgenic CF mouse. Am. J. Physiol. (in press).

Kristidis P, Bozon D, Corey M, et al. (1992). Genetic determination of exocrine pancreatic function in cystic fibrosis. Am. J. Hum. Genet. 50: 1178-1184.

Lingard JM, Al-Nakkash L, Argent BE. (1994). Acetylcholine, ATP, bombesin and cholecystokinin stimulate ^{125}I efflux from a human pancreatic adenocarcinoma cell line (BxPC-3). Pancreas (in press).

Marino CR, Matovcik LM, Gorelick FS, et al. (1991). Localization of the cystic fibrosis transmembrane conductance regulator in pancreas. J. Clin. Invest. 88: 712-716.

Novak I, Greger R. (1991). Effect of bicarbonate on potassium conductance of isolated perfused rat pancreatic ducts. Pflügers Archiv 419: 76-83.

Plant S, Gray MA, Argent BE. (1993). Ionomycin-activated chloride conductance in isolated rat pancreatic duct cells. J. Physiol. 459: 239P.

Ratcliff R, Evans MJ, Cuthbert AW, et al. (1993). Production of a severe cystic fibrosis mutation in mice by gene targeting. Nature Genet. 4: 35-41.

Snouweart JN, Brigman KK, Latour AM, et al. (1992). An animal model for cystic fibrosis made by gene targeting. Science 257: 1083-1088.

Takahashi H. (1984). Scanning electron microscopy of the rat exocrine pancreas. Arch. Histol. Jpn. 47: 387-404.

Warth R, Greger R. (1993). The ion conductances of CFPAC-1 cells. Cell Physiol. Biochem. 3: 2-16.

Acute Regulation of $Na^+:HCO_3^-$ Cotransport System In Kidney Proximal Tubules

Manoocher Soleimani¶, Gwen L. Bizal¶, Yolanda J. Hattabaugh¶, Peter S. Aronson§, and Julia A. Bergman¶

¶Department of Medicine
Indiana University School of Medicine and Veterans Administration Medical Center,
Indianapolis, IN
§Departments of Medicine and Cellular and Molecular Physiology
Yale University School of Medicine, New Haven, CT.

The reabsorption of the filtered load of HCO_3^- in the kidney proximal tubule is mediated via a Na^+/H^+ exchanger and a H^+-translocating ATPase in the lumen and the $Na^+:HCO_3^-$ cotransporter in the basolateral membranes (3,7,32). The Na^+/H^+ exchanger is an electroneutral, amiloride-sensitive transporter that regulates blood pH by secreting H^+ in exchange for Na^+. The H^+-translocating ATPase is an electrogenic, Na-independent transporter that is inhibited by N-ethylmaleimide (NEM). Secretion of H^+ into the lumen via Na^+/H^+ exchange and H^+-translocating ATPase leads to HCO_3^- reabsorption in the proximal tubule. The $Na^+:HCO_3^-$ cotransporter mediates the transport of HCO_3^- from the kidney proximal tubule cell to blood. Although this transporter was first described in the kidney proximal tubule, recent studies have shown its presence in numerous types of cells (3,7,32). The purpose of this article is to review the functional characteristics and acute regulatory mechanisms of the $Na^+:HCO_3^-$ cotransport system.

1. Electrogenicity and Stoichiometry:

Boron and Boulpaep demonstrated that a stilbene-sensitive, electrogenic, Na^+-dependent transporter mediates the exit of HCO_3^- across the basolateral membrane of the amphibian renal proximal tubule cell (11). The coupled transport of Na^+ and HCO_3^- was found to be associated with a net negative charge movement in the same direction. Subsequent studies (1,4,5, 9,10,18,19,20,37,38,39,52) have shown that this pathway is the major mechanism responsible for mediating the exit of HCO_3^- across the basolateral membranes of the mammalian kidney proximal tubule cells. These studies demonstrated that this transporter is electrogenic and carries a net negative charge. Taken together, these results suggest a stoichiometry of more than one HCO_3^- per

NATO ASI Series, Vol. H 89
Molecular and Cellular Mechanisms
of H+ Transport
Edited by Barry H. Hirst
© Springer-Verlag Berlin Heidelberg 1994

each Na$^+$ ion. The precise knowledge of the stoichiometry of the Na$^+$:HCO$_3^-$ cotransport system is important in studying the function and the structure of the transport protein involved in this process. Based on the rate of initial fluxes of Na$^+$ and HCO$_3^-$ in response to sudden reduction of peritubular HCO$_3^-$ concentration in the intact proximal tubule of the rat kidney, Yoshitomi et al. (52) showed that the Na$^+$:HCO$_3^-$ cotransport has a stoichiometry of three equivalent of base per one Na ion. Using a thermodynamic approach and measuring the ^{22}Na flux in basolateral membrane vesicles isolated from rabbit kidney cortex, Soleimani et al. (38) showed that the stoichiometry of Na:HCO3 cotransport was equivalent to three HCO3 per each Na ion. These latter two studies suggest that with such a stoichiometry, the inside negative membrane potential is sufficient to drive HCO$_3^-$ exit against the inward concentration gradients of HCO$_3^-$ and Na that are present across the basolateral membrane of the intact proximal tubule cell under physiologic conditions.

2. Ionic Base Species:

It should be emphasized that although the results of the above studies are consistent with a stoichiometry of 3 HCO$_3^-$ per each Na ion, they are equally consistent with any transport process in which there is the net transfer of three equivalents of base and one Na. Thus, for example, the cotransport of Na with 2 HCO$_3^-$ and one OH ion, or the cotransport of Na with one CO$_3^=$ and one HCO$_3^-$, or the cotransport of Na with 2 HCO$_3^-$ in exchange for one H$^+$ are thermodynamically equivalent processes that can not be distinguished by the approaches that were employed.

Understanding the ionic base species of this transport process is important for defining the number and types of substrate sites that must be present on the transport protein(s). To define the ionic base species of the Na:HCO$_3^-$ cotransport, Soleimani et al. (39) measured the influx of ^{22}Na in cortical basolateral membrane vesicles in the presence of varying external pH, [CO$_3^=$], or [HCO$_3^-$]. The results demonstrated that at constant [HCO$_3^-$], increasing [CO$_3^=$] stimulated Na$^+$ influx, suggesting the existence of a transport site for CO$_3^=$. Sulfite (SO$_3^=$), a structural analog of CO$_3^=$, stimulated Na$^+$ influx in the presence but not the absence of HCO$_3^-$. The SO$_3^=$-stimulated Na$^+$ uptake increased as a saturable function of [HCO$_3^-$], consistent with the existence of a distinct HCO$_3^-$ site. The sulfite-stimulated Na influx was disulfonic stilbene-sensitive, suggesting that it did occur via the stilbene-sensitive Na$^+$:HCO$_3^-$ cotransport system. No other divalent anions showed any affinity for Na$^+$:HCO$_3^-$ cotransport system. Taken together, the experiments suggested that the actual ionic mechanism of the Na$^+$:HCO$_3^-$ cotransport system involves the cotransport of Na$^+$, CO$_3^=$, and HCO$_3^-$ in a 1:1:1 ratio. The cotransport of Na$^+$ and CO$_3^=$ can occur either via distinct Na$^+$ and CO$_3^=$ binding sites or via the NaCO$_3^-$ (sodium carbonate) ion pair, as has been proposed for the Na-dependent Cl$^-$/HCO$_3^-$ exchanger in the red cell (31). SO$_3^=$ can also form an ion pair with Na$^+$. Indeed, NaCO$_3^-$ and NaSO$_3^-$ (sodium sulfite) ion pairs appear to be substrates for transport via the erythrocyte Band 3 anion exchanger (34). One piece of evidence against the ion pair formation in the cortical basolateral membranes was that oxalate and phosphite, two anions that form ion pairs with Na^+ in erythrocytes, did not interact with $Na^+:HCO_3^-$ cotransport. A second argument against the ion pair hypothesis concerned the relative affinities of Na and Li. The association constant of Li^+ for ion pair formation with $CO_3^=$ is 4-fold greater than that of Na^+. However, the affinity of Li^+ for $Na^+:HCO_3^-$ cotransport was found to be five fold less than Na^+ (39). The difference in relative affinities for Na^+ and Li^+ for the $Na^+:HCO_3^-$ cotransporter is opposite to that expected if Na and $CO_3^=$ were transported via ion pair. Another piece of evidence against the ion pair formation in the cortical basolateral membranes comes from the effect of the alkaloid harmaline. In the ion pair hypothesis, cation inhibitors should not affect the transporter, since the transported cation is shielded by the accompanying anion. Harmaline is an organic cation known to compete at the Na^+ site of several Na-coupled transport systems in the kidney. When tested in cortical basolateral membrane, harmaline caused a dose-dependent inhibition of the HCO_3^--stimulated Na^+ influx in a saturable manner (39). Kinetic studies of the inhibitory effect of harmaline on $Na^+:HCO_3^-$ cotransport demonstrated that the degree of inhibition was reduced in the presence of increasing concentration of Na^+ suggesting that there is a competitive interaction between Na^+ and harmaline. These findings, all taken together, suggest that the HCO_3^- exit across the basolateral membrane of the kidney proximal tubule occurs via a cotransport of Na^+, $CO_3^=$, and HCO_3^- in a 1:1:1 ratio on distinct sites. $SO_3^=$ can act as an alternative substrate at the $CO_3^=$ site and Li^+ can likely compete at the Na^+ site.

3. Inhibitor specificity:

Several investigators have studied the inhibitor specificity of $Na^+:HCO_3^-$ cotransport in rat and rabbit basolateral membrane vesicles and intact tubules (1,4,5,9,10,19,20,37, 38,39,52). In these experiments, DIDS, harmaline, acetazolamide, and furosemide were found to inhibit the $Na^+:HCO_3^-$ cotransporter significantly.

One interesting aspect of the inhibitor specificity of $Na^+:HCO_3^-$ cotransport system is its interaction with the carbonic anhydrase inhibitor acetazolamide (ACTZ). Many studies have demonstrated that ACTZ reduces the rate of H^+ secretion and HCO_3^- reabsorption in the proximal tubule by at least 80% (26,27). Historically, major attention has been focused on the role of the luminal membrane carbonic anhydrase in catalyzing the breakdown of intratubular carbonic acid formed by the titration of filtered HCO_3^- with secreted H^+ (16,17,28,35). Two recent reports suggest that acetazolamide prevents the exit of HCO_3^- from the proximal tubule cell when added to the peritubular fluid (10,13). Histochemical and immunocytochemical studies have demonstrated that carbonic anhydrase is localized at the luminal membrane, the basolateral membrane, and in the cytoplasm of the proximal tubule cell (18). Moreover, carbonic anhydrase activity can be measured in both luminal and basolateral membrane vesicles isolated from the renal cortex (49). The effect of ACTZ on the $Na^+:HCO_3^-$ cotransporter has been studied in basolateral membrane vesicles of rabbit kidney cortex (41). These studies showed that ACTZ does not directly inhibit the $Na^+:HCO_3^-$ cotransport system. To determine if the inhibitory effect of ACTZ on HCO_3^- exit from the proximal tubule cell might result from its ability to inhibit the carbonic anhydrase-mediated generation of HCO_3^- at the basolateral membrane of the cell, the effect of ACTZ on $Na^+:HCO_3^-$ cotransport was measured under conditions of the NH_4^+ pre-pulse method adapted for vesicles (41). The results showed that ACTZ can indeed inhibit the $Na^+:HCO_3^-$ cotransport system by blocking the carbonic anhydrase-mediated generation of HCO_3^- at the inner surface of the basolateral membrane. Inhibition of this process by acetazolamide would be expected to alkalinize the proximal tubule cell, as has actually been observed. This in turn would result in inhibition of luminal membrane H^+ secretion. Thus, the indirect interaction of acetazolamide with the basolateral membrane $Na^+:HCO_3^-$ cotransport system may be an important mechanism underlying inhibition of proximal tubule acid secretion by this agent.

4. Acute Regulation of $Na^+:HCO_3^-$ Cotransport:

Bicarbonate transport across the basolateral membranes of the proximal tubule cells is regulated by a number of factors. In particular, acute and chronic changes in pH and HCO_3^- are associated with alterations in the $Na^+:HCO_3^-$ cotransport system. In addition, several investigators have studied the signal transduction pathways involved in regulating the $Na^+:HCO_3^-$ cotransporter. In the following section we will focus on acute regulatory processes affecting the $Na^+:HCO_3^-$ cotransporter.

a. Regulation of $Na^+:HCO_3^-$ cotransport by cell pH (allosteric interaction). The allosteric regulation of $Na^+:HCO_3^-$ cotransporter by cell pH has been studied by Soleimani et al. (42). The basolateral membrane vesicles isolated from rabbit kidney cortex were pre-equilibrated with 1 mM ^{22}Na at pH_i 6.8-8.0 and known concentrations of HCO_3^-. The net flux of ^{22}Na was then monitored over 5 sec. The results demonstrated that the rate of Na^+ efflux was significantly higher at pH_i 7.2 than pH_i 8.0 despite much higher $[HCO_3^-]$ and $[CO_3^=]$ at pH_i 8.0. Increasing the concentration of intracellular HCO_3^- and $CO_3^=$ at constant pH did not inhibit the rate of HCO_3-dependent Na efflux. Increasing the concentration of intracellular HCO_3^- at constant $CO_3^=$ or increasing the concentration of intracellular $CO_3^=$ at constant HCO_3^- did not inhibit the rate of HCO_3-dependent Na efflux. Further studies have shown that $Na^+:HCO_3^-$ cotransport is less active in very acidic pH_i (6.3) compared to mild acidic pH_i (7.0-7.2) (43). Taken together, the results of these studies are compatible with the presence of an internal pH modifier site which regulates the activity of the $Na^+:HCO_3^-$ cotransporter. This modifier site inhibits the activity of the cotransporter at alkaline and acidic pH and displays a maximal functional activity around physiologic pH. This is in contrast with the pH sensitivity of the other HCO_3^- extruding processes, particularly the Cl^-/HCO_3^- exchanger. Two recent studies evaluating the pH sensitivity of the latter exchanger in lymphocytes and intestinal luminal membrane vesicles demonstrated that an alkaline pH stimulated this exchanger (30,31).

The $Na^+:CO_3^=:HCO_3^-$ cotransporter displays an inhibitory profile at alkaline or acidic pH and is more active at physiologic pH. The luminal Na^+/H^+ exchanger and basolateral $Na^+:CO_3^=:HCO_3^-$ cotransporter in the renal proximal tubule act in series to reclaim filtered HCO_3^-. It has been suggested that the increased activity of the luminal Na^+/H^+ exchanger leads to alkalinization of the cell providing more substrate for the basolateral $Na^+:CO_3^=:HCO_3^-$ cotransporter. The inhibitory effect of an alkaline pH on HCO_3^- transport across the basolateral membrane does not fit into this scheme. Intracellular pH measurements in rat proximal tubule cells show a baseline pH of 7.1-7.2 under normal conditions (52). The results of our experiments demonstrate that the internal modifier site of the basolateral $Na:CO_3^=:HCO_3^-$ cotransporter will be maximally functional at pH 7.0 to 7.2. This suggests that the activity of the luminal Na^+/H^+ exchanger is coupled to the activity of the basolateral $Na^+:CO_3^=:HCO_3^-$ cotransporter via cell pH rather than by the substrate concentration. According to this hypothesis, an increase in the activity of the luminal Na^+/H^+ exchanger at physiologic or mildly acidotic (7.0) cell pH is associated with an increase in the activity of the basolateral $Na^+:CO_3^=:HCO_3^-$ cotransporter. This maintains the cell pH in a narrow range and provides maximal capacity for reabsorption of HCO_3^- from the luminal fluid.

Studies in brush border membrane vesicles have shown that the activity of the Na^+/H^+ exchanger increases with decreasing pH_i (8). Those studies also demonstrated the presence of a H^+ stimulatory modifier site on the Na^+/H^+ exchanger protein. One can speculate that pH-sensitive sites located on the Na^+/H^+ and $Na^+:HCO_3^-$ transport systems enhance the ability of these two transport processes to extrude intracellular acid and base loads and thereby regulate both cell pH and HCO_3^- reabsorption. The molecular mechanism of the stimulatory effect of acidic pH and inhibitory effect of alkaline pH on the activity of the $Na^+:HCO_3^-$ transporter remains speculative. One possible explanation is that this cotransporter contains titrable amino groups that are facing inward. Titration of amino groups may lead to conformational changes in the transport protein which affect its activity. Such a candidate is histidine which has a pK_a in the range of 6.8-7.2. Studies of the luminal Na^+/H^+ exchanger are suggestive of the presence of a histidine amino group in the exchanger. It is conceivable, therefore, that one or more histidyl groups on the luminal Na^+/H^+ exchanger and basolateral $Na^+:HCO_3^-$ transporter may represent pH sensitive sites that coordinate and regulate the activity of these two transport proteins. According to this scheme, increased activity of the luminal Na^+/H^+ exchanger at physiologic or mildly acidic pH is matched by maximal activity of the basolateral $Na^+:HCO_3^-$ transporter. This will mitigate against any significant changes in cell pH and provide maximal capacity for reabsorption of filtered HCO_3^- from the tubule lumen. The physiologic significance of the pH sensitivity of the $Na^+:CO_3^=:HCO_3^-$ cotransporter and its role in cell alkalosis has not been examined in intact tubules.

b. Regulation of $Na^+:HCO_3^-$ cotransport by hormones and protein kinases-The role of hormones and protein kinases in regulating $Na^+:HCO_3^-$ cotransport has been studied in rabbit kidney proximal tubules. In the studies that were performed in basolateral membrane vesicles isolated from rabbit kidney cortex, Ruiz et al. found that cyclic AMP-dependent protein kinase and Ca-calmodulin-dependent protein kinase inhibited the $Na^+:HCO_3^-$ cotransporter (36). Protein kinase C was found to stimulate the $Na^+:HCO_3^-$ cotransporter (36).

The effect of protein kinases on the brush border membrane Na^+/H^+ exchanger has been studied by several investigators. In liposomes reconstituted with solubilized brush border membrane proteins, Weinman et al. have found that cyclic AMP-dependent protein kinase and Ca-calmodulin-dependent protein kinase inhibit whereas protein kinase C stimulates the Na^+/H^+ exchanger(51). These results indicate that the regulation of the $Na^+:HCO_3^-$ cotransport system by protein kinases is simultaneous with and independent of their effects on the Na^+/H^+ exchange system (36).

Angiotensin II has been demonstrated to increase Na^+/H^+ exchange and $Na^+:HCO_3^-$ cotransport systems (21). Simultaneous regulation of Na^+/H^+ exchange and $Na^+:HCO_3^-$ cotransport systems by hormones or protein kinases should minimize significant changes in proximal tubule cell pH and maximize cells ability to reabsorb the filtered HCO_3^-.

c. Effect of acute acid base disorder on Na^+/H^+ exchange and $Na^+:HCO_3^-$ cotransport systems-Since the bulk of HCO_3^- reabsorption in the proximal tubule occurs via the luminal Na^+/H^+ exchanger and basolateral $Na^+:CO_3^=:HCO_3^-$ cotransporter, it might be predicted that alterations in proximal tubular acidification in pathologic states should ultimately result from changes in the activity of these transporters. Two frequent clinical conditions affecting the pH homeostasis of blood are metabolic acidosis and alkalosis. Primary metabolic acidosis, a condition manifested by decreased serum $[HCO_3^-]$ and pH, has been shown to be associated with an increased ability of the renal tubules to reabsorb HCO_3^- (15,23). Primary metabolic alkalosis, a condition manifested by increased serum $[HCO_3^-]$ and pH, has been shown to be associated with a decreased ability of the renal tubules to reabsorb HCO_3^- (6). Studies which have evaluated the luminal Na^+/H^+ exchanger and/or basolateral $Na^+:HCO_3^-$ cotransporter in animals with acid-base disorders suggest that the activity of these two transport processes are increased in metabolic acidosis and decreased in metabolic alkalosis (2,33).

To determine whether the alteration in the activity of the luminal Na^+/H^+ exchanger and basolateral $Na^+:HCO_3^-$ cotransporter in metabolic acidosis and alkalosis is due to systemic effects or is mediated locally at the level of the kidney, Soleimani et al. studied the activity of these two transport processes in an *in vitro* model. Proximal tubular suspensions were prepared and incubated in acidic, normal, and alkaline media. Following the incubation of the proximal tubules in the experimental media for forty five minutes, brush border and basolateral membrane vesicles were isolated and studied for the activity of these transporters. The results demonstrated that exposure of proximal tubules to an acidic pH resulted in adaptive increases in the luminal Na^+/H^+ exchanger and basolateral $Na^+:CO_3^=:HCO_3^-$

cotransporter activities (44). Exposure of proximal tubules to an alkaline pH caused an adaptive decrease in the luminal Na^+/H^+ exchanger but had no effect in the basolateral $Na^+:CO_3^=:HCO_3^-$ cotransporter activities (22). The effects are apparently due to changes in the Vmax of the transporters (22,44). The results also demonstrated that the effect of acidic or alkaline pH on these two transporters is mediated directly at the level of the renal tubule and is independent of systemic factor(s).

The cellular mechanisms responsible for these adaptive changes in *in vitro* acid base disorders are presently unknown. Given the brief time of exposure of the tubular suspensions to acidic and alkaline pH, synthesis of new transport proteins seems unlikely. Indeed, cyclohexamide failed to prevent the adaptive increases in Na^+/H^+ exchange and $Na^+:HCO_3^-$ cotransport in metabolic acidosis (44), lending support to this conclusion. Other possibilities, including activation of currently inactive membrane proteins (phosphorylation) or incorporation of intracellular proteins into the membrane (exocytosis), are potential mechanisms for the observed changes.

Studies examining the role of phosphorylation on the luminal Na^+/H^+ exchanger and basolateral $Na^+:HCO_3^-$ cotransporter have demonstrated that PKC-mediated phosphorylation increases the activity of these two transporters (36,51). Inhibition of protein kinase C in proximal tubular suspension partially prevented the adaptive increases in the luminal Na^+/H^+ exchange and basolateral $Na^+:HCO_3^-$ cotransport systems in metabolic acidosis (22) suggesting that PKC may play a role in this process.

The role of exocytic/endocytic processes in mediating some of the adaptive changes in Na^+/H^+ exchange and $Na^+:HCO_3^-$ cotransport that are observed in acid-base disorders remains undetermined. Alteration in the rate of incorporation of intracellular proteins into membranes has been shown to occur in response to several stimuli. This process, which occurs via recruitment of endosomal vesicles to the membrane, is involved in mediating the action of ADH (50), insulin (24,25), and CO_2-induced acidosis (45,46). Recently, Sabolic et al. demonstrated the presence of an amiloride-sensitive Na^+/H^+ exchange system in endosomal vesicles isolated from rabbit kidney proximal tubule cells (47). In these cells, lowering cell pH by increasing pCO_2 stimulated incorporation of acid-extruding endosomes into the cell membrane (46). It is possible, therefore, that an acidic medium lowers proximal tubule cell pH which in turn stimulates the exocytosis of endosomal vesicles to the luminal membrane. This would lead to extrusion of acid via luminal Na^+/H^+ exchange. Whether the increased activity of the basolateral $Na^+:HCO_3^-$ cotransporter in acute *in vitro* metabolic acidosis is via increased exocytosis remains to be seen.

The adaptive changes in pH-regulating processes in response to *in vitro* alkalosis are more complex. The luminal Na^+/H^+ exchanger is down-regulated and the basolateral $Na^+:HCO_3^-$ cotransporter remains unaffected in metabolic alkalosis. The absence of adaptive changes in $Na^+:HCO_3^-$ cotransport in response to metabolic alkalosis might suggest that proximal tubule cells are not well prepared to defend intracellular pH against alkalosis or that longer incubation time is needed to elicit an adaptive response.

In summary, *in vitro* metabolic acidosis and alkalosis cause complex adaptive changes in the luminal Na^+/H^+ exchanger and basolateral $Na^+:HCO_3^-$ cotransporter in the rabbit proximal tubule. The signal(s) responsible for evoking these adaptive changes reside at the level of the renal tubule and is (are) independent of systemic factor(s). Some of these changes are partially dependent on protein kinase C. Further studies will be required to determine the nature of the signal(s) involved in these adaptive changes.

References:
1. Akiba, T., Alpern, R.J., Eveloff, J., Calamina, J. and Warnock, D.G. 1986. *J. Clin. Invest.* 78:1472-1478.
2. Akiba, T., V.K. Rocco., and D.G. Warnock. 1987. *J. Clin. Invest.* 80:308-315.
3. Alpern R.J. 1990. *Physiol Rev.* 70: 79-114
4. Alpern, R.J. 1985. *J. Gen. Physiol.* 86:613-636.
5. Alpern, R.J. and Chambers, M. 1986. *J. Clin. Invest.* 78:502-510
6. Alpern, R. J., D.G. Warnock., and F.C. Rector, Jr. 1986. In the *Kidney*. B.M. Brenner and F.C. Rector, Jr., editors. W.B.Saunders & Co., Philadelphia. #rd ed. 206-249
7. Aronson, P.S., M. Soleimani., and S.M.Grassl. 1991. *Seminars in Nephrology.* 11:28-36
8. Aronson, P.S., Nee, J., and Suhm, M.A. 1982. *Nature*(London) 299:161-163
9. Biagi, B.A. 1985. *J. Membr. Biol.* 88:25-31.
10. Biagi, B.A. and Sohtell, M. 1986. *Am. J. Physiol.* 250:F267-F272.
11. Boron, W.F., and E.L. Boulpaep. 1983. *J. Gen. Physiol.* 81:53-94.
12. Burg, M., and N. Green. 1977. *Am. J. Physiol.* 233:F307-F314.
13. Burckhardt, B.-Ch., K. Sato, and E. Fromter. 1984. *Pflügers Arch.* 401:34-42.
14. Cogan, M.G., D.A. Maddox, D.G. Warnock, E.T. Lin, and F.C. Rector, Jr. 1979. *Am. J. Physiol.* 237:F447-F454.
15. Cogan, M. G., Alpern, R.J. 1984. *Am J. Physiol.* 247:F387-395
16. DuBose, T.D., Jr., L.R. Pucacco, and N.W. Carter. 1981. *Am. J. Physiol.* 240:F138-146.
17. DuBose, T.D., Jr., L.R. Pucacco, D.W. Seldin, N.W. Carter, and J.P. Kokko. 1979. *Kidney Int.* 15:624-629.
18. Dobyan, D.C., and R.E. Bulger. Renal carbonic anhydrase. 1982. *Am. J. Physiol.* 243: F311-F324.
19. Grassl, S.M. and Aronson, P.S. 1986. *J. Biol. Chem.* 261:8778-8783.
20. Grassl, S.M., Holohan, P.D. and Ross, C.R. 1987. *J. Biol. Chem.* 262:2682-2687
21. Giebel, J., Giebisch, G., and Boron, W. F. 1990. *Proc. Natl. Acad. Sci.* 87:7917-7920
22. Hattabaugh, Y.J., and Soleimani, M. 1992. *J. Am. Soc. Neph.* 3:778A
23. Kunau, R. T. Jr., Hart, J. I., Walker, K. A. 1985. *Am. J. Physiol.* 249:F62-68
24. Kono, T., Robinson, F. W., Blevins, T. L., Ezaki, O. 1981 *J. Biol. Chem.* 257:10942-10947
25. Karnieli, E., Zarnowski, M. J., Hissin, P. J., Simpson, I. A., Salana, L. B., Cushman, S. W. 1981. *J. Biol. Chem.* 256:4772-4777
26. Lucci, M.S., L.R. Pucacco, T.D. Dubose, Jr., J.P. Kokko, and N.W. Carter. 1980. *Am. J. Physiol.* 238:F372-F379.
27. Lucci, M.S., D.G. Warnock, and F.C. Rector, Jr. 1979. *Am. J. Physiol.* 236: F58-F65.
28. Lucci, M.S., L.R. Pucacco, T.D. DuBose, Jr., J.P. Kokko, and N.W. Carter. 1980. *Am. J. Physiol.* 238:F372-F379.
29. McKinney, T.D., and M.B. Burg. 1977. *Kidney Int.* 12:1-8.
30. Mugharbil, A., Knickelbein,R.G., Aronson, P.S., and J.W.Dobbins. 1990. *Am. J. Physiol.* G666-G670
31. Mason, M.J., Smith, J.D., Garcia-Sato, J.D., and S.Grinstein. 1989. *Am.J.Physiol.* 256:C428-433
32. Preisig, P. A., and R.J. Alpern. 1989. *Am. J. Physiol* 256:F751-F756
33. Preisig, P. A., and R. J. Alpern. 1988. *J.Clin.Invest.* 82:1445-1453.
34. Passow, H. 1986. *Rev. Physiol. Biochem. Pharmacol.* 103:62-223.
35. Rector, F.C., Jr., N.W. Carter, and D.W. Seldin. 1965. *J.Clin.Invest.* 44: 278-290.
36. Ruiz, O., and JAL Arruda. 1990. *Kid. Int.* 37:544 A
37. Sasaki, S., Shiigai, T. and Takeuchi, T. 1985. *Am. J. Physiol.* 249:F417-F423.
38. Soleimani, M., Grassl, S.M. and Aronson, P.S. 1987. *J. Clin. Invest.* 79:1276-1280.
39. Soleimani, M., and P.S. Aronson. 1989. *J. Biol. Chem.* 264:18302-18308.

40. Soleimani, M., Lesoine, G. A., Bergman, J.A., and Aronson, P.S. 1991. *J. Biol. Chem* 266:8706-8710
41. Soleimani, M., and Aronson, P.S. 1989. *J.Clin.Invest* 83: 945-951.
42. Soleimani, M., Lesoine, G.A., Bergman, J.A., and T.D. McKinney. 1991. *J.Clin.Invest.* 88:1135-1140
43. Soleimani, M., Hattabaugh, J.A., and Bizal, G.L. 1992. *J. Biol. Chem* 267: 18349-18355
44. Soleimani, M., Bizal, G.L., McKinney, T.D., and Hattabaugh, J.A. 1992. *J.Clin.Invest* 90:211-218
45. Stetson, D. L., Steinmetz. P. R. 1983. *Am. J. Physiol.* 245:C113-120
46. Schwartz, G. J., and Q.Al-Awqati. 1985. *J. Clin.Invest.* 75:1638-1644.
47. Sabolic, I., and D. Brown. 1990. *Am. J. Physiol.* 258:F1245-1253
48. Vieira, F. L., and G. Malnic. 1968. *Am. J. Physiol.* 214:F710-F718.
49. Wistrand,P.J., and R. Kinne. 1977. *Pflügers Arch.* 370:121-126.
50. Wade, J. 1980. *Curr. Top. Membr. Transp.* 15:123-147
51. Weinman, E. D., Dubinsky, W., Shenolikar, S. 1989. *Kidney Int.* 36:519-525
52. Yoshitomi, K., Burckhardt, B.-C. and Fromter, E. 1985. *Pflügers Arch.* 405:360-366.

Potassium dependent H+/HCO3- transport mechanisms in cells of medullary thick ascending limb of rat kidney.

Pascale Borensztein, H. Amlal, M. Froissart, F. Leviel, M. Bichara, and M. Paillard
Département de Physiologie et INSERM U 356
Université Pierre et Marie Curie
Hôpital Broussais
Paris, France

The medullary thick ascending limb (mTAL) of rat kidney has been recently shown to absorb bicarbonate at substantial rate and to contribute to acidification of the final urine (5).

At least two special features of mTAL should be stressed. First, the rate of bicarbonate absorption by mTAL may modulate the bicarbonate concentration and pH of the medullary interstitium because of the low effective blood flow in this area (9). Since the interstitial pH is known to influence proton secretion into medullary collecting duct (6), the rate of bicarbonate absorption by mTAL may indirectly modulate the net acid excretion. Second, large variations of potassium concentration may occur in the medullary interstitium and in the lumen of mTAL due to the potassium recycling which depends on the state of potassium balance (7). For example, during potassium loading, potassium leaves the medullary collecting duct, accumulates in the interstitium, and is secreted into the pars recta and descending limb (7).

The aim of this paper was to describe the cellular mechanisms of H+/HCO3- transport in rat mTAL cells that we have recently found in our laboratory, with a special attention to the potassium-dependent transport mechanisms.

All the methods used have been previously described by our group (3). In brief, we used a suspension of freshly isolated mTAL fragments from inner stripe of outer medulla of rat kidney. Cell pH was estimated from the fluorescence ratio of BCECF loaded cells. Variations in cell membrane potential were estimated from the

NATO ASI Series, Vol. H 89
Molecular and Cellular Mechanisms
of H+ Transport
Edited by Barry H. Hirst
© Springer-Verlag Berlin Heidelberg 1994

degree of quenching of DIS-C-3-5 fluorescence after addition of constant amount of cells in a probe containing medium.

Transepithelial bicarbonate absorption by mTAL implies a net proton secretion at the luminal side and a bicarbonate transport at the basolateral side.

Na^+/H^+ antiporter and H^+-ATPase. These two transport mechanisms have been shown in mTAL cells by our group (3). In rat mTAL cells preincubated at 37°C in HCO_3^-/CO_2 free, Hepes-buffered medium containing 67 mM sodium, cell pH recovery after sudden 40 mM K-acetate-induced acidification was inhibited 50% by 2×10^{-3} M amiloride and 36% by 200 nM bafilomycin A1. The doses of these two drugs were sufficient to completely inhibit the Na^+/H^+ antiporter and H^+-ATPase. These two H^+ transport mechanisms were operational at normal external pH 7.4, since the resting cell pH values were lower in the presence of amiloride or bafilomycin. In addition, we have shown that the Na^+/H^+ antiporter activity was inhibited by 8 Bromo-cAMP, which supports the location of this transporter in the luminal membrane (2). It should be noted however that Sun et al. have recently described an additional Na^+/H^+ antiporter in the basolateral membrane of mouse mTAL cells, which is not operational under normal conditions (11).

K^+/HCO_3^- cotransporter. This new transporter mechanism has been recently described by our group in rat mTAL cells (10). The general approach to characterize this transporter was the following : mTAL cells, preincubated in medium containing 25 mM HCO_3^-, 5% CO_2, pH 7.4, were suddenly diluted in a nominally HCO_3-CO_2 free medium, Hepes-buffered, pH 7.4. After an initial cell alkalinization due to immediate CO_2 exit from cell, the cell pH recovery (i.e. acidification) reflects the HCO_3^- efflux. It should be noted that when cells were alkalinized to the same extent by preincubation at pH 7.8 in a HCO_3^--CO_2-free Hepes-buffered medium, and then diluted in a similar medium pH 7.4, cells did not acidify rapidly, which eliminates a significant H^+ influx.

A first series of experiments were performed to characterize the HCO_3^- efflux. 10^{-4} M DIDS inhibited 50% the HCO_3^- efflux. 2 mM Barium, which inhibits potassium conductance and depolarizes mTAL cells, did not modify the initial rate of cell acidification, which suggests that the HCO_3^- efflux mechanism in mTAL is electroneutral. This electroneutrality of HCO3⁻ efflux has been further confirmed by additional experiments measuring cell potential (10). A Cl⁻-free medium (Cl replaced by gluconate) did not modify the cell acidification rate, which excludes a Cl:HCO3⁻ exchanger. A sodium-dependence of the HCO3⁻ efflux was tested in the presence of 2×10^{-3} M amiloride to inhibit Na⁺/H⁺ antiporter which remains operational at alkaline cell pH values and thus may modify the cell acidification rate independently of HCO3⁻ efflux. Under these conditions, creating an outward transmembrane Na⁺ gradient by diluting cells in sodium free medium did not increase the cell acidification rate, compared to sodium-containing medium, which rules out a sodium-dependent HCO3⁻ efflux.

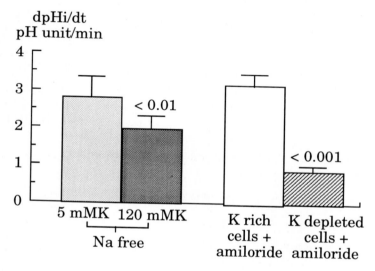

Fig. 1 : Potassium dependence of cell pH recovery from alkalinization (HCO3⁻ efflux) in rat mTAL cells preincubated in HCO3⁻/CO2-containing medium (see text for explanations).

The potassium dependence of HCO3⁻ efflux is shown in Fig. 1. First, cell acidification rate was reduced when cells were diluted in a potassium-rich medium (120 mM potassium) compared to a normal potassium-containing medium (5 mM potassium). In both cases, the dilution medium was Na⁺-free to avoid any difference in the transmembrane Na gradient.Thus, increasing external potassium reduced HCO3⁻ efflux. Second, in cells made potassium-depleted by several washings in potassium-free media, cell acidification rate was strongly inhibited compared to potassium rich cells. Amiloride was present in both cases to inhibit any interference of the Na⁺/H⁺ antiporter Thus, decreasing cell potassium concentration inhibits HCO3⁻ efflux.

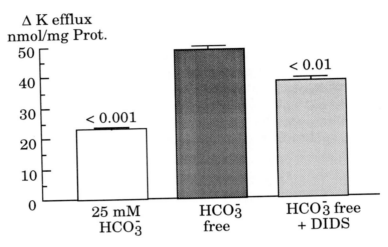

Fig. 2 : Bicarbonate dependence of potassium efflux in rat mTAL cells (see text for explanations).

Finally, the HCO3⁻ dependence of potassium efflux is shown in Fig. 2. The net efflux of potassium was measured with an external potassium-sensitive electrode, after diluting cells in a potassium-free medium containing 1 mM ouabain, 1 mM furosemide, 10 mM barium to inhibit the known potassium transport mechanisms in these cells. Potassium efflux was higher in cells diluted in HCO3⁻-free media compared to 25 mM HCO3⁻-containing medium. Thus, creating an outward HCO3⁻ gradient increased potassium efflux. In addition, this stimulated potassium efflux was reduced by DIDS.

Taken together, these results show clearly that an electroneutral K⁺:HCO3⁻ cotransporter takes place in rat mTAL cells. This transporter is probably located in the basolateral membrane where

HCO3- exit normally occurs This transporter is inhibited by DIDS and insensitive to Barium.

To assess the sensitivity of the transporter to the internal HCO3-, the absolute values of the initial rate of cell pH recovery after sudden dilution of cells in HCO3-/CO2 free media were plotted against a corresponding large range of intracellular HCO3- concentrations, calculated from cell pH and PCO2 measured at the 6th second. We observed a relationship of first order kinetics, with a calculated Km of about 16 mM, which is similar to the normal intracellular HCO3- value for external pH 7.4 and PCO2 40 mmHg. This value of Km HCO3- suggests a role of cellular HCO3- concentration in controlling the activity of the K+:HCO3- cotransporter under physiological conditions.

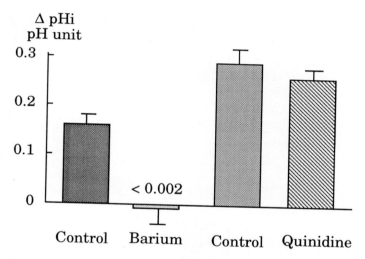

Fig. 3 : Potassium dependence of cell acidification (H+ influx) in rat mTAL cells incubated in a HCO3-/CO2-free medium (see text for explanations).

K+/H+ antiport. We have very recently found that 10 mM Barium, which does not influence the K+/HCO3- cotransporter, alkalinized by 0.10 pH unit mTAL cells incubated in a HCO3-/CO2-free Hepes-buffered medium (1). To further investigate this finding, we looked for the effect of a potassium outward gradient on cell pH in mTAL cells incubated in a HCO_3^--CO_2-free Hepes-buffered medium. Cells were preincubated in a 140 mM K- and Na-free medium, and suddenly diluted in a K- and Na-free medium. As shown in Fig. 3, a rapid cell acidification was observed, which was suppressed by barium (1). The simplest explanation could be that potassium efflux causes hyperpolarization of cells, which in turn could acidify cells by H^+ influx through H^+ conductance, and barium, which blocks potassium conductance and thus suppresses cell hyperpolarization, inhibits the cell acidification. However, in the presence of quinidine, which also blocks potassium conductance and thus prevents also cell hyperpolarization, cell acidification was not suppressed (1) (Fig. 3). Thus barium may have a specific effect on this potassium-dependent H^+ transport independently of its effect on potassium conductance. We have checked that the cell hyperpolarization induced by potassium efflux is actually suppressed by barium or quinidine (not shown). Finally, the cell acidification induced by outward potassium gradient was not affected by 2 mM amiloride, 0.2 mM DIDS, or 0.1 mM SCH 28080 (H^+/K^+-ATPase inhibitor).

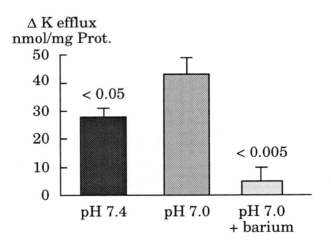

Fig. 4 : External pH dependence of K^+ efflux in rat mTAL cells incubated in HCO_3^-/CO_2-free medium(see text for explanations).

Next, we tested external pH dependence of potassium transport in mTAL cells. The net efflux of potassium was measured by external potassium electrode in the presence of an outward potassium gradient, at two different external pH values, 7.4 and 7.0, in Hepes-buffered medium containing 1 mM ouabain, 1.5 mM furosemide, and 0.1 mM quinidine to block the known potassium transport mechanisms. The net potassium efflux observed was higher at pH 7 than at pH 7.4. In addition, barium reduced the H^+-stimulated potassium efflux (Fig. 4).

Taken together, these results in HCO_3^--CO_2-free Hepes-buffered medium suggest the presence of an electroneutral K^+:H^+ antiporter sensitive to barium and not affected by DIDS. This mechanism is clearly different from the K^+:HCO_3^- transporter which is inhibited by DIDS and insensitive to barium. Although the luminal or basolateral location of the K^+/H^+ antiporter cannot be assessed in our preparation of mTAL fragments, we suggest that it takes place in the luminal membrane of cells. Indeed this barium-sensitive K^+/H^+ antiporter can operate as K^+/NH_4^+ (1), and a luminal barium-sensitive NH_4^+ transport has been shown in the isolated and perfused mTAL tubule (8).

To conclude, we can only speculate on the physiological role of the potassium-dependent H^+:HCO_3^- transport mechanisms in rat mTAL (Fig. 5). It has been shown recently by Good that raising potassium concentration, in both the lumen and bath in experiments with mTAL fragments perfused in vitro, did not modify transepithelial bicarbonate absorption (4). One possibility is that these potassium-dependent H^+:HCO_3^- transport mechanisms help maintain a normal transepithelial bicarbonate absorption in vivo when potassium concentrations vary simultaneously in medullary interstitium and tubular lumen during potassium recycling phenomenon. Indeed, raising peritubular potassium concentration should reduce the activity of basolateral K^+:HCO_3^- transporter, which results in alkalinization of the cells and thus should reduce H^+ secretion across the luminal membrane by Na^+/H^+ and H^+-ATPase. However, raising potassium in the lumen should have two effects : first, it reduces potassium efflux across luminal membrane through potassium conductance and thus depolarizes the luminal membrane, which would enhance the voltage dependent H^+-ATPase activity. Second, increasing potassium reduces the K+:H+ antiporter activity in the luminal membrane, and thus limits H+ entry into cells across the luminal membrane. Thus the net effect of increasing luminal potassium concentration is a stimulation of net proton secretion into the lumen which counterbalances the inhibiting effect of elevated interstitial potassium concentration on HCO3- exit across the basolateral membrane.

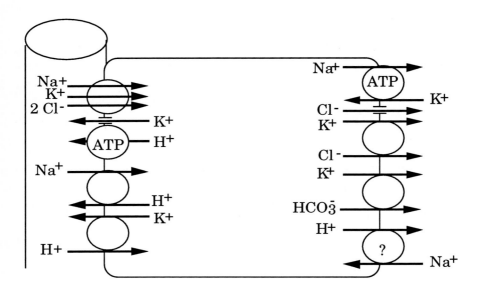

Fig. 5 : Tentative model for H+/HCO3- transport mechanisms in rat mTAL cells.

References
1. Bichara M, Amlal H, Paillard M (1993) Barium-sensitive K+/NH4+(H+) antiport in medullary thick ascending limb (MTAL) cells. J A S N (in press).
2. Borensztein P, Juvin P, Vernimmen C, Poggioli J, Paillard M, Bichara M (1993) cAMP-dependent control of Na+/H+ antiport by AVP, PTH, and PGE2 in rat medullary thick ascending limb cells. Am J Physiol 264:F354-F364.

3. Froissart M, Borensztein P, Houillier P, Leviel F, Poggioli J, Marty E, Bichara M, Paillard M (1992) Plasma membrane Na+:H+ antiporter and H+-ATPase in the medullary thick ascending limb of rat kidney. Am J Physiol 262:C963-C970.

4. Good DW (1988) Active absorption of NH4+ by rat medullary thick ascending limb : inhibition by potassium. Am J Physiol 255:F78-F87.

5. Good DW, Knepper MA, Burg MB (1984) Ammonia and bicarbonate transport by thick ascending limb of rat kidney. Am J Physiol 247:F35-F44).

6. Jacobson HR (1984) Medullary collecting duct acidification. Effects of potassium, HCO3 concentration, and PCO2. J Clin Invest 74:2107-2114.

7. Jamison RL (1987) Potassium recycling. Kidney Int 31:695-703.

8. Kikeri D, Sun A, Zeidel ML, Hebert SC (1989) Cell membranes impermeable to NH3. Nature (London) 339:478-480.

9. Knepper M, Burg M (1983) Organization of nephron function. Am J Physiol 244:F579-F589.

10. Leviel F, Borensztein P, Houillier P, Paillard M, Bichara M (1992) Electroneutral K+/HCO3- cotransport in cells of medullary thick ascending limb of rat kidney. J Clin Invest 90:869-878.

11. Sun AM, Kikeri D, Hebert SC (1992) Vasopressin regulates apical and basolateral Na+:H+ antiporters in mouse medullary thick ascending limb. Am J Physiol 262:F241-F247.

Characterization of Parietal Cell Functional Morphology Utilizing Isolated Gastric Gland Perfusion, Atomic Force and Confocal Microscopy

Irvin M. Modlin[1], Steven J. Waisbren[2], Walter F. Boron[2], Carol J. Soroka[1], James R. Goldenring[1,2], John Geibel[1,2]

Gastrointestinal Surgical Pathobiology Research Unit
Departments of Surgery[1] and Cellular and Molecular Physiology[2]
Yale University School of Medicine
333 Cedar St., New Haven, CT 06510

INTRODUCTION

In recent times, significant advances have taken place in regard to the evaluation and delineation of parietal cell function. The characterization of the proton pump and the elucidation of the mechanisms relating to acid secretion have been of considerable physiological and clinical relevance. An area which remains unresolved has been the delineation of parietal cell apical membrane function. In particular, the characteristics which enable cells to withstand the low pH and proteolytic content of the gastric lumen have yet to be defined. Similarly little is know about the mechanism by which membrane movement takes place within the parietal cell during the secretory process. To date, there exist a number of limitations regarding the study of the functional morphology of the parietal cell. The development of the isolated gastric gland and the isolated parietal cell preparations have been useful. However, alternative methodologies and technologies are required to evaluate further membrane function and to relate function to morphology. With this in mind, we have applied three novel techniques to the study of parietal cell function. Firstly, we have developed an isolated, perfused gastric gland preparation to evaluate apical versus basolateral membrane function. Secondly, we have utilized a parietal cell culture system and scanning confocal fluorescence microscopy to evaluate vectorial membrane movement and cytoskeletal transformation during secretion. Thirdly, we have employed atomic force microscopy to study the apical membrane surface of living parietal cells both at rest and during stimulation. The development of methodologies

NATO ASI Series, Vol. H 89
Molecular and Cellular Mechanisms
of H[+] Transport
Edited by Barry H. Hirst
© Springer-Verlag Berlin Heidelberg 1994

useful in studying the dynamic changes that occur during parietal cell stimulation may provide information of use in further delineating the functional topography of these cells.

METHODS

A)   Isolated perfused gastric glands

New Zealand White rabbits (1-2 kg) were used. A 1 cm² section of the gastric fundus was excised and placed in a Ringer solution at 4°C. Individual glands were hand dissected and immediately transferred to the stage of an inverted microscope. Typically, the individual glands had a length of ~ 1 mm. After transfer to the chamber the glands were attached to a series of concentric glass micropipettes and perfused in a manner similar to that described in detail for isolated perfused renal tubules (Geibel,et al)(Fig. 1). After establishing cannulation, the luminal and basolateral perfusates could be exchanged. We used this technique to examine acid secretion mechanisms located on the apical membrane of the parietal cell. In these studies the fluorescent marker BCECF was used to monitor intracellular pH following various ionic substitutions.

ISOLATED PERFUSED GASTRIC GLAND

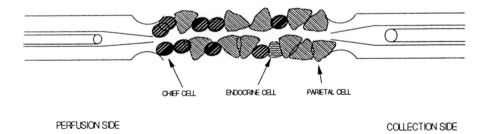

CHIEF CELL   ENDOCRINE CELL   PARIETAL CELL

PERFUSION SIDE   COLLECTION SIDE

Figure 1. Schematic drawing of the isolated perfused rabbit gastric gland. Please note that both sides of the gland are canulated using a series of concentric glass micropipettes.

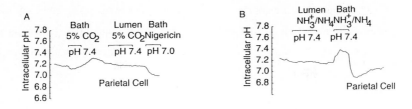

Figure 2. A) Intracellular pH tracing of a perfused gastric gland exposed to both luminal and bath CO_2. B) A separate experiment showing luminal and bath exposure of NH_4Cl.

CO_2/HCO_3

Individual gastric glands were exposed to a non-bicarbonate control solution in both the apical and basolateral perfusate following dye loading. After recording the resting steady-state pH^i values. Either the bath or luminal perfusates was exchanged for a solution containing $5\%CO_2/22$mM HCO_3. Following the perturbation the gland was again bathed in a non-bicarbonate buffer.

NH_4Cl

In a separate series of glands the luminal and or basolateral perfusate was exchanged for a solution of 20 mM NH_4Cl. Glands were initially bathed and perfused in a non-bicarbonate buffer solution. Following the addition of 20 mM NH_4Cl, a 5 minute equilibrium period was established before washing into the control solution.

B)   Parietal Cell Isolation and Culture

In order to investigate the direct effect of various modifiers upon secretagogue-stimulated alterations in the gastric parietal cell, an enriched population of cells is a primary requirement. We have utilized an isolation and culture system for rabbit gastric cells previously described by Chew (Chew) yielding enrichments of >95% parietal cells. Briefly, gastric mucosa was subjected to sequential digestions with pronase (0.5 mg/ml) and collagenase A (0.75 mg/ml). Subsequently, parietal cells were enriched by Nycodenz gradient separation, followed by centrifugal elutriation. Contaminating fibroblasts were removed by a 30 minute pre-attachment step to wells coated with 10% fetal calf serum. The cells were then placed in sterile DMEM:F12 medium containing growth factors, bovine serum albumin, and antibiotics. The cells were cultured in wells or on glass coverslips at 37°C.

Confocal Immunofluorescence

Immunofluorescence was used to establish the distribution of specific protein and membrane markers following the culture procedure. We were especially interested in determining whether the tubovesicular marker, H/K-ATPase,was translocated to the canalicular secretory surface upon stimulation with a secretagogue. Furthermore, we wished to establish whether membrane polarity and cytoskeletal integrity was maintained during the isolation, and culturing protocols. The immunofluorescence studies were carried out on cells that had been fixed for 15 min with 3% paraformaldehyde in 80 mM PIPES, pH 6.8, 5 mM EDTA, 2 mM $MgCl_2$. Appropriately diluted primary antibodies were applied for 2-4 hours at room temperature or overnight at 4°C. In the case of the double labeling, both primary and secondary antibodies were applied simultaneously. Texas Red-conjugated goat anti-mouse IgG or FITC-conjugated goat anti-rabbit IgG were used as the secondary antibody, incubations times were 1 hr at room temperature. Following incubation cover slips were washed and stored in the dark at 4°C until use. In the case of the double-labeling for F-actin, the Bodipy- phallacidin (1:20) was incubated simultaneously with the Texas Red-secondary antibody. Cells were evaluated for positive immunofluorescent responses using the PHOIBOS 100 laser scanning confocal microscope (Molecular Dynamics) equipped with the argon laser. The personal Iris computer with the PHOIBOS 4D Image Analysis software was used to digitize the optical images and to preform the 3-D reconstruction of the cells. The image data were collected with a 100X objective (N.A. 1.4). Optical sections of approximately 0.95 μm were used for all studies except the microtubule series in which 0.1 μm sections were taken. The sequential images were then processed using the VoxelView volume rendering software.

Electron Microscopy

Electron microscopy was preformed to verify the presence of tubulovesicles in the resting cultured cells, and to examine vesicular trafficking following stimulation with the acid secretagogue. Cells were washed in PBS and fixed with 3% glutaraldehyde in 0.1 M cacodylate. Cells were post-fixed with 4% OsO_4 dehydrated in ethanol and Epon embedded. Thin sections were stained with uranyl acetate and lead citrate and viewed with a Philips 300 transmission electron microscope.

[^{14}C]-aminopyrine Accumulation

The functional integrity of the cultured cells was determined by measuring the acid secretory activity by accumulation of [^{14}C]-aminopyrine (AP). Cultured cells were rinsed with control media and pre-incubated for 20 min with media containing [^{14}C]-AP. The cells were then exposed to either histamine(10^{-4} M), carbachol (10^{-4} M) and cimetidine (10^{-5} M) or cimetidine alone. In some series, cytoskeletal modifiers (cytochalasin or colchicine) were added in addition to the above compounds. Thiocynate (10mM) was added to separate wells to correct for nonspecific trapping of AP. After incubation the cells were rapidly washed in PBS and lysed with 1% Triton X-100. Aliquots of cell lysates and supernatants were counted in a Beckman liquid scintillation counter.

C)   Atomic Force Microscopy

Microscopes use light waves and applicable optical components to create a two-dimensional projection of the specimen. The resolution of the system is limited purely by the mechanical properties of the substrates of the specimen, the interface surface, and the relay and system optics of the microscope. However, the examination of living cells under a conventional microscope is limited to a resolution of roughly half the wavelength of the light. A major advance in microscopy was implemented by the development of the *near field* microscope(Radmacher, et al), or the scanning probe microscope. These instruments pass a probe in close proximity to the specimen and collect spatial information about the surface properties, such as the physical topography, ion conductance or even temperature from a specimen. This novel form of microscopy has enabled the *light barrier* to be surpassed. Thus, measurements of ultra-high resolution surface topography of cells and tissues can be undertaken.

We have utilized this novel technique to examine the apical membrane of the parietal cell. Cells were studied during resting and induced acid secretion states using a specially modified chamber allowing for continuous perfusion of biological preparations with physiological solutions maintained at 37 °C.

The specimen is attached to a stable support and the chamber is then lowered into position and sealed with an O-ring assembly. Using two injection ports it is possible to superfuse the surface of the specimen with a specific solution. Included in the system is a thermistor for temperature regulation connected to a feedback heating system. Utilizing the present system the temperature can be maintained to within 0.5 °C at flow rates from 1-15 ml/min, allowing rapid exchange of solutions.

Preparation of Isolated gastric Glands

In order to observe the apical membrane of the gastric gland epithelia, we devised a technique allowing access to the apical surface while maintaining epithelial polarity of the remaining cells. Individual gastric glands were hand dissected at 37°C, using a series of sharpened forceps and glass knives. After dissection, the apical membrane of the gastric gland comprising of segments of approximately 6-8 cells was exposed and populations of both chief and or parietal cells were evident. The gland was then transferred to a cover-slip prepared with CELL-TAK and was inserted at the base of the atomic force microscope (AFM). The chamber was then installed, the superfusion system activated and the probe positioned over an area of cells prior to activation of the scanning procedure. Typically a 20-30 μm area was initially scanned so that both parietal and chief cells were visible. The apical surface was exposed to fluorescent probes to ensure proper cell identification. Following the initial scan, the field of view was reduced so that only the apical surface of a single cell was in focus.

RESULTS

Isolated Perfused Gastric Glands

Apical and Basolateral Membrane permeabilities

Addition of 5% CO_2/ 22 mM HCO_3 to the apical perfusate for periods greater than 5 minutes had no effect on pH^i in the parietal cells (n=52) of the perfused glands. Following removal of the 5% CO_2/ 22 mM HCO_3 solution pH^i remained constant. Conversely, addition of the same solution to the bath gave a marked acidification of the cell, this effect was reversible upon removal of the 5% CO_2/ 22 mM HCO_3 solution. In separate studies, the order of 5% CO_2/ 22 mM HCO_3 was reversed (bath followed by lumen), and again the bath pulse had a significant effect while luminal addition had no effect on pH^i.

In a separate series of glands 20 mM NH_4Cl was added to either the luminal or basolateral perfusate of perfused gastric glands in a non-bicarbonate buffer solution. Luminal 20 mM NH_4Cl had no effect on pH^i, whereas bath 20 mM NH_4Cl caused a typical pronounced acid load to the parietal cell. As was the case with CO_2 the order of presentation of the experimental solution had no effect on the permeability. These results indicate that the apical surfaces of parietal cells are impermeable to both NH_3 and CO_2.

Confocal Microscopy of Cultured Parietal Cells

Following isolation and gradient separation we were able to obtain a population of parietal cells with > 95% enrichment. After three days in culture the cells were checked for functional integrity as well as maintenance of membrane and cytoskeletal domains. The cells were responsive to histamine and carbachol stimulation as measured by [^{14}C]-AP accumulation and morphological alteration. Electron microscopy demonstrated that the resting cells contained an extensive tubulovesicular compartment. Immunofluorescence microscopy showed that the cultured parietal cells maintained membrane polarity as detected by restricted basolateral localization of the Na/K-ATPase. F-actin microfiliments were associated with the intracellular canaliculi and the lateral cortex of the cells. Ezrin, an actin-binding protein co-localized with F-actin on the intracellular canaliculi. In resting cells, H/K-ATPase-containing tubulovesicles were localized to a region deep to the F-actin-rich intracellular canaliculus. Microtubules were evident and scattered throughout the cytoplasm. After stimulation with an acid secretagogue, electron microscopy demonstrated that the tubulovesicular compartment largely disappeared. H-K-ATPase was detected by confocal microscopy at the surface of the expanded secretory canaliculi, along with the ezrin and F-actin. Conversely, no alteration in the localization of the Na-K-ATPase or the microtubules was evident following the stimulation. These results indicate that stimulated vectorial movement of membrane in the parietal cell precedes through the processing of membrane vesicles through an F-actin-rich target secretory surface.

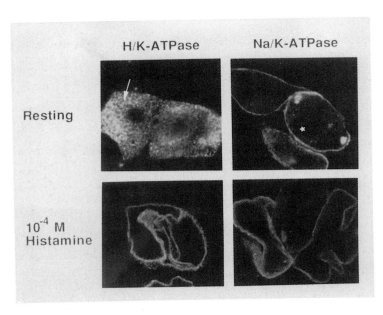

Figure 3. Fluorescent visualization of the proton pump in a vesicular compartment in the resting cell, but at the secretory surface after stimulation. The polarity marker NA/K-ATPase is maintained at the lateral surface after stimulation.

Atomic Force Microscopy

In resting glands, parietal cells were distinguished by their relatively smooth surface with irregular depressions. In contrast, the surface of the chief cells was covered with microvilli. When isolated superfused gastric glands were exposed to 10^{-4} M histamine we were able to demonstrate a change in the apical morphometry of the rabbit parietal cell. It was evident that the interior of the caniliculus projected toward the surface of the cell and, in some cases, was visible reaching the surface of the apical membrane. The diameter of the orifice did not change, however. Continued observation of these cells following removal of the histamine enabled us to note that the morphometry returned to pre-stimulated values after 10 minutes.

Figure 4. Apical Surface of a Parietal cell using atomic force microscopy.

DISCUSSION

The cells of the gastric mucosa must withstand the unusually caustic milieu of the gastric juice. It is, therefore, understandable that these cells have developed unique strategies in their apical surfaces for maintenance of their intracellular environment. Our data indicate that, although the basolateral membrane of the parietal cell has a normal permeability to either NH_3 or CO_2, the apical surface is impermeant to these substances. The observation that the apical surface is impermeant to CO_2 is not only novel, but rather surprising as CO_2 is far less polar than NH_3, which is only slightly less polar than H_2O. The basis for this novel impermeability property is unknown. It is generally believed that both NH_3 and CO_2 traverse the membrane via lipphilic pathways (Golchini et al,Gutknecht et al). An alternative explanation is that the apical surface of the membrane is coated with a molecule that prevents permeation of all such substances. As the apical membrane is constantly exposed to both acid and proteolytic enzyme pepsin, it is not difficult to imagine that a barrier has evolved to protect the cell.

The highly ordered polarization of parietal cell domains is further emphasized in the examination of the structural morphology. Confocal fluorescence microscopy demonstrates the order separation of a basolaterally oriented tubulin-based structural domain, as well as an apically-oriented F-actin-rich secretory zone surrounding the secretory canalicular target membranes. The location of actin microfilaments suggests that they play an important role in vesicular fusion and recycling. Thus, while cytochalasin B can inhibit AP uptake, colchicine treatment had little or no effect on acid secretion in the isolated cells. Thus parietal cells appear to define their functional domains through separation of cytoskeletal domains.

The cytoskeletal domains serve to create the strict polarity required for an epithelial cell such as the parietal cell. With the atomic force microscope, we have been able to demonstrate the morphometric changes that occur during the onset acid secretion and their reversibility after withdrawal of secretagogue. In addition, these studies demonstrate the likely functional separation of the true apical surface from the intracellular canalicular membrane. It is tempting to speculate that the true apical surface may be responsible for the NH_3 and CO_2 impermeability of parietal cells. The combination of the techniques described in this report may lead to greater insights into division and utilization of parietal cell functional and structural domains.

REFERENCES

Chew, C.S., Ljungstrom, M., Smolka, A., Brown,M.R. (1989) Primary culture of secretagogue responses of parietal cells from rabbit gastric mucosa. Am.J. Physiol. 256:G254-263

Geibel, J., Giebeisch, G. Boron,W.F. (1989) Effects of acetate on luminal acidification processes in the S3 segment of the rabbit proximal tubule. Am. J. Physiol.257:F586-F594

Golchini, K. Kurtz, I. (1988) NH_4Cl permeabilites of the rabbit thick ascending limb perfused
in vitro. Am J. Physiol. 255:F135-141

Gutknecht, J., Walter, A. (1981) Transport of protons and hydrochloric acid through lipid bilayer membranes. Biochem. Biophys. Acta. 641:183-188

Radmacher, M., Tillmann, R.W., Fritz, M., Gaub, H.E. (1992) From molecules to cells:imaging soft samples with the atomic force microscope. Science 257:1900-1905

Activation of calpain in gastric parietal cells

Xuebiao Yao and J.G. Forte
Department of Molecular and Cell Biology
University of California,
Berkeley, CA 94720, U.S.A.

Introduction

Ezrin, an 80 kDa protein originally found in chicken intestinal microvilli, is localized to microvilli and other plasma membrane structures in a variety of cell types (Bretscher 1991). In A431 cells, ezrin is constitutively phosphorylated on serine and threonine, and EGF treatment leads to an increase in tyrosine and threonine phosphorylation along with EGF-induced membrane ruffling (Gould et al. 1986; Bretscher 1989). Ezrin is also a prominent cytoskeletal protein in gastric parietal cells, where it has been associated with remodeling of apical cell membrane that occurs with cAMP-dependent protein kinase stimulation (Hanzel et al. 1991).

From deduced amino acid sequence a family of ezrin-related proteins has been defined, including talin and erythrocyte band 4.1, that are involved in interactions between plasma membrane and cytoskeleton (Rees et al. 1990). These three proteins share a high degree of identity within a 200-300 amino acid stretch at their N-termini. The binding site of band 4.1 to glycophorin has been mapped to this homologous region (Anderson & Marchesi, 1984). Talin has been shown to bind to $\alpha_5\beta_1$ integrin with low affinity, but its specific binding domain is not known (Horwitz et al. 1986). No specific ezrin-associated proteins have been reported, although ezrin has been cytolocalized to regions of F-actin filaments and plasma membranes. Recently reported cDNA sequences for two additional proteins, moesin and radixin, show a respective 73 and 75% identity to the entire ezrin sequence (Lankes & Furthmayr 1991; Funayama et al. 1991). There is an even higher identity (~85%) among the first 300 N-terminal amino acids; just beyond this region all three proteins contain an α-helical domain followed by a highly charged region. Radixin is reported as an actin-barbed-end capping protein, localized to cell-to-cell adherents junctions (Tsukita et al. 1989). More recently, Franck et al. (1993) showed similar subcellular localization of ezrin and moesin in A431 cells, though moesin expression was more variable.

Calpain I, a micromolar Ca^{2+}-dependent protease, has been identified in a variety of tissues and has been implicated in many cellular functions

NATO ASI Series, Vol. H 89
Molecular and Cellular Mechanisms
of H⁺ Transport
Edited by Barry H. Hirst
© Springer-Verlag Berlin Heidelberg 1994

(Suzuki *et al.* 1987). Calpain I is a heterodimer composed of an 80 kDa catalytic subunit and a 30 kDa regulatory subunit. It is generally believed that calpain I is maintained in an inactive state by association with calpastatin, an endogenous calpain inhibitor (Murachi 1989). Ca^{2+}-dependent activation of calpain I has been shown to involve autolytic processes in the NH_2-terminal region, and although the detailed mechanism for initiating autolysis is not clear, it was suggested that the 76 kDa autolytic form possesses the hydrolytic activity of calpain (Saido *et al.*, 1993).

Recently, we demonstrated the existence of calpain I in gastric parietal cells (Yao *et al.* 1993). In the present report, we provide evidence that mobilization of intracellular Ca^{2+} ($[Ca^{2+}]_i$) triggers calpain activation, and we correlate the autolysis of calpain I with the hydrolysis of ezrin. In addition, we show the dynamic effects of elevated $[Ca^{2+}]_i$ on actin-based cytoskeletal elements in the parietal cell.

Ezrin is Hydrolyzed In Situ by Mobilization of $[Ca^{2+}]_i$

Using sonicates of gastric glands, we previously showed that activation of calpain led to the proteolysis of ezrin (Yao *et al.* 1993). To gain insight into the calpain-ezrin interaction *in situ*, we used the Ca^{2+} ionophore, ionomycin, to increase Ca^{2+} entry and activate calpain I. Figure 1A shows the rapid hydrolysis of ezrin for isolated gastric glands incubated with 5 µM ionomycin and 1.8 mM extracellular Ca^{2+}. A transient production of a 55 kDa breakdown product of ezrin can also be seen; the immunoreactive 55 kDa product was further hydrolyzed into smaller peptide fragments. Similar degradation of ezrin in gastric glands was achieved by varying external Ca^{2+} in the presence of 5 µM ionomycin (Fig. 1B), or by varying ionomycin in the presence of 1.8 mM Ca^{2+}. Thus, degradation of ezrin was a sensitive index of treatments designed to elevate $[Ca^{2+}]_i$.

Hydrolysis of Ezrin Is Correlated with Autolysis of Calpain

It has been suggested that the activation of calpain involves an autolytic processes toward its N-terminus. To test whether the hydrolysis of ezrin was correlated with the autolysis of calpain, we carried out parallel immunoblot analysis using antibodies against ezrin and calpain I, $3C_6C_3$ (a gift from Dr. Seiichi Kawashima). Figure 2A shows the typical degradation of ezrin in response to adding 5 µM ionomycin to the gland suspension, as well as the blockage ezrin degradation when the incubation included 50 µg/ml calpain inhibitor I (Boehringer Mannheim). In the parallel blot probed for calpain, shown in 2B, autolysis of the 78-80 kDa catalytic subunit of calpain coincided with the addition of ionomycin, producing the 76 kDa activated form within 2 min. Autolysis was blocked by pretreatment with calpain inhibitor I. These data are consistent with the observations of Saido *et al.* (1993), indicating that hydrolysis of endogenous calpain substrate is associated with the autolysis of calpain.

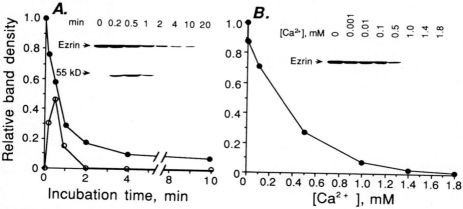

Figure 1. Proteolysis of ezrin *in situ* in ionomycin-treated gastric glands. A. Time-dependent ezrin hydrolysis. Freshly isolated gastric glands were incubated with 5 μM ionomycin and 1.8 mM extracellular Ca^{2+}. Samples were taken at intervals and immunoprobed for ezrin. In addition to the 80 kDa ezrin band that diminished with time of ionomycin-treatment, the antibody also identified the transient appearance of a 55 kDa ezrin-breakdown product. B. Ca^{2+}-dependent ezrin hydrolysis. Gastric glands were incubated with 5 μM ionomycin at the indicated [Ca^{2+}] for 25 min at 37°C. For both experiments, reactions were stopped by centrifugation and samples subjected to SDS-PAGE and immunoblotting with anti-ezrin antibody (6H11). The main graphs show data quantified by densitometry. The insets show actual immunoblots for individual experiments.

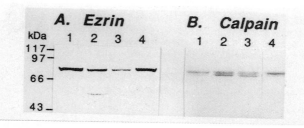

Figure 2. Correlation of calpain autolysis with the hydrolysis of ezrin. Isolated gastric glands were preincubated for 5 min at 37°C without or with 50 μg/ml calpain inhibitor I. The glands were then challenged with 5 μM ionomycin for 2 and 5 min. The reaction was stopped by centrifugation, and the pellets were solubilized and subjected to SDS-PAGE and immunoblotting. Two identical blots were probed: Blot *A* using antibody for ezrin; and Blot *B* using antibody for calpain. In both *A* and *B*, lane 1 represents glands incubated without ionomycin; lanes 2 and 3 show glands treated with ionomycin for 2 and 5 min, respectively; lane 4 shows glands pretreated with calpain inhibitor followed by 5 μM ionomycin for 5 min.

Activated Calpain is a Discriminating Protein Hydrolase

Thus far, ezrin appears to be an exquisite substrate for calpain, but we wondered whether the proteolysis might be more general for other proteins, especially cytoskeletal proteins. Since the actin-cytoskeleton has been implicated in parietal cell secretion, we tested the sensitivity of actin to activated calpain. As shown in Figure 3, a doubly stained Western blot, addition of Ca^{2+} to ionomycin-treated glands caused a Ca^{2+}-related, progressive disappearance of ezrin, while actin remained stable. Therefore, actin is relatively insensitive to activated calpain.

As a further test of the selectivity of calpain in the hydrolysis of parietal cell components, we used similar immunoblot analyses to screen a number of cytoskeletal proteins. Results for several cytoskeletal proteins are summarized in Table 1. Fodrin, like ezrin, is hydrolyzed by activated calpain, but it appears that the majority of proteins tested were stable to the rise of $[Ca^{2+}]_i$, which is consistent with observations made on Coomassie blue-stained preparations (Yao et al. 1993).

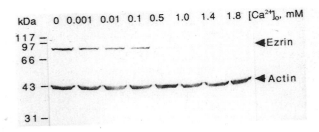

Figure 3. Actin is not hydrolyzed by activation of calpain in gastric glands. Glands were treated with 5 µM ionomycin and the indicated concentrations of external Ca^{2+} for 20 min at 37˚C. The reaction was stopped by centrifugation, and the pellets were solubilized by 2% SDS. The samples were subjected to SDS-PAGE and double immunoblotted, using anti-ezrin and anti-actin antibodies, respectively.

Table 1. Sensitivity of Gastric Cytoskeletal Proteins to Calpain

Protein	kDa	parietal cell	sensitivity to calpain
Actin	43	+	−
Ezrin	80	+	++
Myosin I	110	− (+ gland lumen)	?
Myosin II	200	+	−
α-Fodrin	240	+	++
α-Tubulin	53	+	−
Keratin	55	?	?

Table 2. Effects of ionomycin on proton accumulation, ezrin content and $[Ca^{2+}]_i$. Gastric glands were pretreated with various concentrations of ionomycin in the presence of 1.8 mM extracellular Ca^{2+} for 5 min. A. For estimate of proton accumulation, samples were either taken as resting, or stimulated with 100 μM histamine and 50 μM IBMX for measurement of the AP ratio. Separate samples of stimulated glands were taken for SDS-PAGE, immunoblot, and densitometric analysis of relative ezrin content. Values are mean ±SEM; N = 7, except for 0.5 μM ionomycin where n = 6. B. Samples were loaded with Fura-2 prior to ionomycin treatment, then assayed for $[Ca^{2+}]_i$ or for ezrin content as in A.

[Ionomycin]	A. AP ratio & ezrin content			B. $[Ca^{2+}]_i$ & ezrin content	
	AP accumulation ratio		% Ezrin	% Ezrin	$[Ca^{2+}]_i$
μM	resting	stimulated	remaining	remaining	μM
0	23.7 ±1.9	300.3 ±37.5	100	100	0.19
0.001	24.7 ±2.2	341.0 ±47.8	85.5 ±11.7	n.d.	n.d.
0.01	26.2 ±2.1	349.6 ±53.5	95.5 ±8.9	n.d.	n.d.
0.1	23.5 ±1.9	364.6 ±57.8	92.3 ±8.1	85.2	0.37 ±0.15
0.5	19.1 ±1.8	194.3 ±64.7	64.0 ±9.7	82.8	0.71 ±0.17
1.0	17.1 ±2.1	103.7 ±49.8	56.7 ±7.9	34.0 ±8.0	1.51 ±0.65
2.0	n.d.	n.d.	n.d.	22.0 ±5.0	1.74 ±0.78
5.0	15.9 ±1.4	17.0 ±1.4	10.4 ±2.9	12.0 ±4.0	5.43 ±0.88

Ezrin Content and Secretory Capacity of Parietal Cells

A role for gastric ezrin has been suggested in the cytoskeletal remodeling of the apical membrane when parietal cells are stimulated by the cAMP-dependent pathway. To explore whether there was any correlation between the activation of calpain I and the acid secretory process we monitored the proton secretory capacity of gastric glands simultaneous with the proteolysis of ezrin. Isolated gastric glands were pretreated with a wide range of ionomycin in the presence of 1.8 mM extracellular Ca^{2+}. For the experiments shown in Table 2A, the glands were challenged with either histamine plus IBMX, for stimulation, or cimetidine, for resting. Ezrin content was used to monitor calpain activation; aminopyrine (AP) accumulation was a measure of proton secretory capacity, as described by Berglindh *et al.* (1976). Table 2B shows a separate set of experiments designed to measure free $[Ca^{2+}]_i$ and ezrin content over the same range of ionomycin concentration. Viewing the data collectively, the control preparations, with no ionomycin, showed a good AP accumulation response to stimulation, and the level of $[Ca^{2+}]_i$ was sustained at about 200 nM. At low levels of ionomycin, up to 0.1 μM, there were relatively small changes. We usually noted a slight potentiation of AP accumulation for the stimulated glands, but only in the case of 0.1 μM ionomycin was the AP accumulation significantly increased over the control (P<0.05). At 0.1 μM ionomycin,

$[Ca^{2+}]_i$ was increased to about twice the control level, but there was little activation of calpain, as evidenced by the relative constancy of ezrin. As ionomycin was increased above 0.5 μM, $[Ca^{2+}]_i$ was progressively increased, the content of ezrin diminished, and the AP ratio for stimulated glands decreased. Thus, when conditions were sufficient to cause the steady state level of $[Ca^{2+}]_i$ to exceed 1 μM, there was a correlation between the parietal cell ezrin content and the capacity to accumulate protons.

Effect of increased $[Ca^{2+}]_i$ on F-actin and G-actin

Data above show that actin was not hydrolyzed under conditions where ezrin was proteolyzed by calpain. Because ezrin is co-localized with F-actin to the apical membrane of parietal cells, and may play a role in regulated membrane remodeling, we used cytostaining to test whether there was a change in the state and cellular distribution of actin when ezrin was hydrolyzed by activating calpain. Figure 4 shows cytostaining of control gastric glands and for glands treated with 5 μM ionomycin and sampled at subsequent intervals. Cells were stained for ezrin (anti-ezrin antibody), F-actin (rhodamine-phalloidin), and G-actin (FITC-conjugated DNase I).

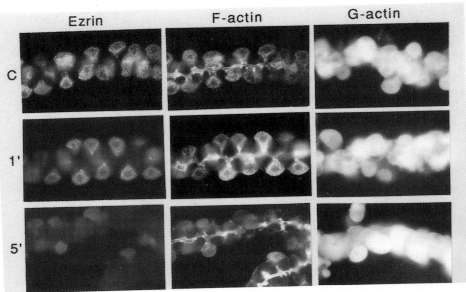

Figure. 4. Effect of elevated $[Ca^{2+}]_i$ on the subcellular localization of ezrin, F-actin and G-actin. Gastric glands were taken as control (C) or incubated with 5 μM ionomycin for 1 (1') or 5 (5') min, then fixed and permeabilized according to standard procedure. Glands were doubly stained for ezrin (visualized using FITC-conjugated secondary antibody following primary antibody, 6H11) and F-actin (rhodamine-phalloidin). Parallel preparations of glands were probed for G-actin using FITC-conjugated DNase I.

As shown in Figure 4, ezrin and F-actin were highly localized in control parietal cells, with typical colocalization to the apical membrane. On the other hand, G-actin was distributed throughout the cytoplasm. F-actin staining was also prominent along the entire glandular lumen as well as localized to the basolateral membrane. For glands treated with ionomycin for 1 min, the infrastructural staining of ezrin became more diffuse, and there was some degeneration of the F-actin signal within the parietal cell. As the incubation was prolonged to 5 min, the ezrin signal disappeared, and the localization of F-actin to parietal cell canaliculi became obscure and/or lost, although staining of F-actin along the gland lumen and at lateral cell aspects remained intense. Throughout the entire period of ionomycin treatment the intensity of G-actin staining did not decrease. In fact, photometric measurements suggested an increase in G-actin staining signal, although more definitive quantification must be carried out.

Thus it appears that gross activation of calpain and loss of ezrin produce more extensive changes in the parietal cell cytoskeleton. There is a diminution of F-actin microfilaments in the apex of parietal cells, but little change in the F-actin signal from the luminal aspects of other cells in which there is not an equivalent pool of ezrin. Throughout the sequence of increased $[Ca^{2+}]_i$ and calpain activation, the G-actin signal does not decay, but in fact appears to increase, consistent with some F- to G-conversion. Figure 5 provides a schematic representation of these changes.

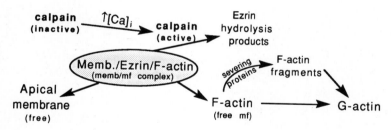

Figure 5. Proposed scheme of cytoskeletal changes when calpain is activated by $[Ca^{2+}]_i$ in the parietal cell. Microfilaments (mf) are ordinarily stabilized by membrane/ezrin/F-actin interactions; destruction of one stabilizing component leads to disruption of the complex and subsequent Ca-mediated F-actin fragmentation.

Discussion of Results

It is generally thought that two distinct intracellular systems can mediate the HCl secretory process in parietal cells: a cAMP-dependent pathway and a Ca^{2+}-dependent pathway. Our experiments suggest that a simple rise in $[Ca^{2+}]_i$ is not sufficient to trigger HCl secretion, since no increase in AP uptake occurred in non-stimulated parietal cells over a broad range of ionomycin. On the other hand, histamine-stimulated gastric glands showed a small potentiation of AP uptake when $[Ca^{2+}]_i$ was slightly elevated in the lower range of ionomycin. Higher doses of ionomycin (0.5-5 µM), that led to more grossly elevated $[Ca^{2+}]_i$, invariably diminished parietal cell capacity to respond to cAMP-dependent stimulation, and this diminution was correlated with calpain activation and ezrin degradation.

The cytoskeleton is known to control cell shape and functional processes in various cell types. Calpain-mediated degradation of cytoskeletal proteins could play a role in these phenomena. From the present work, it is clear that elevation of $[Ca^{2+}]_i$ in gastric parietal cells activates calpain I, which in turn, results in degradation of ezrin and fodrin. Our data for measurement of $[Ca^{2+}]_i$ suggest that the conditions for activating calpain used here clearly exceed the physiological range, and extend to more patho-physiological circumstances, as discussed by Yao *et al.* (1993). Whether, and to what extent, these $[Ca^{2+}]_i$/calpain changes are related to gastric mucosal injury induced mucosal barrier breakers, such as ethanol, remains to be established.

In general, the physiological relevance of calpain in cells has remained unclear despite the body of accumulated knowledge regarding structure-function interrelationships. It has been proposed that the spatial activation of calpain *in situ* may account for its physiological behavior (Saido *et al.* 1993). In parietal cells, it has been noted that carbachol and thapsigargin give a different profile of mobilizing $[Ca^{2+}]_i$ from that of ionomycin-treatment in terms of the pool of Ca^{2+}, the magnitude and sustenance of the Ca^{2+} wave, and a possible compartmentalization. It will be of great interest to determine whether Ca^{2+} mobilizing agents, such as carbachol, thapsigargin and histamine, could initiate the spatial activation of calpain in parietal cells.

The results from parietal cell cytostaining suggest a dynamic interrelationship among ezrin, F-actin and G-actin. In the early phase of ionomycin treatment there was a delocalization of the ezrin signal, followed by a complete loss as ezrin was proteolyzed. Concomitant with the changes in ezrin, there was first a diminution and then a loss of F-actin staining at the apical canalicular surface of the parietal cell. Since calpain-induced proteolysis did not hydrolyze actin, we propose that the observed dynamic changes of F-actin signal may be attributed to the activation of an F-actin severing protein by elevated $[Ca^{2+}]_i$ and/or the loss of actin-based stabilizing proteins, such as ezrin and fodrin (Fig. 5). The maintenance of the G-actin staining signal throughout the treatment is consistent with this explanation.

In summary, calpain activation in gastric parietal cells can be triggered by a rise of $[Ca^{2+}]_i$, and ezrin is an exquisite substrate for activated calpain. Ezrin has been shown to belong to a growing family of membrane-cytoskeletal-associated proteins, including erythrocyte band 4.1, talin, moesin and radixin. The calpain-ezrin-actin interaction in this report occurred at an exaggerated level of $[Ca^{2+}]_i$; however, they suggest that a study of more subtle calpain-ezrin-actin interaction might provide further insight toward the biological relevance of ezrin and its related family of membrane-cytoskeletal linkers.

REFERENCES

Anderson RA, Marchesi VT (1985) Regulation of the association of membrane skeletal protein 4.1 with glycophorin by a polyphosphoinositide. Nature Lond 318:295-298.

Berglindh T, Helander HF, Öbrink KJ (1976) Effect of secretagogues on oxygen consumption, aminopyrine accumulation and morphology in isolated gastric glands. Acta Physiol Scand 97:401-414.

Bretscher A (1991) Microfilament structure and function in the cortical cytoskeleton. Annu Rev Cell Biol 7:337-374.

Conboy J, Kan YW, Shohet SB, Moheandas N (1986) Molecular cloning of protein 4.1, a major structural element of the human erythrocyte membrane skeleton. Proc Natl Acad Sci USA 83:9512-9516.

Croall DE, DeMartino GN (1991) Calcium-activated neutral protease (calpain) system: structure, function, and regulation. Physiol Rev 71:813-847.

Franck Z, Gary R, Bretscher A (1993) Moesin, like ezrin, colocalizes with actin in the cortical cytoskeleton in cultured cells, but its expression is more variable. J Cell Sci 105:219-231.

Funayama N, Nagafuchi A, Sato N, Tsukita S, Tsukita S (1991) Radixin is a novel member of the band 4.1 family. J Cell Biol 115:1039-1048.

Gould KL, Bretscher A, Esch FS and Hunter T (1989) cDNA cloning and sequencing of the protein-tyrosine kinase substrate, ezrin, reveals homology to band 4.1. EMBO J 8:4133-4142.

Hanzel D, Reggio H, Bretscher A, Forte JG, Mangeat P (1991) The secretion-stimulated 80 K phosphoprotein of parietal cells is ezrin, and has properties of a membrane cytoskeletal linker in the induced apical microvilli. EMBO J 10:2363-2373.

Horwitz A, Duggan E, Buck C, Beckerle M, Burridge K (1986) Interaction of plasma membrane fibronectin receptor with talin--a transmembrane linkage. Nature Lond. 320:531-533.

Lankes WT, Furthmayr H (1991) Moesin: a member of the protein 4.1-talin-ezrin family of proteins. Proc Natl Acad Sci USA. 88:8297-8301.

Murachi T (1989) Intracellular regulatory system involving calpain and calpastatin. Biochem Intern 18:263-294.

Negulescu P, Machen TE (1988) Intracellular Ca regulation during secretagogue stimulation of the parietal cell. Am J Physiol, 256(Cell Physiol 23):C130-C140.

Saido TC, Suzuki H, Yamazuki H, Tanoue K, Suzuki K (1993) In situ capture of μ-calpain activation in platelets. J Biol Chem 268:7422-7426.

Suzuki K, Imajoh S, Emori T, Kawasaki H, Minami Y, Ohno S (1987) Calcium-activated neutral protease and its endogenous inhibitor. FEBS Lett 220:271-277.

Yao XB, Thibodeau A, Forte JG (1993) Ezrin-calpain I interactions in gastric parietal cells. Am J Physiol, 265(Cell Physiol 23):C36-C46.

Stimulation of Gastric Acid Secretion Causes Actin Filament Formation and Development of the Secretory Surface in Frog Oxynticopeptic Cells

Susan J. Hagen[1]
GI Division
Brigham and Women's Hospital
75 Francis Street
Boston, MA 02115

INTRODUCTION

Gastric acid secretion is a complicated process that does not occur without dramatic structural changes in parietal cells including the formation of actin-containing microvilli within elaborate intracellular canaliculi after stimulation (2,6). The morphology of inhibited and stimulated parietal cells is well known (6). However, little is known about the structural changes that occur in parietal cells as inhibited cells transform into stimulated cells.

Frog oxynticopeptic (OP) cells, the amphibian counterpart to mammalian parietal and chief cells, secrete both acid and pepsinogen. After stimulation, frog OP cells form long actin-containing surface folds at the apical surface rather than in intracellular canaliculi like mammalian parietal cells (4,6). This apical expansion of secretory membrane and underlying actin filaments after stimulation, as well as the ease in maintaining amphibian mucosal sheets *in vitro* for many hours after isolation, allowed development of a highly controllable model in which to study the coordinated assembly of apical membrane and cytoskeletal components after stimulation. Thus, the present study was

---

[1] and Rudite Jansons, Department of Surgery, Beth Israel Hospital, 330 Brookline Avenue, Boston, MA 02215

NATO ASI Series, Vol. H 89
Molecular and Cellular Mechanisms
of H⁺ Transport
Edited by Barry H. Hirst
© Springer-Verlag Berlin Heidelberg 1994

designed (a) to evaluate morphologically the discrete structural changes that occur in OP cells after stimulation as inhibited cells transform into acid-secreting cells and (b) to determine when actin-filament formation occurs after stimulation. For this, isolated sheets of fundic mucosae were inhibited and then examined by electron microscopy and confocal laser scanning microscopy at sequential time-points after stimulation.

MATERIALS AND METHODS

Animals used for this study were maintained in accordance with the guidelines of the committee on animals at Harvard Medical School and those prepared by the Committed on Care and Use of Laboratory Animals by the National Research Council. Fundic mucosae from bullfrogs (*Rana catesbeiana*) that were caught in the wild were stripped of external muscle layers and submucosa and were mounted in Ussing-type chambers as previously described (4). Acid secretion, measured by the pH-stat method, was inhibited with 1 mM cimetidine or stimulated with 20 µM forskolin.

For electron microscopy, fundic mucosae were fixed and embedded by well-established methods (4). Thin sections cut parallel to the long axis of gastric glands were placed on formvar- and carbon-coated grids and examined with a Philips 300 electron microscope.

For confocal laser scanning microscopy, fundic mucosae were fixed, frozen, and stained with rhodamine-labeled phalloidin to visualize the distribution of F-actin as previously described (4). Sections were analyzed using a BioRad MRC 600 confocal laser scanner and computer. Computer-generated images were photographed with Kodak T-Max professional film.

RESULTS AND DISCUSSION

In our model system, careful stripping of the submucosa resulted in greater inhibition of acid secretion, faster onset and significantly higher rates of acid secretion, and remarkable structural homogeneity of inhibited or stimulated OP cells in gastric glands (4). Inhibited OP cells had short microvilli-like surface folds that projected into the gastric lumen along with expanses of membrane devoid of surface folds; tubulovesicular membranes were abundant in the supranuclear cytoplasm and mitochondria and pepsinogen granules filled the remaining cytoplasm (4). In contrast, stimulated OP cells had long surface folds. Mitochondria and a few pepsinogen granules filled the remaining cytoplasm (4). Few tubulovesicular membranes were present within the subapical cytoplasm of stimulated OP cells, consistent with numerous studies on the translocation of tubulovesicular membranes to the apical surface in oxyntic and parietal cells (6).

To evaluate morphologically the discrete structural changes that occurred after stimulation as inhibited OP cells transformed into acid secreting cells, tissues were prepared for electron microscopy 2, 4, 6, 8, 10, and 30 min after stimulation. When inhibited fundic mucosae were stimulated with forskolin, no acid secretion occurred for the first 10 min. Ten to 16 min after stimulation, acid secretion began and then increased linearly to 30 min. Acid secretion at 30 min was generally one half to one third that of fully stimulated mucosae (4).

Two min after stimulation, OP cells from inhibited and stimulated fundic mucosae were identical. However, by 4 min after stimulation (Fig. 1) the surface of OP cells was highly irregular with many profiles of membrane forming deep clefts in the apical surface. Dark flocculent material was found within the boundaries of infoldings as well as in the lumen of gastric glands. Since no pepsinogen granules were present within the basal cytoplasm, in contrast to the numerous pepsinogen granules in inhibited OP cells (4), our results suggest that pepsinogen granules were released from OP cells after stimulation. Moreover, pepsinogen granules were probably released by compound exocytosis. Images of many apparently fused secretory granules here were similar to those described by Gibert and Hersey (3) who showed that pepsinogen was released by compound exocytosis from chief cells in isolated rabbit gastric glands 2.5 to 30 min after stimulation with CCK-8 or secretin.

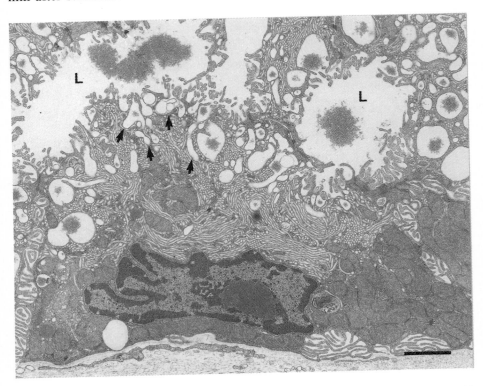

Figure 1. Electron micrograph of an oxynticopeptic cell 4 min after stimulation with forskolin. The surface was highly irregular with many profiles of membrane forming deep clefts in the apical surface (*arrows*). Dark flocculent material was found in the gastric lumen (*L*). Bar, 2 μm.

Ten min after stimulation with forskolin short surface folds were numerous along the length of apical plasma membrane (Fig. 2). Many large vesicles were also present within the subapical cytoplasm next to tubulovesicular membranes. It is not clear whether the lumen of these vesicles is continuous with the gastric lumen although one area of confluence is seen in this micrograph. It is possible that the vesicles represent membrane from pepsinogen granules that was incorporated into the apical plasma membrane after stimulation. If so, whether the pepsinogen granule membrane is required for further development of the secretory surface or is recycled, as has been suggested for mammalian chief cells (5), awaits further investigation.

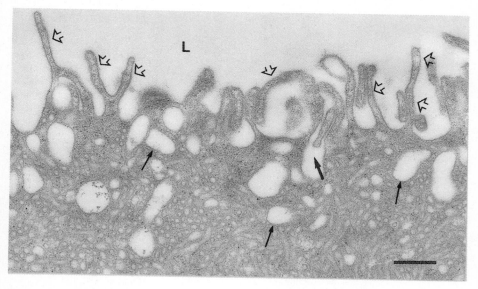

Figure 2. Electron micrograph of the apical surface and subapical cytoplasm of an oxynticopeptic cell 10 min after stimulation with forskolin. Short surface folds (*open arrows*) were prominent along the apical surface. Large vesicles (*thin arrows*) were also present within the subapical cytoplasm. One area where a vesicle is confluent with the gastric lumen is seen in this micrograph (*thick arrow*). Bar, 0.5 μm.

By 30 min after stimulation with forskolin, numerous surface folds were present along the apical plasma membrane (Fig. 3). Most notable was the presence of large and small vesicles within the apical and subapical cytoplasm that were aligned parallel to the long axis of surface folds. Many images suggest that smaller vesicles fuse to create large vesicles. It is not clear if the interior of these vesicles is continuous with the gastric lumen or represent swollen intracellular spaces.

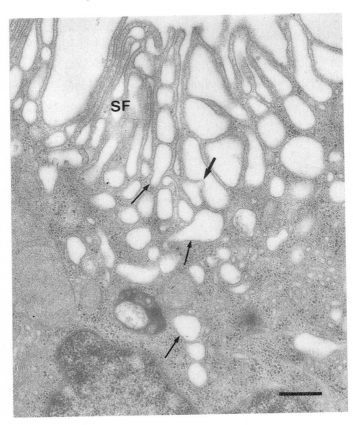

Figure 3. Electron micrograph of the apical surface and subapical cytoplasm of an oxynticopeptic cell 30 min after stimulation with forskolin. Numerous surface folds (*SF*) were present along the apical surface. Large and small vesicles (*thin arrows*) were prominent within the apical and subapical cytoplasm. Many images suggest that small vesicles fuse into larger vesicles (*thick arrow*). Bar, 0.5 μm.

To determine the rate of actin-filament formation after stimulation, frozen sections of gastric mucosae stimulated for 10 and 30 min were stained for F-actin and analyzed by confocal laser scanning microscopy.

In tissues stimulated for 10 min, fluorescence signal for F-actin formed a thin continuous band at the apical surface of OP cells (Fig. 4 *A*). This pattern of staining was different from that of inhibited cells where the fluorescence signal for F-actin was patchy with punctate areas of strong fluorescence signal interposed with areas of no fluorescence signal (4). By 30 min after stimulation, fluorescence signal for F-actin formed a wider continuous band (Fig. 4 *B*), although the width and intensity was not as great as that of fully-stimulated cells.

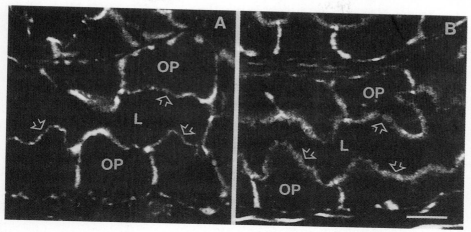

Figure 4. Confocal laser scanning micrographs of fundic mucosae stained with rhodamine-phalloidin to determine the localization of F-actin. (*A*) Fluorescence signal for F-actin formed a thin band along the entire length of apical plasma membrane 10 min after stimulation (*open arrows*). (*B*) By 30 min after stimulation, fluorescence signal for F-actin formed a wide band at the apical surface of OP cells (*open arrows*). OP, oxynticopeptic cells; L, lumen of a gastric gland. Bar, 10 μm.

Our studies demonstrate that elaboration of the secretory surface of gastric OP cells follows a defined progression of events. As early as 4 min after stimulation, pepsinogen granules are released. Next, short surface folds become more numerous at the secretory surface. By 30 min after stimulation, the secretory surface contains prominent surface folds. Large and small vesicles are also present that align parallel to the long axis of surface folds. Our results also show that actin filaments form in concert with surface folds at the secretory surface of frog OP cells. After stimulation, actin filaments increase in concentration as surface folds increase in number and length over time. We previously showed that actin filaments were not formed after stimulation from preformed filaments or by the synthesis of new actin, but that they presumably formed from a large, preexisting pool of non-filamentous actin (4). Taken together, our results suggest that actin binding proteins, which modulate the rate of actin polymerization and depolymerization, actin filament length, and the interaction of actin with components of the apical plasma membrane may be essential for development of the secretory surface and ultimately for the regulation of acid secretion in oxyntic and parietal cells.

ACKNOWLEDGEMENTS

This work was supported by National Institutes of Health grants DK #33827 and AM #15681. We also benefitted greatly from core facilities funded by the Harvard Digestive Diseases Center grant DK #34854.

REFERENCES

1. Black JA, Forte TM, Forte JG (1982) The effects of microfilament disrupting agents on HCl secretion and ultrastructure of piglet gastric oxyntic cells. Gastroenterology 100:395-402
2. Forte JG, Wolosin JM (1987) HCl Secretion by the gastric oxyntic cell. In: Johnson LR (ed) Physiology of the Gastrointestinal trace. Raven Press, New York, pp 853-863
3. Gibert AJ, Hersey SJ (1982) Exocytosis in isolated gastric glands induced by secretagogues and hyperosmolarity. Cell Tissue Res 227:535-542
4. Hagen SJ, Yanaka A, Jansons R (1993) Localization of brush border cytoskeletal proteins in gastric oxynticopeptic cells from the bullfrog, *Rana catesbeiana*. Cell Tissue Res (in press)
5. Helander HF (1978) Quantitative ultrastructural studies on rat gastric zymogen cells under different physiological and experimental conditions. Cell Tissue Res 189:287-303
6. Ito S (1987) Functional gastric morphology. In: Johnson LR (ed) Physiology of the gastrointestinal tract. Raven Press, New York, pp 817-851

Expression of rat gastric H+, K+-ATPase in insect cells using a double recombinant baculovirus

Marianne MARTIN & Paul MANGEAT
CNRS-URA 530
CCIPE, rue de la Cardonille,
34094 Montpellier Cedex 5, France

Introduction

Upon secretagogue stimulation, gastric parietal cells undergo extensive morphological changes linked to the secretion of gastric acid. Secretory membranes are located intracytoplasmically in the resting state and are inserted apically upon stimulation. The mechanism of membrane translocation involves the elongation of apical microvilli, a process dependent on actin and spectrin assembly (Mercier *et al.*, 1989b) and that might be regulated by the phosphorylation state of the cytoskeletal protein ezrin (Hanzel *et al.*, 1991). About 80% of the protein content of these membranes consists of the proton pump complex, H^+, K^+-ATPase. The gastric proton pump is composed of two subunits, an α chain responsible for the catalytic activity of the enzyme and a β chain whose function is still unclear but whose correct assembly with the α subunit is suspected to play a role in the proper structural maturation of the catalytic subunit (Geering, 1991).

In our laboratory, we are trying to reconstitute the different steps leading to the translocation of secretory membranes and we are studying the biogenesis of gastric microvilli. Here we describe the construction of a double recombinant baculovirus enabling high levels of expression of both chains of rat gastric H^+, K^+-ATPase in insect cells.

Methods

Plasmid construction

In order to reconstitute full-length cDNAs coding for the H^+, K^+-ATPase α and β chains with appropriate 5' and 3' restriction sites for cloning respectively in the baculovirus transfer vectors

NATO ASI Series, Vol. H 89
Molecular and Cellular Mechanisms
of H+ Transport
Edited by Barry H. Hirst
© Springer-Verlag Berlin Heidelberg 1994

pGmAc34T and pGm16 (see below), different fragments of these cDNAs were subcloned in derivatives of pEMBL18⁺ plasmid (Dente *et al.*, 1983). For H⁺, K⁺-ATPase α chain (αHK), we used pEMBL18⁺Bg in which the EcoRI site of the polylinker was filled in with Klenow enzyme and a BglII linker was inserted in the SphI polylinker site. For H⁺, K⁺-ATPase β chain (βHK), pEMBL18⁺Bg was further modified by the insertion of an EcoRI linker at the HincII polylinker site, thus generating the pEMBL18⁺BEL plasmid.

<u>Construction of the pGmAc34T-αHK baculovirus transfer vector</u>

Plasmid pRHKA 1-8 (Shull & Lingrel, 1986) which contained the 5' end and the NH2-terminal coding sequence of rat αHK cDNA was digested by NcoI. The proximal 986 bp NcoI fragment was gel-purified, treated by the Klenow enzyme and cloned into the HincII site of the pEMBL18⁺Bg , in the reverse orientation relatively to LacZ. This gave plasmid p-αHK-N. Plasmid pRHKA 3-3 (Shull & Lingrel, 1986), containing all but the six first codons of αHK cDNA, was digested by ScaI, treated by T4 DNA polymerase and a KpnI(Asp718) linker was added at the 3' end. This plasmid was then digested by EcoRI and Asp718 and the resulting 2.62 kb fragment was gel-purified and cloned in the corresponding sites of p-αHK-N. This led to plasmid p-αHK consisting of full-length αHK cDNA .

To construct a baculovirus transfer vector containing αHK coding sequence under the control of the polyhedrin promoter, p-αHK was digested by HindIII, filled in with Klenow, then digested by Asp718. The resulting 3.26 kb fragment was gel-isolated and cloned in the SmaI-Asp718 sites of the pGmAc34T vector, leading to pGmAc34T-αHK construct. In the pGmAc34T vector, the polyhedrin start codon is mutated (ATT instead of ATG) and a cloning SmaI site is introduced at a deletion between nucleotide + 45 and +462. This vector allows the production of high levels of non-fused recombinant protein (Royer *et al.*, 1992).

Recombinant baculoviruses (Ac34T-αHK) expressing αHK were obtained by homologous recombination between pGmAc34T-αHK and wild type viral DNA after cotransfection in insect cells (see below).

<u>Construction of the pGm16-βHK baculovirus transfer vector</u>

Plasmid pBHKA RS25-3 (Shull, 1990) was partially digested by SphI, treated by T4 DNA polymerase and a BamHI linker was added at the 3' end of rat βHK cDNA. The plasmid was digested by EcoRI-BamHI. The 0.96 kb fragment missing the 16 first codons was gel-purified and cloned in the corresponding sites of pEMBL18⁺BEL, leading to plasmid p-βHK-C. Plasmid p-βHK containing full-length βHK cDNA was obtained by inserting a set of overlapping complementary oligonucleotides by shot gun ligation (Grunström *et al.*, 1985) in the BglII - EcoRI restriction sites of p-βHK-C. Oligonucleotides were used as followed :

```
           +1     91A1    /     91A2      /     91A3       EcoRI
    5'  /GATCTATGGCAGCCCTGC/AGGAGAAGAA GTCATG/CAGCCAGCGCATGGCCG/   3'
    3'      /ATACCGTCGGGACG TCCTCTTCTT/CAGTAC GTCGGTCGCGTACCGGCTTAA/5'
        BglII             91A4      /               91A5
```

Oligonucleotide 91A1 created a BglII restriction site upstream of the ATG start codon of βHK cDNA.

To construct a baculovirus transfer vector containing βHK coding sequence under the control of the p10 promoter, p-βHK was digested by BglII and BamHI. The resulting 1.0 kb fragment was gel-isolated and cloned in the BglII site of the pGm16 transfer vector (Blanc *et al.*, 1993), leading to pGm16-βHK construct. In the pGm16 vector, the p10 start codon is mutated and a cloning BglII site is introduced at a deletion between nucleotide +16 and +265.

Recombinant baculoviruses (SLP10-βHK) expressing βHK were obtained by homologous recombination between pGm16-βHK and AcSLP10 viral DNA after cotransfection in insect cells. AcSLP10, a modified virus deleted for the polyhedrin gene, expresses polyhedrin under the control of p10 promoter (Chaabihi *et al.*, 1993), thus allowing selection of recombinant baculoviruses according to the standard procedure described by Summers & Smith (1987).

Insect cell culture, virus obtention and infection

Wild type (AcMNPV clone 1.2, Croizier *et al.*, 1988) and recombinant baculoviruses were propagated in Spodoptera frugiperda (Sf9) cells (ATCC CRL 1711), grown at 28°C in TC100 medium (GIBCO BRL, Cergy Pontoise, France) supplemented with 10% heat-inactivated FCS, antibiotics and components from Grace medium (except yeastolate) that are missing in the TC100 medium.

The different transfer vectors were cotransfected with the appropriate viral DNA preparation (see above) onto Sf9 cells using the transfection-reagent DOTAP (Boehringer Mannheim, Meylan, France). Briefly, plasmid DNA (7µg) and viral DNA (< 1µg) were mixed in 1.5 ml of TC100 medium without serum and left at room temperature for 45 min. 40 µl of DOTAP were added just before use to a second tube containing the same volume of medium. Sf9 cells (3.10^6cells/25 cm^2 flasks) plated one hour before in serum-free medium, were rinsed with the same medium. DNAs and DOTAP were mixed and the mixture added onto the cells. 4h later, fresh complete medium was added, and cells were left 4d at 28°C. Recombinant viruses were then purified by plaque assays as described (Summers & Smith, 1987). For expression experiments, cells, plated at a density of 3.10^6 cells/25 cm^2 flask, were infected 1h later, with viral stocks at a multiplicity of infection of 10.

Results and discussion

Production of a double recombinant baculovirus expressing H^+K^+-ATPase α and β chains

Coinfection of cells by two different viruses theoritically enables expression of two different proteins in the same cell. However, one cannot exclude that some cells are infected with only one of the two viruses. Consequently the level of expression of the two proteins in the same cell is difficult to control. To circumvent this problem, one solution is to contruct a double recombinant virus.

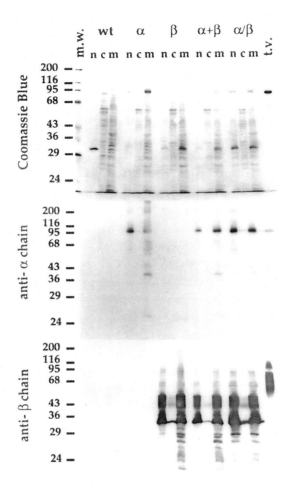

Figure 1: Fractionation of Sf9 cells infected for 40h with either wild type virus (wt), or recombinant virus Ac34T-αHK (α), or SLP10-βHK (β), or infected with both viruses (α+β) or with double recombinant virus Ac34T-αHK/p10-βHK (α/β). Cells were lysed by sonication in 20 mM sodium phosphate pH 7.0, 10 mM NaCl, 0.5 mM PMSF, 0.1% β-mercaptoethanol, 1 mM EGTA, 1 mM MgCl2. Nuclei were pelleted by centrifugation for 5 min at 200 g. Supernatant was collected. The same volume of lysis buffer was added on the nuclear (n) pellets which were treated as above. Pooled supernatants were centrifuged for 1h at 100,000g leading to cytosolic fraction (c) and membrane fraction (m). The fractions were boiled in Laemmli's denaturation buffer and the equivalent of 1.5 x 10^5 cells was analyzed on 12.5% SDS-PAGE containing 0.1% bis-acrylamide. Western blots were performed as described earlier (Mercier et al., 1989a), using either $[^{125}I]$-protein A and autoradiography (anti-αHK, rabbit HK9 antibody, Gottardi and Caplan, 1993) or peroxydase-conjugated sheep anti-mouse IgG (anti-βHK, mouse monoclonal antibody 2G11). m.w. = mol. weigth standards; t.v. = purified rabbit H,K-ATPase-containing tubulovesicules.

Production of such a virus expressing simultaneously αHK and βHK under the control of the polyhedrin and the p10 promoter respectively, required two rounds of cotransfection/selection. First, viral DNA from Ac34T-αHK baculovirus was prepared according to standard procedures (Summers & Smith, 1987) and used to co-transfect insect cells with pGm8022P plasmid, a transfer vector possessing the polyhedrin coding sequence under the control of the p10 promoter (Chaabihi *et al.*, 1993). Thus, homologous recombination at the p10 locus allowed the selection of recombinant baculovirus Ac34T-αHK / p10-Pol which expressed both αHK and polyhedrin and consequently produced polyhedra (OB+ phenotype). Viral DNA from Ac34T-αHK / p10-Pol was prepared and used in a cotransfection experiment with the pGm16-βHK plasmid. The double recombinant Ac34T-αHK / p10-βHK was then screened by the common OB- phenotype.

Expression of rat gastric αHK and βHK in insect cells

The subcellular localization of αHK and/or βHK expressed in Sf9 cells infected with viruses expressing either αHK or βHK or infected with both viruses (α+βHK) or with a double recombinant baculovirus (α/βHK) was analyzed by two complementary procedures. Nuclear, membrane and cytosolic fractions were prepared, separated on SDS-PAGE and western blotted with specific rabbit anti-αHK or mouse anti-βHK antibodies (figure 1). In addition, cells grown and infected on glass coverslips were formaldehyde-fixed, detergent-permeabilized (this step was omitted when indicated) and processed for double labelling immunocytochemistry (figure 2).

Biochemically (figure 1), a similar level of expression of αHK and of βHK was observed when cells were infected with each respective virus or in combination (α+β) or when a double recombinant virus was used (α/β). Recombinant αHK was expressed at a mol. weight similar to the native chain (95 kDa) and fractionated in particulate fractions (nuclear and membrane fractions). Recombinant βHK partionned also as a membrane protein but was expressed under different mol. weigth forms which did not co-migrated with native βHK. Instead of a mature and highly glycosylated form (60-80 kDa), recombinant βHK migrated as unglycosylated protein as well as an underglycosylated moiety. This underglycosylated state of foreign proteins produced in baculovirus-infected insect cells is well documented (Luckow & Summers, 1988; Maeda, 1989). Recombinant glycoproteins with a high mannose oligosaccharide content are produced but conversion to complex N-linked oligosaccharides does not occur.

This explains why only the 32 kDa unglycosylated and the 43-51 kDa core-glycosylated forms of βHK were obtained.

Expression of rat gastric H,K-ATPase α and β chains in Sf9 cells

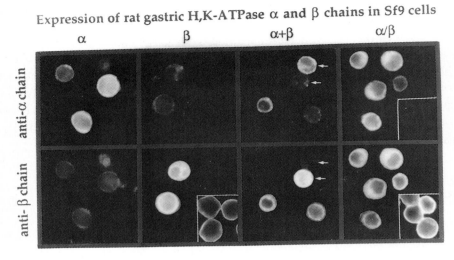

Figure 2: Expression of αHK and βHK in Sf9 cells. Cells plated on glass coverslips and infected with recombinant virus expressing αHK (α), βHK (β), or with both viruses (α+β) or with double recombinant virus (α/β). Cells were fixed 24h post infection with 3.7% formaldehyde in PBS for 30 min, permeabilized with 0.2% Triton X-100, and processed for indirect immunofluorescence microscopy. Double labelling was performed using rabbit anti-αHK HK9 antibody (Gottardi and Caplan, 1993) and mouse monoclonal anti-βHK 146.14 antibody (Mercier et al., 1989a) followed by fluorescein-conjugated goat anti-mouse IgG and rhodamine-conjugated goat anti-rabbit IgG. Note that in cells coinfected with two viruses expressing αHK and βHK (α+β), some cells express both chains, others express only one chain (arrows). In contrast, as expected, all cells infected with the double recombinant virus (α/β) expressed both chains. In inserts non permeabilized cells showing cell surface expression of βHK.

Analysis at the light microscopy level indicated that αHK remained intracellularly expressed in perinuclear structures, presumably in the endoplasmic reticulum, whatever the type of infection used (αHK alone or in combination with βHK). A high proportion of βHK also remained associated with perinuclear structures, but a significant amount was inserted at the plasma membrane as demonstrated by the cell surface staining observed in non permeabilized cells. The amount of βHK expressed at the cell surface did not vary whatever the type of infection (βHK alone or in combination with αHK). This transport competence of the unassembled βHK-subunit was also observed by Horisberger *et al.* (1991) and Gottardi & Caplan (1993).

The fluorescence studies demonstrated that the construction of the double recombinant virus was required to strictly control the coordinate synthesis of both chains of the gastric proton pump in single cells, since when expression was triggered by coinfection with two viruses, a heterogeneous situation occurred where a differential level of expression of both chains was achieved in individual cells. Nevertheless, the coordinate expression of both chains obtained in insect cells infected with the double recombinant baculovirus did not overcome the retention of most newly synthesized proteins in the endoplasmic reticulum.

From these studies one can conclude that, although a coordinate synthesis of αHK and βHK was possible, a correct assembly of the rat gastric proton pump at the plasma membrane of insect cells was not achieved. This failure might be due to a defect in the glycosylation pattern of βHK. Glycosylation of βHK appears therefore important for the proper maturation and intracellular transport of αHK but not of βHK.

Acknowledgements

The authors are indebted to Dr. G. E. Shull for kindly providing αHK and βHK cDNAs, Drs M.J. Caplan and J.G. Forte for antibodies, Drs. G. Devauchelle and M. Cerutti for Sf9 cells, wild type and SLP10 viruses and baculovirus transfer vectors. This work was supported in part by a grant from l'Association pour la Recherche sur le Cancer (n° 6844 to P.M.)

References

Blanc S, Cerutti M, Chaabihi H, Louis C, Devauchelle G and Hull R (1993) Gene II product of an aphid-nontransmissible isolate of Cauliflower mosaic virus expressed in a baculovirus system possesses aphid transmission factor activity. Virology 192 : 651-654

Chaabihi H, Ogliastro MH, Martin M, Giraud C, Devauchelle G and Cerutti M (1993) Competition between baculovirus polyhedrin and p10 gene expression during infection of insect cells. J. Virol. 67 : 2664-2671

Croizier G, Croizier L, Quiot JM and Lereclus D (1988) Recombination of Autographa californica and Rachiplusia ou nuclear polyhedrosis viruses in Galleria mellonella L. J. Gen. Virol. 69 : 177-185

Dente L, Cesareni G and Cortese R (1983) pEMBL : a new family of single stranded plasmids. Nucl. Acids Res. 11 : 1645-1655

Geering K (1991) The functionnal role of the ß-subunit in the maturation and intracellular transport of Na, K-ATPase. FEBS Lett. 285 : 189-193

Gottardi CJ and Caplan MJ (1993) An ion transporting ATPase encodes multiple apical localization signals. J. Cell Biol. 121 : 283-293

Grunström T, Zenke WM, Wintzerith M, Matthes HWD, Staub A and Chambon P (1985) Oligonucleotide-directed mutagenesis by microscale 'shot-gun' gene synthesis. Nucl. Acids Res. 13 : 3305-3316

Hanzel D, Reggio H, Bretscher A, Forte JG and Mangeat P (1991) The secretion-stimulated 80k phosphoprotein of parietal cells is ezrin, and has properties of a membrane cytoskeletal linker in the induced apical microvilli. EMBO J. 10 : 2363-2373

Horisberger JD, Jaunin P, Reuben MA, Lasater LS, Chow DC, Forte JG, Sachs G, Rossier BC and Geering K (1991) The H,K-ATPase ß-subunit can act as a surrogate for the ß-subunit of Na,K-pumps. J. Biol. Chem. 266: 19131-19134

Luckow VA and Summers MD (1988) Trends in the development of baculovirus expression vectors. Biotechnology 6 : 47-55

Maeda S (1989) Expression of foreign genes in insects using baculovirus vectors. Ann. Rev. Entomol. 34: 351-372

Mercier F, Reggio H, Devilliers G, Bataille D and Mangeat P (1989a) A marker of acid-secreting membrane movement in rat gastric parietal cells. Biol. Cell 65 : 7-20

Mercier F, Reggio H, Devilliers G, Bataille D and Mangeat P (1989b) Membrane-cytoskeleton dynamics in rat parietal cells: mobilization of actin and spectrin upon stimulation of gastric secretion. J. Cell Biol. 108 : 441-453

Royer M, Cerutti M, Gay B, Hong SS, Devauchelle G and Boulanger P (1992) Expression and extracellular release of human immunodeficiency virus type 1 gag precursors by recombinant baculovirus-infected cells. J. Virol. 66 : 3230-3235

Shull GE (1990) cDNA cloning of the ß-subunit of the rat gastric H, K-ATPase. J. Biol. Chem. 265 : 12123-12126.

Shull GE and Lingrel JB (1986) Molecular cloning of the rat stomach ($H^+ + K^+$)-ATPase. J. Biol. Chem. 261 : 16788-16791

Summers MD and Smith GE (1987) A manual of methods for baculovirus vectors and insect cells culture procedures. Texas Agricultural Experiment Station Bulletin no.1555.

Influence of F-Actin on Na+/K+/2Cl- Cotransport in Cultured Intestinal Epithelia

Jeffrey B. Matthews, M.D.
Department of Surgery, Beth Israel Hospital
Harvard Medical School
330 Brookline Ave.
Boston MA 02215
USA

It is increasingly recognized that the microfilament-based cytoskeleton of many non-muscle cell types plays a crucial role in diverse cellular functions including locomotion and endo/exocytosis (Condeelis and Hall, 1991). The role of microfilament dynamics in specialized functions of epithelial cells (such as vectorial ion transport) has been less well-studied. The potential importance of the cytoskeleton in epithelial ion transport processes is illustrated by the gastric parietal cell, in which the onset of H+ secretion involves a microfilament-dependent fusion of subapical membrane vesicles containing H+/K+-ATPase units with the apical membrane (Mercier et al., 1989).

In addition to the dramatic microfilament-dependent membrane fusion events observed in the parietal cell, the actin-based cytoskeleton may serve subtler functions in epithelial ion transport. Recently, a number of ion transport proteins have been found to be closely associated with the actin-based cytoskeleton. For example, the alpha subunit of the renal Na+/K+-ATPase has been found to bind ankyrin, thereby forming a metabolically stable complex with actin; this association is likely to be important in the maintenance of cell polarity (Nelson and Hammerton, 1989; Molitoris et al., 1991) and may also influence pump activity (Cantiello and Bertorello, 1991). In general, however, whether dynamic alterations in F-actin can specifically alter the function of associated transport proteins remains largely unstudied.

Electrogenic Cl- secretion is the fundamental means by which many epithelial surfaces are hydrated. The cultured human intestinal cell line T84, which phenotypically and functionally resembles the intestinal crypt cell, serves as a model for electrogenic Cl- secretion (Dharmsathaphorn et al., 1984). Using the T84 model, we reported that cAMP-elicited electrogenic Cl- secretion was paralleled by a rearrangement of F-actin confined to the basal aspect of the cell and demonstrated that prevention of this cytoskeletal remodelling with the microfilament stabilizer phalloidin largely attenuated the Cl secretory response as well (Shapiro et al., 1991).

NATO ASI Series, Vol. H 89
Molecular and Cellular Mechanisms
of H+ Transport
Edited by Barry H. Hirst
© Springer-Verlag Berlin Heidelberg 1994

Our data indicate that the regulation of the diuretic-sensitive $Na^+/K^+/2Cl^-$ cotransporter, which is the integral component of the basolateral membrane Cl^- entry mechanism of Cl^- secreting epithelia, may involve cAMP-elicited F-actin reorganization (Matthews et al., 1992; Matthews et al., in press).

METHODS

T84 cells seeded onto collagen-coated permeable supports were used 7-14 days after plating. Transepithelial transport studies were performed as previously described (Matthews et al., 1992) or using monolayers mounted in low-turbulence modified Ussing chambers. Monolayers were bathed either in a Hepes-phosphate buffered Ringer's solution (HPBR) or a bicarbonate-buffered Ringer's solution gassed with 95% O_2-5% CO_2, pH 7.5. Potential difference was measured by voltage-current clamp; transepithelial resistance (R) and short-circuit current (I_{sc}) were calculated from the voltage response to 25 μA externally applied current. Subsets of monolayers were loaded with the F-actin stabilizer phalloidin by overnight incubation in media containing 33 μM phalloidin (Matthews et al., 1992). This technique does not result in discernable monolayer toxicity (Shapiro et al., 1991).

Electrogenic Cl^- secretion involves the coordinated activities of four distinct transport pathways: 1) the Na^+/K^+-ATPase, which provides a transmembrane Na^+ gradient, 2) the $Na^+/K^+/2Cl^-$ cotransporter, which accumulates Cl^- intracellularly, 3) K^+ channels, which allow K^+ recycling, and 4) apical Cl^- channels, which form the ultimate route for electrogenic Cl^- exit. These specific pathways were studied as described previously (Matthews et al., 1992; Matthews et al., in press). Briefly, apical and basolateral Cl^- and K^+ channels were assessed by the rate constant of efflux of ^{125}I and ^{86}Rb, respectively. The ion-pumping activity of the Na^+/K^+-ATPase was assessed as the Na^+ absorptive current generated in response to apical membrane nystatin-permeabilization using monolayers bathed in nominally Cl^- free (gluconate) solutions. The $Na^+/K^+/2Cl^-$ cotransporter was defined as bumetanide-sensitive ^{86}Rb uptake, 85% of which is ouabain-resistant. Briefly, monolayers were incubated in HPBR with or without ouabain (0.5 mM), bumetanide (10 μM), or agonists for 10 minutes. Monolayers were then moved to wells containing identical uptake buffers plus 1-1.5 μCi-ml^{-1} ^{86}Rb. Uptakes were terminated by rapid washes in ice-cold 100 mM $MgCl_2$-Tris buffer. Filters were excised and radioactivity was counted.

RESULTS AND DISCUSSION

T84 cell monolayers loaded with phalloidin show a markedly attenuated Cl⁻ secretory response to cAMP and cGMP-mediated agonists and to Ca^{+2} ionophore. Interestingly, the receptor-mediated agonists carbachol and histamine, which elicit transient rather than sustained I_{sc} responses in T84 cells, do not appear to be affected by phalloidin (Figure 1).

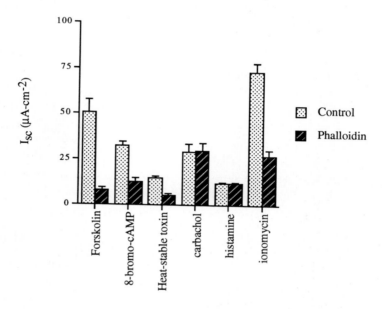

Figure 1. Maximal I_{sc} response to various secretory agonists in control and phalloidin-loaded T84 monolayers in HPBR. The peak response to the Ca^{+2} agonists carbachol and histamine is unaffected by phalloidin-loading; these agonists elicit only transient (*i.e.*, less than 10 min) secretory responses in T84 cells in contrast to the other agonists. Data are mean ± SEM for n=10-22 monolayers.

To confirm that these results were not limited to Hepes buffer conditions, experiments were performed in a bicarbonate-CO_2 system. As shown in Figure 2, T84 monolayers loaded with phalloidin showed a markedly attenuated I_{sc} response to the cAMP agonist forskolin compared to controls. The absence of phalloidin-induced monolayer toxicity is supported by several findings evident in this figure: first, monolayer R was not affected by phalloidin; second, in response to forskolin, transepithelial R rapidly decreased to the same extent in both control and loaded monolayers, indicating that signal transduction was unimpaired; and, third, the I_{sc} response to subsequent addition of carbachol was preserved, indicating that the Cl⁻ secretory apparatus remained intact.

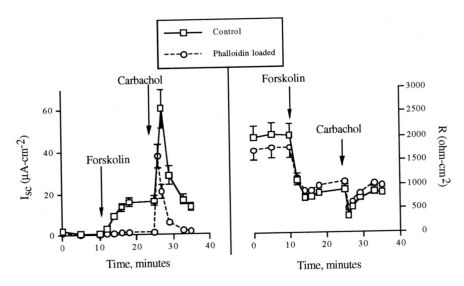

Figure 2. Short-circuit current (I_{sc}, left) and resistance (R, right) responses to forskolin (10 μM) followed by carbachol (0.1 mM) in control (*open squares*) and phalloidin-loaded (*open circles*) T84 monolayers bathed in Ringer's-CO_2 buffer. Agonists added at arrows. The I_{sc} response to forskolin is abolished in phalloidin-loaded monolayers, but the response to the Ca^{+2} agonist carbachol is preserved. Data are mean ± SEM for n=3, each group.

Because morphologic evaluation of cAMP- and cGMP-stimulated T84 monolayers indicated that *apical* and *perijunctional* F-actin does not undergo significant remodelling (Shapiro *et al.*, 1991; Matthews *et al.*, in press), the inhibitory effect of phalloidin on transepithelial Cl⁻ secretion would not be expected to involve impaired activation of *apical* membrane Cl⁻ channels. Furthermore, the observation that the forskolin-induced decrease in transepithelial R is preserved in phalloidin-loaded monolayers (Figure 2) also argues against conductive (channel) pathways being the major site affected by cytoskeletal stabilization. Finally, the maximal rate constant of efflux of ^{125}I evoked by cAMP was no different between control and loaded monolayers (Matthews *et al.*, 1992; Matthews *et al.*, in press).

F-actin in the *basal* pole of T84 cells, visualized by fluorescent microscopy using rhodamine-phalloidin, undergoes striking remodelling during stimulation of monolayers with cAMP or cGMP (Shapiro *et al.*, 1991; Matthews *et al.*, in press). Under unstimulated conditions, F-actin in this region of the cell appears as a fine, randomly dispersed cortical meshwork of microfilaments; after stimulation, F-actin is seen to be displaced toward the periphery of the cells and appears as thick bundles surrounding a central cleared zone. Because the basolateral Na^+/K^+-ATPase has been shown previously to be associated with actin, we investigated the possibility that pump activity was inhibited under phalloidin-loaded conditions. However, we found that the ability of phalloidin-loaded monolayers to transport Na^+ in an absorptive direction in Cl^- free buffer was no different than controls either under unstimulated or forskolin-stimulated conditions (Shapiro *et al.*, 1991; Matthews *et al.*, 1992; Matthews *et al.*, in press). In addition, ouabain-sensitive, bumetanide-resistant ^{86}Rb uptake was only minimally decreased. Similarly, the basolateral K^+ conductance, as assessed by ^{86}Rb efflux, was not affected by phalloidin loading

In current models of electrogenic Cl^- secretion, apical membrane Cl^- channels are viewed as the primary site of regulation. In order to maintain cellular composition, however, the rate of basolateral Cl^- entry must increase to match the augmented rate of Cl^- exit. Indeed, we have shown that both cAMP and cGMP increase basolateral $Na^+/K^+/2Cl^-$ cotransporter activity, as assessed by bumetanide-sensitive, ouabain-resistant ^{86}Rb uptake (Matthews *et al.*, 1992; Matthews *et al.*, in press). Despite the apparently normal function of Cl^- and K^+ exit pathways and the Na^+/K^+-ATPase observed in phalloidin-loaded monolayers, activation of the basolateral Cl^- entry mechanism -- the $Na^+/K^+/2Cl^-$ cotransporter-- was found to be markedly attentuated (Matthews *et al.*, 1992; Matthews *et al.*, in press) (Figure 3). This reduction in cotransporter activity was sufficient to account for the inhibition of the net Cl^- secretory response observed in phalloidin-loaded monolayers.

In avian erythrocytes and in shark rectal gland, activation of $Na^+/K^+/2Cl^-$ cotransport appears to involve phosphorylation of the cotransporter protein and an increase in the number of functional cotransporter units in the plasma membrane, as determined by specific binding of ^{3}H-bumetanide (Pewitt *et al.*, 1990; Lytle and Forbush, 1992). In preliminary experiments with T84 monolayers, we have thus far been unable to demonstrate cAMP-mediated increases in bumetanide binding, although such regulation may have been obscured by high levels of non-specific binding encountered in the T84 system.

Thus, it is intriguing to speculate that stimulation of Cl^- secretion by cAMP leads not only to activation of cotransporters already present in the basolateral membrane but also to the recruitment of new cotransporter units, possibly from cytoplasmic pools, and that such recruitment involves microfilament-dependent insertion. A similar microfilament-dependent insertion event has been observed in

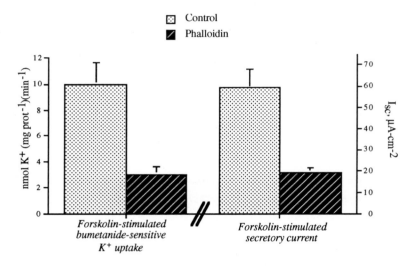

Figure 3. Cyclic AMP-mediated stimulation of $Na^+/K^+/2Cl^-$ cotransporter activity (bumetanide-sensitive ^{86}Rb uptake) is inhibited in T84 monolayers loaded with phalloidin (left-side of figure). The percentage reduction is similar to the inhibition of the overall secretory response (I_{sc}, right side of figure). From (Matthews *et al.*, 1992), with permission.

the toad bladder and mammalian collecting duct involving the vasopressin-regulated water channel (Wade and Kachadorian, 1988). Alternatively, it is conceivable that the association of F-actin with the $Na^+/K^+/2Cl^-$ cotransporter tonically inhibits its function (*e.g.*, through steric hindrance or limitation of dynamic structural changes necessary for ion translocation), and that depolymerization of F-actin enhances activation of membrane-bound cotransporters by releasing such inhibition. Lastly, dynamic changes in the F-actin cytoskeleton may be necesssary for the transduction of secondary regulatory signals for cotransporter activation (*e.g.*, decreased cell volume due to salt loss via activated Cl^- and K^+ channels). We are at present investigating these various possibilities. [1]

[1] The advice and support of Dr. James L. Madara is gratefully acknowledged, as is the technical assistance of Christopher S. Awtrey, Kevin J. Tally, and Jeremy A. Smith. This work was supported by a Glaxo Basic Research Award, the Harvard Digestive Diseases Center, and the Beth Israel Surgical Group.

REFERENCES

Cantiello, HF and AM Bertorello (1991). Actin filaments regulate epithelial sodium, potassium ATPase activity. J Cell Biol **115**: 353a.

Condeelis, J and AL Hall (1991). Measurement of actin polymerization and cross-linking in agonist-stimulated cells. Methods Enzymol **196**: 486-496.

Dharmsathaphorn, K, JA McRoberts, et al. (1984). A human colonic tumor cell line that maintains vectorial electrolyte transport. Am J Physiol **246**: G204-G208.

Lytle, C and B Forbush III (1992). The Na-K-Cl cotransport protein of shark rectal gland II. Regulation by direct phosphorylation. J Biol Chem **267**: 25438-25443.

Matthews, JB, CS Awtrey, et al. (1992). Microfilament-dependent activation of Na+/K+/2Cl- cotransport by cAMP in intestinal epithelial monolayers. J Clin Invest **90**: 1608-1613.

Matthews, JB, CS Awtrey, et al. (in press). $Na^+/K^+/2Cl^-$ Cotransport and Cl^- Secretion evoked by heat-stable enterotoxin is microfilament-dependent in T84 cells. Am J Physiol.

Mercier, FH, H Reggio, et al. (1989). Membrane-cytoskeleton dynamics in rat parietal cells: mobilization of actin and spectrin upon stimulation of gastric acid secretion. J Cell Biol **108**: 441-453.

Molitoris, BA, A Geerdes, et al. (1991). Dissociation and redistribution of Na^+,K^+-ATPase from its surface membrane actin cytoskeletal complex during cellular ATP depletion. J Clin Invest **88**: 462-469.

Nelson, WJ and RW Hammerton (1989). A membrane-cytoskeletal complex containing Na^+K^+-ATPase, ankyrin, and fodrin in Madin-Darby canine kidney (MDCK) cells: implications for the biogenesis of epithelial cell polarity. J Cell Biol **108**: 893-902.

Pewitt, EB, RS Hegde, et al. (1990). The regulation of Na/K/2Cl cotransport and bumetanide binding in avian erythrocytes by protein phosphorylation and dephosphorylation. J Biol Chem **265**: 20747-20756.

Shapiro, M, J Matthews, et al. (1991). Stabilization of F-actin prevents cAMP-elicited Cl- secretion in T84 cells. J Clin Invest **87**: 1903-1909.

Wade, JB and WA Kachadorian (1988). Cytochalasin B inhibition of toad bladder apical membrane responses to ADH. Am J Physiol **255**: C526-C530.

Ca^{2+} Pools and H$^+$ Secretion by the Parietal Cell

Fabian Michelangeli, Giovanna Tortorici, and Marie-Christine Ruiz

Laboratorio de Fisiología Gastrointestinal
Centro de Biofísica y de Bioquímica
Instituto Venezolano de Investigaciones Científicas
Caracas, 1020A
Venezuela

Cellular responses in a number of tissues have been shown to be dependent on increases in cytosolic Ca^{2+} (Berridge 1993, Irvine 1992). The typical byphasic Ca^{2+} response is characterized by an initial spike due to Ins(1,4,5)P$_3$-induced release of Ca^{2+} from intracellular pools, followed by a sustained phase linked to Ca^{2+} entry from the extracellular medium (Berridge 1993, Irvine 1992). Histamine, gastrin and muscarinic agonists have been shown to induce a Ca^{2+} response bearing such a pattern in the gastric parietal cell (Chew 1986; Negulescu and Machen 1988; Michelangeli et al 1989). Ca^{2+} release by cholinergic agonists has been associated to an increase of Ins(1,4,5)P$_3$, whereas for histamine its participation is not clear (Chew and Brown 1986). It has not been possible to measure changes in Ins(1,4,5)P$_3$ in the parietal cell, although in hepatoma cells expressing the canine H$_2$ receptor, Ca^{2+} release in response to Ins(1,4,5)P$_3$ generation has been documented (DelValle et al 1992). The sustained phase of Ca^{2+} increase is dependent on the presence of Ca^{2+} in the extracellular medium suggesting the activation of Ca^{2+} entry pathways during stimulation (Negulescu and Machen 1988a,b). The linkage between receptor activation and Ca^{2+} entry has been the subject of studies in numerous cells but not the parietal cell. Voltage dependent Ca^{2+} channels, receptor-operated Ca^{2+} channels (coupled via G proteins), second-messenger operated channels, and Ca^{2+} store-regulated channels have all been proposed as Ca^{2+} entry mechanisms into different cells (Rink 1990, Putney 1990, Berridge 1993). The pathway(s) for Ca^{2+} influx into the stimulated parietal cell is not well understood. On the other hand, the study of the role of Ca^{2+} increase in the stimulation of acid

NATO ASI Series, Vol. H 89
Molecular and Cellular Mechanisms
of H$^+$ Transport
Edited by Barry H. Hirst
© Springer-Verlag Berlin Heidelberg 1994

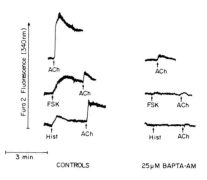

Figure 1. Effect of secretagogues on fura2 fluorescence in isolated gastric glands. In the right-hand panel, the effect of preincubation with BAPTA (25µM, 60 min) is shown.

secretion has been attempted using the intracellular Ca^{2+} chelator, BAPTA-AM (Michelangeli et al 1989; Negulescu et al 1989). Buffering of the intracellular Ca^{2+} changes by incorporated BAPTA induced a variable degree of inhibition of HCl secretion (aminopyrine accumulation ratio, AP ratio) depending on the agonist used. In this paper we attempt to evaluate the different Ca^{2+} pools touched during stimulation and their relationship with H^+ secretion by the gastric parietal cell.

Effect of stimulants on Ca^{2+} pools

Cholinergic agonists, forskolin and histamine all induced an increase in $[Ca^{2+}]_i$ in isolated rabbit gastric glands (Fig.1). In general, the response is characterized by a spike, followed by a decrease to attain a plateau level above resting $[Ca^{2+}]_i$. The cAMP analogue 8Br-cAMP, had no effect (not shown). Acetylcholine (ACh), added after forskolin or histamine during the plateau phase, evoked a new peak in $[Ca^{2+}]_i$. After stimulation with ACh, forskolin or histamine provoked instead a small decrease in the sustained level of Ca^{2+} (not shown). As in Ca^{2+} free medium the spikes are still present, the transient elevation of $[Ca^{2+}]_i$ is thought to be the result of Ca^{2+} release from intracellular deposits (results not shown; Negulescu and Machen 1988a,b). The accumulation of Ca^{2+} in nonmitochondrial stores is dependent on

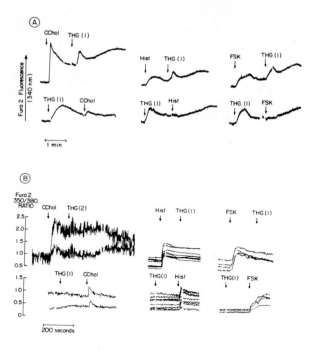

Figure 2. Effect of thapsigargin (THG, 1μM) on fura2 fluorescence in isolated gastric cells **(A)** or different parietal cells on a single gland **(B)** measured with a multiple point microscope photometer system.

the activity of a Ca^{2+}/ATPase (see Thastrup 1990). Inhibition of this pump by thapsigargin (THG), a sesquiterpene lactone, has been shown to deplete stores in a number of cells, resulting in variable increases in $[Ca^{2+}]_i$ and rendering cells insensitive to further release by stimulants (Thastrup 1990). This release is independent of $Ins(1,4,5)P_3$ generation. In isolated glands THG induced an increase in $[Ca^{2+}]_i$ (Fig.2A) that became transient in Ca^{2+} free medium (Chew and Petropoulos 1991). THG prevented the transitory Ca^{2+} response induced by carbachol (Cchol), forskolin or histamine. However, a slow and progressive elevation of $[Ca^{2+}]_i$ was observed after histamine or forskolin, but not with Cchol. On the other hand, in all cases THG always induced a further increase of $[Ca^{2+}]_i$ when added after secretagogues. In unicellular recordings of parietal cells in isolated glands, increases in $[Ca^{2+}]_i$ with THG were observed only at higher concentrations (2 μM), but had the same effect on the further Ca^{2+} response induced by secretagogues (Fig.2B). These results taken together with other published evidence (Chew and Brown 1986, Negulescu and Machen 1988a,b) suggest the participation of distinct intracellular Ca^{2+} pools differentially sensitive to be released by secretagogues. At this point we can propose the existence of at least three different Ca^{2+} pools: one released by histamine (and forskolin) and ACh, another mobilized by ACh alone and the third one insensitive to secretagogues (and $Ins(1,4,5)P_3$) but depleted by THG. This last agent would also deplete the agonist-sensitive pools by virtue of inhibition of the Ca^{2+}-ATPase of ER.

Effect of stimulants on Ca^{2+} entry

The sustained phase of increase of $[Ca^{2+}]_i$ in response to secretagogues and THG was dependent on the presence of Ca^{2+} in the extracellular medium (Fig.3). Ca^{2+} free medium reduced $[Ca^{2+}]_i$

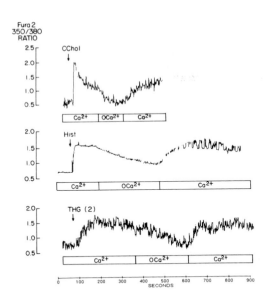

Figure 3. Effect of extracellular Ca^{2+} concentration on fura2 fluorescence response to CChol, histamine and THG (2μM) measured in individual parietal cells in isolated gastric glands.

to resting levels and readdition of Ca^{2+} to the medium, in the continous presence of the secretagogue, provoked a sustained increase in $[Ca^{2+}]_i$. These changes in medium Ca^{2+} have little or no effect on $[Ca^{2+}]_i$ (not shown). This suggests that secretagogues, in addition to the release of intracellular Ca^{2+}, are able to activate the opening of a Ca^{2+} permeation pathway in the plasma membrane. The same pattern of response with media with different Ca^{2+} was observed with THG in parietal cells. This finding supports the hypothesis of a linkage between the state of filling of intracellular stores and the activation of Ca^{2+} entry (Rink 1990; Thastrup 1990; Montero et al 1990). The measurement of Mn^{2+} entry in isolated gastric glands as indicative of Ca^{2+} entry confirms the existence of Ca^{2+} permeation pathways activated by Cchol, histamine and THG (Fig.4). However, the small effect of secretagogues on Mn^{2+} entry observed may be related to the existence of channels permeable to Ca^{2+} but poorly permeable to Mn^{2+}, as it has been shown for parotid acinar cells (Merritt and Hallam 1988).

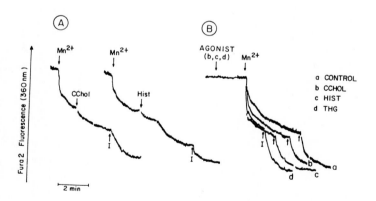

Figure 4. Effect of secretagogues on Mn^{2+} entry and quenching of fura2 fluorescence (at 360 nm) in isolated gastric glands.

As Cchol induces little or no effect on $[Ca^{2+}]_i$ after THG (Fig. 2A,B), it is suggested that Cchol and THG act on the same type of Ca^{2+} channels on the plasma membrane. Therefore the activity of this channel would be regulated by the state of filling of the intracellular Ca^{2+} stores. . On the other hand, histamine (and forskolin) elicits a slow and sustained increase after THG suggesting that it acts on a different type of Ca^{2+} channel linked to receptor activation. This would be either a receptor-operated channel (coupled via G protein) or a second-messenger operated channel.

$[Ca^{2+}]_i/H^+$secretion relationship in the parietal cell.

The relationship between the increase in $[Ca^{2+}]_i$ and H^+ secretion by the parietal cell has been studied using the incorporated Ca^{2+} chelator BAPTA as a buffer (Michelangeli et al 1989; Negulescu et al 1989). BAPTA decreased AP ratio stimulated by ACh, forskolin and histamine. Stimulation by 8Br-cAMP, which does not change $[Ca^{2+}]_i$, was not inhibited by BAPTA (Fig. 5A).

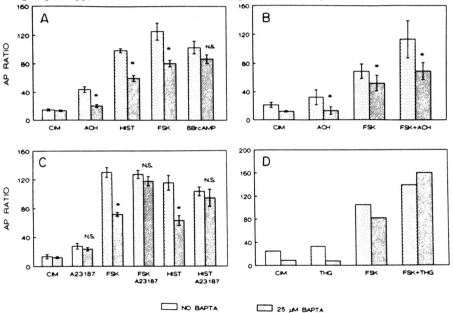

Figure 5. Effect of preincubation with BAPTA (25μM, 60min) on secretagogue stimulation of H^+ secretion (AP ratio) in isolated gastric glands under different conditions.

The Ca²⁺ ionophore A23187 completely reversed the inhibition by BAPTA of histamine and forskolin stimulated H⁺ secretion (Fig. 5C). In this case, the ionophore slowly increases $[Ca^{2+}]_i$ even in the presence of the Ca²⁺ chelator (Michelangeli et al 1989). The potentiated fraction of the forskolin + ACh secretory response was also inhibited by BAPTA (Fig 5B). THG potentiated the forskolin response. However, at difference with ACh, this potentiated fraction was not inhibited by BAPTA (Fig 5D). This may suggest that Ca²⁺ entry during treatment with THG may be able to, at least partially, overcome buffering by BAPTA after a long time. These results indicate that the sustained elevation of $[Ca^{2+}]_i$ induced by secretagogues in the parietal cell is necessary and sufficient to evoke a H⁺ secretory response and is responsible for the potentiated fraction observed with the cAMP-Ca²⁺ dependent secretagogues (histamine and forskolin), or the combination of these and Ca²⁺ dependent ones like ACh.

Figure 6 represents a schematical model depicting the different mechanisms of Ca²⁺ release, Ca²⁺ entry pathways and Ca²⁺ pools discussed in this paper, that play a role in producing an integrated response after secretagogue stimulation of H⁺ secretion.

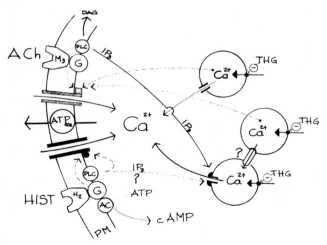

Figure 6. Schematical model depicting the different mechanisms of Ca²⁺ release, Ca²⁺ entry pathways and Ca²⁺ pools discussed in this paper, that play a role in producing an integrated response after secretagogue stimulation of H⁺ secretion.

ACKNOWLEDGEMENTS

This work was supported in part by Grant CI-1*-CT90-0786 (SSMA) from the European Economical Communities (EEC)

REFERENCES

Berridge MJ (1993) Inositol trisphosphate and calcium signalling. Nature 361:315-325

Chew CS (1986) Cholecystokinin, carbachol, gastrin, histamine, and forskolin increase [Ca2+]i in gastric glands. Am. J. Physiol. 250:G814-G823

Chew CS, and Brown M (1986) Release of intracellular Ca2+ and elevation of inositol trisphosphate by secretagogues in parietal and chief cells isolated from rabbit gastric mucosa. Biochim. Biophys. Acta 888:116-125

Chew CS and Petropoulos AC (1991) Thapsigargin potentiates histamine-stimulated HCl secretion in gastric parietal cells but does not mimic cholinergic responses. Cell Regulation 2:27-39

Delvalle JD, Wang L, Gantz I and Yamada T (1992) Characterization of H_2 histamine receptor: linkage to both adenyl cyclase and [Ca2+]$_i$ signaling systems. Am. J. Physiol. 263:G967-972

Irvine RF (1992) Inositol lipids in cell signalling. Current Opinion in Cell Biology 4:212-219

Michelangeli F, Ruiz MC, Fernandez E and Ciarrocchi A (1989) Role of Ca2+ in H+ transport by rabbit gastric glands studied with A23187 and BAPTA, an incorporated Ca2+ chelator. Biochim. Biophys. Acta 983:82-90

Merrit JE and Hallam TJ (1988) Platelets and parotids acinar cells have different mechanisms for agonist-stimulated divalent cation entry. J.Biol. Chem. 263:6161-6164

Negulescu PA., Reenstra WW, and Machen TE (1989) Intracellular Ca requirements for stimulus-secretion coupling in parietal cell. Am. J. Physiol. 256:C241-C251

Negulescu PA and Machen TE (1988) Release and reloading of intracellular Ca stores after cholinergic stimulation of the parietal cell. Am. J. Physiol. 254:C498-C504

Negulescu PA and Machen TE (1988) Intracellular Ca regulation during secretagogue stimulation of the parietal cell. Am. J. Physio. 254:C130-C140

Putney J.W.Jr (1990) Capacitative calcium entry revisited. Cell Calcium 11:611-624

Rink TJ (1990) Receptor-mediated calcium entry. FEBS Letter 268(2):381-385

Thastrup O (1990) Role of Ca2+ -ATPases in regulation of cellular Ca2+ signalling, as studied with the selective microsomal Ca2+-ATPase inhibitor thapsigargin. Agents and Actions 29(1/2):8-15

Acidophylic Cl⁻ and K⁺ Channels of the Gastric Parietal Cell: A New Model of Regulated HCl Secretion

John Cuppoletti, Ann M. Baker and Danuta H. Malinowska
Department of Physiology and Biophysics
University of Cincinnati College of Medicine
231 Bethesda Avenue
Cincinnati, Ohio 45267-0576

The gastric parietal cell is responsible for HCl production. Primary active transport of H^+ occurs through the action of the Mg^{2+}-dependent H^+/K^+ ATPase, which catalyses the ATP-dependent inward movement of one extracytosolic K^+ for each H^+ produced. Pathways for the movement of K^+ and Cl⁻ are also required for continued action of the H^+ pump, to provide luminal K^+ and equivalents of Cl⁻ for HCl production. KCl flux across the apical membrane is under regulation (Sachs et al, 1976; Malinowska et al, 1983). The gastric parietal cell also undergoes a morphological reorganization upon stimulation of HCl secretion, whereupon intracellular vesicles containing the H^+/K^+ ATPase (tubulovesicles) disappear, and elaborate into the apical membrane. Since HCl concentrations in the primary gastric secretion exceed 0.15 M, the H^+/K^+ ATPase and any other proteins (such as ion channels) required for HCl secretion must function in this harsh environment.

The pathways for K^+ and Cl⁻ movement are conductive (Cuppoletti & Sachs, 1984; Reenstra & Forte, 1990). In the present work, we summarize recently published data on an acidophylic Cl⁻ channel (Cuppoletti et al, 1993) and new direct evidence for an acidophylic K^+ channel from electrophysiological (single channel) studies of mammalian (rabbit) gastric apical membranes. Both K^+ and Cl⁻ channels continue to function when low pH bathes the presumed extracytoplasmic surface of the membrane in which they are embedded. Studies have been carried out on the effects of alteration of membrane potential and asymmetric pH changes on the probability of opening (P_o) of the channels. These manipulations mimic the environment of the gastric apical membrane in which the channels must function in concert with the H^+/K^+ ATPase for HCl secretion to occur. Our detailed studies of the Cl⁻ channel suggest a model wherein channel activity (as measured by P_o) is modulated by changes in the environment of the membranes as the channels are recruited from intracellular stores (tubulovesicles) to the apical membrane. Superimposed on these changes due to altered environment, differences in the P_o of the channel of membranes derived from non-secreting (cimetidine-treated) and secreting (histamine-stimulated) rabbit

NATO ASI Series, Vol. H 89
Molecular and Cellular Mechanisms
of H⁺ Transport
Edited by Barry H. Hirst
© Springer-Verlag Berlin Heidelberg 1994

gastric mucosa were also observed, presumably due to covalent modification such as by cyclic AMP dependent protein kinase. Detailed studies of the K^+ channel (described here for the first time) are in progress. Comparisons of the properties of the Cl^- and K^+ channels with the characteristics of HCl accumulation suggest that both of these channels may be involved in regulated HCl secretion.

Materials and Methods

Membrane vesicles enriched in the gastric H^+/K^+ ATPase from secreting (histamine plus diphenhydramine-injected) or non-secreting (cimetidine-injected) rabbits were prepared as previously described in detail (Cuppoletti & Sachs, 1984). Bilayer reconstitution methods, equipment for channel recording and analysis were carried out as described (Cuppoletti et al, 1993). CsCl and K_2SO_4 solutions were used for Cl^- and K^+ channel current recordings respectively and were buffered to pH 7.4 unless otherwise stated. Measurement of HCl accumulation with acridine orange, ATP hydrolysis, para-nitrophenylphosphatase activity, and protein concentration were carried out as described previously (Cuppoletti & Sachs, 1984).

Results

Evidence for a class of Cl^- channels in apical membrane preparations from stimulated rabbit gastric mucosa. Shown in **Figure 1** is a single channel current recording obtained after fusion of stimulated rabbit gastric vesicles to planar lipid bilayers. These channels exhibited a linear current-voltage (I/V) relationship, a conductance of 28 pS when recorded in solutions containing 800 mM CsCl, and 7 pS in 150 mM CsCl. The channels were anion selective as judged by the positive reversal potential of +22 mV obtained with a 5 fold gradient of CsCl (800 mM cis, 160 mM trans). The discrimination ratio was 6:1 (Cl^-:Cs^+) as calculated using the Goldman-Katz-Hodgkin equation and ionic activities. Anion selectivity of the channel measured under near bi-ionic conditions (with permeability relative to Cl^- in parentheses) was I^- (2.0) > Cl^- (1.0) $\geq$ Br^- (0.9) > NO_3^- (0.6) (See Cuppoletti et al, 1993).

Evidence that gastric Cl^- channels continue to function at low pH. When the pH of the solution bathing the trans (but not cis) side of the bilayer, was reduced to pH 3.0, single channels persisted (**Figure 2**). These channels were indistinguishable from those recorded under conditions of symmetrical pH 7.4 solutions with respect to their linear I/V

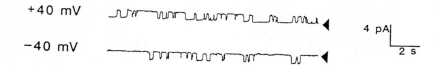

Figure 1. The Gastric Apical Membrane Cl⁻ Channel at pH 7.4. Single channel current recordings were obtained after fusion of stimulated gastric membranes to lipid bilayers. Symmetric 800 mM CsCl solutions at pH 7.4 were used. Holding potentials were as indicated. Arrowheads, closed state of the channel. (From Cuppoletti *et al*, 1993).

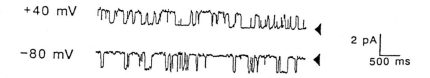

Figure 2. The Gastric Apical Membrane Cl⁻ Channel at pH 3. Single channel current recordings were obtained with the solution on the *trans* side of the bilayer reduced to pH 3 using 10 mM K-citrate buffer. 800 mM CsCl solutions were on both sides of bilayer. The *cis* solution was at pH 7.4. Holding potentials were as indicated. Arrowheads, closed state of the channel. (From Cuppoletti *et al*, 1993).

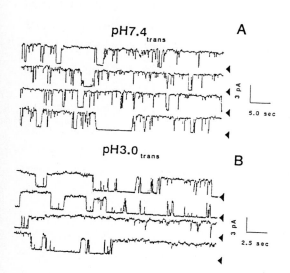

Figure 3. The Gastric Apical Membrane K⁺ Channel at pH 7.4 (A) and pH 3 (B). Single channel current recordings with solution on the *trans* side of the bilayer at pH 7.4 (A) or pH 3 (B). 400 mM K₂SO₄ (800 mM K⁺) solutions were on both sides of the bilayer. In both recordings, solution on *cis* side was at pH 7.4 and the holding potential was -40 mV. Arrowheads, closed state of the channel.

curve, conductance of 27 pS, anion-cation discrimination ratio and anion selectivity. In contrast to the stability of the channels with pH 3 solution on the trans side, reducing the pH of the solution in the cis chamber bathing the presumed cytoplasmic face of the membrane to pH 3.0, always resulted in loss of Cl⁻ channel activity. These Cl⁻ channels were indifferent to Ca^{2+} (see Cuppoletti et al, 1993).

Effects of membrane potential, pH and state of stimulation on Cl⁻ channel opening. In studies presented in detail elsewhere (Cuppoletti et al, 1993), the gastric Cl⁻ channel was shown to be activated at the negative membrane potentials exhibited by the gastric parietal cell, such that a 6-10 fold increase in P_o occurred. Lowering the pH of the trans side to 3 increased the P_o approximately 6 fold relative to that at pH 7.4 at -80 mV. These channels are thus activated by acid at physiologically relevant membrane potential. In addition, when Cl⁻ channels of vesicles from secreting rabbit gastric mucosa (presumably originating from the apical membrane), were compared with those of vesicles from resting gastric mucosa (presumably originating from the tubulovesicles), the stimulated vesicle channel P_o was found to be increased. Cl⁻ channels of both resting and stimulated vesicles exhibited similar increases in P_o upon lowering the trans solution pH or imposing more negative membrane potentials.

Evidence for a class of K⁺ channels which are active at pH 7.4 and pH 3 in apical membrane preparations from stimulated rabbit gastric mucosa. In preliminary studies presented in abstract form (Cuppoletti, 1993), single channel current recordings of K⁺ channels were obtained in 400 mM K_2SO_4 solutions at pH 7.4 (**Figure 3A**) and with trans solution at pH 3.0 (**Figure 3B**). These K⁺ channels exhibited a linear I/V curve, with a conductance of approximately 15 pS at 800 mM K⁺, and a strict selectivity of K⁺ over Na⁺ or Cs⁺. The $K^+:SO_4^{2-}$ discrimination ratio as determined from the magnitude of the negative reversal potential obtained from an I/V curve measured under five-fold gradient conditions was high and approached that predicted from the Goldman-Katz-Hodgkin equation. These K⁺ channels were Ca^{2+}-indifferent. Detailed studies of the effects of membrane potential, pH, and state of stimulation of the tissue on the P_o of the K⁺ channel described in this study have not yet been completed, but are the subject of ongoing studies in this laboratory.

Effects of inhibitors of epithelial Cl⁻ and K⁺ channels on single channel characteristics and HCl secretion. Diphenylamine-2-carboxylate (DPC) is known to inhibit a wide variety of Cl⁻ channels (Gogelein, 1988). In order to use DPC as a tool to relate the acidophylic gastric Cl⁻ channel activity to HCl secretion, it is necessary to demonstrate a direct inhibitory effect of DPC on the Cl⁻ channel, with no effect on the other components of the secretory unit such as the H^+/K^+ ATPase and the K^+ channel. **Figure 4** shows that 2 mM DPC inhibited the gastric Cl⁻ channel. In contrast to the reported inhibitory effects of DPC on non-specific cation channels (Gogelein & Pfannmuller, 1989), the gastric K^+ channel was unaffected by 2 mM DPC. In a typical experiment the K^+ channel open time was 148 msec for control channels and 169 msec for channels recorded in the presence of 2 mM DPC. DPC inhibited HCl accumulation in stimulated gastric vesicles with an apparent K_I of 0.25 mM, and at 1 mM concentrations which fully inhibited HCl accumulation, did not increase the H^+ leak rate from the vesicles. At 2 mM DPC (8 times K_I) only 20% inhibition of para-nitrophenyl phosphatase activity occurred. Maximal H^+/K^+ ATPase activity measured in permeable rabbit parietal cells has been shown to be unaffected by DPC (Malinowska, 1990). Since we have shown that only Cl⁻ channels and not K^+ channels or the H^+/K^+ ATPase are sensitive to DPC, inhibition of HCl accumulation in gastric vesicles by DPC must be due to inhibition of the essential Cl⁻ channel. However to make further progress in this area, it will be necessary to develop ligands with higher affinity and to relate microscopic measures of function (single channel studies) to macroscopic measures of function (net KCl and HCl accumulation). Tetraethylammonium ion (TEA⁺), an inhibitor of K^+ channels from a variety of sources, inhibits HCl accumulation (Cuppoletti & Malinowska, 1988; Reenstra & Forte, 1990). **Figure 5** shows that 10 mM TEA⁺ inhibited gastric K^+ channel activity. TEA⁺ had no effect on the gastric Cl⁻ channel. The effects of peptide inhibitors and other tight binding inhibitors of other K^+ channels have not yet been fully investigated.

Discussion

We have presented evidence from single channel studies for the presence of K^+ and Cl⁻ channels in rabbit gastric vesicles enriched in H^+/K^+ ATPase, which uniquely are active at the low pH at which channels involved in HCl accumulation must function. The Cl⁻ channel, which has been most intensely studied, is activated by low pH, negative membrane potential, and by stimulation of gastric HCl secretion in the animals from which apical membrane vesicles are derived. The properties of the K^+ channel, which is also active at low pH, but which has been less extensively characterized, are also consistent with it playing a role in regulated HCl accumulation. These findings and the characteristic

Control

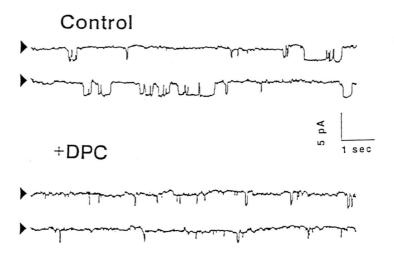

+DPC

Figure 4. Effect of the Cl⁻ Channel Blocker DPC on the Gastric Cl⁻ Channel.
Single Cl⁻ channel current recording at -80 mV holding potential (control) was
inhibited by 2 mM DPC. Arrowheads, closed state of the channel.

CONTROL

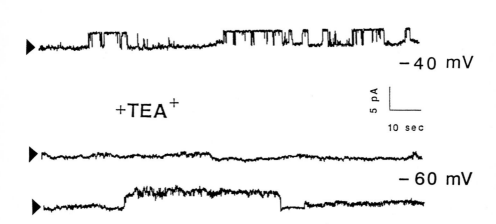

Figure 5. Effect of the K⁺ Channel Blocker TEA⁺ on the Gastric K⁺ Channel.
Single K⁺ channel current recording was inhibited by 10 mM TEA⁺. Arrowheads,
closed state of the channel.

rearrangement of cellular components of the gastric parietal cell which accompany stimulation of HCl secretion have led us to present the model shown in **Figure 6**. In this model, the parietal cell responds to stimulants of secretion by altering the distribution of the gastric H^+/K^+ ATPase and the ion channels from intracellular membranes to the apical membrane. The ion channels thus experience the altered pH and membrane potential of the apical membrane, as well as direct effects of covalent modification such as by cyclic AMP dependent protein kinase, and become more active. The multiplicative sum of these changes results in the massive increases in HCl accumulation observed in intact tissue and in various models of HCl accumulation by the gastric parietal cell.

Experimental support for this model includes direct electrophysiological evidence obtained using the bilayer insertion technique, which allows drastic modification of the environment of the channel, and correlative studies using classic channel inhibitors. Whereas more is now known regarding the properties of these acidophylic gastric Cl channels as a result of our recent studies (Cuppoletti et al, 1993), additional evidence for and consistent with the testable model has been presented in **Figures 3 & 5** in which an acidophylic K^+ channel was identified. Direct electrophysiological evidence for acidophylic apical membrane K^+ channels and Cl channels is new. Quantitative measures of changes in P_o of reconstituted Cl channels from secreting and non-secreting gastric mucosae upon changes of membrane potential and pH have been formalized into a testable model for regulated HCl secretion which may be useful for investigation of regulation of other secretory processes such as Cl secretion by the cystic fibrosis transmembrane regulator protein (CFTR).

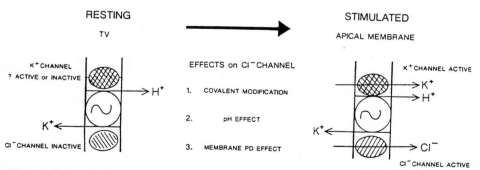

Figure 6. Proposed Model for Regulated HCl Secretion by the Gastric Parietal Cell. Cross-hatched area, K^+ channel; hatched area, Cl channel; H^+/K^+ ATPase. (From Cuppoletti et al, 1993).

The major challenge of our studies in development of a useful model of regulated HCl secretion by the gastric parietal cell will be to definitively relate microscopic measures ion channel activity (single channel studies) to macroscopic measures of transport, such as net KCl flux and HCl accumulation. Clearly, tight binding inhibitors which affect single channel activity and HCl accumulation through specific and direct effects upon their respective targets, such as Cl^+ or K^+ channels, without affecting the gastric H^+/K^+ ATPase, will be essential to further delineation of the mechanism of regulated gastric HCl secretion.

Acknowledgements

This work was supported by NIH grants DK43816 and DK43377 (JC & DHM) and NSF Research Opportunities for Women Career Advancement Award DCB9109605 (DHM). AMB was supported by NIH Training Grant HL07571.

References

Cuppoletti J (1993) K^+ channels in the gastric parietal cell which are active at low pH. Gastroenterology 104 : A60

Cuppoletti J, Baker AM, Malinowska DH (1993) Cl⁻ channels of the gastric parietal cell that are active at low pH. Am J Physiol 264 (Cell Physiol 33) : C1609-C1618

Cuppoletti J, Malinowska DH (1988) K^+ channel in the gastric parietal cell apical membrane. Gastroenterology 94 : A82

Cuppoletti J, Sachs G (1984) Regulation of gastric acid secretion via modulation of a chloride conductance. J Biol Chem 259 : 14952-14959

Gogelein H (1988) Chloride channels in epithelia. Biochim Biophys Acta 947 : 521-547

Gogelein H, Pfannmuller B (1989) The nonselective cation channel in the basolateral membrane of rat exocrine pancreas. Pfluegers Arch 413 : 287-298

Malinowska DH (1990) Cl⁻ channel blockers inhibit acid secretion in rabbit parietal cells. Am J Physiol 259 (Gastrointest Liver Physiol 22) : G536-G543

Malinowska DH, Cuppoletti J, Sachs G (1983) Cl⁻ requirement of acid secretion in isolated gastric glands. Am J Physiol 245 (Gastrointest Liver Physiol 8) : G573-G581

Reenstra WW, Forte JG (1990) Characterization of K^+ and Cl⁻ conductances in apical membrane vesicles from stimulated rabbit parietal cells. Am J Physiol 259 (Gastrointest Liver Physiol 22) : G850-G858

Sachs G, Chang HH, Rabon E, Schackman R, Lewin M, Saccomani G (1976) A nonelectrogenic H^+ pump in plasma membranes of hog stomach. J Biol Chem 251: 7690-7698

Uncoupling of H⁺ and Pepsinogen Secretion in the Amphibian Oxyntopeptic Cell

Marie-Christine Ruiz, Bernardo Gonzalez, Maria Jesus Abad and
Fabián Michelangeli

Laboratorio de Fisiología Gastrointestinal
Centro de Biofísica y de Bioquímica
Instituto Venezolano de Investigaciones Científicas
Caracas, 1020A,
Venezuela

Pepsinogen and HCl secretion in the stomach of inferior
vertebrates, are performed by one single cell type, the
oxyntopeptic cell. This cell is present in the stomach of
elasmobranchs, teleosts, amphibians, reptiles and birds (Smit
1968; Barrington, 1942). From the first studies of Langley
(1881), the presence of pepsinogen has been detected in the
gastric mucosa of amphibians. In contrast to the frog, pepsinogen
in the toad esophagus represents only a small quantity compared
with that contained in the stomach (Inoue et al 1985; Ruiz et al
1993). In the gastric mucosa, pepsinogen has been found to be
more concentrated in the proximal region where the glands are
longer and with a much higher density of zymogen granules (Ruiz
et al 1993). Electron microscopy of the oxyntopeptic cell has
permitted the identification of specialized structures related
to both secretions (Geuze 1971; see Fig.1). The apical pole
contains the tubulo-vesicular membrane system (TVM) in which the
H^+/K^+ pumps, responsible for acid secretion, are localized. During
a stimulatory cycle, the apical surface is increased by a fusion
process of the TVM with the apical membrane (Forte et al 1977;
Jiron et al 1984). This cell also contains organelles involved
in pepsinogen synthesis and export. Particularly, we can observe
Golgi apparatus, rough endoplasmic reticulum and secretory
granules which are stocked in the apical pole before their
release by exocytosis.

NATO ASI Series, Vol. H 89
Molecular and Cellular Mechanisms
of H⁺ Transport
Edited by Barry H. Hirst
© Springer-Verlag Berlin Heidelberg 1994

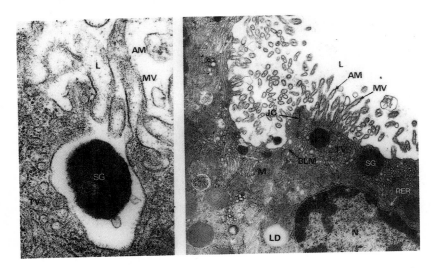

Figure 1. Electron microscopy of the oxyntopeptic cell of *Bufo marinus* stimulated by histamine (10^{-4}M) and exocytosis of secretory granule. L,lumen of the gland; AM, apical membrane; MV, microvillus; TV tubulovesicular system; BLM, basolateral membrane; RER, rough endoplasmic reticulum; SG, secretory granule; M, mitochondria; N, nucleus; LD,lipid deposit;

Effect of Agonists on HCl and Pepsinogen Secretion

The control of HCl secretion has been widely studied. The pepsinogen component, and its relation with acid secretion, has been recently analyzed in our laboratory (Ruiz et al 1993). Pepsinogen release was sensitive to the same secretagogues as acid secretion as shown in Fig.2. Histamine, which is a potent stimulant of acid secretion, also induced the stimulation of pepsinogen secretion. It is known that histamine induces an activation of the adenylate cyclase (AC) in the mammalian parietal cell as well as in the amphibian oxyntopeptic cell (Chew et 1980; Ekblad et al 1980; Soll and Wollin 1979). Forskolin, a direct activator of AC, (Fig.2) and 8Br-cAMP, a cAMP-analogue, also induced the stimulation of both parameters (Fig.3). It has been described that acetylcholine has a double action to stimulate HCl secretion, one mediated by endogenous histamine and another by a direct effect on the oxyntopeptic cell. Moreover, an interaction between the two pathways induces a potentiated response on acid secretion (Ruiz et al 1984). The pattern of stimulation of pepsinogen secretion by cholinergic secretagogues

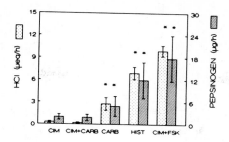

Figure 2. Maximal rate of acid and pepsinogen secretion attained by mucosae stimulated by carbachol (10^{-4}M) in the presence or absence of cimetidine (10^{-3}M), histamine (10^{-4}M) and forskolin (5×10^{-4}M). For details on materials and methods see Ruiz et al 1993.

appears similar to that of the acid component: 1) carbachol stimulates both secretions , 2) in the presence of cimetidine to eliminate the participation of endogenous histamine, carbachol does not induce a response in any of the two parameters (Fig.2), and 3) in the presence of an increase of cAMP induced by exogenous histamine (Ruiz et al 1993), carbachol is able to induce an over effect in both pepsinogen and HCl secretion. This effect was also observed with 8Br-cAMP and carbachol on pepsinogen, but not HCl secretion (Fig.3)

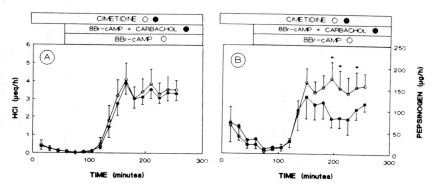

Figure 3. Time course of acid (A) and pepsinogen (B) secretion by paired resting hemi-mucosae of *Bufo marinus* stimulated by joint addition of 8Br-cAMP (10^{-3}M) and carbachol (10^{-4}M). Values are means ± SEM for 4 experiments. *, $p < 0.01$ in paired t test comparing the values of each parameter respectively in the 2 groups at each time point.

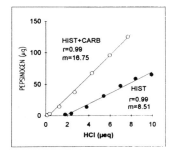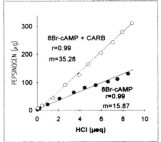

Figure 4. Relation between acid and pepsinogen secretion stimulated with either 8Br-cAMP (10^{-3}M) or histamine (10^{-4}M) (filled circles) and their combination with carbachol (10^{-4}M), (open circles). The values correspond to the accumulated secretions at each time for a single representative experiment. m, slope of the linear regression; r correlation coefficient.

Relationship between HCl and Pepsinogen Secretion

In spite of the two secretions being sensitive to the same secretagogues, the degree of sensitivity seems to be different for each agent and it is possible to dissociate the two parameters under special conditions. Pepsinogen secretion presents a significantly higher stimulation by simultaneous addition of 8Br-cAMP and carbachol than that in the presence of 8Br-cAMP alone, whereas the response in acid secretion attained the same level in the two cases (fig.3). Pepsinogen secretion was still sensitive to carbachol, whereas HCl was fully stimulated with 8Br-cAMP alone. The study of the relationship between the two accumulated secretions at each time revealed an uncoupling between the two secretions for each condition of stimulation. The relation was linear but the slope was higher when the mucosae were stimulated by 8Br-cAMP plus carbachol or histamine plus carbachol than for the group in which 8Br-cAMP or histamine alone were present (Fig.4). Furthermore, addition of carbachol to histamine or forskolin stimulated mucosae changed the slope of this relationship (Ruiz et al 1993). We could interpret that carbachol in the presence of a cAMP-dependent secretagogue, is more efficient to produce an increase of pepsinogen response in relation to acid secretion. Moreover, pepsinogen secreted during stimulation with secretagogues does not seem to be related to

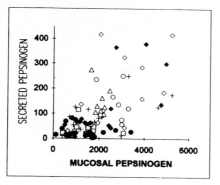
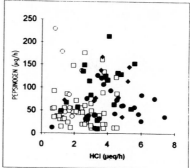

Figure 5. Relation between secreted pepsinogen and mucosal pepsinogen content (left panel) and between the maximal responses of pepsinogen and acid secretion during the stimulation induced by different secretagogues. open squares, 8Br-cAMP (10^{-3}M); open circles, 8Br-cAMP (10^{-3}M) + carbachol (10^{-4}M); filled squares, 8Br-cAMP (1.4×10^{-3}M); filled diamonds, 8Br-cAMP (1.4×10^{-3}M) + carbachol (10^{-4}M); filled circles, forskolin (5×10^{-4}M); open triangles, forskolin (5×10^{-4}M) + carbachol (10^{-4}M); open diamonds, 8Br-cAMP (10^{-3}M) + OAG (10^{-5}M); crosses, 8Br-cAMP (10^{-3}M) + A23187 (10^{-5}M) + OAG (10^{-5}M).

mucosal pepsinogen content (Fig.5A). It may be possible that secretagogues induce the release of newly synthetized pools. The uncoupling between the two secretions can be also evidenced by the lack of correlation between maximal secretion obtained for the two parameters under different conditions of stimulation (Fig.5B).

In Search of Second Messengers

Cholinergic stimulation in gastric mucosa and in other systems, has been associated to an activation of phospholipase C (PLC) with the formation of diacylglycerol (DAG) and inositol 1,4,5 phosphate (IP_3) that induces Ca^{2+} release and an increase of $[Ca^{2+}]_i$ (Chew and Brown 1986, Chiba et al 1989). In this sense, Ca^{2+} release has been evidenced with acetylcholine and histamine in the toad gastric mucosa, as it occurs in mammalian species (Michelangeli and Ruiz, 1984). However, DG or Ca^{2+} alone seems to be a poor stimulant since the Ca^{2+} ionophore A23187, induced a little but significant response in pepsinogen secretion without modifying acid secretion. The DAG analogue, octanoilacylglycerol (OAG) was not able to stimulate any of the two parameters (Fig.6). However, the analysis of the response obtained with

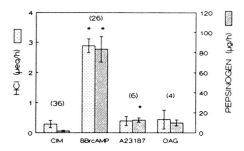

Figure 6. Maximal rate of acid and pepsinogen secretion attained by mucosae stimulated by 8Br-cAMP (10^{-3}M), A23187 (10^{-5}M) and OAG (10^{-5}M) in the presence of cimetidine (10^{-3}M).

joint addition of 8Br-cAMP, OAG and A23187 shows potentiation between the three different pathways (Fig.7). Joint addition of OAG + 8Br-cAMP induced a potentiated pepsinogen response but not on acid secretion. A23187 also induced a new stimulation of pepsinogen secretion over the level attained with 8Br-cAMP. The joint addition of 8Br-cAMP, OAG and A23187 induced an over effect on acid and pepsinogen secretion. These types of interaction were similar of those obtained with carbachol when it was used in combination with cAMP-dependent secretagogues. The difference of sensivity to secretagogues can be explained by a difference of sensitivity to the second messengers involved in the stimulation of two functions. The potentiated response of pepsinogen secretion obtained with the combined stimulation of histamine and carbachol may be due to the interaction between different second messengers generated by the secretagogues, cAMP, Ca^{2+} and diacylglycerol.

In conclusion, cAMP seems to be a potent stimulant of both functions and can activate the chains of events required to produce the two types of secretion (Fig.8). The second messenger generated through the activation of the phospholipase C seem to be more efficient to stimulate pepsinogen secretion than acid secretion. Both of them, involve fusion mechanism. Fusion of the apical membrane with the tubulo-vesicular system, which contains H^{+}/K^{+} pumps, and with the pepsinogen granules. These processes may be differentially sensitive to the second messengers, but at this

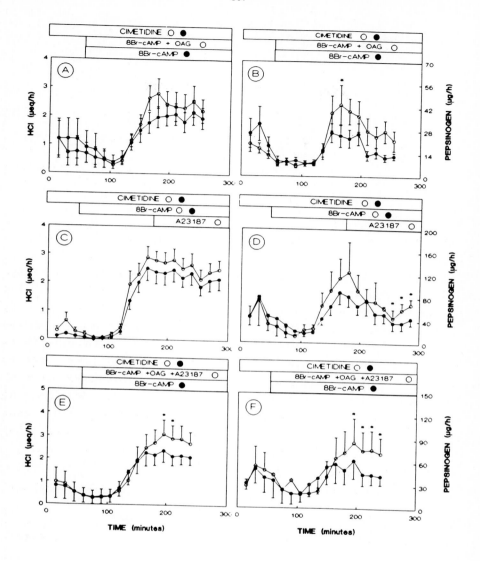

Figure 7. Time course of acid and pepsinogen secretion by groups of paired resting hemi-mucosae of *Bufo marinus* stimulated by joint addition of 8Br-cAMP (10^{-3}M) + OAG (10^{-5}M)(A and B), 8Br-cAMP (10^{-3}M) + A23187 (10^{-5}M)(C and D), or 8Br-cAMP (10^{-3}M) + OAG (10^{-5}M) + A23187 (10^{-5}M)(E and F). Values are means ± SEM for 7 to 9 experiments in each series. * p< 0.01 in paired t test comparing the values of each parameter respectively in the 2 groups at each time point.

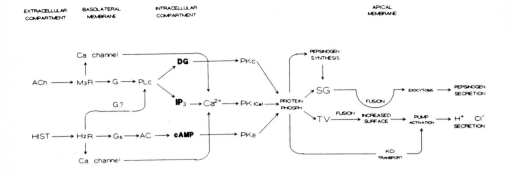

Figure 8. Model of the possible extracellular and intracellular pathways involved in acid and pepsinogen secretions by the oxyntopeptic cell.

point we do not know if the fusion processes occur parallely, independently or sequentially. However, the partial inhibition of pepsinogen secretion, when the acid secretion is totally inhibited by an inhibitor of $H^+/K^+ATPase$, omeprazol, seems to indicate a partial interdependence of the two secretions (Ruiz et al 1993). The interactions at the extracellular level between the secretagogues and at the intracellular level may have an important role in physiological regulation. The uncoupling of the two functions in the oxyntopeptic cell can explain the success, at the evolutionary level, of the separation of these, in two cellular types in mammals.

BIBLIOGRAPHY
Barrington EJW (1942) Gastric digestion in the lower vertebrates. Biol. Rev. 17:1-27
Chew CS and Brown MR (1986) Release of intracellullar Ca^{2+} and elevation of inositol trisphosphate by secretagogues in parietal and chief cells isolated from rabbit gastric mucosa. Biochem. Biophys. Acta. 888:116-125
Chew CS, Hersey SJ, Sachs G and Berglindh T (1980) Histamine responsiveness of isolated gastric glands. Am. J. Physiol. 238:G312-G320
Chiba T, Fisher S, Agranoff B and Yamada T (1989) Autoregulation of muscarinic and gastrin receptors on gastric parietal cells. Am. J. Physiol. 256:G356-G363
Ekblad EBM (1980) Histamine and cAMP as possible mediators of acetylcholine induced acid secretion. Am. J. Physiol. 239:G255-G260
Forte TM, Machen TE and Forte JG (1971) Ultrastructural changes in oxyntopeptic cells associated with secretory function: A membrane recycling hypothesis. Gastroenterology 73:782-786

Geuze JJ (1971) Light and electron microscope observations on a gastric mucosa of the frog (*Rana esculenta*). I.Normal structure. Z. Zellforsch. 117, 87-102

Jirón C, Romano M. and Michelangeli F (1984) A study of dynamic membrane phenomena during the gastric secretory cycle: Fusion, retrieval and recycling of membranes. J. Memb. Biol. 79,119-134

Michelangeli F and Ruiz MC (1984) Ca^{2+} as modulato of gastric secretion and vesicular H^+ transport In: "Hydrogen ion transport in epithelia" (J.G. Forte, D.G. Warnock y F.C. Rector,Eds). Jr. New York: Wiley Interscience, pp 149-159

Langley JN (1881) On the histology and physiology of the pepsin-forming glands. Phil. Trans. Roy. Soc., London, Ser. B 72:663-713

Inoue, M., J. Fong, G. Shah and B.I. Hirschowitz (1881) Mediation of muscarinic stimulation of pepsinogen secretion in the frog. Am. J. Physiol. 248:G79-G86

Ruiz MC and Michelangeli F (1984) Evidence for a direct action of acetylcholine on the gastric oxyntic cell of the amphibian. Am. J. Physiol. 246:G16-G25

Ruiz MC and Michelangeli F (1986) Stimulation of oxyntic and histaminergic cells in gastric mucosa by gastrin C-terminal tetrapeptide. Am. J. Physiol. 251:G529-G537

Ruiz MC Acosta A Abad MJ and Michelangeli F (1993) Non parallel secretion of pepsinogen and acid by the gastric oxyntopeptic cell of the toad (*Bufo marinus*). Am. J. Physiol(in press)

Chew CS, Hersey SJ, Sachs G and Berglindh T (1980) Histamine responsiveness of isolated gastric glands. Am. J. Physiol. 238: G312-G320

Soll AH and Wollin A (1979) Histamine and cyclic AMP isolated parietal cells. Am. J. Physiol. 237:E444-E450

Smit H (1968) Gastric secretion in the lower vertebrates and birds. In: Handbook of Physiology, 6: V. Edited by Code CF, American Physiological Society, Washington p 2791-2805

ACKNOWLEDGEMENTS This work was supported by CONICIT (Venezuela) Grant S1-2221.

The Canine Gastric Muscularis Mucosae and Acid Secretion: A Basket Case?

P.K. Rangachari, M.J. Muller, T. Prior and R.H. Hunt
Intestinal Diseases Research Programme
Dept. Medicine, McMaster University
Hamilton, ON L8N 3Z5

INTRODUCTION

The gastrointestinal tract is designed to digest and absorb ingested nutrients. Appropriate digestion requires the provision of an appropriate milieu for the functioning of secreted enzymes and the co-ordinated movement of digested materials through the length of the gut. Thus motility and secretion need to be interlinked and co-ordinated (Greenwood and Davison, 1987). How this is brought about is still a matter of debate.

With specific reference to the stomach, the frequent association between gastric motor activity and secretion of acid and pepsin has been established (Greenwood and Davison, 1987). The relevance of circular muscle contractions to secretion has also been demonstrated (Yates et al., 1978). However, the precise function of the muscularis mucosae, a thin band of smooth muscle that lies at the base of the gastric mucosa is still uncertain, though several physiological roles have been postulated. Since the arterioles and venules pass through the muscle before branching, contractions and relaxations may significantly affect blood flow to the mucosa. The tone of the muscle could, also, modulate the surface area of the mucosa and thus alter the absorptive surface available.

A more interesting possibility stems from recent studies by Seelig et al., (1985). Using careful three-dimensional reconstructions they have shown "the existence of a unique association between the mucosal smooth muscle strands and the gastric glands. The muscle strands arose from the muscularis mucosa at regular intervals and became branched to form an intricate wrap around a series of gastric glands that empty into one pit." They suggested that nerve-stimulated contractions of this muscle network may cause a compression of the glands and the ultimate ejection of

NATO ASI Series, Vol. H 89
Molecular and Cellular Mechanisms
of H⁺ Transport
Edited by Barry H. Hirst
© Springer-Verlag Berlin Heidelberg 1994

the glandular contents. That acid and pepsin may be transported across the mucus gel only at restricted sites was suggested by Holm and Flemstrom (1990), who performed *in vivo* microscopy of rat gastric surface using pH-sensitive dyes. Later studies by Holm et al., (1992) showed that gastric mucosal gland luminal pressure measured *in vivo* increased during stimulated acid secretion. Thus the combination of an increase in volume secretion and an increase in muscle activity may lead to an increased hydrostatic pressure which forces volume secretion out of the mucus gel (Bhaskar et al., 1992).

Alterations in the functioning of the muscularis mucosae could result in accumulation of acid and/or pepsin in the mucosa. Accumulation of aggressive factors such as acid and pepsin as well as alterations in mucosal blood flow and/or mucosal folding may be involved in the pathogenesis of gastric ulceration (Wallace, 1990).

Our interest in the properties of this muscle stemmed from an earlier study, where we found that stimulation of acid secretion by pentagastrin in an isolated preparation from the canine stomach was brief but explosive (Rangachari and McWade, 1989). Increases in acid secretion that attained rates as high as 16 $\mu Eq \cdot cm^{-2} \cdot h^{-1}$ were noted, although these rates lasted for only a brief period (6 min.). We suggested that this could result from a stimulation of the smooth muscle fibres underlying the glandular mucosa.

We therefore, undertook a systematic study of the pharmacological properties of this muscle strip, paying special attention to the effects of nerve stimulation and the well-known gastric secretagogues (acetylcholine, histamine, and gastrin).

MATERIALS AND METHODS

Preparation

Mongrel dogs of either sex were killed by intravenous injection of phenobarbital. The stomach was removed and placed in oxygenated (95% O_2 / 5% CO_2) Krebs buffer solution of the following composition (in mM): 120.9 NaCl, 1.2 NaH_2PO_4, 14.4 $NaHCO_3$, 5.9 KCl, 2.5 $CaCl_2$, 1.2 $MgCl_2$, 11.5 D-glucose. The pH of the Krebs buffer solution was 7.4. The lesser curvature of the stomach was isolated and the submucosal plexus plus muscularis mucosae was obtained by first removing the external longitudinal and circular layers of smooth muscle and then peeling off the mucosa. This preparation of submucosal plexus plus muscularis mucosae was cut into strips of approximately 2 mm by 10 mm in the direction of the external longitudinal muscle layer.

Recording of Responses to Drugs and Field Stimulation

The muscle strips were placed in conventional organ baths containing continuously oxygenated (95% O_2 / 5% CO_2) Krebs buffer solution at 37°C with one end tied to a hook and the other end to a Grass FT03C force transducer. Isometric contractions were recorded on a Grass 7D polygraph recorder. The strips were placed under 1 g tension (which gave maximum responses to carbachol in preliminary experiments), and allowed to equilibrate for at least 60 min before the addition of drugs. Agonist concentration-response curves were obtained by addition of increasing concentrations of the agonist to the muscle strips in a cumulative manner. Antagonists were applied to the preparation 30 min before addition of the agonist.

To obtain responses to field stimulation, tissue strips were passed between two concentric ring electrodes with one end tied to a hook and the other end to a Grass FT03C force transducer. The strips were placed under 1 g tension and allowed to equilibrate for at least 1 hour before electrical field stimulation was begun. This was done by passing a current of 0.5 ms duration and frequencies ranging from 1-32 Hz and supramaximal voltage through the ring electrodes. These responses were abolished by pretreatment with tetrodotoxin (TTX).

Analysis of Data

The results are expressed as the mean ± standard error of the mean (SEM). N and n denote the number of tissues and the number of dogs, respectively. The student's paired t-test was used to determine the statistical significance of the difference between groups. The groups were considered different when $p < 0.05$.

Materials

Atropine, carbachol, histamine and pentagastrin were purchased from Sigma Chemical Company, USA. Mepyramine maleate was purchased from Baker Ltd., UK. 2-Pyridylethylamine (2-PEA) and dimaprit were gifts from Dr. M. Parsons, Smith, Kline and French, UK. Ranitidine was a gift from Glaxo, Canada. All other chemicals were purchased from Sigma Chemical Company, USA, and were of the highest chemical grade available.

RESULTS:

Responses to Field Stimulation

Three distinct response patterns were seen when tissue strips were subjected to field stimulation. Contraction was the dominant response, being seen in 74% of 66 strips from 11 animals. It was observed that 9% of the strips exhibited a relaxation whereas 17% responded with an initial relaxation followed by a contraction (see Fig. 1). All the responses elicited were inhibited by tetrodotoxin, suggesting that these were nerve-mediated.

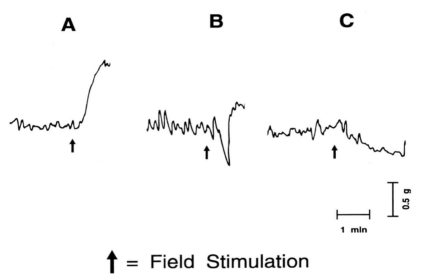

↑ = Field Stimulation

Fig. 1. Shows the three pattern of responses noted to field stimulation (20Hz. and 0.5ms duration, supramaximal voltage) of 66 muscle strips from 11 animals. The first pattern (A) was seen in 74% muscle strips tested. Pattern B and C were seen in 17% and 9% of strips respectively.

To define the putative neurotransmitters involved, we used a pharmacological approach by adding selective antagonists. The contractile responses to field stimulation were inhibited by pretreatment with either atropine (1 μM) or phentolamine (1 μM) (see Fig. 2). When both were present, contractions were abolished and field stimulations elicited only relaxations.

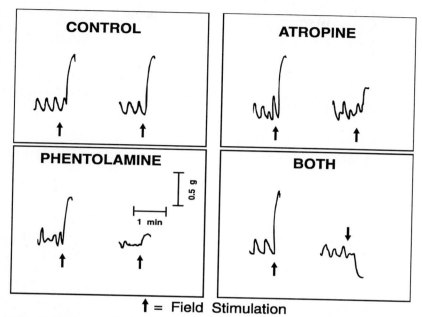

↑ = Field Stimulation

Fig. 2. Representative traces from individual experiments showing the inhibitory effects of pretreatment with either atropine alone, phentolamine alone, or a combination of both. The contractile responses are reduced by either of the antagonists used separately. In the presence of both, nerve stimulation elicits a relaxation.

The relaxations obtained under such conditions were frequency dependent as shown in Fig. 3. This suggested that functional antagonism existed between contractile and inhibitory neurotransmitters in this tissue. Thus the non-adrenergic, non-cholinergic component (NANC) was predominantly inhibitory. To search for the putative neurotransmitter involved, we used tissue strips that had been pretreated with both atropine and phentolamine.

The relaxant responses obtained under such conditions were inhibited by pretreatment with the inhibitor of nitric oxide synthesis, N-omega-L arginine (1 mM) (as shown in Fig. 4). This effect was reversed by L-Arginine (10 mM) but not by an equivalent concentration of D-arginine (Fig. 5a and 5b, respectively).

ATROPINE AND PHENTOLAMINE

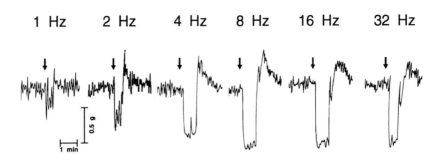

↓ = Field Stimulation

Fig. 3. A trace from a single experiment showing that in the presence of atropine (1 μM) and phentolamine (1 μM), nerve stimulation elicits a frequency-dependent relaxation. Thus the non-adrenergic non-cholinergic response (NANC) is a relaxant one.

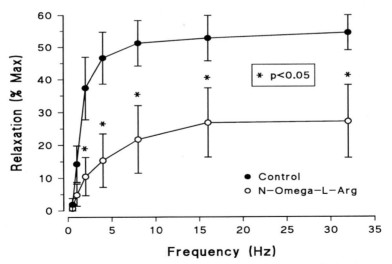

Fig. 4. The NANC response obtained in the presence of atropine and phentolamine is inhibited by pretreatment with N-omega-L-Arginine (1mM). Data are taken from 6 animals.

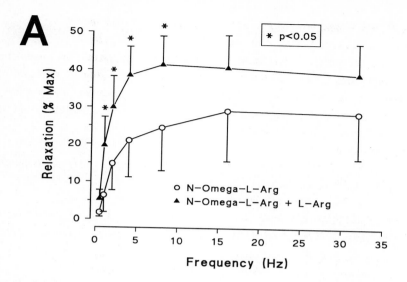

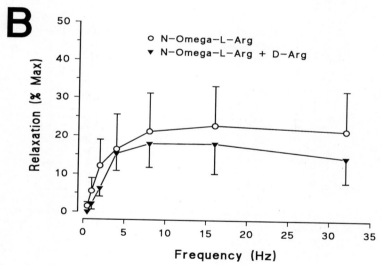

Fig. 5. The inhibitory effects of (A) N-omega-L-Arginine are selectively reversed by L-arginine (10 mM), (B) but not by D-Arginine (10mM). Data taken from 8 animals.

Tissue strips also relaxed when treated with sodium nitroprusside. These effects were not blocked by the inhibitor of NO synthesis (Sanders and Ward, 1992).

<u>Responses to Gastric Stimulants</u>

Gastric acid secretion is markedly stimulated by histamine, gastrin and acetylcholine. Given our general hypothesis that the muscularis mucosae play a significant role in propulsion of acid from the glandular lumen, it was essential that we test the responses of the muscle strips to those three agonists. We chose to use carbachol and pentagastrin for acetylcholine and gastrin.

There were two distinct responses to histamine (Muller et al., 1993). Sixty percent of the strips tested responded to histamine (10 µM) by contracting. The same concentration produced relaxation in the remaining tissues. At higher concentrations, only contractile responses were seen. The possibility that these responses were elicited by stimulation of different receptor subtypes was tested.

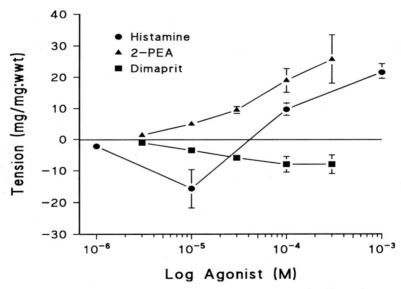

Fig. 6. A comparison of muscle strip tissue responses for histamine, a selective H_1 agonist (2-pyridylethylamine, PEA), and a selective H_2 agonist (dimaprit). Data from 4 animals. Reprinted from Life Sciences 52:PL49-53, 1993 with permission, see acknowledgements.

An H_1 selective agonist (2-Pyridylethylamine or 2-PEA), elicited concentration-dependent contractions, whereas an H_2 selective agonist (dimaprit), elicited only relaxations (Fig. 6). The responses to 2-PEA were selectively antagonised by the H_1 antagonist, mepyramine, and unaffected by the H_2 antagonist, ranitidine. The converse was true for dimaprit. The responses elicited to either histamine, 2-PEA, or dimaprit were found to be unaffected by TTX (Fig. 7).

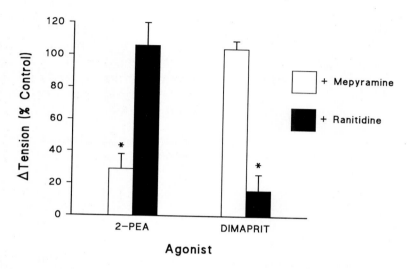

Fig. 7. Tissue responses to 2-PEA are antagonised by a selective H_1 antagonist, mepyramine but not by a selective H_2 antagonist, ranitidine. The converse is true for dimaprit. Data taken from 9 animals. Reprinted from Life Sciences 52:PL49-53, 1993 with permission, see acknowledgements.

Carbachol caused a concentration-dependent contraction of the gastric muscle strips. Atropine pretreatment shifted the concentration response curve to the right and increased the ED_{50} from 3.7 μM to 190 μM (Fig. 8).

Pentagastrin had only marginal effects on the contractility of the muscle strips. In a small number of strips (8 out of 42), a weak relaxant effect was seen.

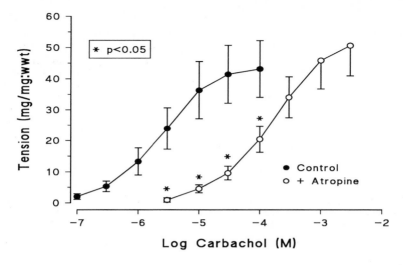

Fig. 8. Tissue responses to carbachol are antagonised by atropine.(10^{-7}M). Data taken from 6 animals.

DISCUSSION:

The precise functions of the gastric muscularis mucosae are uncertain. Given its location, it is possible that it may play a role in the expulsion of acid and pepsin from the gastric glands. The arguments for such a role have been mentioned above.

Our studies were designed to obtain some information regarding the functional innervation and pharmacological properties of this muscle strip and to assess the potential for its responsiveness to gastric stimulants. The results obtained can be summarised as follows:

1. Neural stimulation evoked both contractile as well as relaxant effects.

2. The contractile effects were inhibited by cholinergic and adrenergic receptor antagonists.

3. The relaxant effects were prevented by inhibiting the synthesis of nitric oxide.

4. Histamine evoked both a contractile and a relaxant effect with different receptor subtypes being involved. Contractile effects involved H_1 receptors whereas relaxant effects involved H_2 receptors.

5. Carbachol-induced contractile effects were antagonised by atropine.

6. Pentagastrin occasionally elicited a weak relaxant effect.

We confined our attention to the muscularis mucosae in the region of the lesser curvature. This is the region where ulcers are most commonly found. However the contractile properties of this muscle could vary not only between species but also between different regions of the same species. In humans, Walder (1953) found that whereas acetylcholine caused contractions of all strips, the responses to adrenaline were dependent on the site examined. Thus pieces from the lesser curvature relaxed in responses to adrenaline, whereas those from the greater curvature exhibited contractions. Noradrenaline caused a transient relaxation in all preparations "followed by a period of slow rhythmic contractions of augmented amplitude." In that study, histamine caused only a slight contraction. In the canine stomach, the muscularis mucosae from the antral region of the dog stomach appears to behave differently towards neural stimulation eliciting only a relaxation (Angel et al., 1983). VIP appeared to be the inhibitory neurotransmitter involved.

Several of the properties that we have observed in our study distinguish this muscle from other muscles in the gastrointestinal tract.

Thus in general gastrointestinal muscles respond differentially to cholinergic and adrenergic stimulation. Cholinergic responses are contractile and adrenergic responses are relaxant. Sphincters on the other hand respond to adrenergic stimulation by contracting. Thus this muscle strip appears to possess some of the properties of sphincteric smooth muscle (Phillips, 1992).

The dual effects of histamine are also distinctive. In most other regions of the gastrointestinal tract, histamine elicits only a contractile response, the relaxant effects being much less common. It is also interesting that the relaxant effects involve H_2 receptors which induce gastric acid secretion.

The effects of adenosine on this tissue were also found to be distinctive, since the contractile effects involved A_2 rather than A_1 receptors as in most other smooth muscles (Muller et al., 1992).

It is interesting in our experiments, that the inhibition of nitric oxide synthesis does not completely block the relaxations produced by nerve stimulation. We have not conclusively excluded a role for VIP in producing further relaxation. However, preliminary experiments suggested that the effects of VIP were slight and inconsistent.

How the properties described in our studies relate to a possible role for this muscle in gastric function is not entirely clear. We have mentioned at the outset the studies by Seelig et al., (1985) on the organisation of the muscle fibres in relation to the architecture of the gastric gland. The authors had suggested that these fibres could be stimulated to propel acid out of the glandular lumen (see Fig. 9). Later studies have provided further support for such a possibility.

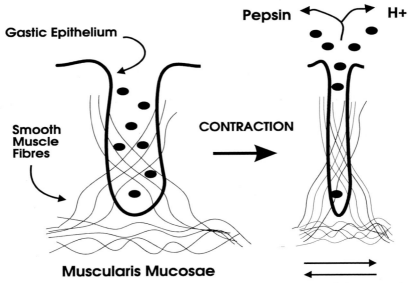

Fig. 9. A schematic diagram showing the organization of the smooth muscle "baskets" around the gastric glands. Smooth muscle strands arising from the muscularis mucosae form "an intricate wrap around a series of gastric glands that empty into one gastric pit" (Seelig et al., 1985). The muscularis mucosae is supplied by cholinergic, adrenergic and NO-ergic fibres. In addition, there are receptors for histamine and adenosine. Contractions of these fibres in response to neuronal stimulation or inflammatory mediators can modulate expulsion of acid and pepsin from the gastric gland.

Our results point to certain unique aspects of this muscle. These include the observation that both cholinergic and noradrenergic stimulation elicit contractile responses. A relaxant nitroxidergic pathway also exists. The coordination between these neural pathways could lead to a fine regulation of the propulsion of acid and pepsin from the lumen. In the presence of inflammatory mediators such as histamine or adenosine, these regulatory mechanisms could be disrupted leading to retention of acid and pepsin with consequent deleterious effects.

ACKNOWLEDGEMENTS:

This work was supported by operating grants to PKR from the Medical Research Council (MRC) of Canada. T.P. is a technical assistant funded by MRC. Figures 6 and 7 are reprinted from Life Sciences, Vol. 52, Muller et al., H1 contractile and H2 relaxant receptors in canine gastric muscularis mucosae, PL49-52, Copyright (1993), with kind permission from Pergamon Press Ltd., Headington Hill Hall, Oxford OX3 0BW, UK

REFERENCES·

Angel F, Schmalz PF, Morgan KG, Go VLW, Szursewski JH (1982) Innervation of the muscularis mucosa in the canine stomach and colon. Scand. J. Gastroenterol. 71:71-75.

Angel F, Go VLW, Schmalz PF, Szursewski JH (1983) Vasoactive intestinal polypeptide: a putative transmitter in the canine gastric muscularis mucosa. J. Physiol. 341:641-654.

Bhaskar KR, Garik P, Turner BS, Bradley JD, Bansil R, Stanley HE, Lamont JT (1992) Viscous fingering of HCL through gastric mucin. Nature 360:458-461.

Greenwood B and Davison JS (1987) The relationship between gastrointestinal motility and secretion. Am. J. Physiol. 252(Gastrointest. Liver Physiol.15): G1-7.

Holm L, Ågren J, Persson AEG (1992) Stimulation of acid secretion increases the gastric gland luminal pressure in the rat. Gastroenterology 103:1797-1803.

Holm L, Flemstrom G (1990) Microscopy of acid transport at the gastric surface in vivo. J. Intern Med 228(Suppl 1)91-95.

Muller MJ, Donoff B, Prior T , Rangachari PK, Hunt RH (1992) Adenosine receptor-mediated contraction of canine gastric muscularis mucosae. (Abstract) Gastroenterology 102:A490.

Muller MJ, Prior T, Hunt RH, Rangachari PK (1993) H_1 contractile and H_2 relaxant receptors in canine gastric muscularis mucosae. Life Sciences 52: 49-53.

Phillips, SF (1992) Sphincters of the gastrointestinal tract: functional properties. In Sphincters: Normal Function — Changes in Diseases. eds. Daniel E.E., Tomita, T., Tsuchida, S., Watanabe, M.. CRC Press Inc., Boca Raton, Florida, pp. 7-27.

Rangachari PK, McWade D (1989) An "epithelial" preparation from the canine gastric mucosa: responses to secretagogues. Am. J. Physiol. 257 (Gastrointest.Liver Physiol.20); G46-51.

Sanders KM, Ward, SM (1992) Nitric oxide as a mediator of non-adrenergic noncholinergic neurotransmission. Am. J. Physiol. 262 (Gastrointest. Liver Physiol. 25): G379-G292.

Seelig LL, Schlusselberg DS, Smith WK, Woodward DJ (1985) Mucosal nerves and smooth muscle relationships with gastric glands of the opossum: an ultrastructural and three dimensional reconstruction study. Am. J. Anat 174: 15-26.

Walder DN (1953) The muscularis mucosae of the human stomach J.Physiol.120:365-372.

Wallace, JL (1990) Mucosal defence: new avenues for treatment of ulcer disease. In: Gastroenterology Clinics of North America, Peptic Ulcer Disease, vol. 19:1, ed. R.H. Hunt, Saunders, Philadelphia, PA, pp.87-100.

Yates JC, Schofield B and Roth SH (1978) Acid Secretion and motility of isolated mammalian gastric mucosa and attached muscularis externa. Am. J. Physiol. 234(3): E319-326.

THE RECEPTORS REGULATING ACID SECRETION ON AND OFF THE PARIETAL CELL

K. C. Kent Lloyd, D.V.M., Ph.D.
Andrew H. Soll, M.D.
CURE/UCLA Digestive Disease Core Center,
University of California, Los Angeles
Department of Veterans Affairs,
VA Medical Center,
West Los Angeles, California, 90073. USA

Summary

Using a combination of *in vitro* and *in vivo* experimental techniques has made possible the identification of receptors that regulate gastric acid secretion by the parietal cell. Some of the newest discoveries in the area of regulation of acid secretion are related to the cellular localization of physiologically relevant receptors for acid secretagogues and acid inhibitors. *In vitro* techniques have enabled the ability to isolate, culture, and study the function of parietal cells, gastrin-containing "G" cells, histamine-containing ECL cells and somatostatin-containing "D" cells, and the ability to clone genes encoding for specific receptors. *In vivo* techniques have enabled the testing of hypotheses generated in a physiological model from what we learn from *in vitro* studies. This experimental strategy has greatly enhanced our understanding of the physiological role and the regulation of parietal cell secretion by secretagogues and inhibitors produced by various cell types within the gastric mucosa.

NATO ASI Series, Vol. H 89
Molecular and Cellular Mechanisms
of H⁺ Transport
Edited by Barry H. Hirst
© Springer-Verlag Berlin Heidelberg 1994

Introduction

A simple balance exists between stimulatory and inhibitory factors that regulate acid secretion by the parietal cell. These factors function as endocrine, neurocrine, paracrine, and autocrine mediators and may either be chemicals or polypeptides. The interaction of these factors with specific receptors on and off the parietal cell provide for fine-tuning of the acid response to stimulation and for sufficiently redundant control mechanisms.

Pathways regulating acid secretion

Today we recognize three, and possibly four, pathways for the regulation of acid secretion. These pathways are endocrine, paracrine, neurocrine, and possibly autocrine. Each of these pathways also can be categorized by stimulatory and inhibitory mechanisms. Additionally, the pathways are interdependent. While it is important to know the existence of these pathways, it is even more important to understand which of these pathways is(are) physiological.

Mechanisms of regulation of the parietal cell

The classic debate over the mechanism of secretagogue action on the parietal cell centers around the identification of the site of physiologically important receptors for the three acid secretagogues, histamine, gastrin, and acetylcholine. This debate can be understood if one considers the effect of histamine type-2 receptor (H2) antagonists on secretagogue action. H2 antagonists inhibit pentagastrin, cholinergic, and histamine-stimulated acid secretion. When only these results are considered, it suggests that gastrin and acetylcholine act to stimulate acid secretion via histamine. However, we know also that atropine inhibits pentagastrin-stimulated acid secretion, but gastrin probably does not act via release of acetylcholine. To resolve these issues, we now hypothesize that the action of gastrin depends upon histaminergic and muscarinic input to the parietal cell. Further, the action of gastrin is direct on the parietal cell, and is indirect through the release of histamine.

A recent article by Mezy and Palkovits (1992) further adds to the complexity of multiple receptors on multiple cell types having a physiologically relevant effect on acid secretion. They found mRNA for gastrin, histamine, and acetylcholine receptors on immunocytes in the lamina propria of the stomach, but not on gastric epithelial cells. They concluded that the targets of antiulcer drugs seem to be cells of the immune system in the gut, and not parietal cells, as is generally believed. Clearly, *in vitro* and *in vivo* studies to date support and establish the presence of physiologically important secretagogue receptors on parietal cells, so that further work to understand the role of other cell types with similar receptors is necessary.

Physiological receptors for acid secretagogues: The Gastrin Story

Despite the intensity of investigations, "Where are the physiologically important receptors regulating acid secretion?" remains an important question. Are these receptors on parietal cells, histamine-containing enterochromaffin-like (ECL) cells, somatostatin ("D") cells, or other endocrine, paracrine, and/or immune cells, or all of the above?

We have applied the reductionist approach, using dispersed and separated canine fundic mucosal cells, to study the effect of agents regulating acid secretion and to localize their physiologically-relevant receptors. In the case of gastrin, morphological transformation of the parietal cell from a quiescent state to a stimulated state can be seen in response to activation by gastrin. A similar effect also would be seen upon stimulation by cholinergic agonists and by histamine. These results indicate that the parietal cell is responding in a "physiological" fashion to each of the three secretagogues.

A second example of experimental results using the reductionist approach is that demonstrating a functional response of cultured parietal cells. Gastrin, as do histamine and cholinergic agonists, induces intracellular accumulation of ^{14}C-aminopyrine by parietal cells, indicating functional activation of the parietal cell.

The reductionist approach has been successful in providing substantial evidence for the existence of gastrin receptors on parietal cells. After elutriation of dispersed cells, the fraction containing principally parietal cells is largely devoid of histamine-containing ECL cells. Gastrin receptors are localized to parietal cells in cell separation studies and on autoradiography. In addition, the potentiation between gastrin and histamine is significantly greater than the maximal response to either secretagogue alone. Finally, the ultimate proof of a parietal cell gastrin receptor has been the cloning of this receptor by Alan Kopin, Larry Miller, and coworkers (1992). They conclusively demonstrated that the parietal cell gastrin receptor is related to the G-protein associated, adrenergic family of receptors that contain 7 transmembrane spanning domains.

Receptors for acid secretagogues on non-parietal cells

While the evidence for gastrin, as well as histamine and muscarinic receptors, on parietal cells is overwhelming, there also is increasing evidence for acid secretagogue receptors on other cell types as well. When the various cell types of the canine fundic mucosa are separated by elutriation, gastrin-binding to cells in each fraction closely parallels the percentage of parietal cells within each fraction. However, a small but significant amount of gastrin binds to non-parietal cells in fractions containing small cells, which include mast cells, somatostatin cells, and histamine ECL-cells. These hypotheses that gastrin acts on histamine ECL- cells are supported by data from Gerber (Gerber JG & Barnes JS, 1987), who has shown that, in the dog, gastrin induces release of histamine from stores in the gastric mucosa. If gastrin stimulates acid secretion principally by releasing histamine, these data indicate that it would do so by releasing histamine from mast cells or ECL-cells.

Attempting to establish the cellular origin, either mast cells or histamine ECL-cells, of histamine released by binding of gastrin receptors has been difficult because these two cell types co-elute in overlapping cell fractions. Therefore, our approach in fundic mucosal cells from dog has been to discriminate between these two cell types, mast cells and ECL-cells, using density gradient separation of cells which elutriate in the small cell fractions. Chuang *et al* (1992) has successfully separated mast cells from ECL-cells using this methodology, confirming previous evidence that ECL-cells are gastrin responsive, while mast cells are not. Functional studies of ECL-cells in short-term culture reveal that histamine is released into the cell culture supernatant by gastrin and by CCK in a dose-dependent fashion. Along with this evidence, data from *ex vivo* studies by Sandvik and Waldum (1987) demonstrate that in the isolated, perfused rat stomach, gastrin induction of histamine release coincides with the acid secretory response to gastrin.

All of this evidence indicates that, returning to the classic debate of where the receptors for acid secretagogues are localized, both hypotheses probably are true in dog, rabbit, rat, and man. However, the real question remains: "Which of these gastrin receptors exerts physiological control over secretion?"

Physiological receptors for acid inhibitors: The Somatostatin Story

Another important point of control is the role of somatostatin. Somatostatin is contained in "D" cells of the gastric fundic and antral mucosa. The functional studies of small cells containing somatostatin-like immunoreactivity cells in short-term culture were conducted by Soll and Yamada and coworkers (1984, 1985). Their studies demonstrate that gastrin is a poorer stimulant for the release of somatostatin than is CCK. This result also might explain why, *in vivo*, CCK is a poorer stimulant of acid secretion in dogs. The tendency for CCK to tip the balance in favor of acid inhibitory mechanisms could be explained by a greater efficacy of CCK than gastrin at somatostatin cell receptors. The selective type "A" CCK receptor antagonist, MK-329 or devazepide, inhibits CCK release of somatostatin from "D" cells *in vitro* and converts CCK into an almost full gastrin agonist *in vivo* (Lloyd KCK, 1992). These results lead to the question, "What mechanisms integrate the secretory response."

Clearly there is a need for integrating the acid secretory response. Along with gastrin, histamine, and acetylcholine, somatostatin is a leading candidate as an integrator of the acid secretory response. One site of action of somatostatin is the histamine ECL-cell. The somatostatin monoclonal antibody, S607, developed by Wong *et al* (1990), when added to short-term cultures of ECL-cells, markedly enhances gastrin-stimulated histamine release. This result implies that endogenous somatostatin tone causes inhibition of basal and gastrin-induced histamine release from ECL-cells. The mechanism of action of somatostatin is probably paracrine, since these cells possess elongated cytoplasmic processes that extend to within reach of gastric epithelial and subepithelial cells.

Conclusion

By returning to our theme of recurrent and overlapping circuits regulating acid secretion, we can see that numerous questions remain unresolved. A current area of research centers around the question, "What is the relative physiological importance of gastrin receptors on different cell types?" We have at least two cell types to investigate, the parietal cell and histamine ECL-cell, and several key physiological functions to study, including secretory activation, trophism, and cellular maturation.

References

Chuang C-N, Tanner M, Chen MCY, Davidson S, Soll AH (1992) Gastrin induction of histamine release from primary cultures of canine oxyntic mucosal cells. *Am J Physiol Gastrointest Liver Physiol* 263:G460-G465.

Gerber JG, Barnes JS (1987) Histamine release in vivo by pentagastrin from the canine stomach. *Journal of Pharmacology & Experimental Therapeutics* 243:887-892.

Kopin AS, Lee Y-M, McBride EW, et al (1992) Expression cloning and characterization of the canine parietal cell gastrin receptor. *Proc Natl Acad Sci USA* 89:3605-3609.

Lloyd KCK, Maxwell V, Kovacs TOG, Miller J, Walsh JH (1992) Cholecystokinin receptor antagonist MK-329 blocks intestinal fat-induced inhibition of meal-stimulated gastric acid secretion. *Gastroenterology* 102:131-138.

Mezey E, Palkovits M (1992) Localization of targets for anti-ulcer drugs in cells of the immune system. *Science* 258:1662-1665.

Sandvik AK, Waldum HL, Kleveland PM, Schulzesognen B (1987) Gastrin produces an immediate and dose-dependent histamine release preceding acid secretion in the totally isolated, vascularly perfused rat stomach. *Scand J Gastroenterol* 22:803-808.

Soll AH, Yamada T, Park J, Thomas LP (1984) Release of somatostatin-like immunoreactivity from canine fundic mucosal cells in primary culture. *Am J Physiol Gastrointest Liver Physiol* 247:G558-G566.

Soll AH, Amirian DA, Park J, Elashoff JD, Yamada T (1985) Cholecystokinin potently releases somatostatin from canine fundic mucosal cells in short-term culture. *Am J Physiol Gastrointest Liver Physiol* 248:G569-G573.

Wong HC, Walsh JH, Yang H, Taché Y, Buchan AM (1990) A monoclonal antibody to somatostatin with potent in vivo immunoneutralizing activity. *Peptides* 11:707-712.

Post-Receptor Signaling Mechanisms in the Gastric Parietal Cell

Catherine S. Chew, Ph.D.
Department of Medicine
Institute for Molecular Biology and Genetics
Medical College of Georgia
Augusta, GA 30912-3100
U.S.A.

Introduction

Because gastric parietal cell acid secretion is associated with peptic ulcer disease, there has been considerable interest in defining the mechanisms that regulate proton transport within this fascinating cell type. Three major agonists, histamine, acetylcholine and gastrin are involved in the stimulation of HCl secretion, which is ultimately initiated by insertion of the H,K-ATPase or proton pump into the mostly internalized, apically oriented canalicular membrane. At the level of the parietal cell, the histaminergic and cholinergic receptor subtypes have been identified respectively as histamine H_2 and muscarinic M_3 (cf Chew, 1991 for a recent review). A gastrin receptor has also been cloned from a canine parietal cell cDNA library and characterized as a CCK-B receptor subtype (Kopin et al, 1992). In contrast to the cellular membrane-associated receptors which allow the cell to communicate with other gastrointestinal cells via endocrine/paracrine/neural mechanisms, intracellular receptor-modulated events are less well characterized. Our present knowledge of the latter mechanisms are briefly described in the following sections.

Second messengers and receptor coupling mechanisms

The specific proteins linking membrane-bound receptors to intracellular activation events have not been unequivocally identified; however, most evidence suggests that parietal cells like other cell types possess a complement of G proteins, some of which enhance second

NATO ASI Series, Vol. H 89
Molecular and Cellular Mechanisms
of H^+ Transport
Edited by Barry H. Hirst
© Springer-Verlag Berlin Heidelberg 1994

messenger production and others of which are inhibitory. Cholera toxin, which increases NAD-dependent ADP ribosylation of Gsα (a G protein subunit that stimulates adenylyl cyclase activity with a resultant increase in cellular cAMP content) increases cAMP levels in isolated parietal cells and increases acid secretion (measured indirectly as accumulation of the weak base, ^{14}C-aminopyrine (AP)), with a similar time course. This toxin also stimulates incorporation of ^{32}P from labelled NAD into a parietal cell protein with a Mr similar to that reported for other cell types (Brown and Chew, 1987). The presence of a member or members of the Gi family is based mainly on experiments with pertussis toxin which initiates NAD-dependent ADP ribosylation of Giα, an event that reduces receptor-mediated inhibition of adenylyl cyclase. When isolated parietal cells are pre-incubated with pertussin toxin, the inhibition of histamine-stimulated AP accumulation by prostaglandins, somatostatin and epidermal growth factor is reversed (Atwell and Hanson, 1988; Brown and Chew, 1987; Lewis et al, 1990; Rosenfeld, 1986). Pertussis toxin has also been shown to increase the incorporation ^{32}P from labelled NAD into a parietal cell protein with a Mr similar to the pertussis-toxin sensitive protein detected in other cell types (Brown and Chew, 1987).

Gq-type proteins have not yet been localized to parietal cells. In other cell types, muscarinic receptors have been shown to be linked to either Gi or Gq proteins with M$_3$ receptor subtypes typically linked to the Gq species. When there is Gi linkage, pertussis toxin blocks cholinergic responses. Prolonged preincubation of cultured parietal cells with maximal concentrations of pertussis toxin (250 - 500 ng/ml, 4-24 hours) has no effect on carbachol-stimulated AP accumulation or on carbachol-induced increases in free intracellular calcium concentrations (Chew, unpublished observations). Thus, evidence to date suggests that Gq rather than Gi-like proteins modulate cholinergic receptor function in parietal cells.

The second messenger signaling events that are initiated upon agonist binding to specific membrane-bound receptors are now reasonably well documented. Histamine elevates cAMP content via H$_2$-receptor-linked activation of adenylyl cyclase. Histamine also elevates free intracellular calcium concentrations ($[Ca^{2+}]_i$) by an undefined mechanism which does not appear to involve generation of cAMP as this action is not mimicked by addition of cAMP analogs (Chew and Brown, 1986: Negulescu et al, 1989; Chew, unpublished observations). In contrast to the cholinergic agonist, carbachol, and the peptide, gastrin, histamine does not appear to increase significantly parietal cell $[Ca^{2+}]_i$ through a phospholipase-dependent breakdown of phosphatidylinositol 4,5-bisphosphate (PtdIns(4,5)P$_2$) (Chew and Brown, 1986). However, when cloned histamine H$_2$-receptor cDNA is expressed in COS cells, histamine addition increases InsP$_3$ as well as cAMP concentrations (DelValle et al, 1992). Since G protein coupling mechanisms in cell lines such as COS may be different from those in the parietal cells, data from these cells may not accurately reflect parietal cell functions. Another possibility is that histamine does increase PtdIns(4,5)P$_2$ breakdown in parietal cells but the response is transient and, perhaps, oscillatory. In approximately 30% of parietal cells, oscillations in the calcium signaling pattern have been observed. Since only ~70% of parietal cells respond to histamine as compared to a 100% response to carbachol, it could be that the combination of transient and/or oscillatory responses along with fewer responding cells make it difficult to demonstrate histamine-stimulated increases in InsP$_3$ concentrations. This latter argument seems unlikely because gastrin/CCK-stimulated increases in InsP$_3$ in enriched parietal cell populations are readily detected despite the fact that gastrin initiates calcium responses a similar proportion of the total parietal cell population as compared to histamine and the calcium responses to this agonist are also transient in contrast to carbachol (Chew and Brown, 1986; Chiba et al, 1988; Roche and Magous, 1989. See also figure 1 below).

Most evidence to date indicates that histamine stimulates HCl secretion via a cAMP-dependent mechanism with histamine-stimulated increases in $[Ca^{2+}]i$ playing either a minor or no role in the activation of the acid secretory response. In brief, data supporting this conclusion are as follows: 1) Chelation of extracellular calcium with EGTA inhibits carbachol- but not histamine-stimulated AP accumulation in isolated gastric glands and parietal cells (Berglindh et al, 1980; Chew, 1985; Soll, 1981); 2) Buffering of $[Ca^{2+}]i$ with BAPTA along with brief chelation of extracellular calcium with EGTA under conditions that completely suppress both carbachol and histamine-induced elevation of $[Ca^{2+}]i$ inhibits carbachol-stimulated AP accumulation in isolated gastric glands but has minimal to no effect on histamine-stimulated AP accumulation(Chew and Petropoulos, 1991; Negulescu et al, 1989); 3) Thapsigargin, a plant alkaloid that stably elevates parietal cell $[Ca^{2+}]i$, potentiates histamine- but not carbachol-stimulated AP accumulation (Chew and Petropoulos, 1991); 4) In single cultured parietal cells in which acid secretory-related responses are assessed simultaneously with intracellular calcium measurements, the correlation between histamine-stimulated changes in $[Ca^{2+}]i$ and secretion are inexact in that 25-30% of cells respond to histamine with an increase in secretion but not $[Ca^{2+}]i$; 5) Histamine does not detectably increase phosphorylation of a 28 kDa phosphoprotein that is responsive to carbachol and calcium ionophores (Brown and Chew, 1987 and unpublished observations). Since histamine does not uniformly elevate calcium in parietal cell populations as compared to carbachol which does, it might be argued that calcium does modulate histamine-stimulated secretion in a subpopulation of parietal cells. However, the "physiological" evidence suggests that if this is the case, the calcium-responsive sub-population does not significantly influence the response of the population as a whole. Another possibility that has not yet been systematically explored is that histamine-stimulated increases in parietal cell $[Ca^{2+}]i$ modulate cellular activities that are unrelated to the acute stimulation of HCl secretion.

Whether or not histamine releases calcium from the same intracellular calcium pool(s) as carbachol is uncertain. If the pools are shared, however, it is likely that histamine is unable to release as much calcium from these pools as carbachol because prestimulation of single parietal cells with maximal concentrations of histamine does not prevent subsequent release of calcium by thapsigargin. In contrast, pre-stimulation of these cells with maximal doses of carbachol almost completely abolishes the response to thapsigargin (Chew and Petropoulos, 1991).

In comparing the calcium signaling patterns to histamine, gastrin and carbachol in single parietal cells in primary culture, clear differences emerge. The two panels in figure 1 depict data obtained from digitized video image recordings of single parietal cells in primary culture that were loaded with the calcium-sensitive fluorescent indicator, fura 2. In the left panel, repetitive doses of a single maximal concentration of gastrin were administered which lead to the development of a rapid tachphyllaxis. In contrast, repetitive administration of maximal doses of carbachol did not alter the response to this agonist. There is also no apparent desensitization in the calcium response to histamine (not shown). Data in the right panel of figure 1 compares calcium signaling responses to histamine, carbachol and gastrin in a different cell. With gastrin, the initial rapid rise in calcium is characteristically followed by a decline to a steady state level that is only slightly above basal. Carbachol induces a substantially higher steady-state elevation in $[Ca^{2+}]_i$ whereas histamine induces a transient rise in $[Ca^{2+}]_i$ that is frequently followed by smaller and irregular spiking activity.

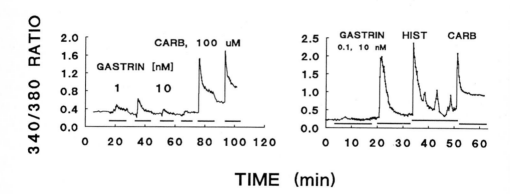

Figure 1. Intracellular calcium recordings from single cultured parietal cells loaded with the fluorescent calcium-sensitive indicator, fura 2. In the left panel, the cell was repetitively stimulated with gastrin then carbachol as indicated. In the right panel, a different cell was stimulated sequentially with gastrin, histamine and carbachol. (Modified from Chew et al, 1992 with permission).

It is not known whether the rapid diminution of the calcium response to repetitive gastrin stimulation is due to down-regulation of the gastrin receptor or to other events downstream from the receptor. However, this phenomenon may help to explain why the AP accumulation response to gastrin in acutely isolated and cultured parietal cells is so weak relative to carbachol and histamine. Another possible factor is the observation that only ~ 70% of cells that respond to carbachol also respond to gastrin with a rise in $[Ca^{2+}]_i$ (Chew et al, 1992).

Although it seems assured that parietal cells possess some type of InsP₃ receptor that modulates release of calcium from intracellular pool(s), the receptor has not yet been identified nor have agonist-sensitive calcium pools been localized to a specific membranous area within this cell type. Several cDNA clones coding for InsP₃ receptors have recently been sequenced and partially characterized. One of these isoforms, designated IP3R-3, was found to be expressed at high levels in small intestine, pancreas and kidney (Blondel et al, 1993). Whether IP3R-3 or a different receptor subtype is present in parietal cells remains to be determined. There is no evidence to support the presence in parietal cells of ryanodine receptors, which have been proposed to be involved in the mediation of calcium-induced calcium release in brain, skeletal muscle, and sea urchin but not frog eggs (Galione et al., 1993), for example. Data from experiments with single parietal cells loaded with fura-2 also does not support the presence of non-selective cation entry channels within the plasma membrane. Such channels have been postulated to serve as agonist-sensitive calcium entry pathways in some cell types based on experiments in which increased quenching of intracellular fura 2 in the presence of extracellular manganese has been detected following agonist addition. Since we were unable to detect these channels in parietal cells, we have postulated that agonist-stimulated calcium influx into the parietal cell may involve either an unidentified channel subtype with unique molecular characteristics as compared to the non-selective cation channels proposed for some cell types or to voltage-sensitive calcium channels of excitable cells. Alternatively, calcium entry into parietal cells might be modulated by some type of ion exchange mechanism (Chew et al, 1992).

Agonist-dependent protein kinase activation and intracellular protein phosphorylation

It is now well established that histamine activates cAMP-dependent protein kinase(s) and activation of a cytosolic type I isoform as well as type II isoforms have been reported (Chew, 1985; Jackson and Sachs, 1982; Goldenring et al 1992). Histamine has also been shown to increase in situ phosphorylation of at least four different phosphoproteins with approximate Mr's of 27, 40, 30 and 80 kDa (Chew and Brown, 1987; Oddsdottir et al, 1988; Uruishidani et al, 1987). The 80 kDa phosphoprotein has been identified as ezrin, a cytoskeletal protein that co-localizes with the H,K-ATPase (Hanzel et al 1991). Thus, it is possible that ezrin may be involved in the regulation of this enzyme's activity. As yet, the other phosphoproteins have not been extensively characterized or identified.

With respect to intracellular pathways modulated by calcium-dependent agonists, high levels of calcium-calmodulin dependent kinase II have been detected in parietal cells and indirect evidence based on the use of inhibitors of this enzyme suggest that it may be involved in the modulation of carbachol-stimulated HCl secretion (Funasaka et al, 1992). Although both carbachol and gastrin/CCK are thought to activate protein kinase C, the specific protein kinase C isozymes that are stimulated are unknown. Protein kinase C activity has been detected in parietal cell extracts and there is evidence based on the use of the protein kinase C activator, 12-O-tetradecanoylphorbol-13-acetate (TPA), that two phosphoproteins or a single protein present in different isoforms with Mr of 66 kDa, pI 5.7-5.9 are phosphorylated in response to carbachol and gastrin via a protein kinase C-dependent mechanism (Chew, 1985; Brown and Chew, 1989). Another protein of Mr 28 kDa, pI ~5 (pp28) exhibits increased phosphorylation when intracellular calcium is elevated with the calcium ionophore, ionomycin, or by carbachol. When the rise in intracellular calcium is prevented through the use of the intracellular calcium chelator, BAPTA, no increase in pp28 phosphorylation is detected (Brown and Chew, 1989). Thus, there is good evidence to show the calcium-dependency of pp28 phosphorylation. Preliminary data obtained from internal amino acid sequences of this protein indicate no clear homology with any known protein although there is 36% homology between this fragment and an internal seqence segment of the skeletal muscle cytoskeletal protein, dystrophin (Chew and Petropoulos, unpublished observations). There is presently a major effort being made to identify pp28 as well as the other unknown agonist-responsive phosphoproteins identified thus far in parietal cells. Clearly their identification and characterization will represent a major step forward in defining the regulated intracellular activation events involved in HCl secretion.

Post-receptor signaling events that may be involved in long-term regulation of parietal cell function

The recent development of a primary cultured parietal cell model (Chew et al, 1989), has made it possible to initiate studies of more long-term effects of a variety of agents including growth factors. Preliminary results with epidermal growth factor (EGF) and transforming growth factor α have suggested the intriguing possibility that these factors, which are known to inhibit acutely agonist-stimulated HCl secretion in vivo and in vitro, may also play a positive regulatory role. For example, when parietal cells are cultured overnight in serum-free media then exposed to EGF, there is an initial suppression of AP accumulation which is followed within approximately six hours by a reversal of this inhibition and, within twenty-four hours, by a significant enhancement in AP accumulation (Nakamura and Chew, 1992). Recent studies indicate that the enhancement is suppressed by the putative tyrosine kinase inhibitors, genistein and tyrphostin B56. Interestingly, when added acutely, genistein potentiates histamine-stimulated AP accumulation in isolated parietal cells (Tsunoda et al, 1993; Nakamura and Chew, 1993). Both genistein and tyrphostin B56 also potently inhibit the phosphorylation of a phosphoprotein with a Mr of ~45 kDa, pI ~6.8-7 which has a similar migration pattern to one of the MAP kinase isozymes, ERK1, on two-dimensional gels. In related experiments, both ERK1 and ERK2 isozymes have been detected in immunoblots of extracts from enriched parietal cells and EGF shown to activate both isozymes. Of potential importance is that both carbachol and the protein kinase C activator, TPA, also activate both isozymes (Nakamura and Chew, 1993). MAP kinases have been most studied in fibroblasts and cancer cell lines where they are thought to form part of an intracellular signaling pathway involved in the mediation of cell division and differentiation. Their apparent presence and activation in a highly differentiated cell type such as the gastric parietal cell suggests that these enzymes may also be involved in the long-term regulation of other cellular activities.

References

Atwell, M.M. and P.J. Hanson (1988) Effect of pertussis toxin on the inhibition of secretory activity by prostaglandin E_2, somatostatin, epidermal growth factor and 12-*O*-tetradecanoylphorbol 13-acetate in parietal cells from rat stomach. Biochim Biophys Acta 971: 282-288.

Berglindh T, Sachs G, Takeguchi N. (1980) Ca^{2+}-dependent secretagogue stimulation in isolated gastric glands. Am J Physiol 239: G90-94

Blondel O, Takeda J, Janssen H, Seino S, Bell GI (1993) Sequence and functional characterization of a third inositol trisphosphate receptor subtype, IP3R-3, expressed in pancreatic islets, kidney, gastrointestinal tract, and other tissues. J Biol Chem 268: 11356-11363

Brown MR and Chew CS (1987) Multiple effects of phorbol ester on secretory activity in rabbit gastric glands and parietal cells. Can J Physiol Pharmacol 65: 1840-1847

Brown MR, Chew CS (1989) Carbachol-induced protein phosphorylation in parietal cells: regulation by $[Ca^{2+}]i$. Am J Physiol 257: G99-G110

Chew CS (1985) Differential effects of extracellular calcium removal and non-specific effects of Ca^{2+} antagonists on acid secretory activity in isolated gastric glands. Biochim Biophys Acta 846:370-378

Chew CS (1991) Intracellular mechanisms in control of acid secretion. Curr Opin Gastroenterol 7:856-862

Chew CS, Brown MR (1986) Release of intracellular Ca^{2+} and elevation of inositol trisphosphate by secretagogues in parietal and chief cells isolated from rabbit gastric mucosa. Biochim Biophys Acta 888: 116-125

Chew CS, Brown MR (1987) Histamine increases phosphorylation of 27- and 40- kDa parietal cell proteins. Am J Physiol 253: G823-G829

Chew CS, Ljungström M, Smolka A, Brown MR (1989) Primary culture of secretagogue-responsive parietal cells from rabbit gastric mucosa. Am J Physiol 256:G254-263

Chew CS, Nakamura K, Ljungström M (1992) Calcium signaling mechanisms in the gastric parietal cell. Yale J Biol Med 65: xx-xx

Chew CS, Petropoulos AC (1991) Thapsigargin potentiates histamine-stimulated HCl secretion in gastric parietal cells but does not mimic cholinergic agonists. Cell Regul 2:27-39

Chiba T, Fisher SK, Park J, Seguin EB, Agranoff B, Yamada T (1988) Carbamoylcholine and gastrin induce inositol lipid turnover in canine gastric parietal cells. Am J Physiol 255: G99-G105

DelValle J, Wang L, Gantz I, Schäffer M, Yamada T (1992) Characterization of the H$_2$ receptor: linkage to both adenylate cyclase and Ca^{2+} signalling systems. Am J Physiol 263: G967-972

Funasaka M, Fox LM, Tang LH, Modlin IM, Goldenring JR (1992) The major calcium-binding protein in rabbit parietal cells is Ca^{2+}/calmodulin-dependent protein kinase II. Biochem Int 27: 1101-1109

Galione A, McDougall A, Busa WB, Willmott N, Gillot I, Whitaker M (1993) Redundant mechanisms of calcium-induced calcium release underlying calcium waves during fertilization of sea urchin eggs. Science 261: 348-352

Goldenring JR, Asher VA, Barreuther MF, Lewis JJ et al (1992) Dephosphorylation of cAMP-dependent protein kinase regulatory subunits in stimulated parietal cells. Am J Physiol 262: G763-773

Hanzel D, Reggio H, Bretscher A, Forte JG, Mangeat P (1991) The secretion-stimulated 80 K phosphoprotein of parietal cells is ezrin, and has properties of a membrane cytoskeletal linker in the induced apical microvilli. EMBO J 10: 2363-2373

Jackson RJ, Sachs G (1982) Identification of gastric cyclic AMP binding proteins. Biochim Biophys Acta 717: 453-458

Kopin AS, Lee Y-M, McBride EW, Miller LJ, Lu M, Lin HY, Kolakowski LF Jr, Beinborn, M: Expression cloning and characterization of the canine parietal cell gastrin receptor. Proc Natl Acad Sci 89:3605-3609, 1992

Lewis JJ, Goldenring JR, Asher VA, Modlin IM (1990) Effects of epidermal growth factor on signal transduction in rabbit parietal cells. Am J Physiol 258: G476-G483

Ljungström M, Chew CS (1991) Calcium oscillations and morphological transformations in single cultured gastric parietal cells. Am J Physiol 260: C67-C78

Nakamura K, Chew CS (1992) Role of tyrosine phosphorylation in mediation of epidermal growth factor effects on the gastric parietal cell. Molecular Biol Cell 3: 151A

Nakamura K, Chew CS (1993) Parietal cells contain an EGF-sensitive MAP kinase that is inhibited by a subclass of tyrosine kinase inhibitors. Gastroenterol 104: A155

Negulescu PA, Reenstra WW, Machen TE (1989) Intracellular calcium requirements for stimulus-secretion coupling in parietal cells. Am J Physiol 256: C241-51

Oddsdottir M, Goldenring JR, Adrian TE, Zdon MJ, Zucker KA, Modlin IM (1988) Identification and characterization of a cytosolic 30 kDa histamine stimulated phosphoprotein in parietal cell cytosol. Biochem Biophys Res Commun 154: 489-496

Roche S, Magous R (1989) Gastrin and CCK-8 induce inositol 1,4,5-trisphosphate formation in rabbit gastric parietal cells. Biochim Biophys Acta 1014:313-318

Rosenfeld GC (1986) Prostaglandins E$_2$ inhibition of secretagogue stimulation of [^{14}C] aminopyrine accumulation in rat parietal cells: a model for its mechanism of action. J Pharmacol Exper Ther 233: 513-518

Soll AH (1981) Extracellular calcium and cholinergic stimulation of isolated canine parietal cells. J Clin Invest 68:270-278

Tsunoda Y, Modlin IM, Goldenring JR (1993) Tyrosine kinase activation in the modulation of stimulation of parietal cell acid secretion. Am J Physiol G351-356

Urushidani T, Hanzel DK, Forte JG (1987) Protein phosphorylation associated with stimulation of rabbit gastric glands. Biochim Biophys Acta 930: 209-219

Relative Sensitivity to Acid Injury of Different Regions of the Intestinal Tract

A. Garner, G. Abu-Hijleh, R. Dib, S.M.A. Bastaki, A.C. Hunter,[*] A. Allen[*]
Departments of Pharmacology and Anatomy
Faculty of Medicine and Health Sciences
United Arab Emirates University
P.O. Box 17666
Al Ain
U.A.E.

Summary

We compared the development of mucosal damage in the duodenum, jejunum, ileum and colon of anesthetised rats following luminal exposure to relatively low concentrations of isotonic HCl. In the distal duodenum, lowering luminal pH by perfusion with 10 mM HCl for 10 min enhanced mucosal permeability as evidenced by an increase in clearance of ^{14}C-urea from blood to lumen. Histologically, injury ranged from villus oedema after instillation of 1 mM HCl for 30 min to widespread necrosis after 25mM HCl for 3 hr. Similar changes in mucosal permeability and/or histological appearance were observed in jejunum and terminal ileum. Transmucosal flux of urea in the proximal colon was lower and this tissue was far more resistant to HCl compared with small intestine. On the evidence of microscopic examination, the colon was unaffected by all except the most severe acid treatment (25 mM for 3 hr). There was no relationship between acid resistance and pre-epithelial protective factors in these various regions suggesting that tissue acidification is the primary determinant of HCl-induced mucosal injury in the intestine.

Introduction

The stomach displays a remarkable ability to withstand the deleterious effects of luminal acid and the majority of gastric ulcers are probably drug-induced. The duodenum is also regularly subjected to low luminal pH and, in this region of the GI tract, acid has a primary role in the genesis of ulceration. There are few studies of

[*]Department of Physiological Sciences, University of Newcastle Upon Tyne, Medical School, Newcastle Upon Tyne NE2 4HH, UK

NATO ASI Series, Vol. H 89
Molecular and Cellular Mechanisms
of H⁺ Transport
Edited by Barry H. Hirst
© Springer-Verlag Berlin Heidelberg 1994

the effects of low luminal pH in more distal regions of the intestine. Surgical diversion of gastric contents into the jejunum in dogs or the ileum in pigs has been shown to induce ulcers. These findings led Florey *et al.* (1939) to propose that the presence of Brunner's glands in the proximal portion of the small bowel conferred a degree of protection against proteolytic digestion. Interestingly, however, even these early studies reported evidence of damage in apparently 'acid-resistant' regions of the intestine when tissues were examined histologically.

Considerable progress has been made over the past few years in defining those elements which are important in protection of the stomach against autodigestion (see Garner *et al.*, 1990 for review). Recent investigations of acid resistance in the intestine have focused on the role of mucosal alkaline secretion, particularly in regions of the duodenum which are devoid of Brunner's glands. There studies reveal that cellular transport combined with paracellular diffusion of HCO_3^- is capable of maintaining neutraility in the surface mucus layer during exposure to low luminal pH (see Allen *et al.*, 1993 for review). In the present study, we compared resistance to luminal acid loads in different regions of the small intestine and the colon. Previous work has demonstrated that dilute HCl ($\geq$10mM) increases mucosal permeability and HCO_3^- efflux in the duodenum *in vitro* and *in vivo* (Feil *et al.*, 1987; Wilkes *et al.*, 1988). Here we have attempted to correlate responses to luminal acid in the various regions with passive permeability of the epithelium and dimensions of the mucus-bicarbonate barrier.

Methods

Experiments were performed on the distal duodenum, mid jejunum, terminal ileum and proximal colon. Morphological appearance was assessed by light microscopic examination. Mucosal permeability (Wilkes *et al.*, 1988), luminal alkalinization (Flemström *et al.*, 1982) and mucus thickness (Kerrs *et al.*, 1982) were measured using methods described in detail in previous publications. With the exception of mucus thickness measurements, which were obtained from rats sacrificed by survical dislocation, all other parameters were determined in anaesthetized animals. Wistar rats (180-250 g) were fasted for 18-24 hr with drinking water supplied *ad libitum*. Anaesthesia was induced by halothane inhalation and maintained with urethane (1.5 g/kg i.m.). Intrarectal temperature was kept at 36-37°C with the aid of thermostatically controlled heated tables. The abdominal cavity was accessed via a midline incision then a 1-2 cm segment of intestine with intact blood supply was located. Fluid at 37°C was either recirculated through the segment from a water-jacketed chamber by means of a 100 % oxygen gas lift or continuously perfused through the segment from a heated reservoir by means of a peristaltic pump to allow collection of discrete aliquots.

To determine the relative susceptibility of different regions to luminal acid, segments of intestine were exposed to 0.3 M mannitol or isotonic solutions containing 1, 5 or 25 mM HCl for 30 min or 3 hr. Tissues were removed, fixed in Bouin's solution then paraffin-embedded section (5 µm) stained with haematoxylin & eosin and examined by light microscopy. Passive mucosal permeability in these regions was determined from the flux of ^{14}C-urea into the lumen. Radiolabelled urea was infused i.v. in a dose of 100 µCi/kg and its appearance in the luminal perfusate measured at 10 min intervals by β scintillation counting. Rates of mucosal alkalinisation were measured by recirculation of unbuffered saline through the segments and continuous back-titration of the bathing solution to pH 7.40 with 50 mM HCl using a pH-stat.

Thickness of the surface mucus gel layer was measured in fasted rats immediately after sacrifice. Segments of intestine (2 cm) were excised, washed in saline and opened by cutting along the anti-mesenteric border. Tissues were placed flat on a Millipore filter with the luminal surface uppermost. Full thickness strips of intestine were cut between parallel razor blades spaced 1.6 cm apart then transferred to a microscope slide. These thick, unfixed sections were viewed with an inverse microscope and dimensions of the adherent gel layer recorded at 500 µm intervals using a graduated eye-piece. Three strips of tissue and about 10 readings per section were obtained from each piece of intestine.

Results and Discussion

Microscopic appearance of the intestinal mucosa following luminal acid exposure was determined on a scale of 0 to +++ (Table 1). Despite the criticism which can be levelled at qualitative assessment, it is abundantly clear that the small intestine is much more sensitive to acid-induced mucosal injury than the colon. Thus even exposure for 30 min to concentrations of HCl as low as 5 mM induced evidence of submucosal oedema in the duodenum and jejunum when compared with control tissues. Increasing the concentration of H^+ in the perfusate to 25 mM caused quite severe necrotic changes in both duodenum and jejunum. The terminal ileum appeared somewhat less sensitive to the deleterious action of acid compared with other parts of the small intestine. However, there was no evidence of colonic injury after 30 min of acid exposure.

Table 1. Microscopic assessment of mucosal damage induced by luminal acid

Region	HCl for 30 min			HCl for 3 hr		
	1 mM	5mM	25mM	1mM	5mM	25mM
distal duodenum	0	+	+++	++	+++	+++
mid jejunum	0	+	+++	++	++	+++
terminal ileum	0	0	++	++	+	++
proximal colon	0	0	0	0	0	+

Scale of damage; 0 = indistinguishable from control, + subepithelial oedema with minimal surface cell distruption, ++ necrotic changes confined to the superficial mucosa, +++ extensive damage including deep necrosis and loss of villi.

When the duration of exposure was increased to 3 hr, pathological changes were noted in all three regions of the small intestine even at the lowest acid concentration of 1 mM. Widespread necrotic changes occurred after 3 hr exposure to 5 mM HCl in the duodenum and 25 mM HCl in the jejunum. Again the terminal ileum appeared least susceptible. In contrast, the colon was unaffected after 3 hr exposure to 1 and 5 mM HCl. Only the most severe conditions tested (25 mM HCl for 3 hr) produced colonic damage and, even after this treatment, pathological changes were confined to oedema with only minimal disruption of the surface cell layer. The dramatic difference in sensitivity between the small and large intestine is illustrated by photomicrographs taken after 3 hr of luminal perfusion with 5 mM HCl (Figure 1).

Blood to lumen fluxes of small inert probe molecules have been employed as a sensitive means of monitoring enteropathy (Bjarnason et al., 1985). We have previously used the clearance of ^{14}C-urea to detect acid-induced increases in duodenal mucosal permeability (Wilkes et al., 1988). A similar approach has been reported using ^{51}Cr-EDTA to compare changes in duodenal and jejunal permeability following luminal acidification (Nylander et al., 1989). The pathological changes induced by exposure of the small intestine to HCl for 30-180 mins could be detected after much shorter periods of acid exposure using ^{14}C-urea clearance.

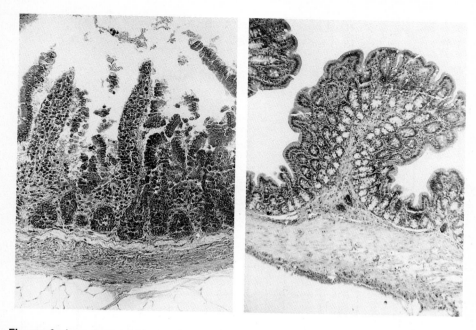

Figure 1. Low power photomicrographs of the jejunum and colon after luminal exposure to 5mM HCl for 3 hr. While the jejunum was severely damaged as evidenced by deep necrosis and loss of villi, appearance of the proximal colon was indistinguishable from controls perfused with mannitol (magnification x 200).

In the ileum, for example, urea flux increased two-fold after luminal perfusion with 10 mM HCl for 10 min (Figure 2). It seems reasonable to conclude that the increase in clearance of urea from blood to lumen reflects an increase in permeability of diffusional pathways rather than an increase in transcellular pathways. This could result from either enhanced permeability of the paracellular route or creation of pores by exfoliation of entire surface cells. Acidification of the colon also caused an increase in urea clearance but the response was far smaller both in terms of magnitude and duration when compared to the small intestine.

In the ileum, the return of urea clearance towards control over a 60 min period following acid exposure is consistent with recovery of electrical integrity observed in studies of mucosal restitution after acute injury. Indeed loss of barrier function as evidenced by loss of tissue electrical resistance and transmucosal PD in the duodenum has been reported in response to extremely small concentrations of HCl 10-100 fold less than those used in the present study (Vattay *et al.*, 1988).

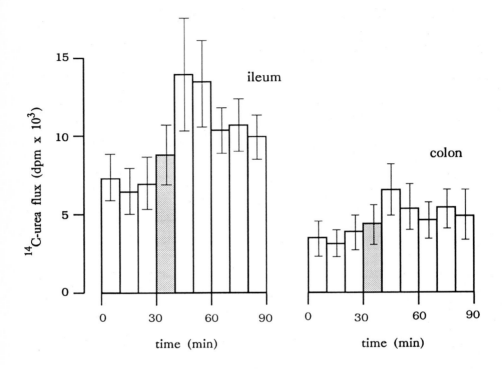

Figure 2. Urea clearance from blood to lumen in the terminal ileum and proximal colon of anaesthetised rats. The lumen was acidified with 10 mM HCl for a 10 min period indicated by the shaded columns (mean ± sem, n=6 for both regions).

The increase in mucosal permeability accompanying acidification of the duodenal lumen allows alkaline interstitial fluid to diffuse into the layer of adherent mucus thereby buffering acid at the epithelial surface. In order to determine whether any correlation could be discerned between these parameters in the basal state and the relative susceptibility of different regions of the intestine to acid injury, we compared urea clearance, alkaline secretion and mucus thickness. There was no relationship between acid resistance and the 'mucus-bicarbonate' barrier. For example, alkaline secretion in the acid-resistant colon was comparable to that in the distal duodenum and far lower than in the terminal ileum. Moreover, dimensions of the surface mucus layer in the colon and ileum were comparable and about twice those recorded in the duodenum or jejunum (Table 2).

Table 2. Mucosal permeability and the mucus-bicarbonate barrier in the intestine

Region	Permeability (dpm x 10^3)	Alkaline Secretion ($\mu eq/cm/hr$)	Mucus Thickness (μm)
distal duodenum	19.2 ± 0.6 (6)	3.1 ± 0.44 (29)	79.4 ± 8.6 (8)
mid jejunum	not tested	1.2 ± 0.24 (6)	86.2 ± 11.5 (8)
terminal ileum	13.0 ± 1.5 (6)	13.2 ± 1.08 (32)	174.2 ± 38.6 (8)
proximal colon	4.8 ± 0.7 (6)	3.2 ± 0.56 (30)	162.5 ± 15.9 (8)

Values are means (urea permeability and alkaline secretion) or median (mucus thickness) ± one standard error with the number of animals given in parenthesis.

Regional differences in permeability in the intestine based on measures of shunt conductance were evidenced in this study by comparing rates of urea clearance. Thus fluxes of urea across the 'leaky' small intestine were considerably higher than encountered in the colon. Susceptibility to acid-induced injury was directly related to epithelial permeability. It therefore seems reasonable to conclude that diffusion of acid/protons into the mucosa leads to acidification and ultimately to injury whenever the rate of influx exceeds the capacity of the tissue to buffer/dispose of H^+ ions. In the duodenum we have observed that H_2SO_4 is less toxic than HCl at equimolar concentrations (Garner et al., 1993). Whilst HCl is a permeable acid and may even cross membranes in the molecular (undissociated) form, H_2SO_4 is relatively impermeant having a non-transported divalent anion. These findings support the contention that cytoplasmic acidification is the major mechanism underlying mucosal injury by luminal acid. In view of the much greater sensitivity of the small intestine to acid damage when compared with the colon, it is tempting to speculate that cellular acidification by luminal HCl occurs via the basolateral side rather than the more accessible apical membrane of the enterocyte.

We conclude that the colon, like the stomach, is effectively able to withstand a luminal acid load. The small intestine, responds to luminal acidification with an initial increase in permeability. This response could serve a functional role inasmuch as diffusion of interstitial fluid would dilute and neutralise acid thereby allowing recovery of mucosal integrity and barrier function. If however the rate of acid influx exceeds its rate of clearance then mucosal acidification will lead to superficial damage and eventually to necrosis as evidenced by the effect of increasing luminal acid concentration or prolonging the time of acid exposure.

References

Allen A, Flemström G, Garner A, Kivilaakso E (1993). Gastroduodenal mucosal protection. Physiol Rev 73: 823-858.

Bjarnason I, Smethurst P, Levi AJ, Peters TJ (1985). Intestinal permeability to ^{51}Cr-EDTA in rats with experimental induced enteropathy. Gut 26: 579-585.

Feil W, Wenzl E, Vattay P, Starlinger M, Sogukoglu T, Schiessel R (1987). Repair of rabbit duodenal mucosa after acid injury *in vivo* and *in vitro*. Gastroenterology 92: 1973-1986.

Flemström G, Garner A, Nylander O, Hurst BC, Heylings JR (1982). Surface epithelial HCO_3^- transport by mammalian duodenum *in vivo*. Am J Physiol 243: G348-G358.

Florey HW, Jennings MA, Jennings DA, O'Connor RC (1939). The reactions of the intestine of the pig to gastric juice. J Pathol Bacteriol 49: 105-123.

Garner A, Allen A, Gutknecht J, Yanaka A, Goddard PJ, Silen W, Lacy ER, Bauerfeind P, Starlinger M, Wallace JL (1990). Mechanisms of gastric mucosal defence. Eur J Gastroenterol Hepatol 2: 165-188.

Garner A, Hunter AC, Wilkes JM, Tanira MOM, Dib R, Bastaki SMA, Abu-Hijleh G (1993). The mechanism of acid-induced duodenal mucosal damage. Gut 34: S16.

Kerrs S, Allen A, Garner A (1982). A simple method for measuring thickness of the mucus gel layer adherent to rat, frog and human gastric mucosa: influence of feeding, prostaglandin, N-acetylcysteine and other agents. Clin Sci 63: 187-195.

Nylander O, Kvietys P, Granger DN (1989). Effects of hydrochloric acid on duodenal and jejunal mucosal permeability in the rat. Am J Physiol 257: G653-G660.

Vatttay P, Feil W, Klimesch S, Wenzl E, Starlinger M, Schiessel R (1988). Acid stimulated alkaline secretion in the rabbit duodenum is passive and correlates with mucosal damage. Gut 29: 284-290.

Wilkes JM, Garner A, Peters TJ (1988). Mechanisms of acid disposal and acid-stimulated alkaline secretion by gastroduodenal mucosa. Dig Dis Sci 33: 361-367.

"FIRST-PASS" PROTECTION AGAINST LUMINAL ACID IN NECTURUS ANTRAL EPITHELIUM.

E.Kivilaakso, T.Kiviluoto , H.Mustonen and H.Paimela
II Department of Surgery
Helsinki University Central Hospital
00290 HELSINKI, FINLAND

ABSTRACT

The mechanisms contributing to the "first-pass" protection against intracellular acidosis in acid exposed gastric mucosa are reviewed. The studies were performed with isolated Necturus antral mucosa using microelectrode technique. The following observations are discussed: **1.** The pre-epithelial mucus-HCO_3 buffer layer has a buffer capacity strong enough to maintain a pH gradient between the luminal bulk solution and the epithelial surface in acid-exposed gastric mucosa. Perturbation of this layer by HCO_3 inhibition or mucolytic agents leads to dissipation of the pH gradient, with subsequent acidification of intracellular pH. **2.** Luminal acid (pH 3.0 - 2.0) increases the apical cell membrane resistance of surface cells by ca 100%, thus decreasing its permeability to luminal H^+ (and other ions). **3.** In resting antral mucosa, with weak pre-epithelial buffer capacity and permeable apical cell membranes, acute exposure to luminal acid leads to transient acidification of surface cells, which is controlled by activation of basolateral Na^+/H^+ antiport. After more prolonged acid exposure, the enhanced pre-epithelial buffer capacity (due to stimulated epithelial HCO_3 secretion) and the tightened apical cell membranes form a diffusion barrier efficient enough to prevent the   access of luminal H^+ inside surface cells, and the action of Na^+/H^+ antiport is not needed. However, if these diffusion barriers are deranged,   the basolateral Na^+/H^+ antiport is again activated and becomes the critical regulator of pH_i to maintain it within physiological ranges.

INTRODUCTION

The gastric mucosa is continuously exposed to high luminal acid, the H^+ gradient across the surface epithelium being as high as 10^5-10^6. This implies efficient protective mechanisms to maintain physiological intracellular pH (pH_i) in the surface epithelial cells. Since these mechanisms are located in the most superficial part of the mucosa, their action may be called as "first-pass" protection, in distinction from the defence system protecting the mucosa against deeper damage. This paper reviews these protective mechanisms, which include, at least, three components: (i) the pre-epithelial mucus-HCO_3 buffer layer (ii) the diffusion barrier generated by the apical cell

NATO ASI Series, Vol. H 89
Molecular and Cellular Mechanisms
of H^+ Transport
Edited by Barry H. Hirst
© Springer-Verlag Berlin Heidelberg 1994

membranes (and intercellular tight junctions) and (iii) the regulatory mechanisms of intracellular pH in surface cells.

MATERIAL AND METHODS

Necturus (Necturus maculosus) antral mucosa was stripped of its seromuscular coat and mounted mucosal side up in a perfusion chamber. Both sides of the mucosa were perfused individually with Ringer's solutions, having serosal side buffered to pH 7.3 with HCO_3-CO_2 or $HEPES/O_2$. Mucosal side was always gassed as was the serosal side. Intracellular pH and Na^+ acitivity in surface epithelial cells was measured with double-barrelled liquid sensor pH/PD and Na^+/PD microelectrodes (Kivilaakso & Kiviluoto 1988). Epithelial surface pH was measured with similar double-barrelled pH/PD microelectrodes positioned in close vicinity but not in contact with the apical cell membrane of a surface cell (Kiviluoto et al. 1993).

Intraepithelial resistances were measured using 2D-cable analysis (Frömter 1972). Four conventional microelectrodes were impaled into surface cells at measured distances from each other. Current pulses were passed through one of the electrodes, measuring the voltage deflections in the others. The 2D-cable equation $V(r) = AKo (r / \lambda)$ was fitted to the measurements to obtain A and λ .Apical (R_a), basolateral (R_b) and paracellular shunt (R_s) resistances were calculated from R_a/R_b, transmucosal reistance (R_t), A and λ

RESULTS AND DISCUSSION

Pre-epithelial mucus-HCO_3 layer

The gastric mucosa is known to actively secrete (or transport) HCO_3. Since luminal acid is a potent stimulator of the secretion, it has been proposed that the secreted HCO_3 has a protective function against luminal acid by forming together with surface mucus gel an "alkaline" buffer layer at the epithelial surface (Flemström 1987). The role of this layer in the protection against intracellular acidosis in acid-exposed mucosa was assessed by measuring intracellular pH (pH_i) and extracellular epithelial surface pH (pH_s) simultaneously with two double-barrelled pH-PD-microelectrodes and perturbating the functional integrity of this layer by inhibiting HCO_3 secretion and with mucolytic agents.

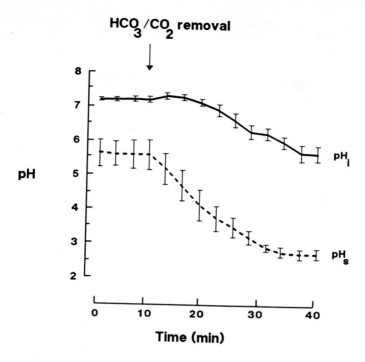

HCO$_3$/CO$_2$ removal

Fig. 1. Simultaneous measurement of intracellular pH (pH$_i$) and extracellular epithelial surface pH (pH$_s$) in acid-exposed Necturus antral mucosa. The mucosa was continously exposed to luminal pH 2.5. In the presence of serosal HCO$_3$/CO$_2$, a distinct pH-gradient exists between the luminal bulk solution (pH 2.5) and epithelial surface (pH$_s$ 5.6). Removal of HCO$_3$/CO$_2$ from the serosal perfusate dissipated this gradient, followed by acidification of pHi (mean ± SEM, N=9)(Adapted from Kiviluoto et al. 1993).

When the mucosas were exposed to luminal pH 2.5, a pH-gradient of ca 3 pH-units was observed between the luminal bulk solution and epithelial surface (Fig. 1). However, when HCO$_3$ secretion was inhibited by removal of serosal (systemic) HCO$_3$/CO$_2$, this pH-gradient rapidly dissipated, followed after a 5 - 10 min delay by acidification of also pH$_i$ (Fig.1). When the other component of this layer, mucus gel, was digested with luminal pepsin (5 mg/ml) or decomposed with N-acetyl-L-cystein (NAC, 5%) during acid exposure, the outcome was essentially similar, but occurred much slower (Figs 2 and 3). These findings indicate that the pre-epithelial mucus-HCO$_3$ layer has, indeed, a protective function against luminal acid and that the two components of it, HCO$_3$ and the mucus gel, are equally important for this function.

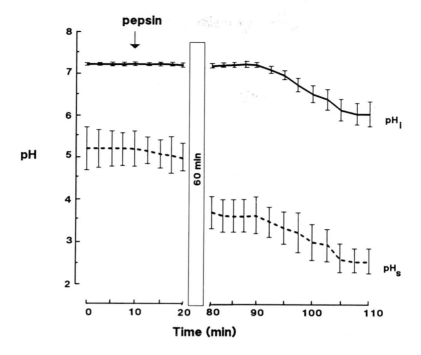

Fig. 2. Simultaneous measurement of pH_i and pH_s in acid-exposed (pH 2.5) Necturus antral mucosa. Digestion of the surface mucus gel by luminal pepsin (5 mg/ml at pH 2.5) provoked slow, progressive dissipation of the pre-epithelial pH-gradient, followed after a delay of ca 80 min by acidification of also pHi (mean ± SEM, N=7)(Adapted from Kiviluoto et al. 1993).

Apical cell membrane permeability

In order to elucidate the role of the apical cell membranes (and intercellular tight junctions) of the surface epithelial cell as a diffusion barrier to luminal acid, the behaviour of intraepithelial resistances

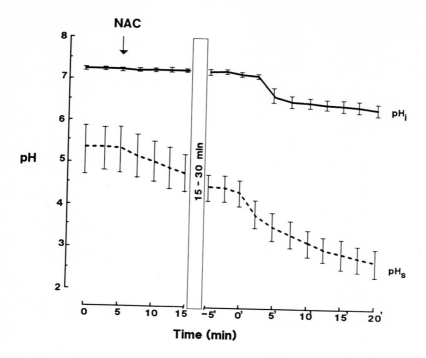

Fig. 3. Simultaneous measurement of pH_i and pHs in acid-exposed (pH 2.5) Necturus antral mucosa. Decomposition of the surface mucus gel by 5% NAC provoked progressive dissipation of the pre-epithelial pH gradient, followed after a delay of 30-40 min by acidification of also pH_i (mean ±SEM, N=7). (Adapted from Kiviluoto et al. 1993).

(and conductances) during acid exposure was investigated using 2D-cable analysis (Mustonen et al. 1992). When the luminal solution was acidified from pH 7.0 to pH 3.5, no changes in intraepithelial resistances occurred (Fig. 4). However, when the luminal pH was acidified to 3.0, and further to 2.0, a sudden, ca 100% increase in apical cell membrane resistance (R_a) occurred. Stronger acidities led to disruption of the epithelium. In shunt resistance (R_s) no such increase occurred, and the basolateral cell membrane resistance (R_b) and transepithelial (R_t)likewise remained unchanged. Thus, also the apical cell membrane of the surface cells seems to have an important protective role as a diffusion barrier against luminal acid, being able to decrease its conductance to H^+ in face of enhanced acid load.

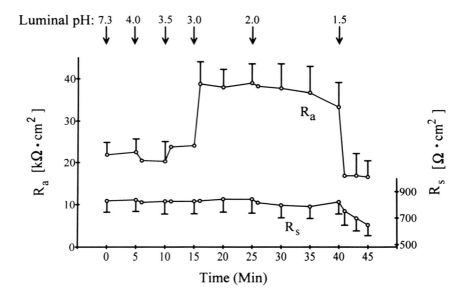

Luminal pH: 7.3 4.0 3.5 3.0 2.0 1.5

Fig. 4. Effect of luminal acid on apical cell membrane resistance (R_a) and intercellular shunt resistance (R_s) in Necturus antral epithelium. Acidification of the luminal perfusate from pH 7.3 to pH 3.5 had no effect on R_a and R_s. Further acidification to pH 3.0 and 2.0 provoked a rapid increase in R_a, while R_s remained unchanged. Stronger luminal acidities caused disruption of the mucosa (mean $\pm$ SEM, N= 8).

Regulatory mechanisms of Intracellular pH

When the Necturus antral mucosa is acutely exposed to relatively strong (considering in vitro conditions) luminal acid, pH 2.0, a rapid intracellular acidification of 0.2 - 0.3 pH-units occurs, whereafter a reversal and partial recovery of pH_i occurs, followed by a new steady state at a somewhat lower level than the baseline pH_i (Kiviluoto et al. 1990). This suggests that in a resting mucosa, with low epithelial HCO_3 secretion and high apical cell membrane conductance, some luminal H^+ gets inside the surface cells, but the regulatory mechanisms of intracellular pH are activated with resultant normalization of pH_i.

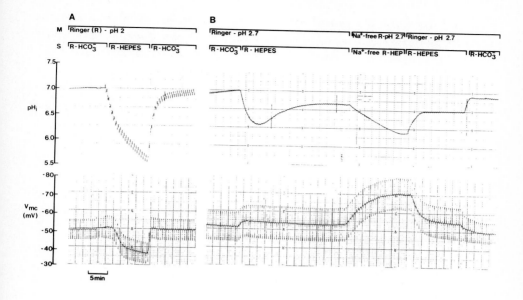

Fig. 5. Effect of HCO$_3$/CO$_2$ and Na$^+$ removal on pH$_i$ in Necturus antral mucosa exposed to luminal acid (pH 2.0-2.7).   A. Removal of serosal HCO$_3$/CO$_2$ provoked a profound but reversable acification of pH$_i$. B. Removal of, first , HCO$_3$/CO$_2$ from the serosal perfusate provoked an initial acidification of pH$_i$ followed by steady state pH$_i$ at somewhat lower level than the baseline. Subsequent removal of also Na$^+$ from the serosal perfusate provoked a progressive but still reversible acidification of pH$_i$ (N=5)(From Kiviluoto et al. 1990).

In all eukaryotic vertebrate cells so far studied the main regulator of intracellular pH (against acidosis) is Na$^+$/H$^+$ antiport (exchanger), which extrudes intracellular H$^+$ out of the cell electrochemically uphill by the force of the strong inward-directed electrochemical gradient of Na$^+$ (Roos & Boron 1981). Obviously, this is the case also in the gastric mucosa, since the initial acidification and reversal of pH$_i$ following exposure to luminal acid was accompanied by a concomitant rapid and transient accumulation of Na$^+$ inside the cell ($[Na+]_i$ from 16.3 to 39.1 mmol/L), while blocking of the function of Na$^+$/H$^+$ antiport by removal of serosal Na$^+$ or addition of serosal amiloride (1 mM) abolished both the increase of intracellular $[Na+]$ and the reversal of acidification of pH$_i$,   pH$_i$ rapidly falling below pH 6.0 (Kiviluoto et al. 1991).

The outcome was different after more prolonged acid exposure, when pH$_i$ had obtained its new steady state level. Blocking of the function of Na$^+$/H$^+$ antiport   in this situation had no influence on pH$_i$, pHi remaining stable despite continuous exposure to luminal acid, indicating that Na^+/H^+ antiport is no more needed for the maintenance of steady state pH_i (Kiviluoto et al. 1990). In contrast, when serosal HCO_3^-/CO_2 was removed, a rapid and profound acidification of pH_i occurred (Fig. 5a).

In order to elucidate the mutual relationship of these two pH_i regulatory mechanisms, the two ions were removed successively, HCO_3/CO_2 first and Na+ thereafter. In these experiments a weaker luminal acid, pH 2.7, had to be used in order to get a stable pH_i in the absence of serosal HCO_3/CO_2. Removal of serosal $HCO_3/CO2$ in this situation induced an initial acidification of pH_i, which was, however, rapidly reversed and partially recovered, followed by a new steady state (Fig. 5b).This suggests that, as during acute acid exposure, again a pH_i regulator, presumably Na^+/H^+ antiport, was activated. This was confirmed by the finding that when the function of Na^+/H^+ antiport was blocked by removal of serosal Na^+, a progressive acidification of pH_i occurred (Fig. 6b). An essentially similar finding was obtained with addition of serosal amiloride (Kiviluoto et al. 1990).

Our intepretation of the above findings is that the initial intracellular acidification following exposure to luminal acid stimulates epithelial HCO_3 secretion (luminal acid is a potent stimulator of epithelial HCO_3 transport) with resultant increase in the buffering capacity of the pre-epithelial mucus-HCO_3 layer. It also decreases the conductance of the apical cell membrane of surface cells. These two intensified diffusion barriers are efficient enough to completely prevent the entry of luminal H^+ inside the surface cells, and therefore the action of the basolateral Na^+/H^+ antiport is not needed to maintain a stable pH_i during prolonged luminal acid exposure. However, when the buffering capacity of the pre-epithelial HCO_3-mucus layer is reduced by inhibition of the epithelial HCO_3 secretion, luminal H^+ gets again inside the surface cells activating basolateral Na^+/H^+ antiport, which becomes the critical regulator of pH_i to restore and maintain a stable and physiological pH_i.

References

Flemström G (1987) Gastric and duodenal mucosal bicarbonate secretion. In: Johnson KR, ed. Physiology of the Gastrointestinal Tract. 2nd ed. New York; Raven Press, pp 1011-1029.

Frömter E (1972) The route of passive ion movement through the epithelium of Necturus gallbladder. J Memr Biol 8: 259-301

Kivilaakso E, Kiviluoto T (1988) Intracellular pH in isolated Necturus antral mucosa in simulated ulcerogenic conditions. Gastroenterology 85: 1198-205.

Kiviluoto T, Ahonen M, Bäck N, Häppölä O, Mustonen H, Paimela H, Kivilaakso E (1993) Pre-epithelial mucus-HCO_3 layer protects against intracellular acidosis in acid-exposed gastric mucosa. Am J Physiol 264: G57-G63.

Kiviluoto T, Paimela H, Mustonen H, Kivilaakso E (1990). Intracellular pH in isolated Necturus antral mucosa exposed to luminal acid. Gastroenterology 98: 901-908.

Kiviluoto T, Paimela H, Mustonen H, Kivilaakso E (1992). Intracellular Na^+ and regulation of intracellular pH in gastric mucosa during acute exposure to luminal acid. Gastroenterology 100: A98.

Mustonen H, Kiviluoto T, Kivilaakso E (1992) Luminal acid increases apical cell membrane resistance in isolated Necturus antral mucosa. Gastroenterology 102: A130

Roos A, Boron WF (1981) Intracellular pH. Physiol Rev 61: 297-435.

Regulation of pHi in Oxyntic Cells in Intact Sheets of Frog Gastric Mucosa

Akinora Yanaka, M.D.
Department of Gastroenterology
Institute of Clinical Medicine
University of Tsukuba
1-1-1 Tennodai, Tsukuba City
Ibaraki 305, JAPAN

William Silen, M.D.
Department of Surgery
Beth Israel Hospital
330 Brookline Avenue
Boston, MA   02215
U.S.A.

Gastric mucosal epithelial cells are regularly exposed to enormous concentrations of luminal H^+, with gradients at times as high as 1,000,000:1. The ability of the mucosa to withstand such gradients has properly focused considerable attention on the mechanisms by which the mucosal epithelial cells regulate pHi. While several studies have previously examined the regulation of pHi in surface epithelial cells in intact amphibian antrum[7,9,2,3,8,6], or in monolayers of chief cells[11], experiments to assess oxynticopeptic cells directly have been limited to isolated cells[10,13,12] until recently. Since isolated cells invariably lose polarity, and because the interaction of the different types of cells within the complex fundic mucosa may be important,[5] we have developed methods to measure pHi in intact amphibian fundic mucosa using the fluorescent dye, 2',7'-bis-(2-carboxy-ethyl)-5(6)-carboxyfluorescein), (BCECF)[4,18,1,15,14,17,16]. Our results have been presented in several previous publications[4,18,1,15,14,17,16] and therefore this communication represents an up to date synthesis and review of this material.

NATO ASI Series, Vol. H 89
Molecular and Cellular Mechanisms
of H^+ Transport
Edited by Barry H. Hirst
© Springer-Verlag Berlin Heidelberg 1994

METHODS AND RESULTS

Initial studies showed that sheets of frog (Rana Catesbeiana) fundic gastric mucosa exposed to the esterified form of BCECF on the luminal side for 3 hours in an Ussing chamber serendipitously concentrated the fluorescent BCECF in the oxynticopeptic cells[4]. An enhanced fluorescent signal from the oxynticopeptic cells and even more selective loading was obtained when both the muscularis propria and the muscularis mucosa were stripped from the mucosa and the tissue was exposed to the BCECF-AM from the basolateral side.[18] The results reported herein are a compilation of the findings in the latter preparation.

Basolateral anion exchange

Using the doubly stripped mucosa we have confirmed the existence of a HCO_3^-/Cl^- exchanger on the basolateral membrane of oxynticopeptic cells[18], the presence of which was suggested by others in isolated gastric glands[10] and by us in isolated rabbit parietal cells[12]. This exchanger was readily blocked by DIDS (an inhibitor of anion exchange) in the nutrient solution or by removal of Cl^- from both the nutrient and luminal solutions[18]. The activity of the HCO_3^-/Cl^- exchanger was greater in stimulated than in inhibited tissues as evidenced by a greater rise in pHi of stimulated oxynticopeptic cells exposed to serosal DIDS than in resting cells. Of considerable physiologic interest was the finding that the presence of Cl^- in the luminal solution prevented the rise in pHi observed when Cl^- was removed from both bathing solutions, indicating that luminal Cl^- is available to the basolateral HCO_3^-/Cl^- exchanger[18]. Our studies do not discern whether movement of the luminal Cl^- to the basolateral side is paracellular or through the cells.

In addition to stimulation of HCO_3^- extrusion in exchange for Cl^- induced by acid secretion we have recently demonstrated that dm PGE_2 also stimulates HCO_3^- extrusion through the basolateral Cl^-/HCO_3^- exchanger in resting oxynticopeptic cells and that the relatively high $[HCO_3^-]$ on the basolateral surface afforded by dm PGE_2 protects oxynticopeptic cells from acidification during exposure to luminal or serosal acid[17].

Disposal of H^+ by oxynticopeptic cells

Forskolin-stimulated mucosae bathed in nutrient HCO_3^- solutions maintain a pHi in oxynticopeptic cells of about 7.03 when the luminal pH is 7.2. This pHi was unchanged by luminal acidification to pH 1.5, but further luminal acidification to 1.0 caused a decrease in pHi to ~ 6.85. The decrease in pHi was somewhat greater in stimulated tissues bathed in HEPES nutrient solutions, but pHi returned to control levels when luminal pH was restored to 7.2. Similar results were obtained in resting tissues, i.e. those pre-treated with the H^+/K^+ ATPase inhibitor, omeprazole. In resting tissues bathed in HEPES, the decrease in pHi induced by increasing luminal $[H^+]$ was exaggerated by serosal amiloride, an inhibitor of Na^+/H^+ exchange, but not by serosal H_2-DIDS. In resting tissues bathed in HCO_3^-, the decrease in pHi induced by lowering luminal pH was not altered by either amiloride or H_2-DIDS alone but was accentuated by a combination of these two agents. Amiloride did not alter the changes in pHi induced by increasing luminal $[H^+]$ in resting tissues bathed in Cl^- free HCO_3^- but H_2-DIDS enhanced the decrease in pHi and partially inhibited the recovery at luminal pH 7.2. The amiloride-insensitive, Cl^- independent recovery of pHi in the presence of serosal HCO_3^- was completely inhibited either by removal of ambient Na^+ or by the addition of serosal H_2-DIDS. We concluded from these studies[15] that pHi in oxynticopeptic cells in intact sheets of frog mucosa exposed to high luminal $[H^+]$ is maintained more readily in the stimulated state than in the resting state, and that serosal HCO_3^- protects oxynticopeptic cells from a major decrease in pHi during exposure to high luminal $[H^+]$. In addition, our data clearly demonstrate that both a basolateral Na^+/H^+ exchanger and basolateral $Na^+-HCO_3^-$ co-transport are involved in the recovery of pHi from exposure to low luminal pH[15].

Since others have shown in antral mucosa that basolateral acidification is much more deleterious to the mucosal cells than is luminal acidification, we studied the effects of acidification of the nutrient solution in intact sheets of fundic mucosa[1]. Lowering the pH of an unbuffered nutrient solution from 7.2 to 3.5 acidified pHi in oxynticopeptic cells from 7.05 only to 6.44 without a change in potential difference or resis- tance, but with a striking increase in acid secretion from 0.86 to 1.88 $\mu Eq/cm^2/hr$. By light and electron microscopy, oxynticopeptic cells appeared normal despite the nutrient pH of 3.5. pHi returned to normal values upon return of the nutrient pH to 7.2 in untreated and forskolin stimulated tissues, whereas pHi was irreversibly lowered to 6.26 and 6.44 by the nutrient pH 3.5 in omeprazole and cimetidine treated tissues respectively. In the latter tissues, oxynticopeptic and smooth muscle cells of the muscularis mucosa showed extensive morphologic damage. These data indicate that per se, a pHi as low as 6.4 is not deleterious in the secreting mucosa, whereas the inhibited fundic mucosa is much more susceptible to a similar degree of basolateral acidification.

In addition to the basolateral Na^+/H^+ exchanger and the basolateral Na^+/HCO_3^- co-transporter, both of which serve to alkalinize the oxynticopeptic cell, the studies on acidification of the nutrient solution described above[1] suggested that acid secretion by the oxyntic cell can also serve as a means of disposal of excessive intracellular protons and thus play a role in the regulation of pHi. Further studies in our laboratory[14] strongly support this concept.

While we have not specifically examined the relative tolerance to acidification of oxynticopeptic and surface cells, our studies have suggested that oxynticopeptic cells may be more susceptible to an acid challenge than surface cells. For example, the decrease in pHi in response to a challenge with change of gassing from 100% O_2 to 95% O_2- 5% CO_2 was much greater in fundic oxynticopeptic cells than in antral surface cells[4]. In addition, we have observed a much more profound morphologic injury of oxynticopeptic than to surface cells during basolateral acidification[1]. Most investigators have naturally focused attention on surface cells as the initial site of ulceration since these cells are most superficially located and are therefore presumably more exposed to luminal acid. Since systemic acidification, however, has been shown to be so injurious, and because luminal acidification is well tolerated by both surface and oxyntic cells, we must begin to focus on the oxyntic cell as perhaps the initial site of damage under conditions of systemic acidification.

DISCUSSION AND SUMMARY

We have shown the presence of a HCO_3^-/Cl^- exchanger, a Na^+/H^+ exchanger, and a co-transporter of $Na^+-HCO_3^-$ on the basolateral membrane of amphibian oxynticopeptic cells. Especially in the stimulated fundic gastric mucosa the HCO_3^-/Cl^- exchanger is poised to extrude HCO_3^- and to alkalinize the basolateral surface when it is exposed to an acidic challenge or to back-diffusion of luminal H^+. The presence of HCO_3^- in the nutrient solution serves a similar purpose, probably because of $Na^+-HCO_3^-$ co-transport. Luminal Cl^- can participate in the operation of the basolateral HCO_3^-/Cl^- exchanger after diffusion of the Cl^- through the mucosa. Strong acidification of the luminal surface of the mucosa is better tolerated than acidification of the nutrient solution, with minimal lowering of pHi, probably because of the relatively impermeant nature of the apical membrane to H^+. It appears that the damaging effect of luminal H^+ occurs mainly after back-diffusion of H^+ to the basolateral side. The response to an alkaline challenge which is mediated solely by the HCO_3^-/Cl^- exchanger is less brisk and less complete than that to an acidic challenge.

453

REFERENCES
1. Arvidsson,S.,Carter,K.,Yanaka,A.,Ito.S.,Silen,W. Effect of basolateral acidification on the frog oxynticopeptic cell. American Journal of Physiology (Gastrointestinal and Liver Physiology) 22:G564-G570, 1990.
2. Ashley,S.W.,Soybel,D.I.,Cheung,L.Y. Measurements of intracellular pH in Necturus antral mucosa by microelectrode technique. Am. J. Physiol. 250 (Gastrointest. Liver Physiol. 13):G625-G632, 1986.
3. Ashley,S.W.,Soybel,D.I.,Moore,D.,Cheung,L.Y. Intracellular pH (pHi) in gastric surface epithelium is more susceptible to serosal than mucosal acidification. Surgery 102:371-379, 1987.
4. Carter,K.J.,Saario,I.,Seidler,U.,Silen,W. Effect of PCO_2 on intracellular pH in in vitro frog gastric mucosa. Am. J. Physiol. 256 (Gastrointest. Liver Physiol. 19):G206-G213, 1989.
5. Curci,S.,Debellis,L.,Frömter,E. Evidence for rheogenic sodium bicarbonate co-transport in the basolateral membrane of oxyntic cells of frog gastric fundus. Pflügers Arch. 408:497-594, 1987.
6. Kivilaakso,E. Contribution of ambient HCO_3^- to mucosal protection and intracellular pH in isolated amphibian gastric mucosa. Gastroenterology 85:1284-1289, 1983.
7. Kivilaakso,E.,Kiviluoto,T. Intracellular pH in isolated Necturus antral mucosa in simulated ulcerogenic conditions. Gastroenterology 95:1198-1205, 1988.
8. Kiviluoto,T.,Paimela,H.,Mustonen,H.,Kivilaakso,E. Intracellular pH in isolated necturus antral mucosa exposed to luminal acid. Gastroenterology 98:901-908, 1990.
9. Kiviluoto,T.,Voipio,J.,Kivilaakso,E. Subepithelial Tissue pH of rat gastric mucosa exposed to luminal acid, barrier breaking agents, and hemorrhagic shock. Gastroenterology 94:695-702, 1988.
10. Paradiso,A.M.,Townsley,M.C.Wentzl,E.Machen,T.E. Regulation of intracellular pH in resting and in stimulated parietal cells. Am. J. Physiol. 257 (Cell Physiol. 26):C554-C561, 1989.
11. Sanders,M.J.,Ayalon,A.,Roll,M.,Soll,A.H. The apical surface of canine chief cell monolayer resist H^+ back-diffusion. Nature 313:52-54, 1985.
12. Seidler,U.,Carter,K.,Ito,S.,Silen,W. Effect of CO_2 on pHi in rabbit parietal, chief, and surface cells. Am. J. Physiol. 256 (Gastrointest. Liver Physiol. 19):G466-G475, 1989.
13. Townsley,M.C.,Machen,T.E. Na^+-HCO_3^- co-transport in rabbit parietal cells. Am. J. Physiol. 257 (Gastrointest. Liver Physiol. 19):G350-G356, 1989.
14. Yanaka,A.,Carter,K.J.,Goddard,P.J.,Heissenberg,M.C.,Silen,W. H^+-K^+-ATPase contributes to regulation of pHi in frog oxynticopeptic cells. American Journal of Physiology. 24:G781-G789, 1991.
15. Yanaka,A.,Carter,K.J.,Goddard,P.J.,Silen,W. Effect of luminal acid on intracellular pH in oxynticopeptic cells in

intact frog gastric mucosa. Gastroenterology. <u>100</u>:606-618, 1991.

16. Yanaka,A.,Carter,K.,Goddard,P.,Silen,W. Regulation of intracellular pH (pHi) in oxynticopeptic cells: Studies in intact amphibian gastric mucosa. In: <u>Mechanisms of Injury, Protection, and Repair of the Upper Gastrointestinal Tract</u>. A. Garner and P.E. O'Brien (eds), John Wiley & Sons, England, pg. 247-251, 1991.

17. Yanaka,A.,Carter,K.J.,Goddard,P.J.,Silen,W. Prostaglandin stimulates $Cl^--HCO_3^-$ exchange in amphibian oxynticopeptic cells. American Journal of Physiology. <u>25</u>:G44-G49, 1992.

18. Yanaka,A.,Carter,K.J.,Lee,H.H.,Silen,W. Influence of Cl^- on pHi in oxynticopeptic cells of in vitro frog gastric mucosa. American Journal of Physiology (Gastrointestinal and Liver Physiology) <u>21</u>:G815-G824, 1990.

Oxyntic cell Na$^+$/H$^+$ and Cl$^-$/HCO$_3^-$ exchangers

U. Seidler
II. Department of Medicine
Technical University of Munich
Ismaninger Str. 22
81675 Munich
F.R.G.

Summary

We have studied the role and regulation of the basolateral Na$^+$/H$^+$ and Cl$^-$/HCO$_3^-$ exchanger of the acid-secreting gastric oxyntic cell in maintenance of ion homeostasis during acid secretion and the maintenance of pH$_i$ during acidification of the gastric lumen. Inhibition of the Na$^+$/H$^+$ exchanger had little effect on the acid secretory rate but strongly compromised the ability of the oxyntic cells to maintain a neutral pH during luminal acidification. As expected, there was a strong increase of transport capacity with decreasing pH, but stimulation of acid formation by histamine or forskolin did not shift the pH$_i$-dependency to higher pH$_i$-values. Thus, the oxyntic Na$^+$/H$^+$ exchanger is involved in the cellular defence against acid damage, but its role in the regulation of acid secretion, if any, is minor. Oxyntic cells express the AE2 isoform of the anion exchanger gene family. The flux through the Cl$^-$/HCO$_3^-$ exchanger increased during acid secretion without a concomitant change in pH$_i$. Modulation of maximal anion flux rates was found by pH$_i$ but not by the secretagogues histamine and forskolin. Inhibition by stilbene derivatives abolished HCl secretion. Thus, Cl$^-$/HCO$_3^-$ is essential for acid secretion but a regulation of its maximal transport rate by secretagogues was not observed. Inhibition of acid secretion did not influence the maintenance of a neutral pH$_i$ during luminal acidification in the presence of serosal 5% CO$_2$/HCO$_3^-$, but compromised pH$_i$-regulation in serosal Hepes/O$_2$. This suggests that the "alkaline tide" produced during acid secretion may help in the prevention of acid damage.

Introduction

The ability of the different types of gastric epithelial cells to actively regulate their intracellular pH may play an important role in gastric mucosal protection against luminal acid (11,22,23). It has been shown that the acid-secreting gastric mucosa is much more resistant to luminal acid than the inhibited mucosa (7). This may be due to several factors, among them the fact that the "alkaline tide", which is the HCO$_3^-$ produced intracellularly during acid secretion, may enhance pH$_i$-regulatory processes in oxyntic and surface cells and may be used for secretion of HCO$_3^-$ into the mucus layer. Thus, during acid secretion, there appears to be relatively little danger for oxyntic cell damage due to acid. However, the lowest intragastral pH is not found postprandially, when acid secretion is maximally stimulated, but at night, during the interdigestive state with low rates of acid secretion. This apparent paradoxon is due to the

NATO ASI Series, Vol. H 89
Molecular and Cellular Mechanisms
of H$^+$ Transport
Edited by Barry H. Hirst
© Springer-Verlag Berlin Heidelberg 1994

fact that the food, whose intake has resulted in the strong stimulation of acid secretion, also serves as an intragastral diluent and buffer for many hours. Therefore, the situation in which luminal acid is present together with a "resting state" in oxyntic cells occurs regularly and the prevention of a low luminal pH during this period is the mainstay of ulcer therapy by H_2-antagonists.

All three major gastric epithelial cell types possess a Na^+/H^+ exchanger, which, according to our studies, is the major pH_i-regulator during an intracellular acid load (20,22). Therefore, we were interested in its protective role during a luminal acid load. During stimulation of acid secretion, the oxyntic cell produces intracellular base. The disposal of base is coupled to the uptake of Cl^- via the parietal cell basolateral Cl^-/HCO_3^- exchanger (5). Thus, an increased flux rate through the Cl^-/HCO_3^- exchanger is to be expected during acid secretion. However, the "tuning" of apical and basolateral ion flux rates during acid secretion is ill understood. Theoretically, a number of mechanisms could account for an increase in anion flux rates during stimulation of acid secretion: A change in driving force, a hormone-induced phosphorylation of the transport protein, resulting in a conformational change with subsequent change in ion affinity, pH_i-sensitivity, or some other factor affecting the transport capacity, or, as suggested by Muallem et al (13), the presence of an internal pH-sensitive modifier site resulting in a steep increase of maximal transport rates over a small pH-range. The moderate pH_i-increase necessary for this activation should be mediated by a secretagogue-induced stimulation of the Na^+/H^+ exchanger (13). Therefore, we were interested in the role and regulation of the oxyntic cell Cl^-/HCO_3^- and Na^+/H^+ exchanger during stimulation of acid secretion.

In this paper, we describe some of our recent attempts to unravel the role and regulation of the parietal cell Na^+/H^+ and Cl^-/HCO_3^- exchanger during acid secretion and in the maintenance of a neutral pH_i during acidification of the gastric lumen.

Methods:

Cell isolation
The principle of isolating a homogenous parietal cell population is that of enzymatic digestion of the gastric mucosa into single cells and separation of the different cell populations by a combination of counterfluw elutriation, which separates the cells predominantly by size, and Percoll and Nycodenz gradients, which separate the cells by density. With a combination of these techniques, a virtually homogeneous parietal cell population can be isolated. The technique is described in detail elsewhere (20,21).

Measurement of pH_i in isolated rabbit parietal cells
The measurement of pH_i in isolated parietal cells is acchieved by loading the cells with the membrane-permeant ester form of the fluorescent dye BCECF/am. The dye is alternatively excited with the wavelenghts 439 and 490 nm and fluorescence intensity measured at 540 nm emission wavelenght. A ratio of fluorescence readings at 490 nm (pH-sensitive wavelength) and 439 nm (relatively pH-insensitive wavelength) is calculated and this ratio is converted to pH_i-values after calibration with the high K^+-nigericin method, where pH_i is equilibrated with pH_o. The methods of measuring pH_i in single cells have been described in detail by us and others (1,3,12,17,20,21).

Determination of intracellular buffer capacity β_i and calculation of proton/base flux rates
When the acid or base extrusion rate at a given pH_i is to be calculated, it is necessary to know the intracellular buffer capacity β_i for this pH_i. The net proton or base flux rate at a given pH_i is $dpH_i/dt \times \beta_i$. The intrinsic buffer capacity β_{int} was determined by an "intracellular titration" by stepwise addi-

tion of a cell-permeable weak acid (CO_2) or salt of a weak base (NH_4Cl) and determination of the resultant change pH_i, as previously described in detail (21). The CO_2/HO_3^--dependent buffer capacity was calculated with the equation $\beta_{CO2} = (ln)10x[HCO_3^-]$, and total buffer capacity β_i ($\beta_{int} + \beta_{CO2}$) calculated for each pH_i.

[14]C-aminopyrine(AP) accumulation measurements

The measurement of the accumulation of radiolabeled aminopyrine is a well-validated method to assess acid formation in isolated parietal cells. The method makes use of the circumstance that during secretagogue stimulation of isolated parietal cells, fusion of tubulovesicles commences as well as formation of secretory canaliculi with an acidic interior. The weak base aminopyrin is membrane-permeant in the nonprotonated form, and the protonated form accumulates in the acid spaces of the parietal cell. Thus, accumulation of aminopyrine reflects acid formation in isolated parietal cells, although a linearity between the two has not been established. We measured ^{14}C-AP accumulation ratios as previously described in detail (20,21)

Measurement of pH_i in oxyntic cells within intact frog gastric mucosa

Recently, we have succeeded in loading BCECF into oxyntic cells within intact frog gastric mucosa. For the experiments described in this paper, we have used the European waterfrog (Rana esculenta). The method for tissue preparation was adapted from that described by Debellis et al (2). After splitting the submucosa from the mucosa by injection of isotonic saline, the mucosa was mounted serosal side up onto a metal frame and placed into a preparation bath with continuous perfusion. The remaining connective tissue and muscularis mucosa was removed with forceps under a stereomicroscope. Thus, a 3 x 3 mm area was prepared and the metal plate was fixed between the two halves of a lucite mini-perfusion chamber. When we managed to completely remove the muscle strands that covered the base of the glands, loading of the oxyntic cells at the base of the glands was acchieved after continuous perfusion of the serosal side with buffer containing 10-30 μM BCECF/am. pH_i was then assessed fluorometrically by standard techniques.

Ussing chamber experiments:

Ussing chamber experiments were performed by standard techniques to assess tissue vitality and measure the acid secretory rate. Potential difference (PD) was measured, electrical resistance was calculated after passing a constant current of 25 or 50 mA through the tissue and measuring the resultant drop in PD. the acid secretory rate was measured by a pH-stat titration technique (27).

Results and Discussion:

I. The parietal cell Na$^+$/H$^+$ exchanger

1. General characteristics

Early experiments had established the presence of a Na$^+$/H$^+$ antiporter in the plasma membrane of the rabbit parietal cell (12,17,20). This Na$^+$/H$^+$ antiporter was activated by decreasing pH_i, with a Hill coefficient of 2.13 (21), and had a k_i for amiloride in the micromolar range (unpublished observations), suggesting that it was of the basolateral type. A recent report indicate that there may also be an apical Na$^+$/H$^+$ exchanger on the parietal cell (5). We are presently in the process of localizing the different Na$^+$/H$^+$ exchanger mRNAs found in the stomach (16) to the individual cell types, and this will hopefully unravel the question how many different Na$^+$/H$^+$ exchangers are expressed in the parietal cell.

When we studied a variety of substances for whom receptors exist on the parietal cell for their effect on the K_m for internal protons and the V_{max} of the Na$^+$/H$^+$ exchanger in isolated parietal cells, we observed a rather unexpected pattern: A shift of the pH_i-dependency to the right, indicative of a change in K_m for H^+_i (6), was observed for

EGF and carbachol, whereas histamine, gastrin, PGE_2 were without apparent effect. The effect of EGF was inhibited by the tyrosine kinase inhibitor genistein, but the effect of carbachol was also blunted by this substance (unpublished). For us, this pattern of activation did not fit into a physiological context. Recently Chew et al reported their observations on the possible presence of MAP-kinases in rabbit parietal cell which are stimulated by EGF and carbachol and which may have long-term effects on the acid secretory pathway (15). We are now searching for a possible link between their observations and ours.

2. Role in parietal cell acid formation

Of particular importance to investigators interested in gastric physiology was the question whether the Na^+/H^+ antiporter was involved in a regulatory fashion in the process of acid secretion. An early report had indicated that this may indeed be the case (13). We have therefore measured the effect of amiloride in concentrations up to 1 mM and dimethyl-amiloride (DMA) in concentrations up to $10^{-4}M$. We found that the secretagogue-stimulated ^{14}C-AP accumulation ratios were inhibited only very weakly, with a maximal inhibition of 11% at 1 mM amiloride for stimulation with $10^{-5}M$ forskolin (21). These results are similar to those of Paradiso et al (18).

The pH_i of isolated parietal cells in suspension did not significantly change during stimulation with forskolin ($10^{-5}M$), indicating that a change in pH_i is not mandatory for acid formation. When we measured the effects of forskolin on the pH_i-dependency (K_m for internal H^+) and V_{max} of Na^+/H^+ exchange, we did not find any, indicating that cAMP-mediated acid formation does not coincide with a change in the K_m for internal protons (H^+_i) or V_{max} of the Na^+/H^+ exchanger.

Recent observations in single cultured parietal cells have shown, however, that some parietal cells responded to histamine with an increase in pH_i, while others did not (unpublished). The same variability in the pH_i response to histamine plus a phosphodiesterase inhibitor had been observed in single parietal cells within intact gastric glands by Paradiso et al (18). It has not been unequivocally established yet what mechanism is responsible for this pH_i-change, whether it is due to Na^+/H^+ exchange (13) or the formation of intracellular base during acid secretion (2), and why it is seen in only a fraction of the parietal cells studied.

The data suggest that a strong inhibition of the Na^+/H^+ exchanger does not interfere significantly with the parietal cells' ability of acid formation due to a strong secretory stimulus. Also, no change in the apparent K_m for $[H^+]_i$ or V_{max} by secretagogues was uniformly observed. These data do not rule out a more subtle influence of Na^+/H^+ exchange on acid secretion, however.

3. Role of the Na^+/H^+ exchanger during luminal acidification

One of our most important issues was the question whether the parietal cell makes use of its Na^+/H^+ exchanger to prevent cytoplasmic acidification during a high luminal acid load. As mentioned above, the first question is whether the parietal cell is ever endangered by luminal protons. Figure 1 shows a pH_i trace obtained from BCECF-loaded oxyntic cells in an intact sheet of mucosa during progressive luminal acidification. The serosal buffer is 20 mM Hepes/Tris and the serosal pH is 7.36. Little change in oxyntic cell pH_i was observed down to luminal pH 2. At luminal pH 2, there

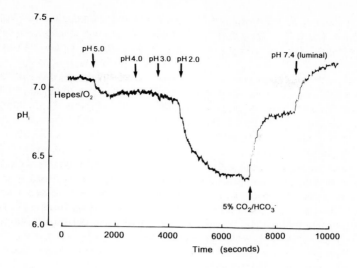

Figure 1: pH$_i$-trace in oxyntic cells within intact frog mucosa during progressive acidification of the luminal perfusate. The serosal buffer was 20 mM Hepes. A significant pH$_i$-drop was observed when the luminal pH was lowered to pH 2.

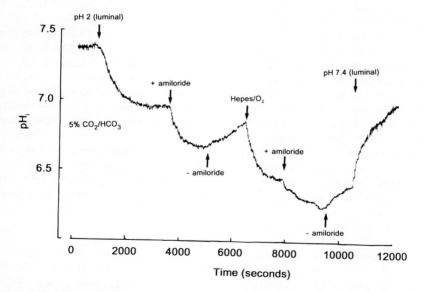

Figure 2.: Effect of 1 mM amiloride on oxyntic cell pH$_i$ during luminal acidification to pH 2 in the presence (1st part of trace) and absence of serosal CO_2/HCO_3^-. Amiloride caused a significant acidification both in the presence and absence of CO_2/HCO_3^-. It is also evident that pH$_i$ is far higher during luminal acidification in the presence than in the absence of serosal CO_2/HCO_3^-.

was a rather rapid acidification with a new plateau at pH_i 6.52 ± 0.09 (n = 5). In serosal 5%CO_2/HCO_3^-, the drop in pH_i during luminal acidification also began at pH 2, but was much less pronounced (pH_i dropped from 7.38 ± 0.09 to 7.13 ± 0.11 at luminal pH 2, n = 4)

The presence of 1 mM amiloride or $10^{-4}M$ DMA resulted in a much more pronounced decrease in pH_i during luminal acidification to pH 2 than in its absence, both in the presence and in the absence of serosal CO_2/HCO_3^- (Figure 2). In serosal 5% CO_2/HCO_3^-, pH_i decreased from 7.06 ± 0.13 to 6.78 ± 0.15 at luminal pH 2 (n = 3). In serosal Hepes/O_2, the pH_i-decrease was from 6.48 ± 0.06 to 6.32 ± 0.05 at luminal pH 2 and from 6.91 ± 0.09 to 6.50 ± 0.11 at luminal pH 3 (data not shown). The results clearly show that the oxyntic cells in the depth of the mucosa do acidify during luminal acidification, and that the oxyntic cell Na^+/H^+ exchanger is involved in pH_i-maintenance during a luminal acid load, both in the presence and absence of serosal CO_2/HCO_3^-. Previous reports had indicated that in the bullfrog fundus (25) and the necturus antrum (8), a Na^+/H^+ exchanger was involved in pH_i-maintenance only in the absence of CO_2/HCO_3^-, which is an unphysiological situation. Recent experiments in the necturus fundus support our findings of an important role of the Na^+/H^+ exchanger in pH_i-regulation in the fundus (9).

II. The Cl^-/HCO_3^- exchanger

1. General characteristics

The existence of an electroneutral Cl^-/HCO_3^- exchanger in the basolateral membrane of the oxyntic cell has been postulated by Rehm and collegues more than 20 years ago (18), and several authors have demonstrated its existance and basolateral location (10,12,16,20,26). The reported kinetic parameters vary significantly, for example the K_m for external Cl^- from 5.5 (26) to 25 mM (12), and the pH_i at which halfmaximal activation is seen ranges from 7.0 (own observations) to far above 7.5 (24). Northern blot analysis and PCR-amplification with AE2 specific primers revealed that parietal cells express the AE2 isoform of the anion exchanger gene family (unpublished results), and a specific antibody for AE2 stained the basolateral membrane of the rat parietal cell (Seth Alper, this volume). We have analyzed the pH_i-dependency of maximal transport rates both in isolated parietal cells (unpublished), parietal cell basolateral membranes (14), and oxyntic cells within intact frog gastric mucosa, and consistantly found the halfmaximal activation at pH_i 6.9 (frog) to 7.15 (rabbit). Figure 3 shows that oxyntic cell pH_i increases during Cl^- removal, indicative of exchange of internal Cl^- for external HCO_3^-. It is evident that both the speed of pH_i-increase and the magnitude decrease with decreasing pH_i.

2. Role in acid secretion

Inhibition of the Cl^-/HCO_3^- exchanger by stilbene derivatives completely inhibits acid formation in isolated parietal cells and in most preparations of intact gastric mucosa from Rana esculenta. This is not due to the increase in pH_i caused by inhibition of Cl^-/HCO_3^- exchange, but the inhibiton of Cl^- uptake (21). The exact coupling mechanism between the proton pump rate and the flux through the apical Cl^--channel, resulting in the secretion of HCl, is not clear. Nevertheless, the logical conclusion is that

461

every base produced intracellularly must be exchanged at the basolateral side for Cl^-. Therefore, unless a second, as yet uncharacterized basolateral Cl^- entry pathway exists, base cannot accumulate intracellularly during acid secretion.

An interesting aspect is the adjustment of the basolateral anion flux rate to that of the apical pump rate. We have not found an secretagogue-induced increase in maximal anion flux capacity (21). The strong pH_i-dependency of anion flux rates with halfmaximal activation at pH_i 7.1, as found in our experiments, could account for an increase in flux rates if the pH_i increases in that pH_i-range during stimulation of acid secretion. However, it is not clear whether the necessary pH_i-increase occurs under physiological conditions. Both Paradiso et al (18) and we (unpublished) have found a variable pH_i-response in individual parietel cells, and in oxyntic cells within intact frog mucosa, we find a variable pH_i-response to different acid secretagogues. This suggests that a pH_i increase is probably not the only signal by which the anion flux rate increases during acid secretion. Another possibility has been suggested by Thomas et al (24). Their data indicated that an initial drop in $[Cl^-]_i$ occured during stimulation. Together with our findings that the pH_i of the stimulated parietal cell is in the pH_i-range for near-complete activation, and the maximal anion exchange capacity is very large, these results suggest that the increase in anion flux rate during acid secretion is mediated primarily by a change in the driving force and not in the ion transport capacity.

3. Role of acid secretion and Cl^-/HCO_3^- exchange in pH_i-maintenance during luminal acidification.

In the presence of serosal $5\%CO_2/HCO_3^-$, isolated frog gastric mucosa withstood a luminal pH_i of 1 for 2 hours without apparent damage. Oxyntic cell pH_i was stable at 6.62 ± 0.12 (n=3) at this pH_i, and it was irrelevant whether acid-secretion is stimulated by forskolin or inhibited by ranitidine. In the presence of serosal Hepes/O_2, however, the tissues could withstand a luminal pH of 2 only, and in this case, the oxyntic cell pH_i was higher in stimulated than in inhibited tissues. One of several possible explanations for these findings is that the intracellular HCO_3^- produced during acid secretion and exchanged for Cl^- at the basolateral membrane, the socalled "alkaline tide" serves as an additional buffer to influxing protons at the basolateral membrane. Yanaka et al have studied this effect in greater detail than we have (25-27) and the findings are summarized in the previous paper (William Silen, this volume).

Acknowledgement:

The author would like to thank all contributors to the work compiled in this manuscript, most of all her doctoral students Georg Lamprecht, Sabine Roithmaier, Michael Hübner, Manuela Nader, Heidi Rossmann, Peter Stumpf, Kristian Siegel, Cornelia Stettner, and Barbara Seidler. She would also like to thank Katherine Carter, Ron Kopito, Rosella Carroppo, Catherine Chew, and Wolfram Nagel for their collaboration, and William Silen and Meinhard Classen for their encouragement and generous support.

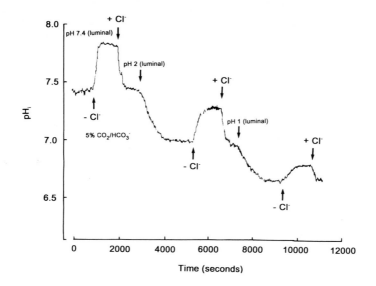

Figure 3.: Effect of Cl^--removal and -readdition on oxyntic cell pH_i at different pH_i. It is evident that both the speed of pH_i-increase and the magnitude decreases with decreasing pH_i. This illustrates the pH_i-dependency of the oxyntic cell anion exchanger.

References:

1. Boyarsky, G., M.B. Ganz, R.B. Sterzel, and W.F. Boron (1988). Intracellular pH-regulation in single glomerular mesangial cells. II. Na^+-dependent and -independent $Cl-HCO3$ exchangers. Am J. Physiol. 255:C844-C856.
2. Debellis, L., Curci, S., and E. Frömter (1992). Microelectrode determination of oxyntic cell pH in intact from gastric mucosa. Effect of histamine. Pflügers Archives 422:253-259
3. Dickens, C.J., J.I. Gillespie, and J.R. Greenwell (1989). Interactions between intracellular pH and calcium in single mouse neuroblastoma (N2A) and rat pheochromocytoma cells (PC12). Q. J. Exp. Physiol. 74:671-9
4. Forte, J.G., and T.E. Machen (1986). Ion transport by gastric mucosa. In: Physiology of membrane disorders. Andreoli, T.E., Hoffman, J.F., Fanestil, D.D., Schultz, S.G. eds. 2nd ed. New York: Plenum Publishing Corporation, pp535-558
5. Geibel, J., and I.M. Modlin (1993). Evidence for Na^+/H^+ exchange activity on the apical membrane of isolated perfused rabbit gastric glands. Gastroenterology 104:A710
6. Green, J. and S. Muallem (1989). A common mechanism for activation of the Na^+/H^+ exchanger by different stimuli. Faseb J. 3:2408-14
7. Kivilaakso, E., D. Fromm, and W. Silen (1978). Effect of the acid secretory state on intramural pH of rabbit gastric mucosa. Gastroenterology 75: 641-648
8. Kiviluoto, T., Paimela, H., Mustonen H., and Kivilaakso, E (1990). Intracellular pH in isolated Necturus antral mucosa exposed to luminal acid. Gastroenterology 1990:98:901-908

9. Kiviluoto, T., H. Mustonen, and E. Kivilaakso (1993). The defence mechanisms against intracellular acidosis induced by luminal acid are different in gastric antral and corpus mucosa. Gastroenterology 104:A119

10. Lamprecht, G., Seidler, U., and M. Classen. pH_i-regulating ion transport mechanisms in isolated parietal cell basolateral membrane vesicles. Am. J. Physiol, in press

11. Machen, T.E., und A.M. Paradiso (1987). Regulation of intracellular pH in the stomach. Ann. Rev. Physiol. 49:19-33

12. Muallem, S, Burnham, C., Blissard, D., Berglindh, T., and G. Sachs (1985). Electrolyte transport across the basolateral membrane of parietal cells. J. Biol. Chem. 260:6641-6653

13. Muallem, S., D. Blissard, E.J. Cragoe, und G. Sachs (1988). Activation of the Na^+/H^+ and Cl^-/HCO_3^- exchange by stimulation of acid secretion in the parietal cell. J. Biol. Chem. 263:14703-14711

14. Nader, M., U. Seidler, and M. Classen (1992). Differences in pH-dependent regulation of the ileal brush border and the parietal cell basolateral anion exchanger. Gastroenterology 102:A1077

15. Nakamura, K., and C.S. Chew (1993). Parietal cells contain an EGF-sensitive MAP-Kinase that is inhibited by a subclass of tyrosine kinase inhibitors. Gastroenterology 104:A155

16. Orlowski, J., R.A. Kundasamy, and G.E. Shul (1992). Molecular cloning of putative members of the Na/H-exchanger gene family. J. Biol. Chem. 267:9331-9339.

17. Paradiso, A.M., R.Y. Tsien, and T.E. Machen (1987). Digital image processing of intracellular pH in gastric oxyntic and chief cells. Nature 325:G524-G534

18. Paradiso, A.M., M.C. Townsley, E. Wenzl, and T.E. Machen (1989). Regulation of intracellular pH in resting and stimulated parietal cells. Am. J. Physiol. 257:C554-561

19. Rehm, W.S., and S.S. Sanders (1975). Implications of the neutral carrier Cl^-/HCO_3^- exchange-mechanism in gastric mucosal. Ann. NY Acad. Sci. 264:442-445

20. Seidler, U. Carter, K., and W. Silen (1989). Effect of CO_2 on pH_i rabbit parietal, chief and surface cells. Am J. Physiol 256:G466-G475

21. Seidler, U., Roithmeier, S., Classen M., and W. Silen (1992). Influence of the acid secretory state on Cl^-/base, Na^+/H^+ exchange and intracellular pH_i in isolated rabbit parietal cells. Am.J.Physiol. 262:G81-G91

22. Seidler, U., S. Roithmaier, V. Schusdziarra, M. Classen, und W. Silen (1991). pH_i-regulating ion transport systems in isolated rabbit parietal, chief and surface cells. In: Mechanisms of injury, protection and repair of the upper gastrointestinal tract. A. Garner und P.E. O'Brien eds. John Wiley und Sons, Chichester 1991, pp 253-271

23. Silen, W (1987). Gastric mucosal defense and repair. In: Physiology of the Gastrointestinal Tract. Johnson, L.R., Christensen, J., Jackson, M.J., Jacobson, E.D., und J. H. Walsh. eds. Raven Press, 2nd edition, pp 1055-1070

24. Thomas, H.A., and T.E. Machen (1991). Regulation of Cl/HCO_3^- exchange in gastric parietal cells. Cell regulation 2:727-737

25. Yanaka, A. K.J. Carter, P.J. Goddard, and W. Silen (1991) . Effect of luminal acid on intracellular pH in oxynticopeptic cells in intact frog gastrid mucosa. Gastroenterology 100:606-618

26. Yanaka, A., K.J. Carter, H.-H. Lee, and W. Silen (1989). Influence of Cl^- on pH_i in oxyntopeptic cells of in vitro frog gastric mucosa. Am. J. Physiol. 258:G815-G824

27. Yanaka, A., K.J. Carter, P.J. Goddard, M.C. Heissenberg, and W. Silen (1992). H^+/K^+ ATPase contributes to regulation of pH_i in frog oxyntopeptic cells. Am. J. Physiol. 261:G815-G824

DUODENAL MUCOSAL BICARBONATE SECRETION IN HUMANS

Jon I. Isenberg and Daniel L. Hogan

Division of Gastroenterology
Department of Medicine
University of California Medical Center
San Diego, CA 92103-8413
USA

The old adage *"Why doesn't the stomach digest itself?"* has intrigued physiologists, clinical investigators and physicians for generations. In humans, pure parietal cell secretion is approximately 160 mM, and while it is buffered within the lumen the gastric and duodenal mucosa are subjected to a continual exposure of high concentrations of acid and pepsin. The ability of the gastroduodenal mucosa to protect itself from acid-peptic injury, as well as damage induced by drugs, other injurious agents, alcohol, ischemia, etc., involve a number of intrinsic pre-epithelial, epithelial and sub-epithelial defensive factors that work in concert to prevent mucosal injury. Indeed, when the aggressive factors overwhelm mucosal defense the result is inflammation, erosion or ulceration with the potential of their inherent complications. A host of defensive factors are under investigation including: mucus and bicarbonate secretion, surface active phospholipids, the "mucoid cap", cellular resistance, acid/base transporters, growth factors, cytokines, leukocyte adhesion molecules and reconstitution.

An important defensive factor, which will be reviewed herein, is mucosal bicarbonate secretion that contributes to the mucus-bicarbonate barrier (Fig. 1). In 1856 Claude Bernard postulated that the mucus layer overlying the surface epithelial cells of the stomach may protect the surface epithelium from damage. Furthermore, Pavlov in 1898 suggested that the "alkaline mucus" served to

NATO ASI Series, Vol. H 89
Molecular and Cellular Mechanisms
of H⁺ Transport
Edited by Barry H. Hirst
© Springer-Verlag Berlin Heidelberg 1994

neutralize gastric acid thereby serving as a protective function. Recent studies have revealed that mucus delays the diffusion of hydrogen ion significantly more than water alone. Mucus, albeit dynamic, adheres to the epithelial cell surface providing a physical barrier between the surface epithelial cells and the luminal contents. In the duodenum, mucus is secreted by Brunner glands and epithelial cells. The mucus layer is thinned by aspirin, non-steroidal antiinflammatory drugs (NSAIDs), bile salts and lysolecithin, while it is thickened by prostaglandins of the A and E type.

Gastroduodenal Mucosal Bicarbonate Secretion. As early as 1892, Schierbeck demonstrated in dogs that feeding not only increased gastric acid secretion, but also induced an increase in luminal CO_2 to levels above those in blood, suggesting that the CO_2 resulted from the reaction of intraluminal H^+ and HCO_3^-. Hollander postulated that gastric alkaline secretion was produced at a constant rate originating from the nonacid-secreting cells. Furthermore, in isolated intestinal loops in dogs, free of pancreaticobiliary secretions, Dorricott et al infused hydrochloric acid and measured loss of hydrogen ions and changes in osmolality in the luminal contents. They observed that hydrogen ion concentration and osmolality decreased ($HCl + NaHCO_3 \rightarrow H_2O + NaCl + CO_2$), and concluded that mucosal bicarbonate secretion was responsible for the decrease in hydrogen ion concentration. During the last 20 years investigators have demonstrated in both animals and humans that gastric and duodenal mucosal bicarbonate secretion (DMBS) is regulated by various agonists and antagonists, and that the secretory mechanism involves specific cellular transport mechanisms including a likely HCO_3^- conductance channel, HCO_3^-/Cl^- and Na^+/H^+ exchangers and a $NaHCO_3$ cotransporter.

There are a number of methods used to measure mucosal bicarbonate secretion which include: 1) pH/PCO_2 utilizing the Henderson-Hasselbalch equation to calculate [HCO_3^-], 2) back-titration, and 3) measurement of pH-osmolality. The two former methods yield similar results, while the latter reveals results that are 3- to 5-fold greater.

<u>Duodenal Mucosal Bicarbonate Secretion (DMBS)</u> In early studies with mammalian duodenum, the origin of alkaline secretion was thought to be mainly from the submucosal Brunner's glands. However, over the past decade studies in species devoid of Brunner glands, as well as in distal duodenum (Brunner gland-free) segments, have revealed that rather high rates of bicarbonate secretion originates from surface epithelial cells. The significance of Brunner gland alkaline secretion requires further investigation. Transport of bicarbonate by the duodenal surface epithelial cells has been studied extensively in bullfrog, rat, rabbit, dog and other species. Only recently has bicarbonate secretion by the human duodenal mucosa been studied.

In spite of rather complicated methods that require isolation of the proximal duodenum from gastric and/or distal duodenal secretions, resting, basal, proximal DMBS has been reported to be approximately 150 to 200 μmol/cm-h in three independent laboratories. Also, DMBS is in large part an active process since bicarbonate is secreted in the absence of a serosal-to-mucosal gradient. Bicarbonate secretion in the third portion of the duodenum is only about 15% of that in the bulb indicating a steep gradient from proximal-to-distal duodenum. Basal bicarbonate secretion by the proximal duodenum is controlled by a number of neural, cellular and subcellular factors. Since atropine decreases basal DMBS by approximately 80%, a major controlling factor is muscarinic cholinergic tone. Also, endogenous prostaglandins contribute significantly to resting DMBS; inhibition of cyclooxygenase with indomethacin reduces resting DMBS by about 60%. Bicarbonate secretion is decreased by the α_2-agonist clonidine. This may be of clinical relevance as plasma concentration of catecholamines and urinary output of their metabolites are increased in some duodenal ulcer (DU) patients, and basal and stimulated (H^+, PGE_2) bicarbonate outputs are impaired in DU patients (see below). Finally, Knutson et al demonstrated that basal DMBS is mediated in part by opioids. Morphine inhibited resting bicarbonate output by about 70%, an effect reversed by naloxone, indicating that this effect is mediated by μ-receptors.

The human duodenal mucosa responds to a variety of agonists and antagonists which are summarized in Table 1. Luminal acidification with a physiologic amount of HCl is a potent stimulus, increasing bicarbonate secretion 2-3 fold. Furthermore, the pH threshold for duodenal mucosal bicarbonate secretion is approximately pH 3.

Luminal acidification of the duodenum stimulates the release of a number of hormones (e.g., secretin, glucagon and VIP). In animals each stimulates duodenal bicarbonate secretion, although the response to secretin is conflicting. Yet, in humans, only VIP significantly increases DMBS.

In addition, prostaglandins of the E class play a major role in regulating duodenal bicarbonate transport (Fig. 1). Selling et al demonstrated that luminal acidification increased endogenous PGE_2 output as well as bicarbonate secretion. And, inhibition of

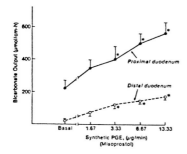

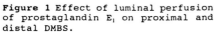

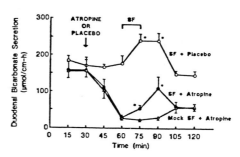

Figure 1 Effect of luminal perfusion of prostaglandin E_1 on proximal and distal DMBS.

Figure 2 Vagal cholinergic control of DMBS. Effect of atropine and sham feeding (SF).

endogenous PGE_2 with the cyclooxygenase inhibitor indomethacin significantly decreased both basal and H^+-stimulated bicarbonate output and luminal PGE_2 release. Thus far, all prostaglandins tested stimulate DMBS. Taken together, these results indicate that

E class prostaglandins act locally to regulate human DMBS.

DMBS is also controlled by vagal-cholinergic factors. Muscarinic blockade with the nonspecific anticholinergic atropine decreased bicarbonate secretion markedly to 20% of basal (Fig. 2). On the other hand, cholinergic-stimulation with bethanachol had no significant stimulatory effect. As untoward side effects occurred in subjects at the higher bethanachol doses, it is possible that cholinergic agonists may affect human DMBS. The marked suppression of DMBS by atropine indicates that cholinergic tone is near maximal during basal bicarbonate secretion. Cephalic-vagal stimulation by sham feeding in humans also significantly increased bicarbonate secretion by 30%. Of interest, cholinergic factors do not regulate the H^+- and vagal-induced bicarbonate responses since atropine did not alter their respective net stimulated bicarbonate outputs. Also, inhibition of local PGs with indomethacin did not impair the response to sham feeding. Therefore, in healthy subjects, whereas acid-stimulated bicarbonate transport is most likely regulated by both PGEs and VIP, vagal-stimulated bicarbonate secretion appears to be mediated by VIP and/or other non-cholinergic (possibly purinergic) mechanisms. Ainsworth et al recently reported that in normal subjects active cigarette smoking diminished H^+-stimulated DMBS but did not alter resting bicarbonate secretion (Fig. 3).

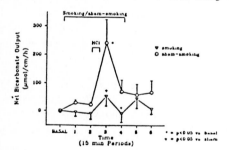

Figure 3 Effect of smoking on acid-stimulated DMBS in normal smoking volunteers.

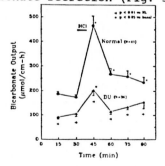

Figure 4 Basal and acid-stimulated DMBS is impaired in DU patients compared to normal controls.

DMBS in duodenal ulcer (DU) disease. In 1987 it was reported that patients with inactive duodenal ulcer (DU) demonstrated impaired proximal, but not distal, DMBS. This has been a consistent observation (Fig. 4). Basal and stimulated DMBS in DU is approximately 50% of that in normal subjects with little overlap. In addition, patients who were studied while their ulcer was active and after their ulcer healed (as determined by upper gastrointestinal endoscopy) revealed that basal and acid-stimulated bicarbonate outputs were similar. This suggests that the impaired proximal duodenal bicarbonate secretion observed in DU patients is independent of the anatomical ulcer defect, and more likely related to a cellular or subcellular abnormality. Further investigations revealed that DU patients released significantly greater amounts of endogenous PGE_2 in response to acidification as compared to normals. Therefore, whereas DU patients secrete less bicarbonate, they release more PGE_2. The mechanism(s) responsible for this observation is unknown. Moreover, when exogenous PGE_2 was perfused into the duodenum there was no significant increase in bicarbonate secretion. Also, cephalic-vagal stimulation in a limited number of DU patients failed to increase DMBS. To explore if the defect in DMBS may be of morphological origin, biopsies were taken from the duodenal bulb in patients with inactive DU and healthy volunteers. There were no differences found in quantitative morphometric histology of any parameter between the two groups as assessed by a "blinded" expert morphologist. The effect of age on DMBS was assessed (younger $\leq$ 44 versus older $\geq$ 44 years) in both normal subjects and DU patients. Although the DU group had significantly decreased basal and stimulated DMBS, the variable of age itself did not influence bicarbonate secretion in either normals or DUs.

The impairment in bicarbonate secretion by the proximal duodenal mucosa in duodenal ulcer patients seems to be a global event including diminished basal bicarbonate production as well as being non-responsive to the agonists H^+, PGE_2 and sham feeding. Finally, although DMBS to all agonists tested is diminished in DU patients, the resting and stimulated transmucosal duodenal potential difference (PD) is similar between groups. This suggests that rather than secreting bicarbonate, DU patients are secreting another anion (possibly Cl^-). These results taken as a whole support the hypothesis that the bicarbonate secretory defect in DU is located at a distal secretory process (i.e., cellular and/or subcellular events). The recent evidence regarding the association of *Helicobacter pylori* (HP) and DU, particularly ulcer recurrence rates, is of interest. Studies are in progress to assess the affect of HP infection on DMBS both in normal subjects and in patients with DU.

Mechanisms of DMBS   Studies evaluating gastroduodenal bicarbonate transport have been performed mainly *in vitro* with mucosal tissue, stripped of serosal and muscle layers, and mounted in Ussing chambers or isolated enterocytes. The proposed mechanism includes transport processes involving: (a) electroneutral Cl^-/HCO_3^- exchange; (b) active electrogenic transport, possibly via the Cl^- conductance (CFTR) channel; and, (c) passive paracellular transport. Furthermore, bicarbonate secretion is an oxygen-dependent metabolic process, and requires Na^+, K^+-ATPase which likely functions with a $NaHCO_3^-$ cotransporter to transport bicarbonate across the basolateral cell membrane. To further delineate specific transport pathways and their regulation in health and disease, studies are currently in progress examining acid/base transporters using isolated enterocytes.

Summary and the Future   The integrity of the gastroduodenal mucosa is maintained by a delicate balance between the so called, "aggressive" and "defensive" factors. When this balance is tipped towards the aggressive factors mucosal damage and ulceration can occur. There is enormous overlap between normal subjects and DU patients in most of the pathophysiological factors studied thus far (e.g., gastric acid secretion, gastric emptying, gastrointestinal hormone content and release, etc.). To date, the only rather consistent pathophysiological observations identifying a defect in a major defensive factor is impaired duodenal mucosal bicarbonate secretion and the presence of *Helicobacter pylori*. Whether or not these two are interrelated requires additional study. In addition, at the clinical level the development of antisecretory drugs is approaching the end of the road with the availability of the H^+/K^+-ATPase inhibitors. The focus in the future will undoubtably be directed at development of effective and safe therapeutic agents that will increase specific defensive factors.

Table 1. Human Gastroduodenal Mucosal Bicarbonate Secretion.

	Gastric HCO_3^-	Duodenal HCO_3^-
Stimulants:	Cholinergic agents Sham-feeding Prostaglandin Es HCl Gastric distension	HCl Prostaglandin Es VIP Sham-feeding Theophylline (cAMP)
Inhibitors:	Cholinergic antagonists Aspirin Indomethacin Taurocholate	Atropine Indomethacin Morphine Clonidine Acetazolamide Smoking
No Effect:	Gastrin Histamine H_2-blockers Omeprazole Acetazolamide	Glucagon Secretin Bethanachol

REFERENCES

Ainsworth MA, Kjeldsen J, Schaffalitzky de Muckadell OB (1990) Morphine inhibits secretion of bicarbonate from the human duodenal mucosa. Scand J Gatroenterol 25:1066-75.

Ainsworth MA, Hogan DL, Koss MA, and Isenberg JI (1993) Cigarette smoking inhibits acid-stimulated duodenal mucosal bicarbonate secretion. Annals Internal Medicine, in print.

Allen A (1981) Structure and function of gastrointestinal mucus. In: Johnson LR (ed). Physiology of the Gastrointestinal Tract. New York, Raven Press.

Ballesteros MA, Wolosin HD, Hogan DL, Koss MA, Isenberg JI (1991) Cholinergic regulation of human duodenal mucosal bicarbonate secretion. Am J Physiol 261:G327-31.

Bukhave K, Rask-Madsen J, Hogan DL, Koss MA, and Isenberg JI (1990) Proximal duodenal prostaglandin E_2 release and mucosal bicarbonate secretion are altered in patients with duodenal ulcer. Gastroenterology 99:951-55.

Feitelberg SP, Hogan DL, Koss MA, and Isenberg JI (1992) The human duodenal mucosa: pH threshold for bicarbonate secretion and diffusion of CO_2. Gastroenterology 102:1252-58.

Flemström G (1987) Gastric and duodenal mucosal bicarbonate secretion. In: Johnson LR, ed. Physiology of the Gastrointestinal tract. 2nd ed. New York: Raven Press.

Hogan DL, Isenberg JI (1988) Gastroduodenal bicarbonate production. In: Stollerman GH (ed). Advances in Internal Medicine. Chicago, Year Book Medical Publishers, Inc.

Isenberg JI, Hogan DL, Koss MA, Selling JA (1986) Human duodenal mucosal bicarbonate secretion: Evidence for basal secretion and stimulation by hydrochloric acid and a synthetic prostaglandin E_1 analogue. Gastroenterology 91:370-78.

Isenberg JI, Selling JA, Hogan DL, Koss MA (1987) Impaired proximal duodenal mucosal bicarbonate secretion in duodenal ulcer patients. N Engl J Med 316:374-79.

Knutson L, Flemström G (1989) Duodenal mucosal bicarbonate secretion in man. Stimulation by acid and inhibition by the $alpha_2$ adrenoceptor agonist clonidine. Gut 30:1708-15.

Odes HS, Hogan DL, Ballesteros MA, Wolosin JD, Koss MA, and Isenberg JI (1990) Human duodenal mucosal bicarbonate secretion: evidence suggesting active transport under basal and stimulated conditions. Gastroenterology 98:867-72.

Selling JA, Hogan DL, Andreas A, Koss MA, Isenberg JI 1987 Indomethacin inhibits duodenal mucosal bicarbonate secretion and endogenous PGE_2 output in human subjects. Ann Intern Med 106:368-71.

Wolosin JD, Thomas FJ, Hogan DL, Koss MA, O'Dorisio TM, Isenberg JI 1989 The effect of vasoactive intestinal peptide, secretin, and glucagon on human duodenal bicarbonate secretion. Scand J Gastroenterol 24:151-57.

Yao BG, Hogan DL, Bukhave K, Koss MA, Isenberg JI 1993 Bicarbonate transport by rabbit duodenum in vitro: Effect of vasoactive intestinal polypeptide, prostaglandin E_2, and cyclic adenosine monophosphate. Gastroenterology 104:732-40.

Gastric bicarbonate secretion

Luca Debellis, Rosa Caroppo, Claudia Iacovelli, Eberhard Frömter*, Silvana Curci
Istituto di Fisiologia Generale
Università di Bari
Via Amendola 165/A
70124 Bari
Italy

Gastric mucosa does not only secrete hydrochloric acid but it is also capable of secreting alkali, probably as part of a defence mechanism against acid-induced ulceration. Alkali secretion can be observed in gastric fundus mucosa if HCl secretion is inhibited by blockers of histamine receptors (Flemström G. 1977, Takeuchi K. 1982). Which cells secrete alkali is still unknown. Amphibian gastric fundus mucosa consists of two major cell types: the oxyntopeptic cells (OC) which form the gastric glands and are known to secrete acid and the surface epithelial cells (SEC) which face the gastric lumen and have been assumed to secrete alkali possibly via an apically or basolaterally located Cl^-/HCO_3^- exchanger (Flemström G. 1977, Takeuchi K. 1982). However, thus far no firm evidence for the localization and mechanism of alkali secretion is available.

In the present study we have investigated the mechanism of alkali secretion and the possible involvement of the OC in Rana esculenta gastric fundus mucosa with the pH-stat technique and with electrophysiological measurements. Specifically we measured the rate of transepithelial alkaline secretion (ASR), serosal membrane potential (V_s) and cytoplasmic pH (pH_i) of OC as well as transepithelial resistance (R_t) and voltage divider ratio (VDR) under the influence of ion substitutions, transport inhibitors and Ca^{2+}- or cAMP-mediated secretagogues.

I. Measurements of alkali secretion rate

For these measurements sheets of gastric fundus mucosa were mounted between two half-chambers, each consisting of a circular fluid canal filled with Ringer solution which was constantly recirculated by means of a bubble lift.

---

* Zentrum der Physiologie, J W Goethe Universität, Theodor Stern Kai 7, 60590 Frankfurt/M 70, F.R.G.

NATO ASI Series, Vol. H 89
Molecular and Cellular Mechanisms
of H+ Transport
Edited by Barry H. Hirst
© Springer-Verlag Berlin Heidelberg 1994

The Ringer solution on the serosal side had the following composition (in mmol/l): 102.4 Na^+, 4.0 K^+, 1.8 Ca^{2+}, 0.8 Mg^{2+}, 91.4 Cl^-, 17.8 HCO_3^-, 0.8 SO_4^{2-}, 0.8 $H_2PO_4^-$ and 11 D-glucose. It was gassed with 5% CO_2 in O_2 and had a pH of 7.36. It also contained ranitidine (10 µmol/l) to suppress HCl secretion. The luminal solution was unbuffered and had the following composition (in mmol/l): 102.4 Na^+, 4.0 K^+, 91.4 Cl^-, 15.0 isethionate, 7.0 mannitol, 11.0 D-glucose. It was gassed with 100% O_2 to prevent CO_2 accumulation. ASR was measured by activating the titration procedure every 15 min with 5 mmol/l HCl as the titrant. Simultaneously also V_t and R_t were measured.

In control conditions in summer season the gastric fundus mucosa of Rana esculenta secretes alkali at a rate of $\approx$ 0.8 µEq·cm^{-2}·h^{-1} and this rate remains usually constant for up to 3 or more hours. ASR was reduced but not abolished when the HCO_3^- on the serosal surface was replaced by HEPES-Tris (pH 7.36), and fell to less than half when an inhibitor of carbonic anhydrase was given (acetazolamide 100 µmol/l). Those observations suggests that ASR reflects primarily secretion of HCO_3^- which may either be transported across the epithelium or generated from metabolic CO_2.

Further experiments indicate that ASR (HCO_3^- secretion) requires metabolic energy. Indeed, inhibition of oxidative metabolism with rotenone (10 µmol/l) or with the "uncoupler" dinitrophenol (100 µmol/l) (which does not interfere with CO_2 production but rather accelerates it) strongly depressed ASR. Moreover, inhibition of the Na^+/K^+ pump by ouabain (100 µmol/l) also depressed ASR. These observations suggest that ASR is probably achieved by some sort of secondary active transport (which requires metabolic energy) in coupling with the flux of Na^+ ions (e.g. driven by a Na^+ gradient which is built up by the Na^+/K^+ pump).

To test for ion flux coupling mechanism that might be involved in secretion we investigated the effect of reducing bath Na^+ and Cl^- concentration. Substitution of serosal Na^+ by N-methyl-D-glucamine significantly reduced V_t and ASR both in the presence (Table 1) and absence of serosal HCO_3^-. Serosal replacement of Cl^- by gluconate nearly collapsed Vt and tended to increase ASR slightly while luminal replacement had the opposite effect: V_t increased and ASR decreased but only slightly (-13% after 1 hour, seeTable 1). Serosal application of amiloride (100 µmol/l) which is known to inhibit Na^+/H^+ exchange, had virtually no effect. On the other hand, serosal application of 4,4-diisothyociano-2,2-disulfonate stilbene (DIDS 200 µmol/l), which is a well known inhibitor of serosal Cl^-/HCO_3^-exchange and serosal $Na^+(HCO_3^-)_n$ cotransport, strongly depressed ASR without greatly affecting V_t both in presence (Table 1) and in absence of serosal HCO_3^-. The latter observations strongly support the conclusion reached above that alkaline secretion is achieved by some kind of secondary active transport of HCO_3^- in coupling with primary active Na^+ absorption. A model that could account for these observations would comprise two steps: 1.) uptake of HCO_3^- from the interstitial fluid into the cell by means of a Na^+ coupled HCO_3^- transporter, such as the DIDS-inhibitable rheogenic

$Na^+(HCO_3^-)_n$ cotransporter and 2.) an apical efflux mechanism, the nature of which is still to be defined.

	n	CONTROL		EXPRIMENTAL	
		Vt [mV]	ASR $[\mu Eq \cdot cm^{-2} \cdot h^{-1}]$	Vt [mV]	ASR $[\mu Eq \cdot cm^{-2} \cdot h^{-1}]$
serosal Na$^+$-free	15	33.9±1.9	0.60±0.06	10.7±1.7	0.20±0.06 p< 0.01
serosal Cl$^-$-free	4	37.0±2.5	0.61±0.06	11.7±0.7	0.70±0.09
mucosal Cl$^-$-free	4	40.6±2.5	0.88±0.15	49.2±3.9	0.78±0.11
200 µmol/l DIDS	10	31.0±3.1	0.64±0.06	27.6±3.2	0.25±0.07 p< 0.01

Table.1 - Effect of ion substitutions (Na$^+$ replaced by NMDG and Cl$^-$ replaced by gluconate) and of DIDS on frog fundus gastric mucosa alkaline secretion rate (ASR) and transepithelial potential difference (V_t). The serosal solution was HCO_3^--CO_2 buffered Ringer (pH 7.36), gassed with 5% CO_2 in O_2. Effects observed after 1 hour treatment are reported.

Regarding the question which cells are responsible for HCO_3^- secretion it is notewort that we have previously detected a rheogenic $Na^+(HCO_3^-)_n$ cotransporter in the serosal cell membrane of the OC of Rana esculenta stomach (Curci et al. 1987). This led us to investigate whether the OC might be involved in HCO_3^- secretion. Since - as it is true in general - steady state measurements are not sufficiently instructive for modelling attempts, we analyzed transients in ASR and in the properties of OC during stimulation of alkaline or acid secretion.

Administration of 100 µmol/l carbachol, a Ca^{2+}-mediated secretagogue, led to a transient stimulation of ASR which relaxed again within 10 min. This stimulation was associated with a transient increase in calculated lumen-negative short-circuit current (SCC), which most probably reflects anion secretion. A quantitative comparison of the increase in ASR and SCC, however, indicates that the change in SCC exceeded the change in ASR and this was also true for the basal values of SCC and ASR. This result is ambiguous. It leaves open the possibility that ASR is a rheogenic process accompanied by flow of negative charges but does not prove this. Any current that would not be carried by HCO_3^- secretion most likely reflects the so called non-acidic Cl secretion (Forte et al. 1980) that appears to be also stimulated by carbachol. To further clarify the mechanism of HCO_3^- secretion and the possible involvement of the OC we have investigated these cells with microelectrodes techniques focusing mainly on the effect of different secretagogues.

II. Microelectrode experiments on oxyntopeptic cells

The techniques applied have been described previously (Debellis L. et al. 1990, 1992). Briefly, sheets of gastric fundus were mounted horizontally between two half chambers with the serosal side facing up. To allow free access to the OC the connective tissue was dissected off and microelectrodes were inserted through the open top of the upper half chamber to impale cells under microscopic observation. Both surfaces of the tissue were continuously perfused with gassed HCO_3^- Ringer solution having the above described composition. The transepithelial potential (V_t) and resistance (R_t) were monitored continuously wile cell membrane potentials (V_s) and intracellular pH (pH_i) were measured with single- and double-barrelled microelectrodes. In some experiments the voltage divider ratio (VDR) was also determined. It represents $\Delta V_m / \Delta V_s$ where ΔV_m and ΔV_s are voltage displacements recorded in response to transepithelial current pulses across the apical and basolateral cell membrane respectively. Double-barrelled pH microelectrodes were constructed from boro-silicate glass tubing and H^+ Ionophore II (Fluka, Switzerland) was used as sensor. The reference channel was filled with 0.5 mol/l KCl while the shank of the sensitive channel contained a reference solution of pH 7.0. All microelectrodes were calibrated before and after the impalement by flushing the serosal chamber with NaCl solutions having pH values between 6.3 and 7.9. The electrodes (tip length $\approx$ 23 mm) showed a response of approximately 58 mV/pH unit. The resistance of the selective channel averaged 390 GΩ and that of the reference channel 160 MΩ. Cimetidine (100 µmol/l in the serosal solution) was used to keep the tissues in the resting state.

Fig. 1 depicts an experiment in which an oxyntopeptic cell was punctured with a double-barrelled pH microelectrode and in which the serosal surface of the mucosa was intermittently perfused with carbachol (100 µmol/l). As can be seen after carbachol treatment the cells transiently acidified, reaching a peak pH shift of 0.09 ± 0.03 pH units (n=6) after approximately 5 min. This time course coincides with the transient increase in ASR described above and hence may reflect an increase in HCO_3^- secretion from the cell into the lumen compartment. In addition, simultaneously with the above described rise in V_t and fall in R_t the voltage divider ratio (VDR) fell, indicating probably a preferential decrease of the apical cell membrane resistance because R_t also decreased. This could explain the observed transient increase in short circuit current described above which may reflect concomitant stimulation of Cl^- and HCO_3^- secretion from OC into gastric lumen probably via activation of an apical anion conductance.

Since we have previously shown that frog stomach oxyntopeptic cells contains a DIDS (SITS)-sensitive $Na^+(HCO_3^-)_n$ cotransporter in their serosal cell membrane (Curci et al. 1987), the possible involvement of this rheogenic cotransporter in the mechanism of alkaline secretion was investigated by studying the effect of carbachol and other secretagogues on the cell potential response to sudden changes in serosal bath HCO_3^- concentration (from 17.8 to 6 mmol/l HCO_3^- replaced by gluconate).

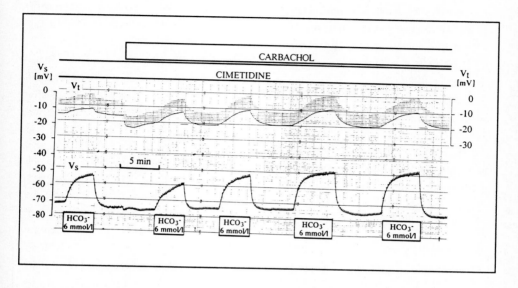

FIG. 1 - Effect of stimulation with carbachol (10^{-4} mol/l) on transepithelial potential (V_t), serosal membrane potential (V_s) and intracellular pH (pH_i) of oxyntopeptic cell in presence of cimetidine (10^{-4} mol/l). Superimposed voltage pulses indicate response to transepithelial constant current pulses of 50 $\mu A \cdot cm^{-2}$ and 1 s duration.

As can be seen in Fig. 2 both the transepithelial and the cell membrane potential response $(\Delta V_t)HCO_3^-$ and $(\Delta V_s)HCO_3^-$ increased under carbachol but with a different time course. $(\Delta V_t)HCO_3^-$ increased immediately from 2.2±0.6 mV (n=5) to 4.7±0.9 mV and remained elevated as long as carbachol was present, while $(\Delta V_s)HCO_3^-$ (17.4±1.7 mV) increased only with a delay of approximately 5 min., reached its maximum (20.8±1.0 mV) after 10 min, after the peak in SCC and ASR was over, and then remained elevated. The delayed rise in $(\Delta V_s)HCO_3^-$ may be a consequence of the electrical properties of the epithelium. If carbachol transiently activates an apical anion conductance as described above, this should attenuate the measurable increase in $(\Delta V_s)HCO_3^-$ until the peak in SCC (and ASR) is over. Alternatively we cannot exclude, however, that carbachol activates the presumed apical anion conductance and the basolateral $Na^+(HCO_3^-)_n$ cotransporter with a different time course.

Since an increase (or decrease) in the potential response to substitution of a permeable ion, i, by a non penetrating ion does not necessarily indicate that the conductance g_i increased (or decreased) but may also result from a decrease (or increase) in other ion conductances in the same membrane, we have also measured the potential responses to steps changes in serosal K^+ concentration from 4 to 13 mmol/l. These potential changes, however, remained constant under carbachol or even slightly increased, suggesting that the observed increase of $(\Delta V_s)HCO_3^-$ after application of carbachol indicates indeed a stimulation of $Na^+(HCO_3^-)_n$ cotransport.

If the OC are capable of secreting HCO_3^- we may expect that this secretion is terminated when the cells are stimulated to secrete acid and if $Na^+(HCO_3^-)_n$ cotransport in the serosal cell membrane is involved in HCO_3^- secretion, we may anticipate that it is down-regulated after stimulation of acid secretion with histamine or other cAMP mediated secretagogues. This was indeed observed. Histamine (100 µmol/l) clearly decreased $(\Delta V_s)HCO_3^-$ from 18.3±0.4 to 13.6±1.0 mV as well as $(\Delta V_t)HCO_3^-$ from 3.7±0.4 to 0.9±0.2 mV and the change in transepithelial resistance after substitution of HCO_3^-, $(\Delta R_t)HCO_3^-$ from 26.3±6.7 to -11.6±3.7 $\Omega \cdot cm^2$ (n=7). Similar results were also obtained with theophylline (10 mmol/l) and dibutyryl cAMP (100 µmol/l).

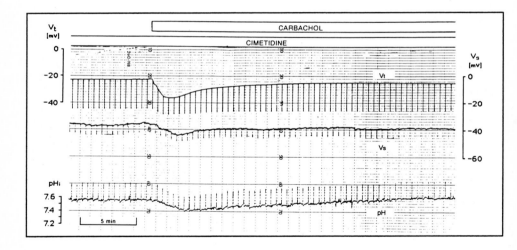

FIG. 2 - Effect of stimulation with carbachol (10^{-4} mol/l) on the response of V_t and V_s to lowering serosal HCO_3^- from 17.8 to 6 mmol/l during marks.

III. Model of HCO₃⁻ secretion

From the data summarized above we conclude that OC of Rana esculenta stomach are capable of secreting HCO_3^- by a mechanism which involves HCO_3^- uptake from serosal fluid into the cell via the basolateral $Na^+(HCO_3^-)_n$ cotransporter and probably conductive HCO_3^- exit across the apical cell membrane into gastric lumen.

From a thermodynamic point of view this model appears reasonable. The first step, HCO_3^- uptake into the cell, could never be achieved by a HCO_3^- conductance, because the electrochemical driving force for HCO_3^- across the basolateral cell membrane favours HCO_3^- efflux. It also could not be achieved by the Cl^-/HCO_3^- exchanger because this would require that intracellular Cl^- concentration rises above extracellular Cl^- concentration. Theoretically Na^+/H^+ exchange could help to alkalinize the cell but its activity is too low at the observed pH of ≈ 7.4 (Debellis et al. 1992) and amiloride did not block ASR in our experiments. However, based on conventional estimates of intracellular Na^+ concentration and on our measurements of intracellular HCO_3^- concentration and of the membrane potential we may conclude that $Na^+(HCO_3^-)_n$ cotransport can import HCO_3^- into the cell provided it operates at a stoichiometry of 2 HCO_3^- to 1 Na^+ ion. Although this stoichiometry contradicts previous estimates of 3 HCO_3^- to 1 Na^+ obtained in renal cells (Yoshitomi et al. 1985; Soleimani et al. 1987) we know today that the stoichiometry is not fixed but may vary between 3 to 1 and 2 to 1 in different preparation as recently observed in renal tubules (Seki et al. 1993).

The second step, HCO_3^- exit from cell to lumen, could indeed proceed via a simple HCO_3^- conductance. The apical membrane potential (and under conditions of apical HCO_3^- free solutions also the concentration gradient) favour HCO_3^- efflux. In addition we know that carbachol stimulates Cl^- secretion in many epithelia, including intestinal epithelia. Since some of the ion channels involved are not exclusively selective for Cl^- ions but allow also HCO_3^- ions to pass (Kunzelmann K. et al. 1991), one might even speculate that such anion channels mediate both non-acidic Cl^- secretion - which also originates from the OC (Curci et al. 1986) - and HCO_3^- secretion.

The question how much the OC contribute to gastric fundus HCO_3^- secretion cannot be answered at present. Our data certainly do not imply that SEC do not secrete HCO_3^-. We do not know enough today about the properties of the SEC to evaluate their transport capabilities. However, the observation that serosal Na^+ replacement abolished ASR almost completely and the observation that luminal Cl^- replacement - which is thought to abolish the Cl^-/HCO_3^- exchanger-mediated HCO_3^- secretion in the apical membrane of SEC (Flemström 1980) - had only a marginal effect, suggests that the contribution from the OC might be substantial. The physiological role of HCO_3^- secretion by the OC might be to protect the gastric gland lumina from prolonged exposure to acid pH when acid secretion has stopped.

REFERENCES

Curci S, Debellis L, Frömter E (1987) Evidence for rheogenic sodium bicarbonate cotransport in the basolateral membrane of oxyntopeptic cells of frog gastric fundus. Pflügers Arch 408: 497-504

Curci S, Schettino T, Frömter E (1986) Histamine reduces Cl activity in surface epithelial cells of frog gastric mucosa. Pflügers Arch 406: 204-211

Debellis L, Curci S, Frömter E (1990) Effect of histamine on the basolateral K^+ conductance of frog stomach oxyntopeptic cells and surface epithelial cells. Am J Physiol 258: G631-G636

Debellis L, Curci S, Frömter E (1992) Microelectrode determination of oxyntopeptic cell pH in intact frog gastric mucosa. Effect of histamine. Pflügers Arch 422: 253-259

Flemström G (1977) Active alkalinization by amphibian gastric fundic mucosa in vitro. Am J Physiol 233: E1-E12

Flemström G (1980) Cl- dependence of HCO_3^- transport in frog gastric mucosa. Uppsala J Med Sci 85: 303-309

Flemström G (1987) Characteristics of gastric bicarbonate secretion. In: Physiology of the Gastrointestinal Tract. New York, Raven Press,1016-1029

Forte JG, Machen TE, Öbrink KJ (1980) Mechanisms of gastric H^+ and Cl- transport. Ann Rev Physiol 42: 111-126

Kunzelmann K, Gerlach L, Fröbe U, Greger R (1991) Bicarbonate permeability of epithelial chloride channels. Pflügers Arch 417: 616-621

Seki G, Coppola S, Frömter E (1993) The Na-HCO_3 cotransporter operates with a coupling ratio of 2 HCO_3^- to 1 Na^+ in isolated perfused rabbit renal proximal tubule. Pflügers Arch 425: 409-416

Soleimani M, Grassl SM, Aronson PS (1987) Stoichiometry of Na^+-HCO_3^- cotransport in basolateral membrane vesicles isolated from rabbit cortex. J Clin Invest 79: 1276-1280

Takeuchi K, Merhav A, Silen W (1982) Mechanism of luminal alkalinization by bullfrog fundic mucosa. Am J Physiol 243: G377-G388

Yoshitomi K, Burckard BCh, Frömter E (1985) Rheogenic sodium-bicarbonate cotransport in peritubular cell membrane of rat renal proximal tubule. Pflügers Arch 405: 360-366

Role of Dopamine in Control of Duodenal Mucosal Bicarbonate Secretion

Gunnar Flemström, Bengt Säfsten and Lars Knutson[1]
Department of Physiology and Medical Biophysics
Uppsala University
S-751 23 Uppsala
Sweden

The bicarbonate secretion of the duodenal mucosa alkalinizes the viscoelastic mucus gel adherent to the mucosal surface and is one major mechanism in the protection of this epithelium against luminal acid (and pepsin). The secretion is influenced by stimulatory and inhibitory neural impulses and by local mucosal production of prostaglandins and other substances (Flemström 1993), and is deficient in patients with chronic and acute duodenal ulcer disease (Isenberg et al. 1987, Mertz-Nielsen et al. 1993). Recent reports have indicated that some dopaminergic compounds are efficacious in the treatment of duodenal ulcer disease in humans (Sikriric et al. 1991), and protective effects of such compounds have been observed in animal models of ulcer disease (Glavin 1989, MacNaughton & Wallace 1989, Glavin & Szabo 1990, Aho & Lindén 1992). It is also well known that dopamine receptor agonists affect the motility in several segments of the gastrointestinal tract and effects on salivary gland and pancreatic exocrine secretions and on small intestinal NaCl absorption have been reported (Donowitz et al. 1982, Willems et al. 1985). These findings made it of interest to investigate the effects of dopamine and some dopaminergic compounds on the duodenal mucosal bicarbonate secretion. The secretion was studied in anaesthetized rats and in human volunteers. Duodenal mucosal production of cyclic AMP was measured in isolated duodenal villus and crypt enterocytes.

Studies in anaesthetized animals

Male Sprague-Dawley rats, weighing 250-315 g (Anticimex, Stockholm, Sweden) were kept under standardized conditions of temperature and light and were allowed to adjust to their new environment for at least four days. They were deprived of food for 20-24 hours before the experiments, but had free access to drinking water up to the beginning of the experiment. The surgical and experimental procedures have been described previously (Säfsten et al. 1991) and a

---

[1]Department of Surgery, Uppsala University Hospital, S-751 85 Uppsala, Sweden

NATO ASI Series, Vol. H 89
Molecular and Cellular Mechanisms
of H⁺ Transport
Edited by Barry H. Hirst
© Springer-Verlag Berlin Heidelberg 1994

brief summary is given here. The animals were anaesthetized with 5-ethyl-5(1'methyl-propyl)-2-thiobarbiturate (Inactin), 120 mg·kg-1, intraperitoneally. The abdominal cavity was opened by a midline incision, the pylorus was ligated and the stomach was drained through a soft catheter placed in the esophagus. The common bile duct was catheterized to prevent exposure of the duodenal segment under study to pancreatic or bile juices. A 12 mm segment of the duodenum, starting 15-18 mm distal to the pylorus and with its blood supply intact, was cannulated in situ between two glass tubes connected to a reservoir containing isotonic NaCl. This fluid was rapidly circulated by a gas lift (100% oxygen) and the bicarbonate secretion into the luminal perfusate titrated continuously at pH 7.4 with 50 mM HCl (NaCl to isotonicity) under automatic pH-stat control. The transmucosal electrical potential difference (PD) between the duodenal lumen and the posterior vena cava was recorded continuously.

Duodenum in the anaesthetized rat spontaneously secreted bicarbonate at a steady basal rate of between 3 and 11 μEq·cm-1·h-1 and developed a transmucosal PD of 2 to 5 mV (lumen negative). The effects of dopamine are illustrated in Fig. 1. Continuous intravenous infusion of dopamine (50 μg·kg-1·h-1) caused a two-fold increase in duodenal mucosal bicarbonate secretion (p<0.01), and a higher rate (250 μg·kg-1·h-1) of infusion resulted in a further increase.

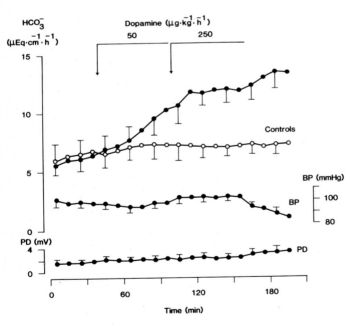

Fig. 1. Intravenous infusion of dopamine increased duodenal mucosal bicarbonate secretion in anaesthetized rats. Means ± SEM of secretion, transmucosal PD and mean arterial blood pressure (BP) in animals receiving dopamine and in control animals receiving vehicle (isotonic saline) alone. (Flemström *et al*. 1993 with permission from the publisher)

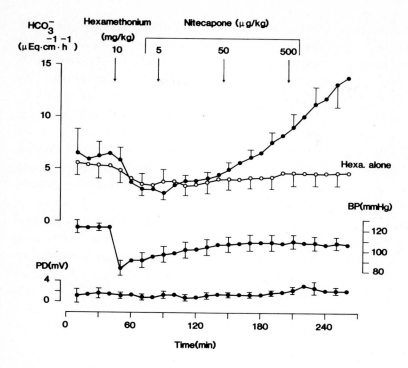

<u>Fig. 2</u>. The ganglion blocking agent hexamethonium decreased duodenal mucosal bicarbonate secretion and the mean arterial blood pressure in anaesthetized rats but did not affect the significant increase in secretion in response to an i.v. bolus injection of the peripherally acting COMT inhibitor nitecapone. Means ± SEM of secretion, PD and mean arterial blood pressure in animals injected with nitecapone and in animals treated with hexamethonium alone. (Flemström *et al.* 1993, with permission from the publisher)

There were no significant changes in PD, but the higher rate of infusion lowered the mean arterial blood pressure. Prostaglandin E_2 (20 μM luminally) added at the end of the experiments as a test of the viability of the duodenal mucosa, stimulated the bicarbonate secretion in all experiments. The dopamine D-1 selective agonist SKF-38393 (10-200 μg·kg-1) caused a dose-dependent increase (p<0.01) in mucosal bicarbonate secretion, similar in magnitude to that produced by dopamine. Dopamine D-2 receptor stimulation by bromocriptine (0.5 mg·kg-1), in contrast, slightly decreased (p<0.05) the secretion. Domperidone (5-100 μg·kg-1), a peripherally acting dopamine antagonist, did not affect basal bicarbonate secretion, PD or blood pressure. Pretreatment with domperidone (100 μg/kg), however, inhibited the increase in secretion in response to the agonist SKF-38393.

Catecholamine-*O*-methyl transferase (COMT) inhibitors decrease tissue degradation of catecholamines, including dopamine. Intravenous bolus injection of the peripherally acting inhibitor 3-[(3,4-dihydroxy-5-nitrophenyl) methylene]-2,4-pentanedione (nitecapone) at a dose of 50 $\mu g \cdot kg^{-1}$ increased duodenal mucosal bicarbonate secretion ($p < 0.05$), and at 500 $\mu g \cdot kg^{-1}$ a further increase was observed. The magnitude of the increases in secretion was similar to that caused by dopamine. Pretreatment with domperidone abolished the stimulatory effect of nitecapone, strongly suggesting that the effect of COMT inhibition was exerted peripherally and due to an increase in the tissue concentration of dopamine.

The ganglion-blocking agent hexamethonium was used to further elucidate the stimulatory effect of peripheral COMT inhibition. Hexamethonium alone (10 $mg \cdot kg^{-1}$) caused a modest decrease ($p < 0.05$) in duodenal mucosal alkaline secretion and a profound fall ($p < 0.01$) in the mean arterial blood pressure. Nitecapone was injected intravenously in hexamethonium-treated animals when the alkaline secretion and blood pressure had attained steady levels (Fig. 2). Hexamethonium did not prevent the ability of the duodenal mucosa to respond to nitecapone with significant ($p < 0.01$) increases in bicarbonate secretion, strongly suggesting that stimulation of secretion by dopamine is independent of enteric as well as extrinsic neural influence.

Studies in humans

The duodenum in healthy volunteers was intubated and the first 3 cm of the bulb isolated using a pear-shaped balloon in the distal part of the stomach and two button-shaped ballons in the duodenum as described in detail previously (Knutson & Flemström 1989). The stomach was drained by a separate sump-type tube and the tubes were positioned under fluoroscopic guidance. Isotonic NaCl solution was infused (2 $ml \cdot min^{-1}$) into the duodenal test segment and the effluent was collected by gravity drainage. The concentration of bicarbonate in the effluent was determined in triplicate (Corning 965 Carbon Dioxide Analyzer, Halstead, U.K.). To remove dissolved CO_2, samples were always gassed with 100% N_2 before analyses. Absence of trypsin activity was used to exclu contamination of the effluent by pancreatic bicarbonate. Studies were carried out after an overnight fast, and the H^+K^+ ATPase inhibitor omeprazole (20 mg) was given orally one hour before the experiments to suppress gastric acid secretion.

The proximal duodenum in humans spontaneously secreted bicarbonate at a basal rate of about 180 $\mu Eq \cdot cm^{-1} \cdot h^{-1}$. Effects of increasing doses of nitecapone and of the prostaglandin E_1 analogue misoprostol were studied after recording steady basal rates of secretion for at least three 15 min periods. The compounds were added to the luminal perfusate as illustrated in Fig. 3, and both significantly ($p < 0.01$) increased the bicarbonate secretion. The secretion after removal of nitecapone decreased relatively rapidly, whereas the decrease after removal of misoprostol was slower.

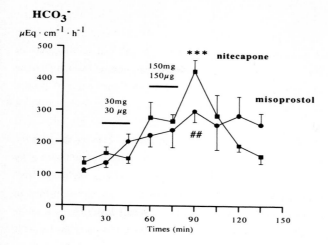

Fig. 3. The peripherally acting COMT inhibitor nitecapone or the prostaglandin analogue misoprostol was added to the duodenal perfusate in healthy volunteers as indicated. Means ± SEM of mucosal bicarbonate secretion are shown. **P<0.01 and ***P<0.001 for difference between drug and basal secretion. (Knutson *et al.* 1993, with permission from the publisher)

Studies of isolated duodenal enterocytes

In order to study the possible involvement of cyclic AMP as intracellular mediator of the response to dopamine receptor stimulation, cells were collected from the duodenum in rats by a combination of enzyme treatment and calcium chelation (Säfsten & Flemström 1993). Two major fractions, one mainly of villus and the other mainly of crypt origin were studied. In the villus fraction, the activity of alkaline phosphatase was 1.6 ± 0.2 μmol·mg protein-1·min-1 and that of sucrase 99±16 nmol·mg protein-1·min-1. In the crypt fraction, activities were 0.7±0.1 and 28±11, respectively.

Dopamine and the D-1 agonist SKF-38393 increased (p<0.05 at 10-5 M with both) the accumulation of cyclic AMP with a maximal response 5-15 min after start of incubation. The response to dopamine is illustrated in Fig. 4. The D-2 agonist quinpirole (10-5 M), in contrast, decreased accumulation of cyclic AMP. There were no significant differences between villus and crypt cell fractions in respect to the effects of dopamine, SKF-38393 or quinpirole. Effects were compared with those of vasoactive intestinal polypeptide, a well established and potent duodenal secretagogue (*cf.* Flemström 1993). The increase in cyclic AMP accumulation in the crypt cell fraction in response to this peptide was maximal 5 min after start of the incubation. In the villus cell fraction, however, the increase did not occur until after 60 min of stimulation.

Conclusions

   Duodenum and in particular the proximal segment transports bicarbonate at a higher rate than does the small distal small intestine. Recently, apical membrane vesicles from guinea pig duodenal enterocytes (Brown *et al*. 1989; 1992) and microfluorospectrophotometry of isolated rat proximal duodenal enterocytes (Isenberg *et al*. 1993) were used to elucidate the acid/base transport process in mammalian duodenal mucosa. An anion conductance pathway with equal conductance for bicarbonate and chloride and inhibited by the channel blocker diphenyl-2-carboxylate was demonstrated in the membrane vesicles. The rat duodenocytes extruded acid by amiloride-sensitive Na^+/H^+ exchange, and base by DIDS-sensitive Cl^-/HCO_3^- exchange. Base was imported by a DIDS-sensitive $NaHCO_3$ cotransporter. Comparison with the results of studies of intact duodenal mucosa (Flemström 1993) strongly suggests that bicarbonate is transported from cell to lumen by Cl^-/HCO_3^- exchange and via an anion conductive pathway and imported by $NaHCO_3$ cotransport. It is possible that, in addition, intracellular hydration of absorbed CO_2 with formation of H^+ and HCO_3^- supplies bicarbonate to the secretory process. Bicarbonate ions would be transported across the luminal cell membrane and H^+ ions would be extruded across the serosal membrane by Na^+/H^+ exchange or by a proton pump (Holm *et al*. 1990).

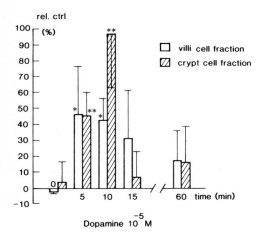

Fig. 4. Dopamine increased cyclic AMP levels in dispersed fractions of rat villus and crypt duodenal enterocytes. Dopamine (10 μM) was added at zero time and cells were incubated for 5, 10, 15 or 60 min in Ringer's bicarbonate solution. Cell concentration was 10^5 cells per incubate of 150 μl. Cyclic AMP was determined by radioimmunoassay and changes are expressed as percentage of simultaneously runs controls. Means ± SEM are presented. *P<0.05 and **P<0.01 vs. control. (Säfsten & Flemström 1993, with permission from the publisher)

Several types of stimuli are known to stimulate or inhibit the secretion. Luminal acidification results in an up to 10-fold increase in duodenal mucosal alkaline secretion, a response mediated by local mucosal production of prostaglandins and neural reflexes. The potent stimulation of the secretion by vasoactive intestinal polypeptide (Nylander *et al*. 1993) and by some sigma-receptor ligands (Pascaud *et al*. 1993) was recently shown to decrease duodenal mucosal susceptibility to luminal acid. Inhibitory stimuli include alpha$_2$-adrenoceptor agonists and cyclooxygenase inhibitors.

Dopamine is present in duodenal cells and neurones (Graffner *et al*. 1985, Dawirs *et al*. 1992). The results of the present studies demonstrate that dopamine D-1 receptor stimulation increases the duodenal mucosal alkaline secretion in the rat. Peripheral COMT inhibition which counteracts tissue degradation of catecholamines, increased the secretion in human volunteers as well as in the rat. The rise in secretion in humans was similar in magnitude to that observed with the prostaglandin E$_1$ analogue misoprostol. The ganglion-blocking agent hexamethonium did not inhibit the stimulatory effect of COMT inhibition on duodenal mucosal alkaline secretion, indicating that stimulation by dopamine is independent of enteric as well as extrinsic neural influence. An effect of dopamine D1-agonists directly on the duodenal enterocytes was confirmed by use of cells isolated from the rat duodenal mucosa. These results, furthermore, indicate that villus and crypts enterocytes are similarly stimulated by dopamine and suggest that cyclic AMP is involved as a messenger in the stimulation of duodenal mucosal bicarbonate secretion by dopamine. Dopamine D1-receptor stimulation did not affect the duodenal trans-mucosal PD, suggesting that dopamine, like vasoactive intestinal polypeptide and beta-endorphin stimulates an electroneutral component of the bicarbonate transport.

Some dopaminergic compounds were recently found to be efficient in the treatment of patients with duodenal ulcer (Sikiric *et al*. 1991) and an increase in dopamine receptor binding to a membrane fraction from duodenal mucosal homogenates was reported in such patients (Hernandez *et al*. 1989). This might relate to the low rates of duodenal mucosal alkaline secretion in duodenal ulcer patients. Dopamine receptor binding to membrane fractions from antral mucosal homogenates did not differ from that in healthy controls. Further studies of the role of mucosal dopamine in the control of gastroduodenal alkaline secretion and in ulcer disease would seem of considerable interest.

References

Aho PA, Lindén I-B (1992) Role of gastric mucosal eicosanoid production in the cytoprotection induced by nitecapone. Scand J Gastroenterol 127:134-138

Brown CDA, Dunk CR, Turnberg LA (1989) Cl-HCO$_3$ exchange and anion conductance in rat duodenal apical membrane vesicles. Am J Physiol 257:G661-G667

Brown CDA, McNicholas CM, Turnberg LA. (1992) A Cl$^-$ conductance sensitive to external pH in the apical membrane of rat duodenal enterocytes. J Physiol (London) 456:519-528

Dawirs RR, Teuchert-Noodt G, Kampen WU (1992) Demonstration of dopamine-immunoreactive cells in the gastrointestinal tract of gerbils (*Meriones unguiculatus*). J Histochem Cytochem 40:1197-1201

Donowitz M, Cusolito S, Battisti L, Fogel R, Sharp GWG (1982) Dopamine stimulation af active Na and Cl absorption in the rabbit ileum. J Clin Invest 69:1008-1016

Flemström G (1993). Gastric and duodenal mucosal bicarbonate secretion. In: Johnson LR, Alpers DH, Christensen J, Jacobson ED, Walsh JH (eds). Physiology of the gastrointestinal tract. Third edition (in press). Raven, New York

Flemström G, Säfsten B, Jedstedt G (1993) Stimulation of mucosal alkaline secretion in rat duodenum by dopamine and dopaminergic compounds. Gastroenterology 104:825-833

Glavin G (1989) Activity of selective dopamine DA_1 and DA_2 agonists and antagonists on experimental gastric lesions and gastric acid secretion. J Pharmacol Exp Ther 251:726-730

Glavin GB, Szabo S (1990). Dopamine in gastrointestinal disease. Dig Dis Sci 35:1153-61

Graffner H, Ekelund M, Håkanson R, Rosengren E (1985) Effect of different denervation procedures on catecholamines in the gut. Scand J Gastroenterol 20:1276-1280

Hernandez DE, Walker CH, Valenzuela JE, Mason GA (1989) Increased dopamine receptor binding in duodenal mucosa of duodenal ulcer patients. Dig Dis Sci 34:543-547

Holm L, Flemström G, Nylander O (1990) Duodenal alkaline secretion in rabbits: effects of artificial ventilation. Acta Physiol Scand 138:471-478

Isenberg JI, Ljungström M, Säfsten B, Flemström G (1993) Proximal duodenal enterocyte transport: evidence for Na^+-H^+ and $Cl^--HCO_3^-$ exchange and $NaHCO_3$ cotransport. Am J Physiol 265:G677-G685

Isenberg JI, Selling JA, Hogan DL, Koss MA (1987) Impaired proximal duodenal mucosal bicarbonate secretion in patients with duodenal ulcer. New Engl J Med 316: 374-379

Knutson L, Flemström G (1989) Duodenal mucosal bicarbonate secretion in man. Stimulation by acid and inhibition by the $alpha_2$-adrenoceptor agonist clonidine. Gut 30:1708-1715

Knutson L, Knutson TW, Flemström G (1993) Endogenous dopamine and duodenal bicarbonate secretion in humans. Gastroenterology 104:1409-1413

MacNaughton WK, Wallace JL (1989) A role for dopamine as an endogenous protective factor in the rat stomach. Gastroenterology 96:972-980

Mertz-Nielsen A, Hillingsø J, Eskerod O, Frøkier H, Bukhave K, Rask-Madsen J (1993) Evidence for increased gastric PGE_2 release and bicarbonate output secondary to a defective duodenal mucosal bicarbonate secretion in patients with duodenal ulcer. Gastroenterology 104:A146

Nylander O, Wilander E, Larson GM, Holm L (1993) Vasoactive intestinal polypeptide reduces hydrochloric acid-induced duodenal mucosal permeability. Am J Physiol 264:G272-G279

Pascaud XB, Chovet M, Soulard P, Chevalier E, Roze C, Junien J-L (1993) Effects of a new sigma ligand, JO 1784, on cysteamine ulcers and duodenal alkaline secretion rats. Gastroenterology104:427-434

Säfsten B, Flemström G (1993) Dopamine and vasoactive intestinal peptide stimulate cyclic adenosine-3',5'-monophosphate formation in isolated rat villus and crypt duodenocytes. Acta Physiol Scand 149:67-75

Säfsten B, Jedstedt G, Flemström G (1991) Effects of diazepam and Ro 15 -1788 on duodenal bicarbonate secretion in the rat. Gastroenterology 101:1031-1038

Sikiric P, Rotkvic I, Mise S, Petek M, Rucman R, Seiwerth S, Zjacic-Rotkvic V, Duvnjak M, Jagic V, Suchanek E, Grabarevic Z, Anic T, Brkic T, Djermanovic Z, Dodig M, Marovic A, Hernandez DE (1991) Dopamine agonists prevent duodenal ulcer relapse. A comparative study with famotidine and cimetidine. Dig Dis Sci 36:905-910

Willems JL, Buylaert WA, Lefebvre RA, Bogaert MG (1985) Neuronal dopamine receptors on autonomic ganglia and sympathetic nerves and dopamine receptors in the gastrointestinal system. Pharmacol Rev 37:165-216

INDEX OF AUTHORS

492

NATO ASI Series H

NATO ASI Series H

NATO ASI Series H

NATO ASI Series H

NATO ASI Series H

Springer-Verlag
and the Environment

We at Springer-Verlag firmly believe that an international science publisher has a special obligation to the environment, and our corporate policies consistently reflect this conviction.

We also expect our business partners – paper mills, printers, packaging manufacturers, etc. – to commit themselves to using environmentally friendly materials and production processes.

The paper in this book is made from low- or no-chlorine pulp and is acid free, in conformance with international standards for paper permanency.